MILLETS–2023

A Transdisciplinary Approach
to its Resurgence and Sustainability

MILLETS–2023

A Transdisciplinary Approach to its Resurgence and Sustainability

Editors

Prof. Anushree Lokur
Prof. Kamini J. Donde
Prof. Jessy Pius

ALLIED PUBLISHERS PVT. LTD.

Noida • Mumbai • Kolkata • Chennai • Bangalore

ALLIED PUBLISHERS PRIVATE LIMITED

D-5, Sector-2, **Noida**–201 301
Ph. Nos.: 0120-4320295/2542557/4352866 • E-mail: delhi.books@alliedpublishers.com

17 Chittaranjan Avenue, **Kolkata**–700072
Ph.: 033-22129618 • E-mail: cal.books@alliedpublishers.com

15 J.N. Heredia Marg, Ballard Estate, **Mumbai**–400001
Ph.: 022-42126969 • E-mail: mumbai.books@alliedpublishers.com

No. 25/10, Commander-in-Chief Road, Ethiraj Lane (Next to Post Office)
Egmore, **Chennai**–600008
Ph.: 044-28223938 • E-mail: chennai.books@alliedpublishers.com

P.B. No. 9932, No. 15, 3rd Floor (Next to Vijaya Bank),
5th Cross, Gandhinagar, Karnataka, **Bangalore**–560009, Ph.: 080-41530285 / 22386239
• E-mail: bngl.journals@alliedpublishers.com / apsabng@airtelmail.in

Sri Jayalakshmi Nilayam, No. 3-4-510, 3rd Floor (Above More Super Market)
Barkatpura, **Hyderabad**–500027
Ph.: 040-27551811, 040-27551812 • E-mail: hyd.books@alliedpublishers.com

Website: www.alliedpublishers.com

ISBN: 978-93-90951-91-8

Published by Sunil Sachdev and printed by Ravi Sachdev
at Allied Publishers Pvt. Ltd., D-5, Sector-2, Noida–201 301

Principal's Message

I write this message with great pleasure and excitement. Millet 2023 is an incredible book that dives into the interesting world of millets.

As we approach a global transition, sustainable and nutritious food options are critical. Millets have emerged as a source of hope and promise for the global need for a healthy planet.

Millets are hardy, resilient crops with a low carbon and water footprint they can withstand high temperatures, and grow on poor soils with little or no external inputs. Thus, Millets not only offer nutrition and resilience but also income and livelihood to small farmers.

According to the reports of United Nations Food and Agricultural Organization, 828 million people go to bed hungry every night globally. Moreover, year 2022 has been identified as 'Year of unprecedented hunger'. With only seven years remaining to attain the goal of zero hunger by 2030, we need very well planned, strategic, and well executed programs to reach the hungry millions with a sustainable solution.

Ramnarain Ruia Autonomous College is a signatory to the United Nations Sustainable Development Goals. Established in 1937, the College is celebrating its 85 glorious years of journey in education and training. To commemorate this occasion, the College had organised an international conference on Millets (Millets 2023) to deliberate on the evident benefits of millets as an alternate and sustainable food resource. The conference was also an attempt to boost up for the achieving the sustainable development goals like "Zero hunger", "Good health" and "Sustainable Consumption". Millets can form an integral part of these efforts due to their nutritional benefits, and precedence of long years of traditional usage to meet the global demands.

As a part of this conference, the research contribution by academicians, researchers, entrepreneurs, Governmental and Non-governmental organisations is collated and presented here in the form of book as "Millets 2023: A Transdisciplinary Approach to its Resurgence and Sustainability".

This is a humble contribution of our institute to the earnest appeal by our Honourable prime minister Shri. Narendra Modiji to bring back millets to the dining table at our homes and proactively help save our planet from an impending food scarcity.

I must extend my heartfelt gratitude to the editors, contributors and reviewers of Millet 2023 for their tireless efforts in bringing this invaluable collection of research papers to realisation.

Prof. Anushree Lokur
Principal
Ramnarain Ruia Autonomous College

Foreword

I have been working on millets for a decade now at the premier food research institute of our country and have always been excited about millet's unique capabilities. This is an opportune time to explore millets and I amelated to write on the topic – a celebration of nature's ancient grains and a journey into the heart of sustainable, nutritious, and delicious food choices.

In the modern age, where we often find ourselves racing against the clock, we tend to miss the profound significance of the food we consume. Our relationship with food has transformed over the years, and with it, our diets. However, as we embrace innovation and global flavors, it is crucial to remember the treasure trove of nourishment that has been cultivated for millennia, often right in our own backyard.

Millets, often referred to as "ancient grains," are the unsung heroes of our culinary heritage. These small but mighty grains have sustained civilizations across the world for thousands of years, offering unparalleled nutrition, versatility, and resilience in the face of environmental challenges. Today, they are making a comeback as people seek healthier, more sustainable food options.

In this book, we invite you to explore the world of millets – from their fascinating history and cultural significance to their nutritional powerhouse and diverse culinary applications. We'll delve into the reasons behind the resurgence of millets in modern diets and their potential to transform our food systems into more sustainable and resilient models. You'll discover inspiring stories of farmers, chefs, and entrepreneurs who are championing millets and driving positive change in the food industry. From traditional recipes that have stood the test of time to innovative, millet-based dishes that cater to contemporary tastes, this book will be your guide to incorporating these remarkable grains into your own culinary adventures.

As you flip through the pages of this book, we hope you'll be inspired to join the millet movement, whether you're a seasoned food enthusiast or a curious newcomer. Millets aren't just grains; they're a testament to the wisdom of our ancestors and a symbol of our commitment to nourishing ourselves and the planet.

I wholeheartedly believe that millets are the answer to our food and nutrition security challenges – where tradition meets innovation, health meets sustainability, and where every bite tells a story of the past and the promise of a brighter future.

As we celebrate the revival of millets, let us also celebrate the wisdom of our ancestors who cherished and nourished themselves with these grains. Let us honor the farmers who work tirelessly to cultivate millets sustainably. And let us all be stewards of change by choosing millets for our health and the health of our planet.

Thank you for joining us on this journey of discovery. Together, we can make millets a prominent part of our diets and contribute to a more sustainable and nutritious future for all.

Suresh D. Sakhare
Principal Scientist, CSIR-Central Food Technological Research Institute

About the Editors

Prof. Anushree Lokur is currently the Principal of S.P. Mandali's Ramnarain Ruia Autonomous College, Matunga, Mumbai.

She has completed her graduation and post-graduation from Abasaheb Garware College of Arts and Science Pune in Microbiology and was a Gold Medalist in both the years. She has completed her M.Phil. and Ph.D. in Microbiology from University of Mumbai. She has also completed her post-graduate diploma in Patent Management and is a Government of India recognized Patent Agent. She is also a Certified Trainer by Life Science Skill and Development Council.

She has total 36 Years' experience of teaching at undergraduate level and 26 years of teaching at postgraduate Level. She is recognized as Research Guide for Ph.D. in Microbiology and Biotechnology by University of Mumbai. Currently, three students are pursuing Ph.D. under her guidance.

Before being appointed as Principal she has been on various administrative positions in the college. She was the Head, Department of Microbiology, Dean Academics Affairs, Vice-Principal, Hon. Director and Nodal Officer UGC, DDU Kaushal Kendra, & Co-coordinator, Steering Committee for 3rd cycle, 2012 (CGPA 3.65/4) and 4th cycle, 2017 (CGPA 3.70/4) for NAAC Accreditation.

She is on the Governing Bodies, Academic Councils and Board of Studies of various education institutions. She has also been a Convener or a Member on LIC Expert Committees. She is on various Advisory committees of University of Mumbai. She is member of Intellectual Property Cell of University of Mumbai.

She has 15 international and 02 national research papers published in Peer Reviewed Journals. She has been granted 01 patent and filed 01 patent. Her area of research is Microfluidics, Diagnostics for Developing Countries, Biofilms etc.

She has been awarded with Gold Medal, Prize for Best Integrated Human Practices at the i-GEM International Competition held in Boston and with Chief Ministers Appreciation Award for success in i-GEM competition 2018.

Currently she is the Principal Investigator for the DBT Builder program by the Department of Biotechnology, Govt. of India.

Prof. Kamini J. Donde is a professor in Chemistry Department of Ramnarain Ruia Autonomous College. She joined the Department of Chemistry, Ruia College in 2004 as an Assistant Professor. In addition to her teaching responsibilities, Dr. Donde also holds the esteemed position of Dean, Research at Ruia College, a testament to her commitment to advancing knowledge and pushing the boundaries of scientific exploration. Her research focuses on the fascinating areas Coordination Chemistry, Polymer Chemistry and Nanotechnology.

Dr. Donde's passion for research is evident in her impressive track record of 23 research publications in various national and international reputed journals. She has also been a guiding force for doctoral students, with four currently pursuing their Ph.D. degrees under her guidance and two already received their degree.

Not only does Dr. Donde excel in her academic pursuits, but she is also committed to promoting and celebrating significant causes. She successfully organised the International Conference on Millets-2023 in January 2023, as a tribute to her commitment to promote healthy food choices and sustainable agriculture.

Prof. Jessy Pius is professor and Head of the Department of Botany at Ramnarain Ruia Autonomous College. She completed her Bachelor's and Master's in Botany with a first rank in M.Sc. from University of Calicut, Kerala. She has been awarded NET, CSIR Fellowship granted by UGC in the year 1986. She has received her Ph.D. in Plant Biotechnology from BARC, University of Mumbai where she has worked on *in vitro* studies of pearl millet and finger millet. To her credit she has 32 research publications in reputed national and international journals.

She has 30 years' experience of teaching at undergraduate level and 25 years of teaching at postgraduate Level. She also has administrative experience of working as Coordinator of Internal Quality Assurance Cell (IQAC) of Ramnarain Ruia Autnomous College for the past 10 years, Course Coordinator to B.Voc., Greenhouse Management course under DDU Kaushal Kendra, 2014–17 and as coordinator of Biological Science Departments for the past five years.

She is a recognized research guide for Ph.D. by the University of Mumbai since 1999 and has successfully guided 6 students for their Ph.D. thesis and one student for M.Sc. by research with three students currently working for their Ph.D. She has worked as a co-investigator on a major research project (DAE-BARC) from 2009–2013 and also as the Principal investigator in a minor research project funded by University of Mumbai in the year 2012. Currently she is one of the co-investigators for the DBT Builder program by the Department of Biotechnology, Govt. of India.

Acknowledgement

It gives me immense pleasure to present this book focusing on the most crucial issue in contemporary times today. Millets has been long served as traditional staple crops for millions of farmers and relatively cheaper source of nutrition especially to the underprivileged population of the world. It was indeed a great pleasure and a wonderful opportunity to organise an international conference and help 'Millets' make a comeback, thanks to the proposal of Government of India to the United Nations, thereby leading them to pass a resolution declaring '2023 as the International Year of Millets'.

Several people contributed to this conference book publication. I would like to take the opportunity to thank all of them. To begin with, I would like to express my heartfelt gratitude to the Principal Prof. Anushree Lokur, who bestowed her confidence and trust in me to become the convenor of the two-day International conference on "Millets-2023: A Transdisciplinary Approach to its Resurgence & Sustainability", organised by S.P. Mandali's Ramnarain Ruia Autonomous College, Mumbai in January 2023.

Any conference and its vision to compile a book can't be imagined until there is faith and commitment instilled for academic excellence. Our management S.P. Mandali has been at the forefront of backing up all such events. We feel extremely happy to have such management and extend our heartfelt thanks, especially to Adv. Shri S.K. Jain (Chairman, Managing Council, SPM, Pune), Adv. Mihir Prabhudesai (Member, Managing Council, SPM, Pune) and all council members for their goodwill and constant support in various forms. I would also like to thank Dr Jessy Pius, Co-convenor of the conference, for her sustained efforts to see to the publication of this book.

I also thank Ms. Varsha Malwade (Co-convenor) for all her support in organising the conference. It is indeed a great privilege for being India's first and one of the only non-agricultural university affiliated colleges, to get support and recognition from the Ministry of Agriculture and Farmers Welfare, Government of India and Government of Maharashtra for conducting this conference. Being a signatory to the 'Sustainable Development Goals' (SDGs) Accord of the United Nations on June 30, 2020, the book also contributes to the SDGs set by the United Nations.

I am thankful to all the research scholars and academicians who presented their research work and traced and connected the phenomenon of millets to the present day needs.

The theme and five diverse sub-themes covered in this book integrates sciences and humanities covering multidimensional topics like business and socio-economic environment for cultivating and marketing millets at national and global level, and reconnecting with long rooted traditional presence of millets in Indian art and culture.

I am thankful to all academic partners, sponsors, entrepreneurs, and NGOs for their contribution towards the publication of this book. I acknowledge my deep gratitude to all the reviewers Prof. (Dr.) Louiza Rodrigues, Dr. Bhavna R Narula, Dr. Urmila Moon, Dr. Madhavi Badole, Dr. Vaishali Dabke, Dr. Mayuri Rege, Dr. Sachin Palekar, Dr. Nandini Girish, Dr. Madhura. Dhavale-Joshi for taking efforts to check all the manuscripts meticulously. My sincere thanks to Dr. Veena Thakare for handling the most important a part of this publication, particularly the plagiarism check.

I am sure, through this book, we will reach out to maximum number of people at pan India level and help in disseminating the knowledge gained through these deliberations to support production and consumption of millets, coupled with enhanced consumer awareness of their nutritional and health benefits; and make millets as a viable and sustainable alternative to modern cash crops. I am thankful to all for contributing to this book and reiterating the significance of millet and achieve the vision of millet becoming the future superfood in India and globally. With deep appreciation, I humbly present this book for wider readership.

Prof. Kamini J. Donde
Department of Chemistry & Dean Research
Ramnarain Ruia Autonomous College

Contents

THEME–2: Millets Sustainable Solution to Food Security

THEME–3: Entreprenerurship, Start-Ups, Product Development and Marketing Strategies

THEME–4: Bioprospecting and Innovative Sustainable Cultivation Techniques for Millets

Theme–5: GO's, NGO's and Policies

THEME-1

Sociocultural, Economic, Geographical and Historical Aspects of Millets

Millets: The Magical Indian Tribal Food

Neeraja Tutakne[1], Sakshi Raj Kumar[2] and Snehal Unde[3]

[1]Department of Botany, SIES (Autonomous) College, Sion (West), Mumbai, India

[2]Department of Botany, Ramnarain Ruia (Autonomous) College, Matunga, Mumbai

[3]Department of Botany, St. Xavier's College (Autonomous), Fort Mumbai, India

ABSTRACT: Millets are India's ancient, heirloom grains that act as a storehouse for proteins, dietary fibres, amino acids, minerals, antioxidants, and vitamins. Indian millets are primarily categorized into two types, major millets which include Sorghum, Finger millet, Foxtail millet, and Pearl millet, and minor millets which incorporate Proso, Kodo, and little millet. The paper reviews various regions across India to find out the usage of millet among tribal communities. The review highlights that millets form an integral part of tribal food. Millets are hardy, rainfed, pest and disease-resistant crops with shorter cultivation periods than other grains. This has made millet a staple grain for the tribals across India. These super grains help the Indian tribal communities lead active and healthier lives. These grains are the best possible solution to battle vitamin and mineral deficiencies. Thus, millets serve as a magical treasure for the well-being of the tribals. Taking some lessons from the tribal communities, the existing food systems in urban life need to be updated. Against this backdrop, millet, one of the earliest domesticated cereal grains, may have a comeback. This will not only help to improve human health but will also be able to survive in the period of global warming and climate change.

Keywords: *Tribal Community, Minor Millets, IYM, Traditional Food, Tribal Food.*

INTRODUCTION

According to the Food and Agriculture Organization (FAO), United Nations, hunger refers to the situation that arises when people lack access to sufficient amounts of safe and nutritious food required for growth, development, and healthy life. As per World Health Organization (WHO) and UN FAO, hidden hunger occurs when the food which people consume is deficient in different nutrients essential for their growth and development (FAO, 2014). Thus, at its most extremities, hunger may lead to severe starvation and famine. In India, the problem of hunger and malnutrition goes hand in hand. In the Global Hunger Index – 2022, India secured a score of 29.1 and ranked 107th out of 121 countries examined. This indicates that India is suffering from a very serious level of malnutrition and hunger problem. Therefore, there is a need to find suitable and sustainable crop substitutes which would not only address the problem of hunger but also improve the income and standard of living of the farmers. In this aspect, the role of millets cannot be ignored. To create awareness throughout the world about the significant role of millets in sustainable agriculture, its utility as a smart superfood and its potential in generating new sustainable market opportunities, the United Nations General Assembly declared 2023 as the 'International Year of Millets' (IYM 2023; FAO, 2022).

Millets are India's ancient, heirloom grains that act as a storehouse for proteins, dietary fibres, amino acids, minerals, antioxidants, and vitamins. This has made millet a staple grain for the tribals across India. These super grains have helped the Indian tribal communities lead active and healthier lives. The tribal knowledge about millets would throw further light on their various utility aspects and health benefits. With the objective of revealing the secrets of the role of millets in tribal food culture leading to healthy and active lifestyles, the present review has been undertaken. The review envisages the Indian tribal regions; their traditional millet-based food practices and products; tribal farming practices for millets; millets as a smart food for future.

MILLETS AND TRADITIONAL MILLET BASED FOOD PRODUCTS

In India, people have enjoyed millets since ancient times. Presently, millet is the sixth most important grain in the world. Millets are the plants showing C_4 cycle and belong to the grass family, Poaceae. They are also known as small millets due to the presence of their small kernels (Paschapur *et al.*, 2021). The review suggests that following are the commonly used millets with their nutritional status and their traditional food products enjoyed across India.

Great Millet (*Sorghum bicolor* L. Moench): Great millet, also known as Jowar, is a major millet. It is a gluten free, nutritious grain which is packed with carbohydrates, proteins, vitamin B, magnesium, potassium, phosphorus, iron, and zinc. It also acts as an excellent source of soluble and insoluble dietary fibers. These insoluble fibers act as a 'prebiotic' agent.

Finger Millet (*Eleusine coracana* L.): Finger millet, also known as Nachni or Ragi, is a nutritional millet that provides people with equal quantities of proteins, minerals, vitamins and dietary fibres like other cereals (Devi *et al.*, 2014; Kumar, 2016). It has the highest calcium content among all cereals. Value added food products prepared from finger millet include khichadi, laddu, batti (Arora and Srivastava, 2002), buns (Tiwari and Srivastava, 2017), bread (Chhavi and Sarita, 2012), muffins (Ulfath *et al.*, 2016) and pizza base (Shrestha, 2017).

Foxtail Millet (*Setaria italica* L. subsp. *italica*): Foxtail millet, commonly known as Kangni or Rala, is a rich source of starch, protein, fat, minerals, and crude fibers. It contains copious amounts of lysine, thiamine, niacin, iron and calcium. Value added food products prepared from foxtail millet include bread (Chhavi and Sarita, 2012), burfi (Bisht and Srivastava, 2013), biscuits (Anju and Sarita, 2010), dosa (Narayanan *et al.*, 2016), etc.

Pearl Millet (*Pennisetum glaucum* L.): Pearl millet, commonly known as Bajra, is a profoundly nutritious grain. It contains proteins, carbohydrates, vitamin E, K, B complex, polyphenols, sodium, zinc, magnesium, phosphorus and potassium. Value added food products prepared from pearl millet include bhakhri, uttapam, cheela, etc. (Kumari *et al.*, 2014).

Proso Millet (*Panicum miliaceum* L. subsp. *miliaceum*): Proso millet, also known as common millet or vari, is a protein and a calcium rich millet. Its protein content is richer in essential amino acids (leucine, isoleucine, methionine) (Kalinova and Moudry, 2006). Value added food products prepared from proso millet include upma, dosa, roti, bhakri, cheela, and pancakes.

Kodo Millet (*Paspalum scrobiculatum* L.): Kodo millet, commonly known as native Paspalum or rice grass or Indian crown grass, is a powerhouse of fibres, protein, vitamins, calcium, iron, polyphenols and various other minerals. Value added food products prepared from kodo millet include idli, sewai, upma, etc. (Yadav *et al.*, 2013).

Little Millet (*Panicum sumatrense* Roth.): Little millet, commonly known as miliare or kutki, is a low glycemic index grain which has impressive amounts of proteins, magnesium, niacin polyphenols, tannins, flavonoids and dietary fibres. Value added food products prepared from little millet include flakes (Patil *et al.*, 2015), cheela, bhakri and pancakes.

Barnyard Millet (*Echinochloa frumentacea* Roxb.): Barnyard millet, commonly known as sanwa rice, is rich in carbohydrates, proteins, calcium, iron, vitamin B complex and dietary fibres. Value added food products prepared from Barnyard millet include khichdi, laddu, batti (Arora and Srivastava, 2002), pizza base (Shrestha, 2017), dosa (Nagalakshmi and Beatrice, 2013) and noodles (Surekha *et al.*, 2013).

MILLETS AS TRIBAL FOOD ACROSS INDIA

An extensive account of literature, regarding the major states across India, exhibiting variety in tribal food culture was reviewed. Tribal millet farming practices, which are the common thread connecting the states across India, have been described below. The prominent regions that have been considered in this review are North-eastern India, Northern India, Eastern India, Central India and Southern India.

North-Eastern India

- *Arunachal Pradesh:* Traditionally, shifting cultivation is carried out in Arunachal Pradesh, which is locally known as Jhum. Fox-tail millet and finger millet are commonly cultivated in the region. In addition to that, proso and pearl millet are also widely grown under Jhum cultivation. The practice of Jhum cultivation provides for easy and secure means of sustenance in an otherwise dynamic climactic environment that prevails in this region. Forage jowar has also been introduced in the region to take care of the unavailability of fodder by Krishi Vigyan Kendra, Lohit (Bhat *et al.*, 2019). Mechuka valley, situated 2000 ft above sea level, is mostly inhabited by the Memba Tribe of Arunachal Pradesh. Finger millet, locally called Tame, is a quite popular crop among these people, due to its adaptability to varying altitudes. It is cultivated by simple farming practices such as broadcasting or direct seeding. Memba people process the finger millet to prepare a local beverage called Chang or Apong, apart from using it for making chapatis, porridges, etc. (Touthang *et al.*, 2022). Monpa tribes also make use of finger millet as a porridge dish called as Zan (Rawat *et al.*, 2021).
- *Assam:* Foxtail millet and barnyard millet are commonly grown in Assam by the practice of Jhum cultivation, especially in the Karbi Anglong district (Bhat *et al.*, 2019). Foxtail millet, locally called Konidhan, is a popular choice for making puddings and deserts in this region. One does find finger millet and proso millet among the crop species that are cultivated here prevalently. Locally, finger millet is referred to as Marua in Assam. It acts as a source of calcium for aging elders and growing children of the tribes residing in this region. It is also a good source of iron, thereby helping in recovery from anaemia (North Eastern States: E-Catalogue for Export of Millets and Value-Added Products, 2022).
- *Manipur:* Like Assam and Arunachal Pradesh, millets are a part of jhum cultivation in Manipur also. Millets are popular among the Paite and Thadou-Kuki tribes as festival food, especially on the day of Mim Kut

(first harvest festival) (Bhat *et al.*, 2019). ICAR has conducted successful cultivation of finger millet, bajra and foxtail millets in this region (E-Catalogue for Export of Millets and Value-Added Products: North Eastern States, 2022).

- ***Meghalaya:*** Finger millet (called Krai-tuh in Khasi Hills) and foxtail millet are cultivated in Meghalaya. Raishan (*Digitaria cruciata* var. *esculenta* Bor) is the millet which is endemic to the Khasi tribe, used as food and fodder, as well as during their harvest festival (Singh and Arora, 1972; North Eastern States: E-Catalogue for Export of Millets and Value-Added Products, 2022).
- ***Mizoram:*** The Mara tribe of Mizoram has been farming millets (collectively referred to as Sato) by jhum cultivation as their chief food. Sorghum (locally called as Faisa) and pearl millet (locally called as Bhutun) are prevalently grown over here (North Eastern States: E-Catalogue for Export of Millets and Value-Added Products, 2022).
- ***Nagaland:*** Jowar, bajra, ragi and small millets are all grown in Nagaland, as they are a part of important survival foods in the state. Millets are grown in a multi-cropping pattern, mainly in the districts of Phek, Tuensang, Kiphire and Kohima. Angami Naga tribals have been known to celebrate a millet harvest festival called 'Tsiinyi'. Millets are commonly prepared as khichdi and apang by the millet consuming tribes of Nagaland (North Eastern States: E-Catalogue for Export of Millets and Value-Added Products, 2022; Bhat *et al.*, 2019).
- ***Sikkim:*** Sikkim, being an organic agriculture certified state, is known mainly for cultivation of small millets, such as finger millet. The Lepcha tribes' traditional agricultural practices predominantly include millet farming (Ghimiray, *et al.*, 2018).
- ***Tripura:*** The tribal communities of Tripura mainly grow foxtail millet in a mixed cropping cultivation system. The three immigrant tribes of Tripura, viz., Oraon, Munda, and Santal have been known to be admirable cultivators of rice as well as millets in the region (Adhikari, 2020).

Northern India

- ***Uttarakhand:*** The moderate climate and fertile soil in this region account for more productivity of millet crops. Barnyard millet, finger millet, foxtail millet, among others are the major crops cultivated here. Sorghum is also cultivated in the region. Millets are grown as kharif crops as a part of a mixed cropping system called baranaja. 'Jhangore ki kheer', a porridge made from barnyard millets is quite a famous dish in this region. 'Maduwa ki roti' is another popular dish which is made with finger millet flour. These are typical Kumaoni and Garhwali fares (Rawat *et al.*, 2020). Proso millet is prepared as 'bhat' by the Bhotiya tribe (Rawat *et al.*, 2021).

- ***Jammu and Kashmir:*** One of the major millet crops of Jammu and Kathua districts, is bajra, although the productivity levels of bajra are low. Jowar millet is quite popular in the Kandi belts of Jammu district and in Kishtwar district. Foxtail millet (locally called Shol or Kangni) has been sown as a substitute for paddy in drier areas of this region. Proso millet (locally called Ping) has been known to be cultivated in areas with moderate rainfall in this region. Both these millets have also been grown as fodder for animals. They are popular in the Doda and Kishtwar districts of Jammu and Kashmir (Bhat *et al.*, 2019). Sheikh and Singh (2013) reported that minor millets such as foxtail millet (locally called Tcha) and proso millet (locally called Tchay tchay) are cultivated in the Ladakh region. The flour of these two millets is used to make bread. The grains are also roasted, dehusked and coarsely ground to make 'satu', which is consumed along with tea. Popularity of these millets is also seen in the Kargil region where they are cultivated for fodder production.
- ***Himachal Pradesh:*** Historically, finger millet, proso millet, foxtail millet and kodo millet have been consistently grown by tribal populations of many districts in Himachal Pradesh. Although, there is a decline in the cultivation of millets in the area, both black grained and white grained types of finger millet (locally called as Mandal) are grown in Kullu, Mandi, Kangra and Sirmour districts. Black grained finger millet is used as fodder. An alcoholic drink called 'Sur' is prepared from ragi in the Lag valley of Kullu and Kangra district (Joshi *et al.*, 2012). Foxtail millet (locally called Kangni) is grown widely in Kangra district mostly for the consumption of poultry and cage birds. For human consumption it is cooked like rice. Similarly, proso millet (locally called as Cheena) is mainly cultivated as a fodder crop, but is also used for human consumption in many areas. It is a popular traditional food of the Gaddi shepherd tribe of Chamba region in Himachal Pradesh. Little millet (locally called as Swank), is grown as a cheap grain, popular among the poorer classes in the region. It is considered sacred and consumed in various forms during the Ekadashi and Lohri festival. Kodo millet (locally called as Kodra) is grown only in Shimla district on a small scale. (Bhat *et al.*, 2019).

Eastern India

- ***West Bengal:*** The Darjeeling district in West Bengal is known for finger millet cultivation on a grand scale. Foxtail millet is also grown in this region along with the Coochbehar district. Gorkha tribe of West Bengal prepares a popular local traditional beer, Tongba, which is brewed by fermenting millets with yeast (Sharma, R., 2021). Kodo millet is also grown in this area (Rawat *et al.*, 2021). Sorghum, pearl millet and finger millet are mostly cultivated in the rainfed uplands of the red and lateritic belt of West Bengal (Brahmachari *et al.*, 2018).

- ***Orissa:*** The Koraput region of Orissa is well known for its mixed cultivation of millets along with other crops like paddy, pulses and vegetables. Finger millet gruel is an integral part of their diet. They are known to prepare country liquor made out of rice, finger millet, mahula (*Madhuca indica*), salpa (*Caryota urens*) and khajuri (*Phoenix sylvestris*). The Paroja, Gadaba and Bhumia communities cultivate a number of minor millets like finger, kodo, foxtail, little, barnyard and pearl millets (Mishra and Chaudhury, 2012).
- ***Jharkhand:*** Finger millet is widely cultivated in the Khunti and Godda districts of Jharkhand by the Munda and Sauria Paharia tribes. Little millet is also grown in various regions of Jharkhand as a staple food (Downs *et al.*, 2022).

Central India

- ***Chhattisgarh:*** Kodo millet has been grown traditionally in this state. Tribal farmers have cultivated kodo millet without differentiating it as varieties, although Sahu (2019) reports that *Os Kosara, Rikki Kosara* and *Bade Kosara* variety of kodo millet are grown in Chhattisgarh. Cooked and fermented Kodo rice (paye) is also prepared here (Sahu and Sharma, 2013). Finger millet is also grown in the state and is used for making a soothing beverage called Pej. The popular varieties of finger millet are *Nani Bitti Madia*, *Bade Madia*, *Laad Madia* and *Bakari Kaan Madia*. In the Bastar plateau zone, kodo millet is not only used as food but for making mud walls and earthen pots. It has also been used as a medicine for various poultry diseases. The prasad of the major festival of Gaadi is the grains of finger millet (Sahu and Sharma, 2013). The Pahadi Korwa community has the tradition of using millet inflorescences to decorate their courtyards and pray for a successful harvest (Pilla *et al.*, 2021).
- ***Madhya Pradesh:*** The Baiga and the Gond Adivasi communities of Madhya Pradesh have been cultivating kodo and kutki millets traditionally (Saju, 2018). The Abujh Maria sub-tribe of Gonds is known to cultivate millets as their main output (State Health Resource Centre, Bhopal, 2020). These millets form a very important source of sustenance in Dindori and Mandla districts. Barnyard, foxtail and finger millet are also

grown in the region but they cover a small area with lower yield. Japanese barnyard millet (*Echinochloa esculenta*) is prepared in the same way as rice, consumed as a popped grain, and used in bakery products as well. The wild counterpart of Japanese barnyard millet is used in the preparation of conventional kheer and puri. It is also known to be used as fasting food in some communities (Yadava, 2018). In Western Madhya Pradesh, the Barela tribe has been farming millets along with other important food crops (Soni and Soni, 2016). The practice of applying turmeric and finger millet flour on the bride and groom during the wedding ceremonies is very common over here (Niyogi, 2018).

Western India

- ***Maharashtra*** **and** ***Gujarat:*** Jowar, bajra, finger millet, shama millet (*Echinocloa colona* L.), kodo millet, etc., are the millets grown in major quantities in Maharashtra as well as Gujarat (Paschapur, *et al.*, 2021). Ambali (thin porridge) is prepared from kodo millet in Maharashtra (Hema, *et al.*, 2022). Kodo, finger millet, and little millet are grown in tribal and hilly Dang district of South Gujarat region (Makadia, 2014).
- ***Rajasthan:*** Rajasthan is one of the largest producers of millets in India. Four varieties of millets such as jowar, bajra and small millets are grown in Rajasthan. Bajra is the major contributor (91%) followed by Jowar (9%) in the production of millet in the Kharif season. In the Rabi season, barley is the largest contributor (almost 100%) in millet production (Rajasthan: E-Catalogue for Export of Millets and Value-Added Products, 2022). In semi-arid regions of India, rabadi is a popular welcome drink prepared by pearl millet fermentation using buttermilk (Modha and Pal, 2011). The Meena tribe of Banswara, Udaipur and Chittorgarh districts of Southern Rajasthan are known to cultivate and consume bajra. Garasia tribes distributed in the Sirohi and Udaipur districts primarily also grow pearl millet. Staple food of tribals in Rajasthan includes millets like bajra and sorghum. They also cultivate minor millets like little millet, proso millet, finger millet, kodo millet and foxtail millet as famine food and fodder for cattle (Sharma, P., 2008).

Southern India

- ***Tamil Nadu:*** The use of millets by the tribals of Tamil Nadu has been documented as long ago as the early 1900s. The Irula tribe of the Nilgiris have been reported to cultivate finger millet—ragi, little millet—samai, and foxtail millet—tenai among the millets, for their own consumption. Eravallars, a small forest tribe in the Coimbatore district and Malabar, have their main occupation as cultivation of proso millet (locally called chama) and great millet (locally called cholam), along with other important food crops. Sowing begins in the month of May and harvesting is done in the month of November (Thurston and Rangachari, 1909). More recently, Venkatesan and Sundaramari (2015) have reported significant farming of finger millet by tribals of Kolli hills of Namakkal district in Tamil Nadu (Malayalis). The Kolli hills region is also synonymous with cultivation of foxtail, barnyard, proso, kodo and little millets. The tribals have inculcated the customary agricultural practices which are deeply rooted in our heritage and culture. These farming practices called the Indigenous Tribal Agricultural Practices (ITAP) have allowed intensive cultivation in a particular area for long periods without being detrimental to the environment. ITAP is now being adopted by many tribal farmers for enhancing the upkeep of soil, water and germplasm. Venkatesan *et al.* (2015), in their survey of the Kolli hill region came across some 21 landraces of the millets which are grown under different land categories. Among the millets that are produced in the region, little (Samai), foxtail (Thinai), kodo (Varagu) and proso (Panivaragu) millets are prevalent.

- ***Telangana:*** Sorghum, ragi and bajra are some of the millet crops grown in the Adilabad district of Telangana. Poshadri *et al.*, 2019, worked on the preparation and evaluation of composite weaning mixes primarily containing the great millet and finger millet. The Gond community of Kumuram Bheem Asifabad district celebrate a harvest festival, 'Atukula Panduga', which centres around the great millet. Dishes prepared from foxtail millet and little millet are offered to local deities. Erra Jonna, Tella Boda Jonna, Konkadala Jonna races of Jowar are popular in the region. Flat bread (roti) and porridge (gatka) made from jowar flour is a staple food of the Gonds (Pilla *et al.*, 2021).
- ***Andhra Pradesh:*** Both major and minor millets are grown in Andhra Pradesh, viz., finger millet (chodi/thaida), foxtail millet (kora), proso millet (variga), little millet (sama), kodo millet (arika), barnyard millet (ooda), great millet (jonna), and pearl millet (sajja) (Andhra Pradesh: E-Catalogue for Export of Millets and Value-Added Products, 2022). According to Byaligoudra *et al.*, (2019), jowar is showing positive growth rate in the Guntur district. Finger and little millets are relished by the tribals in the form of chapati and puddings. They are also grown as forage crops. Foxtail and kodo grains are cooked similar to rice and consumed. Kodo preparations are recommended for patients suffering from diabetes. Proso millet makes for a good poultry feed and fodder for

cattle. It is also used to prepare kheer. Barnyard millet is also considered to be a nutritious food crop by the tribals (Triveni *et al.*, 2020).

- ***Kerala:*** Production of millets in Kerala is mainly done in the Palakkad and Idukki districts. A major share of cultivation of millets in Kerala is done in Attappady taluka in Palakkad. Agali, Sholayur and Pudur panchayats are collectively referred to as the millet cluster in the region. Chief millet crops in the area include great millet, finger millet and little millet (chama). Pearl millet (cumbu) and other minor millets are also grown in small quantities (Kerala: E-Catalogue for Export of Millets and Value-Added Products, 2022). Attappady is also being converted to 'Millet Village' under the Directorate of Agriculture Development and Farmers' Welfare to promote the proper production, marketing, and processing of millets. The Muthuvan tribe of Idukki district grow finger millet extensively by slash and burn method, similar to the jhum cultivation. The most famous dish prepared from the finger millet is 'Katty' which is a pudding-like preparation. These tribals prepare themselves for any famine-like situation by observing seed storage practices (Manithottam and Francis, 2007).

Table 1: Millets Cultivated by Tribal Communities across India

Common Name (English)	*Botanical Name*	*State*	*Name of Tribal Community/Region*
Great Millet	*Sorghum bicolor* (L.) Moench	Mizoram	Mara tribe (North-Eastern States: E-Catalogue for Export of Millets and Value-Added Products, 2022)
		Nagaland	Tribals of Phek, Tuensang, Kiphire and Kohima districts; Angami Naga tribe (Bhat *et al.*, 2019)
		Uttarakhand	Tribals of Garhwal and Kumaon region (Rawat *et al.*, 2020)
		Jammu and Kashmir	Tribals of Kishtwar district and Kandi belt of Jammu district (Bhat *et al.*, 2019)
		Rajasthan	Tribals of Rajasthan (Rajasthan: E-Catalogue for Export of Millets and Value-Added Products, 2022)
		Tamil Nadu	Eravallars tribe of Coimbatore district and Malabar (Thurston and Rangachari, 1909)
		Telangana	Gond community of Kumuram Bheem Asifabad district (Pilla *et al.*, 2021)

Common Name (English)	*Botanical Name*	*State*	*Name of Tribal Community/Region*
		Andhra Pradesh	Tribals of Andhra Pradesh (Triveni *et al.*, 2020)
		Kerala	Muthuvan tribe of Idukki district (Manithottam and Francis, 2007)
Finger Millet	*Eleusine coracana* L.	Arunachal Pradesh	Memba tribe (Touthang *et al.*, 2022), Monpa tribe (Rawat *et al.*, 2021)
		Nagaland	Tribals of Phek, Tuensang, Kiphire and Kohima districts; Angami Naga tribe (Bhat *et al.*, 2019)
		Meghalaya	Khasi tribe (North-Eastern States: E-Catalogue, 2022)
		Uttarakhand	Tribals of Garhwal and Kumaon region (Rawat *et.al*, 2020)
		Himachal Pradesh	Tribals of Kullu, Mandi, Kangra and Sirmour Districts (Joshi *et al.*, 2012)
		Orissa	Paroja, Gadaba and Bhumia tribes (Mishra and Chaudhury, 2012)
		Jharkhand	Munda and Sauria Paharia tribes of Khunti and Godda districts (Downs *et al.*, 2022)
		Chhattisgarh	Tribals of Bastar region (Sahu and Sharma, 2013)
		Gujrat	Tribals of Dang district (Makadia, 2014)
		Tamil Nadu	Irula tribe of the Nilgiris (Thurston and Rangachari, 1909), tribals of Kolli hills of Namakkal district (Venkatesan and Sundaramari, 2015)
		Andhra Pradesh	Tribals of Andhra Pradesh (Triveni *et al.*, 2020)
		Kerala	Muthuvan tribe of Idukki district (Manithottam and Francis, 2007)
Foxtail Millet	*Setaria italica* L. subsp. *Italica*	Assam	Tribals of Karbi Anglong district (Bhat *et al.*, 2019)
		Meghalaya	Khasi tribe (North-Eastern States: E-Catalogue for Export of Millets and Value-Added Products, 2022)

Common Name (English)	*Botanical Name*	*State*	*Name of Tribal Community/Region*
		Tripura	Oraon, Munda, and Santal Tribes (Adhikari, 2020)
		Uttarakhand	Tribals of Garhwal and Kumaon region (Rawat *et al.*, 2020)
		Himachal Pradesh	Tribals of Kullu and Kangra Districts (Joshi *et al.*, 2012)
		Jammu and Kashmir	Tribals of Doda and Kishtwar district, Ladakh region (Bhat *et al.*, 2019, Sheikh and Singh, 2013)
		Orissa	Paroja, Gadaba and Bhumia tribes (Mishra and Chaudhury, 2012)
		Tamil Nadu	Tribals of Kolli hills of Namakkal district (Venkatesan and Sundaramari, 2015)
		Andhra Pradesh	Tribals of Andhra Pradesh (Triveni *et al.*, 2020)
Pearl Millet	*Pennisetum glaucum* L.	Mizoram	Mara tribe (North-Eastern States: E-Catalogue for Export of Millets and Value-Added Products, 2022)
		Jammu and Kashmir	Tribals of Jammu and Kathua District (Bhat *et al.*, 2019)
		Rajasthan	Tribals of Rajasthan (Rajasthan: E-Catalogue for Export of Millets and Value-Added Products, 2022)
		Andhra Pradesh	Tribals of Andhra Pradesh (Triveni *et al.*, 2020)
		Rajasthan	Meena tribe of Banswara, Udaipur and Chittorgarh districts, Garasia tribes of Sirohi and Udaipur districts (Sharma, 2008)
Proso Millet	*Panicum miliaceum* L. subsp. *Miliaceum*	Uttarakhand	Bhotiya tribe (Rawat *et al.*, 2021)
		Jammu and Kashmir	Tribals of Doda and Kishtwar district, Ladakh region (Bhat *et al.*, 2019, Sheikh and Singh, 2013)
		Himachal Pradesh	Gaddi shepherd tribe of Chamba region (Joshi *et al.*, 2012)

Common Name (English)	*Botanical Name*	*State*	*Name of Tribal Community/Region*
		Tamil Nadu	Eravallars tribe of Coimbatore district and Malabar (Thurston and Rangachari, 1909), Tribals of Kolli hills of Namakkal district (Venkatesan and Sundaramari, 2015)
		Andhra Pradesh	Tribals of Andhra Pradesh (Triveni *et al.*, 2020)
Kodo Millet	*Paspalum scrobiculatum* L.	Himachal Pradesh	Tribals of Shimla district (Bhat *et al.*, 2019)
		Orissa	Paroja, Gadaba and Bhumia tribes (Mishra and Chaudhury, 2012)
		Chhattisgarh	Tribals of Bastar region (Sahu and Sharma, 2013)
		Madhya Pradesh	Baiga and the Gond Adivasi (Saju, 2018)
		Maharashtra	Tribals of Maharashtra (Hema, *et al.*, 2022)
		Gujarat	Tribals of Dang district (Makadia, 2014)
		Tamil Nadu	Tribals of Kolli hills of Namakkal district (Venkatesan and Sundaramari, 2015)
		Andhra Pradesh	Tribals of Andhra Pradesh (Triveni *et al.*, 2020)
Little Millet	*Panicum sumatrense* Roth.	Himachal Pradesh	Tribals of Himachal Pradesh (Bhat *et al.*, 2019)
		Orissa	Paroja, Gadaba and Bhumia tribes (Mishra and Chaudhury, 2012)
		Madhya Pradesh	Baiga and the Gond Adivasi (Saju, 2018)
		Gujrat	Tribals of Dang district (Makadia, 2014)
		Tamil Nadu	Tribals of Kolli hills of Namakkal district (Venkatesan and Sundaramari, 2015)
		Andhra Pradesh	Tribals of Andhra Pradesh (Triveni *et al.*, 2020)
		Kerala	Muthuvan tribe of Idukki district (Manithottam and Francis, 2007)

Common Name (English)	*Botanical Name*	*State*	*Name of Tribal Community/Region*
Barnyard Millet	*Echinochloa frumentacea* Roxb.	Assam	Tribals of Karbi Anglong district (Bhat *et al.*, 2019)
		Uttarakhand	Tribals of Garhwal and Kumaon region (Rawat *et al.*, 2020)
		Orissa	Paroja, Gadaba and Bhumia tribes (Mishra and Chaudhury, 2012)
		Tamil Nadu	Tribals of Kolli hills of Namakkal district (Venkatesan and Sundaramari, 2015)
		Andhra Pradesh	Tribals of Andhra Pradesh (Triveni *et al.*, 2020)

MILLETS AS SMART FOOD FOR FUTURE

Millets are C_4 plants with high photosynthetic efficiency, fast-growing pattern, heat, drought, salinity, and toxicity tolerance (Yadav and Rai, 2013). Millets can be grown in semi-arid areas with less rainfall and even in degraded lands with low nutrient content. They support the livelihood of people in famine susceptible regions (Tadele, 2016). They form food even for livestock and birds. All these exceptional characteristics make millets an ideal crop for cultivation and promotion. Besides, the millets come with a package of tremendous health benefits like cardiovascular health protection, gut health management, diabetes prevention, obesity management, mood improvement, and assistance in digestive functions (Paschapur *et al.*, 2021). Hence, millets become a boon for the human population associated with a modern sedentary lifestyle and its several related health issues. Owing to these health benefits, millet farming has made a great comeback in a few states of India. With renewed knowledge and information about millets and their value-added products, cultivation of millets has achieved a new lease of life in Punjab with the help of the Department of Food, Science and Technology, Punjab Agricultural University, Ludhiana (Chaba, 2022). This sets the stage for profitable marketing and export of millets and their products. Millet farming in Bihar is also becoming popular again, especially in the Chakai town of Jamui district (Rawal, 2022). A sustainable farming practice of bajra is carried out in Haryana extensively (Nagaraj *et al.*, 2013). Karnataka is one step ahead with being able to revive the production of millets in the state. It has become one of the chief producers and consumers of millets in India. The demand for millets is high in the state with finger millet, pearl millet, and sorghum being at the top of the list. Bengaluru is not only the IT hub of India but also touted as the 'Millet hub of India'. Products made from millets like pizzas, ice-creams, burgers, breads and a plethora of snacks are being served in restaurants which pride themselves in bringing organic and healthy food to the table (Bhat *et al.*, 2018). Thus, it is observed that millets are seeing an uprising due to the health-conscious population across India and becoming a smart food for the future.

CONCLUSION

On the launch of the 'International Year of Millets 2023' at the headquarters of the FAO Rome, Italy, the honourable Prime Minister of India, Shri Narendra Modi, emphasized that it is the need of the hour to make millets a food choice for the future as they are best for the consumer, cultivator and the climate. In India, tribal populations have already demonstrated the same. Tribals have largely lived on different types of protein-rich millets that have helped them sustain their health. The toughness and flexibility exhibited by the millet crops has facilitated the tribals to choose these crops for rainfed, dry, arid or even hilly regions. Therefore, the protection of these important traditional millets would be necessary for ensuring the food security of the future generations. Learning from the tribal food practices about millets would definitely help to create 'Millet Mindfulness'. It would thereby aid individuals in making health-conscious and planet-friendly food choices.

REFERENCES

[1] Adhikari, O.S. (2020). Four Immigrant Tribes of Tripura: Their Life and Culture. Tribal Research & Cultural Institute, Government of Tripura, ISBN: 978-93-86707-45-1.

[2] Andhra Pradesh: E-Catalogue for Export of Millets and Value-Added Products (2022). Agricultural and Processed Food Products Export Development Authority. Ministry of Commerce and Industry. Govt. of India.

[3] Anju, T. and Sarita, S. (2010). Suitability of foxtail millet (*Setaria italica*) and barnyard millet (*Echinochloa frumentacea*) for development of low glycemic index biscuits. *Mal J. Nutr.* 16:361–368.

[4] Arora, S. and Srivastava, S. (2002). Suitability of millet-based food products for diabetics. *J Food Sci Technol,* 39:423–426.

[5] Bhat, B.V., Rao, B.D. and Tonapi, V.A. (2018). The Story of Millets: Millets were the first crops, Millets are the future crops. Karnataka State Department of Agriculture, Bengaluru, *India with ICAR-Indian Institute of Millets Research,* Hyderabad, India.

[6] Bhat, B.V., Arunachalam, A., Kumar, D., Tonapi, V.A. and Mohapatra, T. (2019). Millets in the Indian Himalaya, Indian Council of Agricultural Research, New Delhi. 84p. ISBN: 81-89335-89-8.

[7] Bisht, A.T. and Srivastava, S. (2013). Efficacy of millets in the development of low glycemic index sweets for diabetics. *Mal J. Nutr,* 19:215–222.

[8] Brahmachari, K., Sarkar, S., Santra, D.K. and Maitra, S. (2019). Millet for food and nutritional security in drought prone and red laterite region of Eastern India. *Int J. Plant Soil Sci.*, *26*(6), 1–7.

[9] Byaligoudra, A.I., Aparna, B., Vani, N. and Naidu, G.M. (2019). Growth and Instability of Major Millets in Andhra Pradesh, India. *Int. J. Curr. Microbiol, App. Sci.* 8(7), pp. 985–993, ISSN: 2319-7706.

[10] Chaba, A.A. (2022). Punjab, the granary of India, lags in millet production: data. *The Indian Express.* Accessed online: https://indianexpress.com/article/cities/jalandhar/punjab-millet-production-data-food-indiustry-8351583/#:~:text=In%20Punjab%2C%20mainly%20two%20major,'jowar%2Fcharri

[11] Chhavi, A. and Sarita, S. (2012). Evaluation of composite millet breads for sensory and nutritional quality and glycemic response, *Mal J. Nutr,* 18:89–101.

[12] Devi, P.B., Vijayabharathi, R., Sathyabama, S., Malleshi, N.G. and Priyadarisini, V.B. (2014). Health benefits of finger millet (*Eleusine coracana* L.) polyphenols and dietary fiber: a review, *J. Food Sci. Technol,* 51(6): 1021–1040.

[13] Downs, S.M., Kapoor, R., Merchant, E.V., Sullivan, T., Singh, G., Fanzo, J. and Ghosh-Jerath, S. (2022). Leveraging Nutrient-Rich Traditional Foods to Improve Diets among Indigenous Populations in India: Value Chain Analysis of Finger Millet and Kionaar Leaves. *Foods, 11*(23), 3774.

[14] Food and Agriculture Organisation of the United Nations (2014). ICN2 Second International Conference on Nutrition. Accessed online: https://www.fao.org/about/meetings/icn2/news-archive/news-detail/en/c/265240/#:~:text=Hidden%20hunger%2C%20or%20micronutrient%20deficiencies,billion%20people%20across%20the%20globe.

[15] Food and Agriculture Organisation of the United Nations (2022). Hunger and Food Insecurity. Accessed online: https://www.fao.org/hunger/en.

[16] Food and Agriculture Organisation of the United Nations (2022). International Year of Millets 2023. Accessed online: https://www.fao.org/millets-2023/en.

[17] Ghimiray, S., Chhetri, B., Dutraj, S. and Sharma, T. (2018). Sustainable Livelihood for Hills: A Case Study of Lepcha. In: (Choudhury, J., S. Goswami and A. Mahato eds.) Tribal Livelihood: Issues, Challenges and Opportunities. Tribal Research and Cultural Institute, ISBN: 978-93-86707-22-2.

[18] Global Hunger Index, India (2022). Accessed online: https://www.globalhungerindex.org/india.html#:~:text=In%20the%202022%20Global%20Hunger,of%20hunger%20that%20is%20serious.

[19] Hema, V., Ramaprabha, M., Saraswathi, R., Chakkaravarthy, P.N. and Sinija, V.R. (2022). Millet Food Products. In: (Handbook of Millets-Processing, Quality, and Nutrition Status). *Springer*, pp. 265–299, Singapore.

[20] Joshi, V.K., Garg, V. and Abrol, G.S. (2012). Indigenous fermented foods. In: (Joshi V.K. and Singh R.S. eds.) Food Biotechnology: Principles and Practices. *IK Publishing House, New Delhi*, pp. 337–374.

[21] Kalinova, J. and Moudry, J. (2006). Content and quality of protein in proso millet (*Panicum miliaceum* L.) varieties. *Plant Foods Hum Nutr,* 61:45–49.

[22] Kerala: E-Catalogue for Export of Millets and Value-Added Products (2022). Agricultural and Processed Food Products Export Development Authority. Ministry of Commerce and Industry. Govt. of India.

[23] Kumar, A., Gaur, V.S., Goel, A. and Gupta, A.K. (2015). De novo assembly and characterization of developing spikes transcriptome of finger millet (*Eleusine coracana*): a minor crop having nutraceutical properties. *Plant Mol Biol Rep.,* 33(4): 905–922.

[24] Kumar, A., Tomer, V., Kaur, A., Kumar, V. and Gupta, K. (2018). Millets: A solution to agrarian and nutritional challenges. *Agric Food Secur,* 7(1):31.

[25] Kumari, N., Gupta, A. and Sheikh, S. (2014). Development of low glycemic foods with the use of pearl millet and finger millet. *Int J. Sci. Res.,* 3:193–195.

[26] Madhya Pradesh: E-Catalogue for Export of Millets and Value-Added Products (2022). Agricultural and Processed Food Products Export Development Authority. Ministry of Commerce and Industry. Govt. of India.

[27] Makadia, J.J., Patel, K.S. and Ahir, N.J. (2014). Marketing of minor millets grown in tribal Dang district of South Gujarat. *International Research Journal of Agricultural Economics and Statistics, 5*(2), 197–200.

[28] Manithottam, J. and Francis, M.S. (2007). Ethnobotany of Finger Millet among Muthuvan tribes of Idukki district, Kerala. *Indian Journal of Traditional Knowledge.* 6(1), pp. 160–162.

[29] Mishra, S. and Chaudhury, S.S. (2012). Ethnobotanical flora used by four major tribes of Koraput, Odisha, India. *Genetic resources and crop evolution, 59*(5), 793–804.

[30] Modha, H. and Pal, D. (2011). Optimization of Rabadi-like fermented milk beverage using pearl millet. *Journal of food science and technology, 48*(2), 190–196.

[31] Nagalakshmi, K. and Beatrice, D.A. (2013). Comparative study of the glycemic response of kodo and barnyard millets in type 2 diabetic subjects, *IJGHC* 2:930–935.

[32] Nagaraj, N., Kundu, K.K., Haldar, S., Bantilan, C. and Chandrakanth, M.G. (2013). Baseline Scenario of Rainy Season Pearl Millet in Haryana. Working Paper Series No. 40. Patancheru 502 324, Andhra Pradesh, India: International Crops Research Institute for the Semi-Arid Tropics, 28 pp.

[33] Narayanan, J., Sanjeevi, V., Rohini, U., Trueman, P. and Viswanathan, V. (2016). Postprandial glycemic response of foxtail millet dosa in comparison to a rice dosa in patients with type 2 diabetes. *Indian J. Med Res.,* 144:712–717.

[34] Niyogi, D.G. (2018). Culture has helped millets survive. DownToEarth. Accessed online: https://www.downtoearth.org.in/news/food/culture-has-helped-millets-survive-61730.

[35] North Eastern States: E-Catalogue for Export of Millets and Value-Added Products (2022). Agricultural and Processed Food Products Export Development Authority. Ministry of Commerce and Industry. Govt. of India.

[36] Paschapur, A.U., Joshi, D., Mishra, K.K., Kant, L., Kumar, V. and Kumar, A. (2021). Millets for Life: A Brief Introduction. In: (Kumar, A., M.K. Tripathi, D. Joshi, and V. Kumar eds.) Millets and Millet Technology. *Springer*, pp. 51–80, ISBN 978-981-16-0675-5, ISBN 978-981-16-0676-2.

[37] Patil, K.B., Chimmad, B.V. and Itagi, S. (2015). Glycemic index and quality evaluation of little millet (*Panicumm iliare*) flakes with enhanced shelf life. *J Food Sci Technol* 52:6078–6082.

[38] Pilla, V.V.S.S., Siripurapu, K.K. and Laxmi, B. (2021). The Bio-Cultural Diversity of Jowar / Sorghum Atukula Panduga of the Indigenous Gond Community of Kumuram Bheem Asifabad district, Telangana State, India. WASSAN—RRA Network.

[39] Poshadri, A., Kumar, Y.P., Charan, G.S., Raghuveer, M., Kumar, M.S. and Devi, A.R. (2019). Energy Rich Composite Millet and Soybean based Malted Weaning Mix: A Complementary Food in Tribal Areas of Adilabad District, India. *Int. J. Curr. Microbiol. App. Sci.*, 8(2), pp. 2058–2064. ISSN: 2319-7706.

[40] Rawal, U.D. (2022). Millet farming brings nutrition, financial security for women farmers in Bihar. *30 Stades: Stories that Inspire*. Accessed online: https://30stades.com/2022/01/17/millet-farming-brings-nutrition-financial-security-women-farmers-empowerment-bihar/

[41] Rawat, L., Karnatak, A.K., Nautiyal, B.P. and Bisht, T.S. (2020). Small millets: Making a potential comeback in state Uttarakhand. Vigyan Varta 1 (2): 11–16.

[42] Rawat, L., Karnatak, A.K., Bisht, T.S. and Kukreti, A. (2021). Minor Millets: Profile and Ethnobotanical Scenario. In: (Kumar, A., M.K. Tripathi, D. Joshi, and V. Kumar eds.) Millets and Millet Technology. Springer, pp. 51–80, ISBN 978-981-16-0675-5, ISBN 978-981-16-0676–2.

[43] Rawat, V.S. (2022). UP eyes bigger pie of millet exports, plans to expand jowar acreage. *Business Standard*. Accessed online: https://www.business-standard.com/article/economy-policy/up-eyes-bigger-pie-of-millet-exports-plans-to-expand-jowar-acreage-122102001054_1.html#:~:text=Some%20of%20the%20major%20millet,jowar%2C%20bajra%2C%20kodo%20and%20sanwa&text=Uttar%20Pradesh%20(UP)%2C%20the,looking%20to%20expand%20their%20exports

[44] Sah U., Dixit, G.P., Kumar, H., Ojha, J., Katiyar, M., Singh, V., Dubey, S.K. and Singh, N.P. (2021). Performance of Millets in Bundelkhand Region of UP State. *Indian Journal of Extension Education,* Vol. 57, No. 4, pp. 120–125, ISSN: 0537-1996, ISSN: 2454-552X.

[45] Sahu, R.K. (2019). Cultivation patterns of small millets in the tribal belts of Chhattisgarh state: An in-depth study. *Journal of Pharmacognosy and Phytochemistry*; 8(12): 2538–2543. E-ISSN: 2278-4136 P-ISSN: 2349-8234.

[46] Sahu, R.K. and Sharma, M.L. (2013). Medicinal and other uses of small millets by the tribal farmers of the Bastar Plateau Zone of Chhattisgarh. *Agric.* Update, 8(4): 596–599.

[47] Saju, S. (2018). Forgotten Kodo-Kutki millets on a comeback trail in this MP tribal belt. *The Indian Express*. Accessed online: https://indianexpress.

com/article/india/forgotten-kodo-kutki-millets-madhya-pradesh-dindori-mal nutrition-tribals-baiga-5395798/

[48] Sharma P. (2008). Minor millets of Tribal Area. Agrotech Publishing Academy, ISBN: 81-8321-082-1.

[49] Sharma R. (2021). Locating food in the cultural history of Sikkim; A study of fermented drinks, Doctoral dissertation submitted to Sikkim University.

[50] Sheikh, S.M. and Singh, O. (2013). Pseudocereals and millets: the lost crops of Kashmir. *Genet Resour Crop Evol,* (2013) 60:1191–1199.

[51] Shrestha, R. (2017). Suitability of finger millet & barnyard millet incorporated pizza base in diabetes and cardiovascular disease. Ph.D. (Human Nutrition) Thesis submitted to the Department of Foods and Nutrition, G.B. Pant University of Agriculture and Technology, Pantnagar.

[52] Singh, H.B. and Arora, R.K. (1972) Raishan (*Digitaria* sp.)—a minor millet of the Khasi Hills, India. *Economic Botany,* 26 (4): 376–380.

[53] Soni, L.N. and A. Soni. (2016). Some Tribes of Western Madhya Pradesh: Their Economy and Change in Livelihood Patterns. In: (Tripathi, B., B.K. Mohanta eds.) Recent Researches on the Tribes of Central India. *Aayu Publications,* pp 455–462, ISBN: 97 8-93-85161-18-6.

[54] State Health Resource Centre, Bhopal. (2020). Tribal Health in Madhya Pradesh: A Roadmap. Atal Bihari Vajpayee Institute of Good Governance and Policy Analysis.

[55] Surekha N., Devi, R. and Naik, R. (2013). Development of value added low glycemic index barnyard millet (*Echinochloa frumentacea* Link) noodles. *IJFANS* 2:20–24.

[56] Tadele, Z. (2016). Drought adaptation in millets. In: (Shanker AK, Shanker C, eds.) Abiotic and biotic stress in plants: recent advances and future perspectives. *InTech, Rijeka,* pp. 639–662.

[57] Thurston, E. and Rangachari, K. (1909). Castes and Tribes of Southern India. Government Press, Madras, Vol. 2: C-J.

[58] Tiwari, N. and Srivastava, S. (2017). Effect of finger millet (*Eleusine coracana*) buns supplementation on serum glucose and serum lipids level in type 2 diabetics. *Asian J Dairy Food Res.,* 36:337–340.

[59] Touthang, L., Kalita, H., Makdoh, B., Angami, T., Tasung, A. and Singh, R.. (2022). Finger Millet Cultivation Technique by Memba Tribe of Mechuka, Arunachal Pradesh. *Agriculture and Food E-Newsletter,* Volume 4, Issue 12. E-ISSN: 2581-8315.

[60] Triveni, U., Patro, T.S.S.K., Rani, Y.S., Rao, P.J., Anuradha, N., Rao, M.M.V.S., Georgia, K.E., Praveen, B., Divyasudha, B., Hemamalini, C., Palanna, K.B., Kumar, M.H., Jamuna, P., Prabhakar and Tonapi, V.A. (2020). Tribal Sub Plan in Small Millets: Bringing hope in the lives of tribal farmers. *Agricultural Research Station, Vizianagaram,* ISBN: 978-93-90028-08-5.

[61] Ulfath, T.K., Nikhitha, M., Praditi, S., Kavitha, G.S., Myrene, R.D. and Sangeetha, P. (2016). Development of value added product to reduce the risk of hypoglycemia in diabetic population. *Eur J Pharm Med Res.,* 3:514–518.

[62] Venkatesan, P. and Sundaramari, M. (2015). Rationality and adoption of indigenous cultivation practices of Finger millet (*Eleusine coracana* (L.) Gaertn.) by the tribal farmers of Tamil Nadu. *International Journal in Management and Social Science*; 03 (02), pp. 970–977. ISSN: 2321-1784.

[63] Venkatesan, P., Sundaramari, M. and Venkattakumar, R. (2015). Ethnotaxonomical classification of little millet (*Panicum sumatrense*) by the tribal people in Tamil Nadu, India. *Eco. Env. & Cons.*, 21, pp. S215–S222, ISSN 0971–765X.

[64] Yadav N., Chaudhary, K., Singh, A. and Gupta, A. (2013). Evaluation of hypoglycemic properties of kodo millet based food products in healthy subjects. *IOSR J Pharm*, 3:14–20.

[65] Yadav, O.P. and Rai, K.N. (2013). Genetic improvement of pearl millet in India. *Agric Res.*, 2(4):275–292.

[66] Yadava, H.S. (2018). Feasibility of promoting bhagar in Mandla and Dindori. In: (Proceedings of Bioversity International and Action for Social Advancement) Kodo and kutki millet in eastern Madhya Pradesh: Local stakeholder workshop on current challenges and opportunities. Rome, Italy.

Millets: A Super Food for a Better Future

Rachana Patil and Vaidehi Choudhari

Rural Management, S.P. Mandali's Prin. L.N. Welingkar Institute of Management Development & Research, Mumbai, India

ABSTRACT: In recent years, the public's attention has shifted from fast food products to healthy food products. This trend is significant due to an increase in the number of patients suffering from various medical conditions. People want to eat healthy, but they don't know which foods are good for them, what their bodies need, or what nutrition their food provides.

The primary goal of the study was to determine the level of knowledge about super foods among various stakeholders. Super foods are foods with a high nutritional density. This means that they are high in nutrients while being low in calories. They are high in minerals, vitamins, and antioxidants. Antioxidants are natural molecules found in certain foods. They help to neutralise free radicals. They are high in minerals, vitamins, and anti-oxidants. Antioxidants are naturally occurring molecules found in certain foods. They aid in the neutralisation of free radicals in our bodies. Free radicals are naturally occurring by-products of energy production that can cause havoc in the body. Millet is one of the most important super foods.

The research is based on both primary and secondary data. Primary research was conducted by interviewing various stakeholders to determine their level of awareness of millet products and consumption patterns. Various research tools were used to analyse qualitative and quantitative data. Millets have a higher nutritive value than wheat and rice, according to the findings of the research.

Millets can aid in the prevention and management of health issues such as diabetes, iron deficiency, and digestive issues. It contains antioxidants and may aid in the prevention of diseases such as cancer. Millets can withstand harsh weather and have higher pest resistance than wheat and rice. Despite their numerous advantages, millets are consumed far less than wheat and rice.

According to the research, a major issue among households and patients is a lack of awareness about the importance of millets. The study recommended that there is a strong need to create awareness among consumers about the benefits of consuming millets through various modes of advertising and promotion.

Keywords: Super Food, Millets, International Year of Millets 2023, Healthy Food.

INTRODUCTION

The Indian food market has undergone tremendous transformation over the years, with grain-based product consumption declining and high-value product consumption increasing. Indian customers have traditionally favoured fresh and raw food over processed and packaged food. Recent shifts in consumption patterns indicate that there is sufficient possibility to boost the country's processed and packaged food segments. According to Euro-monitor data on market volume, the fresh food market accounts for around 90% of the entire food market share. However, between 2000 and 2014, the packaged food industry grew at a rate of 7% each year, while the fresh food market grew at a rate of 5%. Health consciousness will be a major element in both driving initial packaged food consumption and shaping developing demand among Indian consumers. It has been observed that there is a rising trend among businesses to provide healthier options in their product line. Companies selling staples and pulses are introducing polish free pulses, multigrain atta variations, and so on; companies selling juices are introducing cold pressed juices; and companies selling biscuits and baked goods are introducing multigrain biscuits and ragi cookies. Although the healthy cooked food category is new, it is worth more than USD 1 billion and is being driven by today's generation, which values strong food ethics such as fresh or organic goods.

Millets are considered a nutritious food, and the millet market is expected to develop at a 4.5 percent CAGR from 2018 to 2025. The millet market was worth USD 9 billion in 2016 and is predicted to grow to USD 12 billion by 2025. The growing preference of Asia Pacific's urban population for nutritious food, along with the increasing unsustainable nature of rice and wheat production, will increase product demand during the projected period. Diabetes, obesity, and cardiovascular disorders such as heart attack, coronary artery disease, arrhythmias, and so on have all been linked to urban lifestyle choices and related food habits. Millets are high in proteins and minerals like calcium and iron, which can help prevent such disorders. As a result, rising public knowledge of the health benefits of millets consumption will drive industry growth by 2025.

According to FAO, global millet production in 2019 was predicted to be 28.4 million metric tonnes. India is the world's largest producer, accounting for 41.0% of the global market. Millet's prominence as a food staple has declined over the previous two decades, owing to a variety of causes such as rising earnings, increased urbanisation, and government policy. More than 50.0% of millet production is currently going to other applications rather than being consumed as a staple. Millets are high-nutritional-value superfoods. Millets are an excellent source of fibre for patients with chronic conditions such as diabetes, obesity, and heart disease. Millets include fibres, which aid digestion and fight digestive problems. Millets, when consumed on a regular basis, assist to avoid gastrointestinal disorders as well as renal and liver ailments. Thus objectives of the present study are as follows:

- To investigate the role of millets in health and nutrition.
- To determine the extent to which Indians are aware of the nutritional benefits of millets.
- To educate farmers on how producing millets might help them maintain environmental sustainability.

MATERIALS AND METHODS

The research is based on primary and secondary data. Primary research was conducted in Malawa region of MP. Secondary research was carried out by reading several research articles and health websites on the history of millets, their health benefits, and how farming millets can help with climate sustainability. Many articles, websites, and research papers were consulted for secondary research. Various stakeholders were consider for the primary research like Doctors, nutritionists, farmers, gym trainers using in person and telephonic interviews and online surveys. Primary research was conducted by interviewing various stakeholders to determine their level of awareness of millet products and consumption patterns. Various research tools were used to analyse qualitative and quantitative data. Simple tabular analysis approach was used to analyse the data.

FUNCTIONAL FOODS

Also known as nutraceuticals, functional foods are highly nutritious and associated with a number of powerful health benefits. For example, they may protect against disease, prevent nutrient deficiencies, and promote proper growth and development. Functional foods are ingredients that offer health benefits that extend beyond their nutritional value. Some types contain supplements or other additional ingredients designed to improve health.

Cargill India was the leading player in fortified/functional packaged food, and remains so, with over 20% value share.

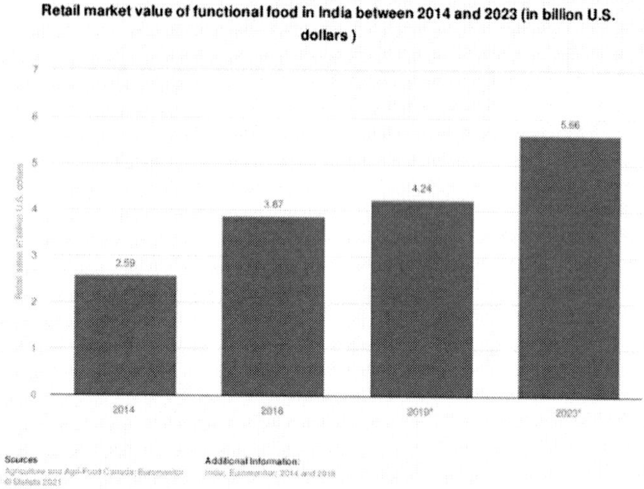

Figure 1: Retail Market Value for Functional Foods in India

FREE FROM FOOD PRODUCTS

Free from foods are made without one or more specific ingredients, making them suitable for people who have a certain allergy, intolerance or health

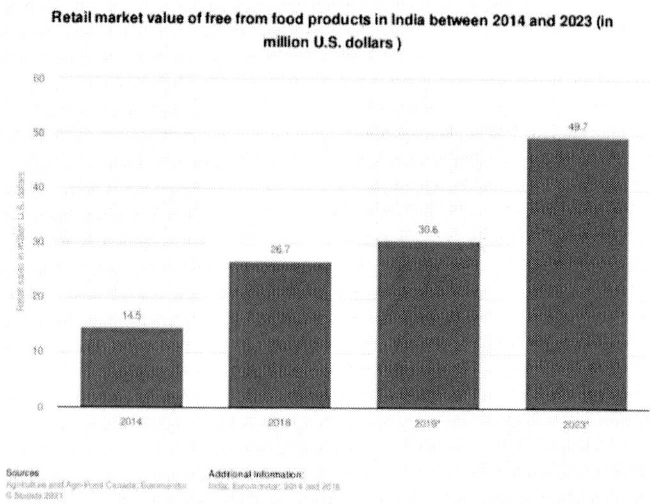

Figure 2: Retail Market Value for Free from Food Products in India between 2014 and 2023

requirement and may need to avoid certain food components. The retail value of the free from food products market in India amounted to approximately 27 million U.S. dollars in 2018. This was forecast to reach nearly 50 million dollars by 2023. The naturally healthy foods category accounted for the highest retail market value within the Indian health and wellness sector during the measured time period.

NATURALLY HEALTHY FOODS

The retail value of the naturally healthy food market in India amounted to approximately 3.8 billion U.S. dollars in 2018. This was forecast to reach nearly 8.5 billion dollars by 2023. This category accounted for the highest retail market value within the Indian health and wellness sector during the measured time period.

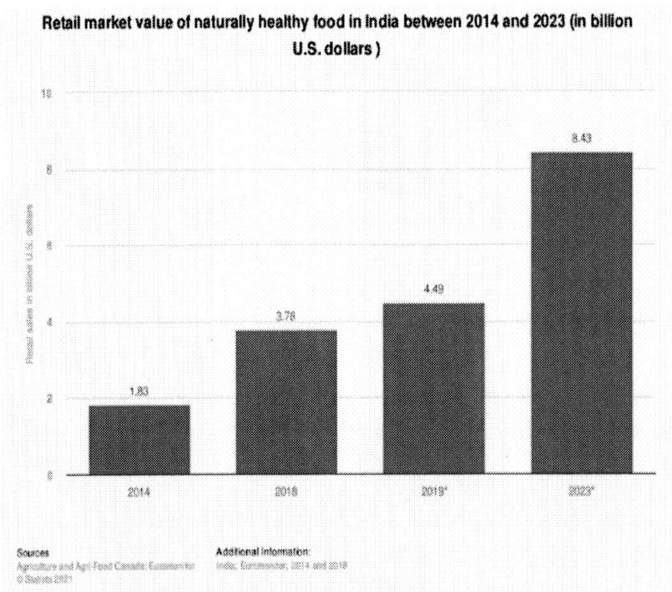

Figure 3: Retail Market Value of Naturally Health Food in India between 2014 and 2023

SUPER FOODS

Superfoods are foods that have a very high nutritional density. This means that they provide a substantial amount of nutrients and very few calories. They contain a high volume of minerals, vitamins, and antioxidants. Antioxidants are natural molecules that occur in certain foods. They help neutralize free radicals in our bodies. Free radicals are natural by-products of energy production that can wreak havoc on the body.

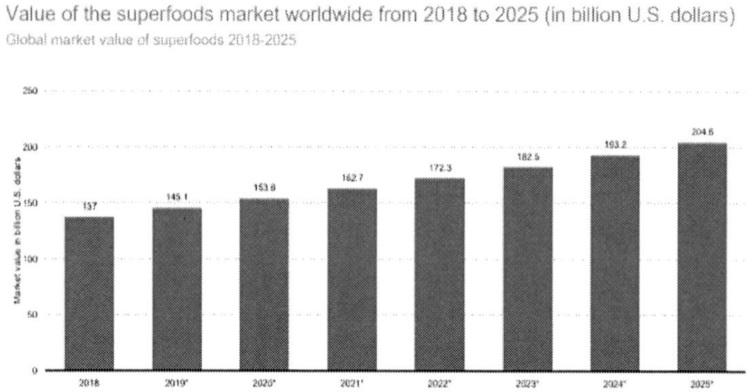

Figure 4: Worldwide Market Value of the Superfoods

MILLETS

Millets are good sources of energy. They provide protein, fatty acids, minerals, vitamins, dietary fibre and polyphenols. Millets are major food source in arid and semi-arid part of the world. Millet protein contains high quantity of essential amino acids (EAAs) which body cannot produce on its own. The bran and germ layers of millets are essential parts and removing them reduces the benefits of millets as they are rich in fibres and can help in digestion related problems. Millets are source of antioxidants, such as phenolic acids and glycated flavonoids. The Millets are characterized to be potential prebiotic and can enhance the viability or functionality of probiotics with significant health benefits.

Millets are important foods in many underdeveloped countries because of their ability to grow under adverse weather conditions like limited rainfall. It has been found that millets have various nutrition and medical functions which make them extremely healthy option. It is a drought resistant crop and can be stored for a long time without insect damage. Due to these qualities' millets can be extremely helpful during famines.

In 2007, global millets production reached about 32 million tonnes with the top producing countries being: India (10,610,000), Nigeria (7,700,000), Niger (2,781,928), China (2,101,000), Burkina Faso (1,104,010), Mali (1,074,440), Sudan (792,000), Uganda (732,000), Chad (550,000) and Ethiopia (500,000) (FAO, 2009).

The use of millet has significantly decreased, according to a report published in The Hindu in 2014. According to the DHAN Foundation, consumption of other small millets has decreased by 83 percent over the past five decades in India, while consumption of finger millets has decreased by 47 percent despite their nutritional benefits and resistance to climate change. The easy supply of rice and wheat through PDS during the 1961–2009 period, which led to food consumption away from the small millets in the producing regions, has been blamed for the drop. Similarly, over the 1961–2009 period, tiny millets' cultivation decreased from 7.22 million hectares to 2.2 million hectares, according to M. Karthikeyan, the foundation's project leader. Consumption of small millets has been restricted by inadequate investments in product development and commercial success, poor social status of small millet food, resistance to dietary habits, and ignorance of the use of small millets in the daily diet, according to Mr. Karthikeyan.

It is observed that millets are often ground into flour. Sometimes millets are prepared as beverages. Roti, made from pearl millets has been the primary food of farmers in Gujarat India. There is an emerging need for the world to feed its growing population, therefore, it is important to explore plants such as millets that are grown locally and consumed by low-income households in places like India and the Sahel zone. Cereals millets-based foods and beverages are known worldwide and are still part of the major diet in most African countries.

NUTRITIONAL COMPOSITION OF MILLET GRAINS

Table 1: Amino Acid Profiles of Different Millet Grains Variety

Amino acids (g/100g)	Foxtail millet (defatted flour) (a)	Proso millet (Dehulled Grain) (b,c)	Pearl millet (true prolamine) (c)	Finger millet (native grain) (d)
Essential Amino Acid				
Isoleucine	4.59	4.1	5.1	4.3
Leucine	13.60	12.2	14.1	10.8
Lysine	1.59	1.5	0.5	2.2
Methionine	3.06	2.2	1.0	2.9
Phenylalanine	6.27	5.5	7.6	6.0
Threonine	3.68	3.0	3.3	4.3
Valine	5.81	5.4	4.2	6.3
Histidine	2.11	2.1	1.7	2.3
Tryptophan	NA	0.8	1.2	NA
Nonessential Amino Acid				
Alanine	9.30	10.9	8.1	6.1
Arginine	3.00	3.2	0.9	3.4
Aspartic acid	7.71	6.2	6.2	5.7
Cystine	0.45	NA	0.8	NA
Glutamic Acid	22.00	21.3	22.8	23.2
Glycine	2.91	2.1	0.7	3.3
Serine	4.56	6.3	5.4	5.3
Tyrosine	2.44	4.0	2.7	3.6
Proline	5.54	7.3	8.2	9.9
*PER[b]	0.80	1.10	1.60	2.00

*Protein Efficiency Ratio (PER). NA: not available. References: (a) Kamara, *et al.* (2009); (b) Bagdi et al., 2011; (c) Saldivar (2003); (d) Devi et al. (2011).

Table 2: Proximate Composition of Millet Grain Varieties

Component (g/100g, dry basis)	Foxtail millet flour	Fonio whole grain	Proso millet dehulled grain	Pearl millet whole grain	Finger millet native grain
Protein	11.50	9–11	11.58	14.8	8.2
Ash	0.47	1–1.1	NA	1.64	2.7
Fat	2.38	3.3–3.8	4.9	4.86	1.8
Total CHO*	75.2	84–86	80.1	59.8	83.3
Crude fiber	NA	NA	0.7	12.19	3.5
References	Kamara et al. (2009)	Vodouhe et al. (2003)	Bagdi et al. (2011)	Taylor et al. (2010)	Devi et al. (2011)

*Carbohydrate (CHO); NA: not available

SOME POTENTIAL HEALTH BENEFITS OF MILLETS

Millets should not be just viewed as potential alternative to common grains as they have many health benefits as compared to common grains. Millet is also rich in phytochemicals, including phytic acid, which is believed to lower cholesterol, and phytate, which is associated with reduced cancer risk. Millet is gluten-free, therefore an excellent option for people suffering from celiac diseases often irritated by the gluten content of wheat and other more common cereal grains. It has a high concentration of the amino acid tryptophan, which helps to improve mood. Aids in the treatment of obesity and excessive cholesterol. It is also useful for people who are suffering from atherosclerosis and diabetic and heart disease. Millets extract from the seed coat were reported to have shown high antibacterial and antifungal activity compared to whole flour extract due to high polyphenols content in seed coat.

MILLET IN THE INDUSTRY

Millets have good grain qualities suitable for processing. Processing of the grain for many end uses involves primary (wetting, dehulling and milling) and secondary (fermentation, malting, extrusion, popping and roasting) operations. All the Millets can be milled by hand grinding or machine milling. Dehulling is not advised as it causes nutrition loss. The emerging principal uses of millets as an industrial raw material include production of biscuits and confectionery, beverages, weaning foods and beer. Grits, flour, and meals from cereals such as millet, sorghum, and corn are now common items in the market. Soft biscuits and cookies are being made using sorghum, maize and wheat composites. Many brands of beer in the underdeveloped countries market have substantial content of local cereal such as millet to reduce the cost of imported barley. The industries are confronted with a number of problems which tend to diminish product qualities and affect overall utilization. In the non-alcoholic beverage and weaning food sectors, storage quality of the grain, nutritional losses after processing, high cost of imported equipment and variation among cultivars are some of the problems militating against improved utilization of millet in the developing countries.

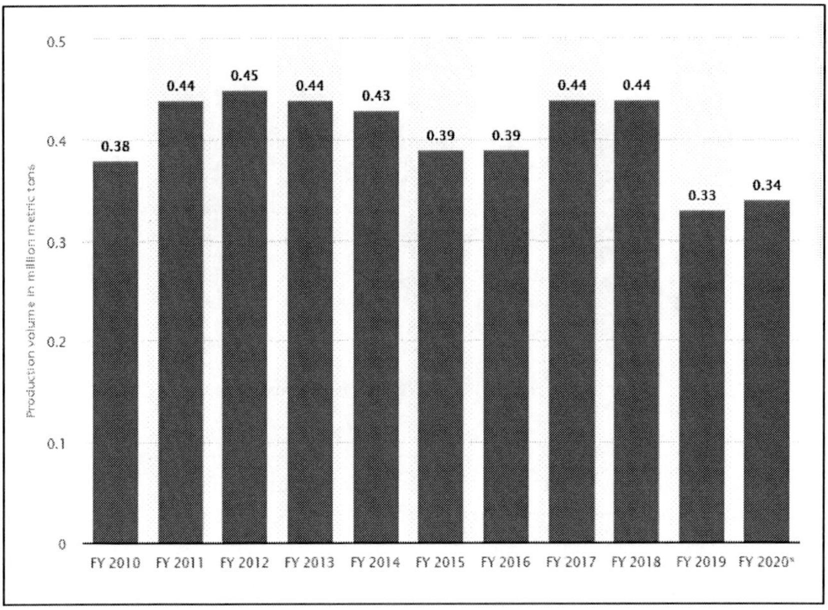

Figure 5: Production Volume of Small Millets during Kharif Season in India from Financial Year 2010 to 2019

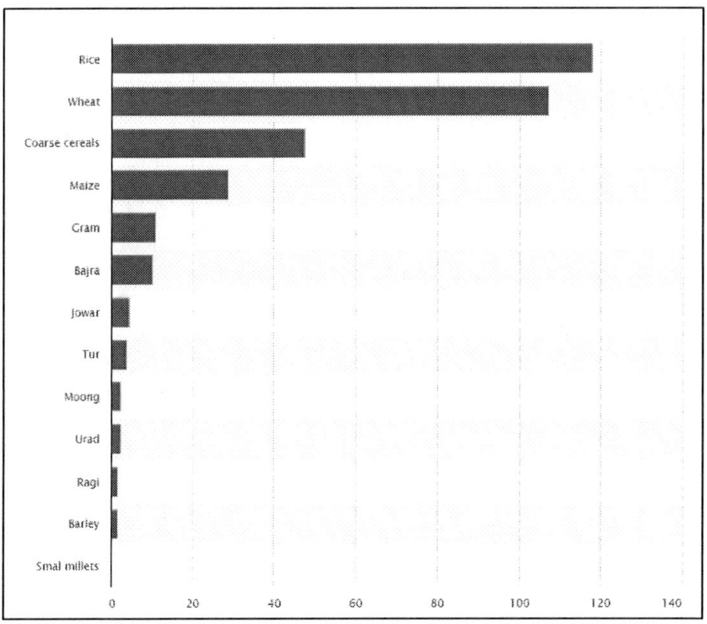

Figure 6: Volume of Food Grains Produced in India in Financial Year 2020 (Volume in million metric tons)

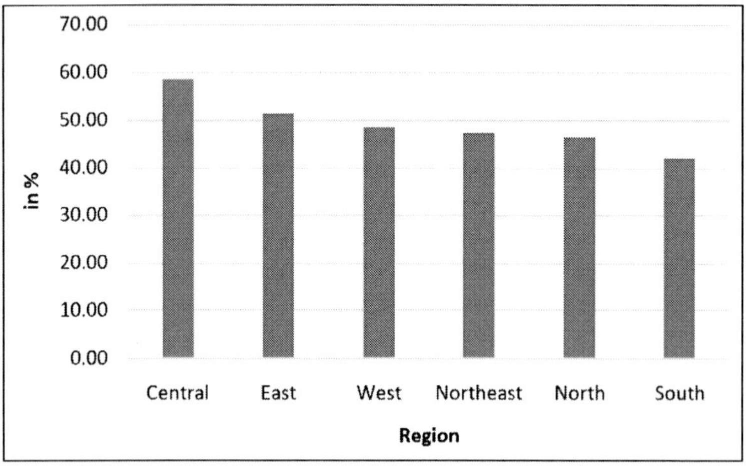

Figure 7: Share of Protein from Cereal and Millets Consumption Urban India FY 2016, by Region

STAKEHOLDERS

Table 3

Category	*Stakeholders*
Producers/ Sellers	Millet Farmers, Farmers involved in Organic farming, Healthy Food Industry, SHGs, FPOs, Organic Food Product Companies, Farm Workers, Farmers
Regulators and Policymaker	FSSAI, Ministry of FPI, Ministry of Agriculture, Ministry of Health
Consumers	Households, Health-conscious community, Gym goers, Cancer patients, Diabetics patients, celiac patients, Obese people, Anemic and calcium deficient patients, Chef, Restaurants
Research	Market Researchers, Food Scientist, Students
Advertisement	Newspapers, Haat, Social Media (organic Reach), YouTube Channel, TV advertisement, exhibition, Health blogs, training and support programs
Ambassadors	Gym Leaders/Trainers, Doctors, SHGs, Nutritionist, Government, Sportsperson, famous Chef, Government Schools and Anganwadis, Hospitals, NGO, Social Media Influencer
Investors	Bank, Insurance companies, Agri Investors, Microcredit institutions
Operations	Retailers, Packaging Material Companies, Transport, Storage, e-commerce sites, exporters/importers

GOVERNMENT POLICIES

Government has recognized the role of millets in the food chain. Under the National Food Security Mission (NFSM) of the preliminary targets for enhancing food grain production by an additional 25 Million Tonnes, the share allocated for millets is 2 Million Tonnes i.e., 8% of the enhanced food grain production. The Indian policymakers refocused their attention towards millet farming systems and enacted policies to create an enabling environment

Table 4: Policy Changes and Possible Outcomes

Sl. No.	*Government Action*	*Possible Effect on Environment/ Agriculture*	*Possible Effect on Health Status*	*Possible Effect on Financial Status*
1.	Compulsory maintenance of landraces by Government and seed producing companies	Helps maintain diversity	Creates possibility for selection for better nutrition and taste qualities	Lowers burden on farmers for landrace maintenance
2.	Limiting acreage under orchard crops/ commercial crops as a percentage of land holding for farmers with larger holdings	Promotes a healthy balance in the environment and ensures diversity	Ensures nutrition.	Helps farmers with more income
3.	Involving private sector companies in promoting millets. E.g., Horlicks to produce millet-based food products	May promote monoculture. Preservation of landraces by the company	Improved health of women/children. With rural areas changing into urban food habits, this ensures millet consumption.	Better market for product
4.	Include millets in MDM scheme and procure from local sources	Maintains local production	Helps to provide better nutrition to kids and helps retain their habit of consumption of millets.	Has a readymade market for the produce. Farmer need not have to struggle for selling his produce

for the farmers. With respect to millets production, some of the existing schemes by the Government of India include:

1. Integrated Cereals Development Programmes in Coarse Cereals (ICDP-CC) based Cropping Systems Areas under Macro Management of Agriculture (MMA).
2. Initiative for Nutritional Security through Intensive Millet Promotion (INSIMP) a part of Rashtriya Krishi Vikas Yojana (RKVY) which is the only comprehensive initiative to support millet production.
3. Rainfed Area Development Programme (RADP): a component of the Rashtriya Krishi Vikas Yojana (RKVY).

The year 2023 will be observed as the International Year of Millets, following India's proposal to the Food and Agriculture Organization, which was approved in at the 160th session of the FAO Council in December 2018. Through decades, the country has enjoyed a rich association with millets, though the Green Revolution favoured rice and wheat. Millets survived, thanks to cultural traditions, but came to be known as "coarse grains". The Government of India's Millet Mission comes under the National Food Security Mission (NFSM), launched in October 2007. NFSM-Coarse Cereals is divided into two parts and one of them is the sub-mission on nutri-cereals to be implemented in 202 districts of 14 states. Before this, millets were being promoted under INSIMP (Nutritional Security through Intensive Millets Promotion) during 2011–12 to 2013–14. Currently, millets are being promoted through technology dissemination, quality seeds through millet seed hubs, awareness generation, minimum support price and inclusion in PDS.

PRIMARY RESEARCH

According to data collected (Table 5) from patients suffering from one or more diseases such as diabetes, blood pressure, and digestive problems, 51.2 percent believe millets can help them fight these diseases and make life much healthier, while 46.5 percent are unsure if daily consumption of millets can help against diabetes, blood pressure, and digestive problems, and 2.3 percent believe millets in daily diet cannot help.

Table 6 reveled, doctors advised to patients to add millets to their diet, 44.2 percent said no, 27.9 percent said their doctor had advised them to add millets to their diet to maintain good health, and another 27.9 percent said their doctor had mentioned millets occasionally while prescribing a healthy diet plan.

Table 5: Patients Opinion on Consumption of Millets to Overcome Diseases such as Diabetes, Blood Pressure and Digestive Problems

Sr. No.	*Opinion*	*Percentage of Respondents (%)*
1.	Yes	51.20
2.	No	46.50
3.	May Be	2.30
Total		**100**

Table 6: Doctors' Advice on Millets Consumption

Sr. No.	*Insisted by Doctor*	*Percentage of Respondents (%)*
1.	Yes	27.90
2.	No	44.20
3.	Some time	27.90
Total		**100.00**

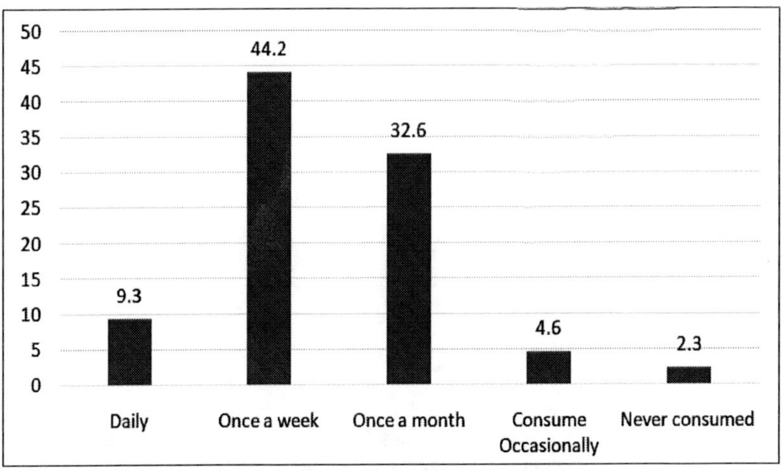

Figure 8: People Consuming Millets (%)

Data shows that 44.2 percent people consume millets once a week, 32.6 percent consume millets once a month and 9.3 percent people consume millets daily in one form or other. 2.3 percent people have never consumed millets and 4.6 percent people consumes millets occasionally.

Many patients chose fruits and eggs for living a healthy lifestyle, with millets coming in second place. The report also identified crucial characteristics that can persuade patients to include millets in their daily diet. The most essential component is guidance from a doctor/nutritionist, followed by influence from family and friends, and the taste of millet-based processed foods. Product availability, pricing, packaging, and advertising were less influential than the prior three criteria.

FARMERS

We discovered that 50% of the farmers in the Malawa region of MP produce millets based on data obtained about them. The majority of them grow maize as millets, and only a few grow Bajra. The region's main crops are wheat for Rabi and soybeans for Kharif. Farmers receive approximately Rs. 1000–1200 per quintal of corn production. There is no government subsidy programme for raising millets. The majority of farmers in the region rely on the monsoon to water their land, and agricultural patterns are also affected by the amount of rain received that year. All farmers sell their produce at the APMC Mandis in their neighbourhood.

GYM TRAINERS

According to the statistics we gathered from gym trainers, just 25% of people or trainers consume millets daily. 16% do not consume millets and do not advise gym goers to consume millet. Millets are occasionally consumed by the remaining trainers. When questioned about healthy food choices recommended by trainers other than supplements, most trainers chose eggs, millets, and meat, with eggs receiving the most support (77%). Few trainers recommend dal, wheat, rice, meat, and soya as healthy foods.

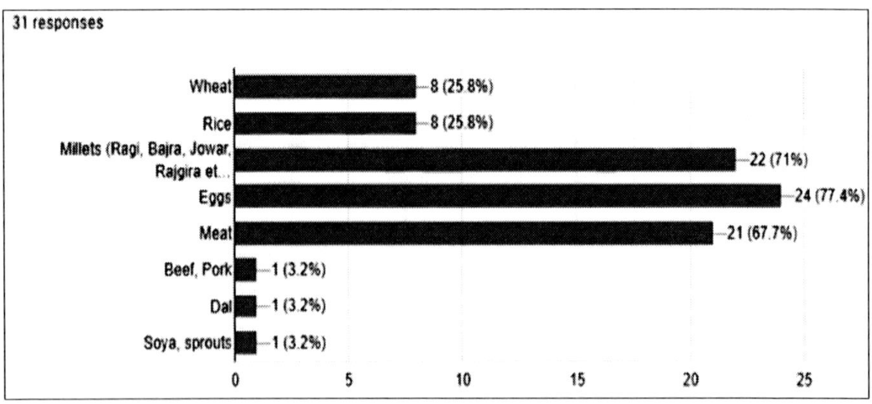

Figure 9: What Healthy Food Apart from Supplements Do You Recommend to Gym Goers?

Pre-workout supplements are recommended by around 57% of trainers. Millets are considered more nutritious by 61.3% of trainers than wheat and rice. 48.4% of trainers are unsure whether millets can aid with weight loss. When questioned about their first impressions of millets, the majority of trainers (about 84 percent) agreed that millets are nutritious.

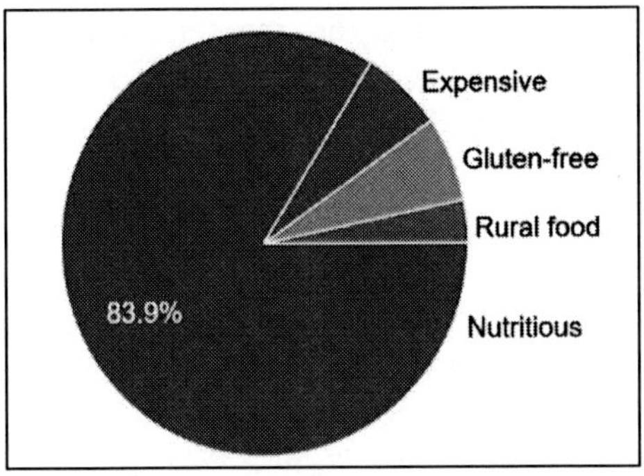

Figure 10: People Perceptions on Millets

HOUSEHOLDS

Approximately 63 percent of households strongly feel that a nutritious diet is a vital aspect in keeping good health. 85% of those polled do not follow a specific diet. The remaining 15% of persons who follow a special diet prefer a low cholesterol diet, a low calorie diet, and a low fat diet. Few people take fruits in addition to their usual diet to preserve excellent health. Supermarkets, street markets, and online businesses are where most people buy healthful food. According to the survey, people prefer Bajra and Jowar in the form of Roti, whilst 25% of ragi consumers prefer ragi in the form of cookies. Data on consumption suggest that wheat and rice are the most often consumed food grains, with the majority of respondents consuming wheat on a daily basis, followed by rice. Whereas Ragi, Maze, Bajara, and Jowar are consumed only a few times per month, a significant number of respondents do not consume millets at all. Taste and appearance are emerging as the most critical factors influencing this segment's decision to purchase the product. Following flavour and appearance in importance order are Gym/Doctor/Nutritionist Influence, Ease of Cooking, Market Accessibility, and Affordability. Packaging is the least of this segment's concerns.

RETAILERS

It was observed that all the retailers surveyed or interviewed were men, if a lady was present at the shop she was accompanied by a male. The interviews were done in Mumbai but most of them that is around 85 percent of the retail shop owners were not from Maharashtra but from other states like UP, West Bengal or other South Indian States. The retail shops that were of two types. One was a general store and other was shops dedicated to selling cereals, pulses and legumes. The small General stores were not selling any of the millets but they sold processed and packed millet-based products like flours of different millets. It was observed that women members of the family were the decision maker when it comes to buying products like flour or instant mix or any other healthy food products. Many of the shops around did not sell millet-based products at all. The number stood at 50% of all the retailers interviewed. Though the shops not selling millets or millet-based products did sell products that were rice or wheat based. When buying processed snack people predominantly look at factors like brand popularity and taste. The millet and millet-based products available at stores were like at some retail chains like Apna Bazaar and Sahakari Bhandar sold both raw millets and processed millet products like flour, instant mix, puffed millet products and some millet processed snacks too. While the smaller shops sold only millet flours. The retailers are most likely to advise the customers to buy a certain product based on his commission and velocity (how fast the product will reach the consumer) of the product which is related to brand popularity of the product.

DOCTORS

From the data that we have collected, it has been observed that majority of the doctors believes that millets are high in nutrition, and they also tend to recommend millets for daily consumption. But when asked about the comparison between the wheat and millets, many of them are not sure about which is more beneficial for health. They believe that millets are beneficial for diabetes and digestion problems but are not very confident about its effects on cancer patients. Many doctors themselves consume millets but not regular, they either consume once or twice a week. About 70% of doctors recommend consuming millets regularly for overall good health.

TRENDS IN MARKET/PRODUCT IDEAS

Protein Supplement: The protein supplement industry is expanding. The growing awareness of fitness and good health has aided the growth of the protein supplement market. When we talk about protein, we usually mean Whey protein, which is produced from milk. However, a new industry of plant-based protein supplements targeting lactose intolerant persons is emerging. The protein powders manufactured from soya, brown rice, and pea are listed below. Millets are also high protein foods, and a product comparable to the ones listed below can be imagined.

Healthy Snacks: According to secondary research, the demand for health food products is rapidly expanding. Primary research has led us to the conclusion that flavour and appearance are the most important factors influencing people's decisions to purchase packaged/processed food goods. Based on these data, we may conclude that the healthy snacking market is a promising one for enterprises. Some millet-based items, such as millet bars, biscuits, and cookies, are shown below.

CONCLUSION AND RECOMMENDATIONS

According to the research study, millets have a higher nutritional value than wheat and rice in terms of proteins, fibre, and micronutrients. Millets can help prevent and regulate health problems such as diabetes, iron deficiency, and digestive disorders. It contains antioxidants and may aid in the prevention of diseases like as cancer. Millets can tolerate hard climates and are more pest resistant than wheat and rice.

According to primary research, a key issue among households and patients is a lack of awareness. Millet product advertising is required to raise awareness. During the epidemic, YouTube's subscriber base increased. Aside from You Tube, Instagram is an excellent medium for connecting with others. Advice from doctors/nutritionists has emerged as one of the most important variables that can add millets to the plates of households. One approach to influence people is through health food festivals. Food festivals presenting various millet recipes can attract consumers to millets, as taste is emerging as a significant factor that can inspire people to consume millets. Protein supplements derived from millets have the potential to become a significant market in the fitness industry. Overall, the protein supplements market is expanding rapidly due to rising consumer protein knowledge, increased purchasing power, and rising cases of lifestyle disorders such as obesity, high blood pressure, diabetes, and malnutrition. Millet protein supplement is a viable choice in the plant-based protein supplement market. Aside from it, different products such as cookies, biscuits, chocolate bars, and ready-to-cook dishes might be introduced to entice customers.

RECOMMENDATIONS

The study recommended that there is strong need to create awareness about the millets ad its nutritive importance among all stakeholders to make changes in the consumption pattern.

All the stakeholders need to take consolidated efforts for this super food to make India the healthy country.

REFERENCES

[1] Amadou, I., Gounga, M.E. and Le, G. (2013). Millets: Nutritional composition, some health benefits and processing—A review. *Emirates Journal of Food and Agriculture, 25*(7), 501–508. Retrieved from https://elibrary.welingkar.org:2051/scholarly-journals/millets-nutritional-composition-some-health/docview/1373193008/se-2?accountid=89041

[2] Best Management Institute and (MBA) college, Mumbai Bengaluru: Weschool. We School—Welingkar Education for Full Time Courses and Part-Time Courses in Mumbai and Bangalore (n.d.). Retrieved January 7, 2023, from https://www.welingkar.org/ https://elibrary.welingkar.org/login?url=https://www.proquest.com/magazines/special-project-health-amp-wellness-foods-fast/docview/2334248508/se-2?accountid=194833

[3] Organic, Sukangan Sardana August 6, R. Priyanka March 24, Sunil Kumar BISOI August 26, Organic August 26, A M Rawat July 7, Mr. Omenta Daniel Dankwa August 6, Pradeep Rao August 6, shailendra Sonekar December 16, Organic December 16, Piyush Tiwary August 12, Best Organic food February 11, Deepak Joshi July 20, Deepak July 20, Organic July 20, Parth Kathuria August 1, Deepak Kumar July 1, Kritika Gupta July 18, Organic July 18, … Name*. (2022, September 30). Best Organic Companies in India in 2022: Pure Eco India—Organic Magazine Organic Directory. Pure Eco India. Retrieved January 7, 2023, from https://www.pureecoindia.in/best-organic-companies-india

[4] India Organic Food Market: Industry trends, share, size, growth, opportunity and forecast 2022–2027. India Organic Food Market Size and Growth Report 2022–2027 (n.d.). Retrieved January 7, 2023, from https://www.imarcgroup.com/indian-organic-food-market

[5] Potential for functional foods in the Indian market: Fi India. Potential for functional foods in the Indian market | Fi India. (n.d.). Retrieved January 7, 2023, from https://www.figlobal.com/india/en/visit/news-and-updates/potential-functional-foods-indian-market.html

[6] BD479E26A0—United States Department of State (n.d.). Retrieved January 7, 2023, from https://www.state.gov/report/custom/bd479e26a0-4/ https://www.mordorintelligence.com/industry-reports/india-gluten-free-foods-beverages-market-industry

[7] Decline in production, consumption of small millets. The Hindu (2021, November 17). Retrieved January 7, 2023, from https://www.thehindu.com/sci-tech/agriculture/The-good-are-not-wanted/article60430393.ece

Understanding Millet Cultivation in Maharashtra

Vandana Panwar, Deepak Deshmukh and Sujitsinh Dubal

PGDM-Rural Management (Emerging Economies), Prin. L.N. Welingkar Institute of Management Development & Research, Mumbai, India

ABSTRACT: India is a sub-tropical country with most of its geography lying in arid and semi-arid regions. Due to scarcity of water, it is imperative that crops requiring least possible water should be grown, possessing all nutrients requirement of the body. And 'Millets' fulfils all these, making them an essential food. All this prompted UNGA (United Nations General Assembly) to declare year 2023 as the 'International Year of Millets' (IYM). This study focuses on 'understanding millet cultivation in Maharashtra'. Starting with global and Indian scenario, the study narrows down on the specifics of three divisions of Maharashtra: Nasik, Konkan and Pune. An exploratory research design was adopted for the study. It examines value chain trends and several public-private partnership (PPP) models. Existence of three public private partnership models namely; Farmer Producer Organisations (FPOs), Contract Farming Method (CFM) and Farmers Associations were observed. The data were collected from nineteen villages covering seven talukas from four districts i.e. Nasik, Thane, Palghar and Satara. With the confidence level of 95%, margin of error of 10% and response distribution of 50%, the recommended sample size was 97. The researchers collected data from total 114 farmers; during November 2022 to December 2022. The collected data were analysed using frequency and percentage analysis. The profile of millet farmers in different PPP models in different studies were examined. The independent variables used in the study includes- age, educational qualifications, occupation, annual income, farm size, cropping pattern, farming experience in years and experience in the PPP mode of farming. The dependent variables are the PPP models-FPOs, contract farming method and Farmer Associations. The study could establish causal relationship in adoption of PPP models and increase in farmer income. Based on the findings of the study, some of the suggestions that may be useful for the functioning of PPP models and millet farmers include- the heterogeneity profile of farmers that can be used for establishing PPP models. Remarkable difference has been observed between the yield and net income of millet farmers; before and after the implementation of PPP models. The study would provide us with an opportunity to raise awareness on and direct our policy attention to their large-scale cultivation and use in our dietary intake, owing to their nutritional and health benefits and their suitability for cultivation under adverse and changing climatic conditions.

Keywords: *Public Private Partnership Models, Millets, Value Chain, Farmers, FPOs, Sustainable Agriculture, SDG.*

INTRODUCTION

The earliest evidence of millets was found in the Indus civilization during 3000 BC. It was the first ancient food grain planted as domesticated food. Millets are a hardy crop, requiring significantly less water and are grown in temperate, subtropical, and tropical dry environments. It is a category of small-seeded annual grasses produced predominantly on marginal land as grain crops (Rao, 2019). Millets, also called Climate-resilient crops known as "nutri-cereals" perform better than wheat and rice under difficult growing conditions and have a higher nutritional value. It can help achieve nutritional security sustainably (Reddy, 2013).

The year 2023 has been announced as the International Year of Millets. To increase the production and consumption of millets, the Indian government has implemented a few measures, including support for Post-harvest Value Addition, enhancing domestic consumption and branding of millets products Nationally and Internationally. Today, it is grown in 131 countries. Millet is a traditional staple food for 59 crore people in Asia and Africa. Table 1 below shows India produces more than 170 lakh tones of millet annually, which is 80 percent of Asia and 20 percent of global production. The global average yield is 1229 kg/ha whereas India's average yield is 1239 kg/ha.

Table 1: Global Scenario of Millets (2020–21)

Regions	*Area (Lakh Hectare)*	*Production (Lakh Tonnes)*
Africa	489	423
America	53	193
Asia	162	215
Europe	8	20
Australia & New Zealand	6	12
India	138	176
World	718	863

Source: FAO Stat 2021.

Sorghum, Pearl millet, finger millet, Little millet, BrownTop Millet, and Foxtail millet are among the millets grown in India according to various agro-climatic conditions. The main millet-growing states are **Rajasthan, Maharashtra, Karnataka, Andhra Pradesh, and Madhya Pradesh.** These states have many millet farmers who grow the grain for both domestic and international markets. Reference?

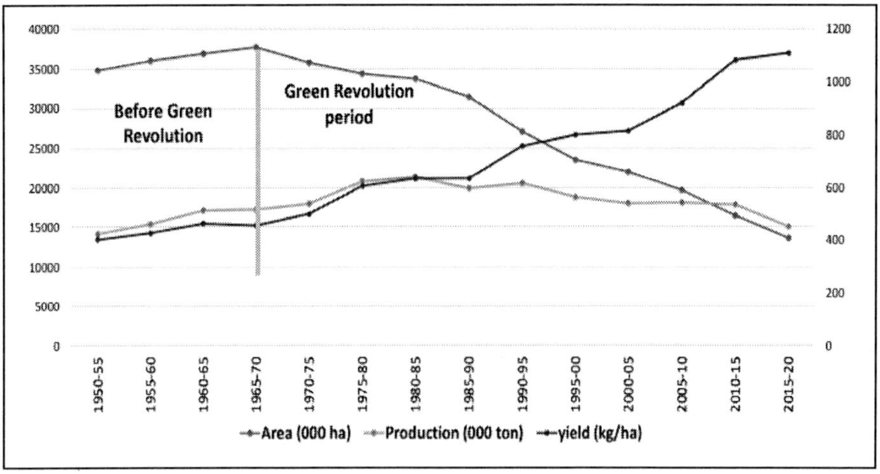

Figure 1: Millets in India (area, production & yield)

Source: Ministry of Agriculture, Government of India.

Millets are susceptible to being impacted by the cultivation of other food crops or cash crops due to the problems of shrinking land holdings, even though they require extraordinarily little water for production and poor soils for survival and growth. The above figure 1 shows the data recurring every five years in India. It is very evident that the area has decreased by 56 percent, and productivity has increased by 228 percent. There has been a high adoption of high-yielding varieties/hybrids. The ineffective pricing strategies brought on by middlemen's involvement in millet marketing were the other factor contributing to the output of millet's downward trend. Up to 1965–70: Millets constituted 20 percent of entire basket of grains for food, now it is only 6 percent. From 21.98 million US (United States) $ in 2017 to 24.73 million US $ in 2020, millets' exports have increased.

To bring back the glory, the government of India has encouraged Public-Private Partnership (PPP) mode, and value chain to empower the farmers and promote the economy to grow overall. It helps the millet farmers to improve their livelihood by utilizing these models effectively.

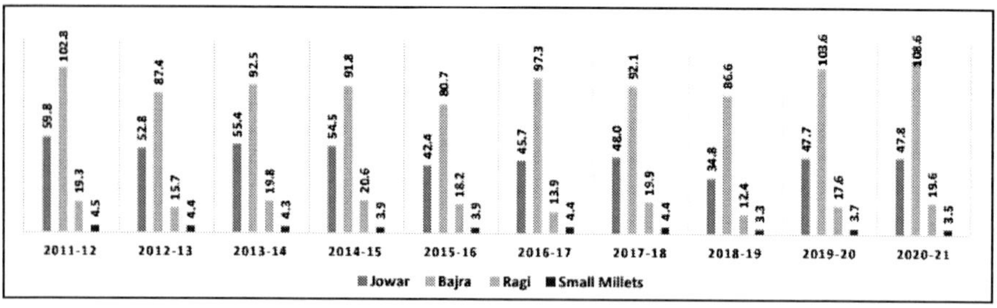

Figure 2: Production of Nutri Cereals during Last 10 Years in India

Source: Ministry of Agriculture, Government of India.

A study to understand the millet cultivation in Maharashtra through PPP models and value chain patterns was deemed relevant considering these aspects. The profile of the millet farmers helps to identify the socio-economic characteristics and perception on performance of implemented PPP model on similar initiatives in agriculture.

MATERIALS AND METHOD

An exploratory research design was adopted for the research study. With the increase in number of FPOs, western India has become the location of choice for these businesses, with Maharashtra taking the lead. Maharashtra has the second-largest population in India, the third-largest area, and the country's biggest economy (Ali 2022). The study examines value chain trends and several public-private partnership (PPP) approaches in millets. The respondents were asked questions based on the reasons for involvement in the selected model, the value chain patterns and the net income.

Maharashtra has a population of approx. 124,904,071 people in the year 2022. It has six revenue divisions, thirty-six districts, 357 talukas, and 41,000 villages. The study was conducted in three divisions i.e. Konkan, Nasik, and Pune. These divisions were purposely selected because they have major areas under millet cultivation. Based on the interaction with officials from the Commissionerate of Agriculture, and State Department of Agriculture, Mumbai, it was found that three public private partnership models namely, Farmer Producer Organisations (FPOs), Contract Farming Method (CFM) and Farmers Associations were functioning effectively.

The data were collected from nineteen villages covering seven talukas from four districts i.e. Nasik, Thane, Palghar and Satara. With the confidence level of 95%, margin of error of 10% and response distribution of 50%, the recommended sample size was 97. The researchers collected data from total

114 farmers who agreed to participate as respondents. Information was gathered through a focused group discussion method and structured interview schedule with an adequate number of questions regarding demographic profile, PPP models, and value chain pattern in Millets. The data were collected in the November-December months of 2022. The collected data were tabulated and analysed using frequency and percentage analysis.

The profile of millet farmers in different PPP models in different studies were examined. The independent variables used in the study includes- age, educational qualifications, occupation, annual income, farm size, cropping pattern, farming experience in years and experience in the PPP mode of farming. These variables are the influencing or contributing variables towards the perception on PPP model. The dependent variables are the PPP models- FPOs, contract farming method and Farmer Associations functioning effectively in the selected districts. The results are discussed in the following paragraph.

RESULTS AND DISCUSSION

The present study analyses the millet cultivation in Maharashtra. The results thus obtained have been organized and explained below:

1. Characteristic profile of millet farmers associated with different PPP models.
2. Reasons for the involvement of the farmers with different PPP models.
3. Value chain patterns followed by farmers.

1. Characteristic Profile of Millet Farmers Associated with Different PPP Models

Maharashtra is frequently praised as a pioneer in the nation's smallholder aggregation movement and is regarded as an agricultural standout with large production of cash crops and staples. It has over 13.6 million farmers, the majority of whom are smallholder and marginal farmers. It is among India's top producers of fruits and vegetables, contributing 5% and 11%, respectively, of the country's total fruit and vegetable production. Maharashtra is also the biggest producer of cotton, pigeon peas, and jaggery. Sorghum, soybeans, wheat, maize, and rice are some of the other important crops raised in the state. The figure 3 below shows the six revenue divisions of Maharashtra i.e. Amaravati, Aurangabad, Konkan, Nagpur, Nashik and Pune divisions. The study was conducted in the three divisions—Konkan, Pune and Nashik.

In six revenue divisions of Maharashtra, the Table 2, below displays the average area, production, and productivity of the three principal millets i.e.

Jowar, Bajra, and Ragi. Production refers to the process of creating, producing, or improving goods and services. It is the quantity produced. Whereas productivity is a term used to describe the efficiency or rate of production in agricultural economics.

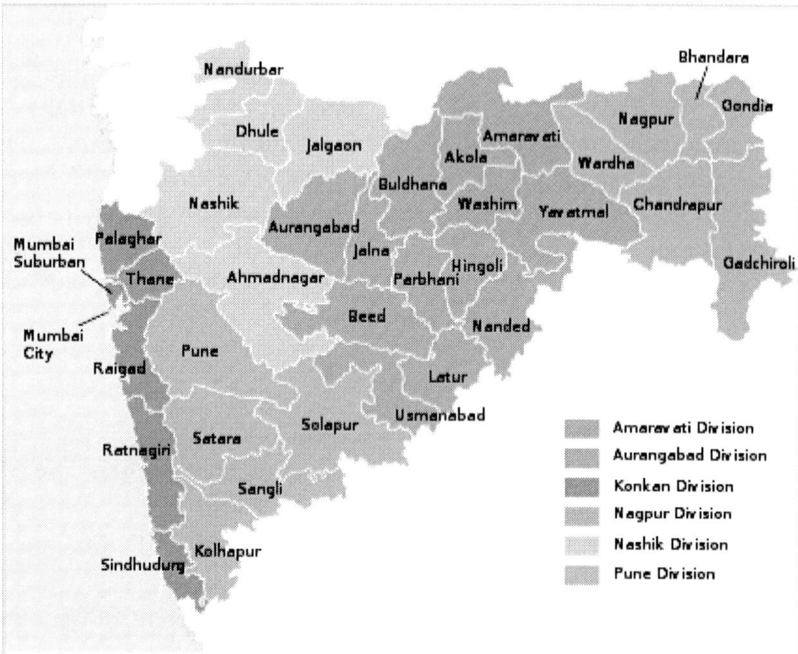

Figure 3: Revenue Divisions and Districts of Maharasthra State

The data reveals that the Nashik division has the largest areas of Jowar (Sorghum) grown at 551.03 hectares, with the largest production at 686.85 tonnes and productivity at 1263.25 kg/ha. The Pune division produces most of the bajra (pearl millet) on an area of 2422.72 ha, producing 2297.72 tonnes and yielding 1888.73 kg/ha. With a production of 329.96 tonnes, Ragi (Finger Millet) has the biggest area of 275.33 ha in the Konkan division, while Pune has the highest productivity with 1769.93 tonnes.

The demographic profile of millet farmers in the above Table 3, was analysed to understand the cultivation of millets. Ninety five percent of the respondents were males and only five percent were female.

1. **Age:** Age represents the maturity of an individual to make decisions. It was observed that maximum farmers in the sample were in the age group of 26–35 years, followed by the age group 36–45 years, more than 55 years and 46–55 years. There were only 1.3 percent respondents in the

Table 2: Average Area, Production and Productivity of Millets during 2021–22–23 in Maharashtra State

Sr. No.	*Division*	*Jowar (Sorghum)*			*Bajra (Pearl Millet)*			*Ragi (Finger Millet)*		
		Area in "00" ha	*Production in "00" Tonnes*	*Productivity in Kg /ha.*	*Area in "00" ha*	*Production in "00" Tonnes*	*Productivity in Kg /ha.*	*Area in "00" ha*	*Production in "00" Tonnes*	*Productivity in Kg /ha.*
1.	Konkan	0.00	0.00	0.00	0.00	0.00	0.00	275.33	329.96	1198.21
2.	Nasik	551.03	686.85	1263.25	1239.01	1341.78	1090.73	182.18	143.40	786.13
3.	Pune	301.07	330.24	1254.83	2422.72	2297.72	1888.73	140.63	224.49	1769.93
4.	Aurangabad	445.19	181.14	873.75	750.69	604.51	1278.21	0.00	0.00	0.00
5.	Amravati	284.31	195.39	700.24	1.13	0.31	282.90	0.00	0.00	0.00
6.	Nagpur	39.52	21.33	537.43	0.00	0.00	0.00	0.06	0.07	638.00

Source: Ministry of Agriculture and Farmers Welfare, Department of Agriculture and Farmers Welfare, Directorate of Economics and Statistics.

Table 3: Profile of Millet Farmers in Maharashtra

Sr. No.	Profile Characteristics		Respondents in Percentage (%)
1.	Age	15–25 Yrs.	1.3
		26–35 Yrs.	48.1
		36–45 Yrs.	34.2
		46–55 Yrs.	3.8
		>55 Yrs.	12.7
2.	Gender	Male	95
		Female	5
3.	Educational Qualification	Below SSC	6.3
		HSC	29.1
		Graduation	38
		Post-Graduation	26.6
4.	Landholding (1 hectares = 100 acres, 1 acre = 100 m^2)	Less than 1 h.a (Marginal farmer)	16.5
		1–2 h.a (Small Farmer)	41.8
		2–4 h.a (Semi Medium Farmer	40.5
		4–10 h.a (Medium Farmer)	1.3
5.	Occupation	Agriculture	98.7
		Any other	1.3
6.	Annual Income	up to Rs. 50,000	2.5
		Rs. 50,000–1 lakh	22.8
		Rs. 1–3 lakh	36.7
		Rs.3–5 lakh	31.6
		More than 5 lakh	6.3
7.	Farming experience	less than 6 years	29.1
		6–10 yrs	32.9
		11–15 yrs	25.3
		More than 15 yrs	12.7

Source: Authors.

age group of 15–25 years. The CEO (Chief Executive Officers) and Director of FPO's were in the age group of 26 –35 years, 36–45 years and more than 55 years. In the farmer society model, the respondent was in the age group of more than 55. This led to the conclusion that most millet farmers in the study area were in their middle and young years of life. They relied solely on agriculture for their livelihood as opposed to other sources. Additionally, because they do not want to incur farming risks, the elderly farmers were not inclined towards participation in PPP model. The results support a report from Thiyagarajan (2014) indicating 57.00% of PPP recipients fell into the group of middle age.

2. **Educational Qualification:** The process of bringing about a desired change in behavior in people is influenced by education. The level of education a person has influenced their knowledge by encouraging them to acquire new things and to comprehend what they have already learned. It is commonly accepted that the amount of confidence to try new things will increase with higher educational qualifications. The data collected on educational status shows the maximum respondents are graduates, followed by SSC, post-graduation, and below SSC. It was also noted that none of the respondents were illiterate. It may be presumed that most of the educated respondents expressed an interest in taking part in the agriculture model of public-private partnerships. The findings suggested that educated farmers were aware of the current extension system, and their attitudes toward partnership models in agriculture were favorable. According to the study by Thiyagarajan (2014), the majority (64.00%) of the beneficiaries in PPP initiatives had education up to the primary or higher level, is corroborated by the finding from this study as well.
3. **Farm Size:** The size of landholding of the millet farmers is categorized as marginal farmers, small farmers, Semi-Medium farmers, medium farmers and large farmers. It was observed that the majority (41.8 percent) of the millet farmers in the study were small farmers, followed by (40.5 percent) semi-medium farmers, (16.5 percent) marginal farmers, and (1.3 percent) medium farmers. According to the data, the landholdings were divided into many small pieces. The small and semi-medium farmers who are interested and have a positive attitude towards new modes of farming are getting more yield. According to Meera *et al.* (2004), 70% of the beneficiaries were small and semi-medium farmers, which is consistent with this conclusion.
4. **Occupation:** A person's occupation is the job they routinely do to receive a set amount of utilitarian compensation. In the present study,

adoption of the PPP model by millet farmers shows that agriculture was the main occupation for 98.7 percent of the respondents. They were solely dependent on agriculture for their livelihood. The remaining 1.3 percent of the respondents were engaged in subsidiary occupations like crane service, electric contractor, goldsmith, hotel owner, salaried employment, teacher, catering agency & vegetable shop, water purifier sales agency. It can be assumed that farming is a key source of livelihood in the study area, having been passed down from ancestors. Additionally, a significant element in the persistently high concentration of labor in agriculture has been the lack of employment opportunities outside of agriculture in and surrounding rural settlements. This conclusion is supported by the findings of Kumar (2011), according to which 95.00% of respondents were dependent solely on agriculture for their living.

5. **Annual Income:** Everyone agrees that income is a significant factor that has a variety of effects on human behavior. It was observed that 36.7 percent of the millet farmers have annual income of Rs. 1–3 lakhs, followed by 31.6 percent with Rs. 3–5 lakh, 22.8 percent with Rs 50,000 to 1 lakh. Only 6.3 percent had annual income of more than Rs. 5 lakh and 2.5 percent had income upto Rs. 50,000 annually. The public-private partnership has improved market connectivity that helps farmers make more money. Adequate training provided by public and commercial partners helps most millet farmers add value to millet grains production, with value addition in logistics, and marketing. Additionally, secondary employment such as raising animals provides the durability of the income if the monsoon fails or the prices of the commodities drop.
7. **Farming Experience:** It is evident from the table that most of the respondents have farming experience of 6–10 years, followed by less than 6 years, 11–15 years and more than 15 years. It is observed that most of the farmers in three separate PPP models were middle-aged or young and had 6 or more years of agricultural experience. The limited percentage of respondents with fewer than six years of farming experience can be attributed to the very few young farmers in the research area who chose agriculture as their primary employment.
8. **Experience in PPP:** Agriculture success is facilitated by years of experience in public-private partnerships, which can also play a significant role in shaping perceptions of these partnerships. In the current study, "farming experience in PPP" refers to the number of years a person has worked in agriculture that involves a public-private collaboration. In our data collected, 41 percent of the respondents have an experience of two years working in the PPP mode, followed by

25 percent with one year of experience, 15 percent have no experience, 13% with three years, and 3 percent with the experience of 1.5 years and 5 years respectively. In Maharashtra, the PPP model was initiated in 2012. As a result, very few beneficiaries have experience of over two years.

9. **Cropping Pattern:** The cropping pattern plays a significant factor in determining how successfully a farmer practices agriculture. Irrigation also changes the cropping pattern in many regions. Millets require little moisture so they can be grown in the rain-fed conditions which require no irrigation. These can be classified as single crop, double crop and mixed crop. The data reveals that 53 percent of the farmers grow mixed crop(sugarcane, bajra, onion, seasonal vegetables, ginger, turmeric, jowar, soyabean, grapes, watermelon, flowers, groundnuts and sweet corn.), 28 percent grow double crop (bajra, onion) and 19 percent growonly singlecrops (bajra or jawar).

2. Reasons for the Involvement of the Farmers with Different PPP Models

In the PPP model, the farmers get linked with FPOs, contract farming method, farmer society etc. FPOs is an organization of farmer-producers that provide support to small farmers with end-to-end services covering all aspects of cultivation from inputs, technical services to processing and marketing. They have direct links with producers, local traders, processing units, local merchants and the marketing of millet grains. For the benefit of millet cultivators, the State department of agriculture has helped with financial and technical requirements for cultivation, storage and training for value addition activities to gain better prices in the market.

The contract farming model in the study area has been implemented to focus on production, marketing and value addition to the traditional variety of millets. The farmers' association on the other hand was started to insulate the farmers from the middleman and dealers and to establish direct marketing linkages. Private processing units and millet-based industries join hands with the State department to educate farmers, supply adequate raw materials and provide solutions for the benefit of everyone.

In the study area there were 178 Farmer producer organizations out of which one from each district was studied. It was observed that 82.3 percent of the farmers were involved with FPOs, 5.1 percent were linked with contract farming method and another 5.1 percent farmers were involved with the farmer society.

Table 4: Factors Affecting Farmers' Involvement in Different PPP Models

Sr. No.	*Factors*	*Ranking for FPOs*	*Ranking for Contract Farming*	*Ranking for Farmer Society*
1.	Better pricing in PPP models	I	I	II
2.	Long lasting middlemen intervention in the marketing of harvested millet	III	I	II
3.	Opportunities for value addition in PPP model	V	I	II
4.	Inadequacy of the inputs for millet cultivation in conventional method	IV	I	I
5.	Storage and Infrastructure facilities.	IV	I	I
6.	Benefits of PPP model approach	II	I	III

The above Table 4 shows the factors affecting the involvement in the PPP models. For the farmers involved with FPO (Farmer Producer Organizations) model, it was observed that for about 73 farmers, the reason for involvement with PPP model was better price realisation for their harvested produces. Besides, 62 and 61 millet farmers found solution to the problem of middlemen intervention in marketing as the reason for joining PPP model. Another 54 farmers found inadequacy of inputs for millet cultivation in conventional methods and availing storage and infrastructure facilities as the reasons for involving with PPP model. Lastly, 45 farmers were involved in PPP to avail the opportunities for value addition to get more prices in the market.

Farmers involved with contract farming model expressed all the reasons being important. Farmers involved with the Farmers association model expressed the inadequacy of the inputs for millet cultivation in conventional method to be the most important reason. Better pricing in PPP models, long lasting middlemen intervention in the marketing of harvested millet and the opportunities for value addition was ranked second reason for participating in Farmer Association. Last reason to join the model was to avail the benefits of PPP model.

3. Value Chain Patterns followed by Farmers

The complete set of activities starting from selection of land, variety of produce and upto the delivery of the processed finished produce to the end consumer represents the value chain. What and how much to produce is determined by the mechanism of feedback between producer farmer, market forces and end consumer. The capturing of this whole defines the efficient value chain analysis.

The value chain pattern followed at Nasik, Pune and Konkan; the major millet production areas of Maharashtra, in different PPP models, were studied and presented as below:

1. VCP-I (Value Chain Pattern I) involving Grading and Packaging.
2. VCP-II (Value Chain Pattern II) involving Grading, Dehulling and packaging.

The only actions included in the VCP (VALUE CHAIN PATTERN) I are grading and packaging. Farmers carried out the grading and packaging tasks at the farm level. The government has set up a drying yard in each community to aid in this procedure. Farmers were utilising gunny bags to pack the graded produce after the drying process, as advised by the state department of agriculture. Intermediaries like wholesaler/trader/retailer etc are involved in extending the channel up to the consumer. Grading and Packaging without the involvement of middlemen helps farmers get more profit. The net income earned was on an average Rs. 2,237 per acre.

The farmers in the Farmer Producer Company (FPO) followed VCP II, as suggested by agricultural marketing department. Grading was done at the field level by the farmer himself after the process of drying of harvested millet grains in the common drying yards. Warehousing and packaging are done by Farmer Produce Companies at their premises and then the packaged produce is given to distribution channel partners to reach it to the end consumers. Thus, the FPO profited from processors and whole sellers by the well-established distribution channels. This profit helped members of FPO in adopting modern inputs and technologies for profitable millet cultivation. The net income earned was on an average Rs. 1,202 per acre.

In the study area, 63.3% of the farmers were involved in the VCP I and 36.7% were involved in the VCP II for enhancing the profitability opportunities for millet farming. The reason ranked first was the better acceptance of products in the market, second-easy way of marketing for their harvested produces. Third important reason was the better pricing for value added products, fourthly efficient storage to enable marketing at appropriate time and lastly it enhances the self-life of the product. The VCP I benefitted the farmers with increased income of Rs. 5/kg of millets.

According to the study, connecting all stakeholders through value addition throughout the value chain system would renew and uplift the diminishing millet area and production, with the ultimate economic benefits going to farmers and other stakeholders in the value chain.

LIMITATION OF THE STUDY

The findings are based on the sample collected from selected area among selected farming communities only and hence applicable in similar situations only. This had the same financial, time, and resource constraints as any other scientific investigation, making it challenging to take up the subject across a vast area. The conclusions of this study cannot be generalized to other farmers or to the entire region because it only included a sample of 114 respondents. However, some of the findings may be helpful in generating insightful recommendations.

CONCLUSION & IMPLICATIONS

The study could establish causal relationship in adoption of PPP and increase in farmer income. With the increasing population pressure and consequent reduction in arable land, it is of utmost importance to increase productivity and profitability of farmers to ensure food security. The symbiotic relation between producer farmer and the PPP associate is becoming a key to the enhanced productivity and reduction of wastage, which meets twin objectives of food for all and increased income to the farmers. The present study aimed at understanding the millet cultivators in Maharashtra. Based on the findings of the study, some of the suggestions that may be useful for the functioning of PPP models and millet farmers include- the heterogeneity profile of farmers can be used for establishing PPP models. Remarkable difference has been observed between the yield and net income of millet farmers; before and after the implementation of PPP models. Since the millets are rich in nutrients and minerals, awareness among the consumers should be created, possibilities of making variety of dishes can be exhibited to promote millet production. So, the year 2023 is celebrated as international year of millets to provide us with an opportunity to raise awareness of and direct our policy attention to their large-scale cultivation and use in our dietary intake, owing to their nutritional and health benefits and their suitability for cultivation under adverse and changing climate conditions.

For future research; action research on effectiveness of PPP model in the development of modern technologies for the promotion of millet cultivation may be attempted. An exploratory study on constraints encountered by the stakeholders for successful partnership models in agriculture may be attempted.

ACKNOWLEDGEMENT

Authors greatly acknowledge time and support provided for this research project by S.P. Mandali's Prin. L.N. Welingkar Institute of Management Development & Research (WeSchool), Mumbai.

REFERENCES

[1] Rao Dayakar, B. (2019). Sorghum Value Chain for Food and Fodder Security. *Breeding Sorghum for Diverse End Uses*, 409–419. https://doi.org/10.1016/B978-0-08-101879-8.00023-1

[2] Reddy, A. and Amarender, A. (2013). Training Manual on Value Chain Analysis of Dryland Agricultural Commodities Patancheru 502 324, Andhra Pradesh, India: International Crops Research Institute for the Semi-Arid Tropics (ICRISAT). 88 pp., Available at SSRN: https://ssrn.com/abstract=2281677 or http://dx.doi.org/10.2139/ssrn.2281677

[3] Ilahi, Ali (2022) Maharashtra Leads the Way in FPO Promotion. https://tci.cornell.edu/?blog=maharashtra-leads-the-way-in-fpo-promotion

Community Awareness and Knowledge Related to Millets and its Application in Society—A Survey Based Article

Ranjitsing Bayas[1], Saleemuddin Panwale[2], Manish Singh[1] and Jagruti Mahadik[1]

[1]Department of Life Science, Ramnarain Ruia Autonomous College Mumbai, India

[2]Department of Statistics, Ramnarain Ruia Autonomous College Mumbai, India

ABSTRACT: Millets are known to be a part of our Basic Indian everyday diet. A survey was conducted based upon the idea of awareness of millets and International Millets Year which is going to be held in January 2023. The general purpose of this survey/questionnaire was to collect data for research purposes and with the objective to spread awareness regarding millets and its nutritional aspects.

A variety of aspects on millets was considered while preparing a questionnaire regarding the awareness of millets. Millets can be considered in terms of various aspects such as nutritional values, disease-related, biofortification, consumption and production, and even on the future replacement of cereals. When we analyze the questionnaire flow it depicts people's opinions on various aspects of millets like in case of disease-related outlook many agree on the fact that millets play a major role in terms of cardiovascular diseases (CVD) and malnutrition as well. Furthermore 98.1%of people agreed on millets being part of entrepreneurship. Having looked at the trend of consumption of millets, we got to know weekly consumption, and also most cultivated millets in native places were found to be Ragi (nachni). Around 96.2% accept the fact that millets maintain a healthy lifestyle.

So through the survey conducted it can be concluded that according to the data collected, still to this day people are not yet aware of the general idea of what millets are. We need to have an Initiative to spread awareness or basic knowledge regarding millets and their benefits in various aspects.

Keywords: *Millets, Awareness, Consumption, Nutritional Value.*

INTRODUCTION

According to recent studies, the need for identifying healthy and environmental Sustainable diets and enhanced usage of underused plant species, such as quinoa, millets, Sorghum, etc. due to their climate resilience and dense nutritional content is increasing significantly. Millets are highly preferable than cereals due to their rich nutritional composition which can solve malnutrition problems thriving in India's more than half of the population. Millets are recognized for their resilience, ability to survive under high temperatures and in degraded Soils, and minimum requirements of water, pesticides, and fertilisers. Millets have high fiber content, antioxidants, gluten-free proteins, low glycemic index, and are fulfilled with biologically active compounds. By understanding the significance of the millets and its role in solving various problems, a survey was conducted on the basis of the awareness regarding millets In Maharashtra. There is a growing interest in reviving millets in India, owing to their nutritional content and ability to grow in harsh climatic conditions due to their climate smart traits. Millets become not handiest of our traditional meals it turned into wealthy in electricity wealthy compounds, and has first rate effect in nutrients. No food has a nutrient level like millets. The only food that has no side effects is millets, from beginning child to grandparents everybody can devour. Millets and Sorghum took 0.33 place in east and critical Africa. This plant has a separate importance in food [10]. Describing about millets and in health benefits, millets are store house of nutrition, by any nutritional parameter; millets are miles ahead of rice and wheat in terms of their mineral content, compared to rice and wheat India, lack of dietary diversity is one of the key factors behind malnutrition and the prevalence of non-communicable diseases such as diabetics, cardiovascular diseases and obesity as such. Small millets grown as a complement to existing crops contribute to an answer [7]. One of the maximum commonplace sicknesses affecting millennials, diabetes is the gateway to a host of illnesses. It has been found that human beings who have included millet into their diets are less likely to be affected by this ailment. Millet contains a very good quantity of magnesium which allows the frame to produce insulin efficiently and additionally prevents the occurrence of diabetes. Eating millet has also shown to preserve the sugar stages in test [10]. An analysis of low GI millets (including sorghum) and their effects on managing and reducing the risk of developing type 2 diabetes shows that millets have a beneficial] effect on various outcomes, such as the fasting and post meal blood glucose levels and insulin index. Consumption of whole grains like millets increases the intake of unsaturated fatty acids and dietary fiber, which enhances the satiety value, which, in turn, could have a beneficial effect on reducing energy intake and hyperlipidemia, and reducing the risk of

Cardiovascular diseases (CVD) and Non-alcoholic fatty liver disease (NAFLD) [8]. Millets awareness and their knowledge amongst people was being collected to understand their perceptions towards it. Low productivity, high labor intensity, drudgery of agricultural operations and lack of alternative farm gate prices, easy availability of rice and wheat through public distribution system, inadequate investment in product development and commercialization, inadequate availability of small millets in local markets and high Prices, etc. can be reason for the decline in the production and consumption of millets which was seen through this survey [7].

The goal of the present study is to examine the degree of knowledge amongst Indian men and women about Millet grain and its nutritional advantages. As a result, the study's primary objective is to determine the level of awareness and consumption pattern of millet amongst Indian men and women regarding millet grains. [1] additionally, a survey is carried out to find out the awareness and purchase of millets by the respondents in their day to day life and to analyze the reasons for usage of millets among women consumers [7].

MATERIALS AND METHODOLOGY

Sample Collection

The survey was being conducted among various kinds of people in Maharashtra states of India. A Google form questionnaire was being circulated throughout the areas and answers to the questions were being collected to study the awareness regarding millets. Overall 162 responses were collected from the survey conducted in this study to analyze data considering various aspects and factors using the Statistical tools.

The Composition of samples with respect to gender and occupation is being represented in the below mentioned table.

Gender	*Men*	*Women*	*Prefer Not to Say*
Percentage of the participants	24.1%	74.1%	1.8%

Occupation	*Self Employed*	*Salaried*	*Student*	*Housewife*
Percentage of the participants	7.7%	17.3%	67.3%	7.7%

Data Collection

The survey was conducted through Google form considering questionnaire patterns for the same along with subjective and multiple choice questions with the help of literature and experts in this field which include as follows.

- Awareness based on gender, highest qualification, occupation of sample collected.
- Awareness regarding international millet year 2023.
- Production and consumption of millets along with the type of millets known and its benefits.
- Kinds of millets cultivated at the native place and recommendations regarding millet recipes.
- Application of millets in terms of diseases and Role in maintaining a healthy lifestyle.
- Effect on the inflation trend in millets.
- Nutritional properties and millets in entrepreneurship.

Methodology

The questionnaire was developed keeping in mind the information collection needs. The study was conducted among people of Maharashtra. Knowledge, Attitude and Practice were the basis of this questionnaire. All the questions designed were to gather data ranging from use of millets in household to the awareness and knowledge about it in Society. The survey form was circulated for about 30 days in the month of Nov 2022 and carried out with the help of pre-tested questionnaires and under the guidance of experts and professors of Ramnarain Ruia College.

The information given by the respondent was checked before filling into the sheet. Data validation checks were performed at regular intervals. Collected data was fed in Microsoft Excel and R Software [11].

RESULT AND DISCUSSION

This study's main aim is to know up to what extent subjects are aware of millets and the consumption pattern of millets.

1. Percentage Distribution of Gender Against Awareness

	Don't Know	*Know*	*Grand Total*
Female	0.111111	62.96296	74.07
Male	0.037037	20.37037	24.07
Prefer not to say	0	1.851852	1.85
Grand Total	**14.81**	**85.19**	**100**

Sample is 74.07% female and 24.07% male. In total 85.19% of the people are aware of pulses.

2. Awareness Regarding International Millet Year

Year	*2023*	*2024*	*2025*	*2030*	*Total*
Percentage	77.78	12.96	7.41	1.85	100.00

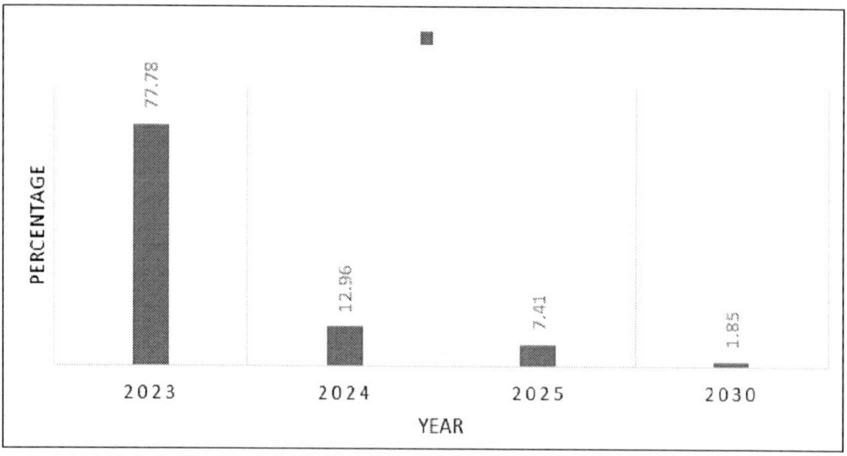

Figure 1: International Millet Year

The Government of India had proposed to the United Nations for declaring 2023 as the 'International Year of Millets' (IYOM). In the presentation, it is observed that 77.78% are aware about the millet year 2023 while others 12.96% selected 2024, 7.41% selected 2025 and 1.85% selected 2030 and are unaware about the Millet year. Through this data obtained it can be interpreted that sufficient awareness has been carried out regarding the Millet year.

3. Awareness about Millets Types and Variety

Responses	*Percentage*
All	1.85185185
Don't Know	1.85185185
Jowar	81.48
Maize	5.556
Rice	3.7
Wheat	5.556
Total	**100**

81.48% of the participants reported that jowar is millet and 18.5% could not recognize the millet name amongst others options. Among the participants, many of them were not aware of the fact that jowar is one of the millets and what millets actually are.

4. Frequency of Consumption against the Pattern of Consumption

Row Labels	*All of the Above*	*Bajra*	*Jowar (Sorghum)*	*Ragi*	*Wheat*	*Grand Total*
Daily	0.00	13.33	29.17	15.38	100.00	22.22
I don't consume	0.00	0.00	0.00	7.69	0.00	1.85
Monthly	0.00	13.33	16.67	7.69	0.00	12.96
Weekly	100.00	66.67	50.00	61.54	0.00	57.41
Yearly	0.00	6.67	4.17	7.69	0.00	5.56
Grand Total	**1.85**	**27.8**	**44.44**	**24.07**	**1.85**	**100**

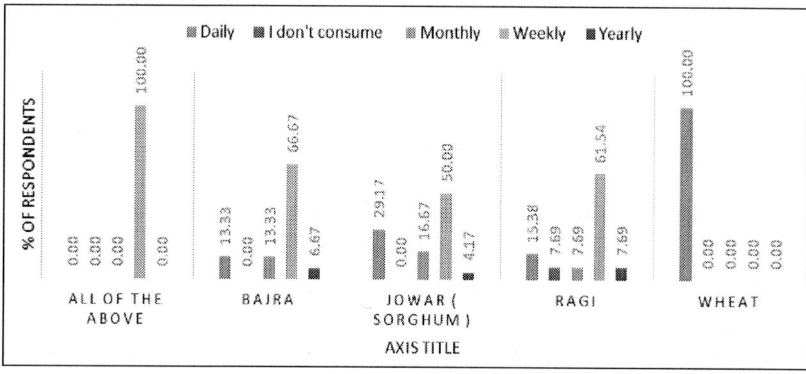

Figure 2: Consumption Pattern of Millet

Considerable proportion of people consumed millets i.e. 1.85% (all of the above), 27.8% (Bajra), 44.44% (Jowar), 24.07% (Ragi), 1.85% respectively. It can also be concluded that consumption of millet and wheat on a weekly basis is almost the same. On a weekly basis Bajra (66.67%) is being most consumed and jowar (50.00%) is being least consumed while taking daily basis into consideration wheat is consumed in the most amount than millets.

5. Consumption and Cultivation

Cultivation	*All of the above*	*Bajra*	*Jowar (sorghum)*	*Ragi*	*Wheat*	*Grand Total*
All	0.0	0.0	8.3	0.0	0.0	3.7
Jowar	0.0	35.7	25.0	14.3	0.0	24.1
Bajra	0.0	21.4	8.3	0.0	0.0	9.3
Ragi	100.0	14.3	8.3	35.7	0.0	18.5
Wheat	0.0	0.0	8.3	0.0	0.0	3.7
Others	0.0	21.4	12.5	0.0	100.0	13.0
No cultivation	0.0	0.0	20.8	35.7	0.0	18.5
Don't know	0.0	7.1	8.3	14.3	0.0	9.3
	1.9	25.9	44.4	25.9	1.9	100.0

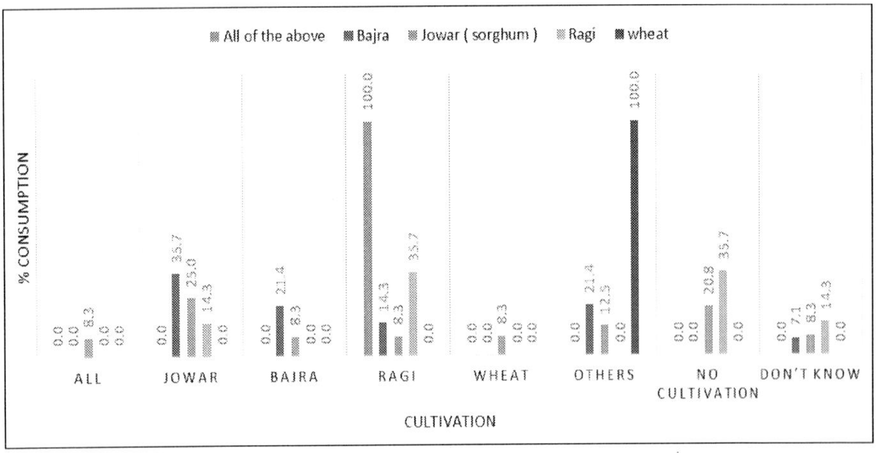

Figure 3: Consumption *vs.* Cultivation of Millets

In this statistics it is observed that in the area where jowar is cultivated we have 35.7%, 25%, 14.3% of bajra, jowar and ragi respectively. In a similar way, for areas where bajra is being cultivated, 21.4% bajra and 8.3% jowar is consumed. For the ragi cultivation areas, 100% of wheat, 14.3% of bajra, 8.3% of jowar and 35.7% of ragi is consumed. In cultivated areas in terms of cereals such as wheat we got 100% consumption in the other areas whereas in 'all cultivation' we have 8.3% jowar and for 'no cultivation' we have 20.8% of jowar and 35.7% of ragi is consumed.

6. Effect of Inflation on Millets

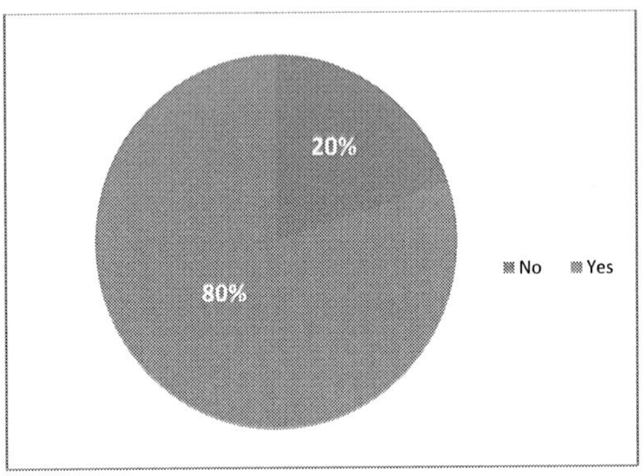

Figure 4: Effect on Inflation

H0: Proportion of people who feel that an inflation affects the millets is 0.5.

H1: Proportion of people who feel that an inflation affects the millets is more than 0.5.

Data: x out of n, null probability 0.5
X-squared = 55.71, *df* = 1, p-value = 4.2e–14 = 0
Alternative hypothesis: true p is greater than 0.5

From total participants, 79.63% of them agree to the fact that there is an effect in the inflation of millet production i.e. while 20.37% of them disagree to the fact.

7. Millets—A Disease Prevention Crop

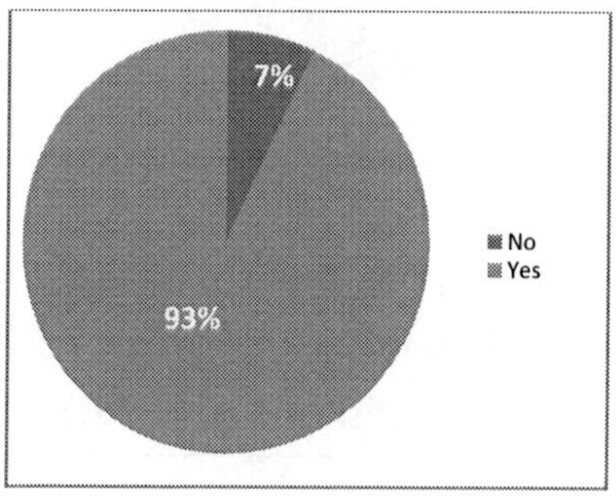

Figure 5: Millets Effect on Diseases

H0: Proportion of people who feel that consumption of millet in our daily diet keeps unsafe from diseases is 0.5.

H1: Proportion of people who feel that consumption of millet in our daily diet keeps unsafe from diseases is more than 0.5.

Data: x out of n, null probability 0.5
X-squared = 115.86, df = 1, p-value < 2.2e–16
alternative hypothesis: true p is greater than 0.5

Millets have potential health benefits and epidemiological studies have shown that consumption of millets reduces the risk of various diseases [Nutritional and health benefits of millets, June 2017]. Around 92.59% participants think of the contribution of consumption of millets on a daily basis keeping us safe from diseases while 7.41% disagreed.

8. Millets a Cure for Obesity and Cardiovascular Disease

H0: Proportion of people who feel that millets are a cure for obesity and cardiovascular diseases is 0.5

H1: Proportion of people who feel that millets are cure for obesity and cardiovascular diseases is more than 0.5.

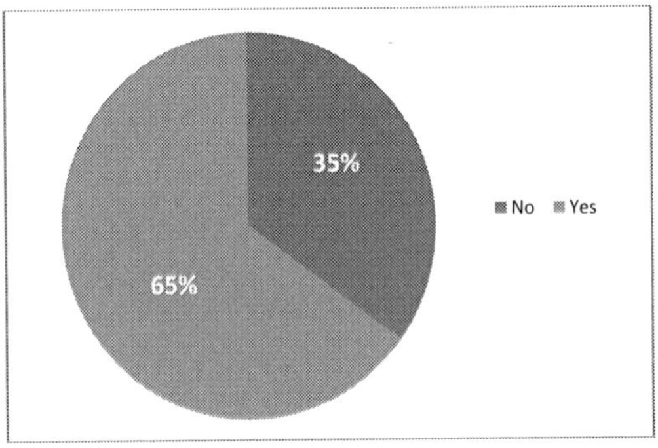

Figure 6: Millets on Obesity and Cardiovascular Diseases

Data: x out of n, null probability 0.5
X-squared = 13.636, df = 1, p-value = 0.000111
alternative hypothesis: true p is greater than 0.5

64.81% people think millets can tackle the problem of obesity or cardiovascular diseases while 35.19% people replied' No' to the question. Millets being high in fiber, antioxidants and complex carbohydrates are potential candidates for having beneficial effects against diseases like CVD, cancer and aging in general. [The story of millets ICAR-Indian Institute of Millets Research, Hyderabad, India, 2018].

9. Millets—A Solution to Malnutrition

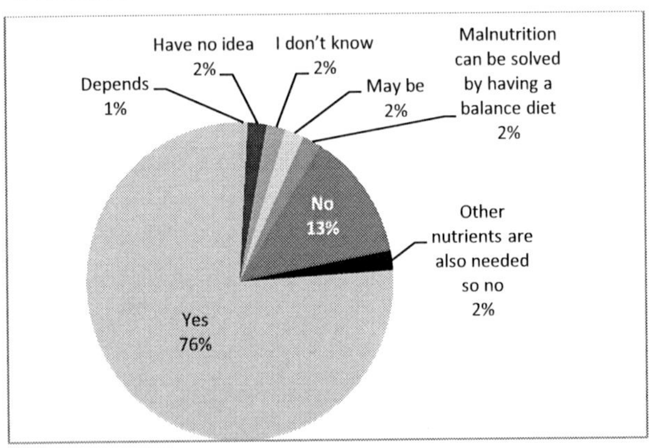

Figure 7: Millets on Malnutrition

H0: Proportion of people who feel that millet is a solution to malnutrition is 0.5.

H1: Proportion of people who feel that millet is a solution to malnutrition is more than 0.5.

Data: *x* out of *n*, null probability 0.5
X-squared = *42.525, df = 1, p-value* = 3.49e–11
alternative hypothesis: true p is greater than 0.5

In this data, 75.93% commented on the millet's contribution to malnutrition problems whereas 12.96% denied. Individually 1.85% had comebacks as 'depends', 'have no idea', ' i don't know', 'maybe ' 'Malnutrition can be solved by having a balanced diet' and 'other nutrients are also needed so no'.

10. Availability of Millets in Fair Price Shop

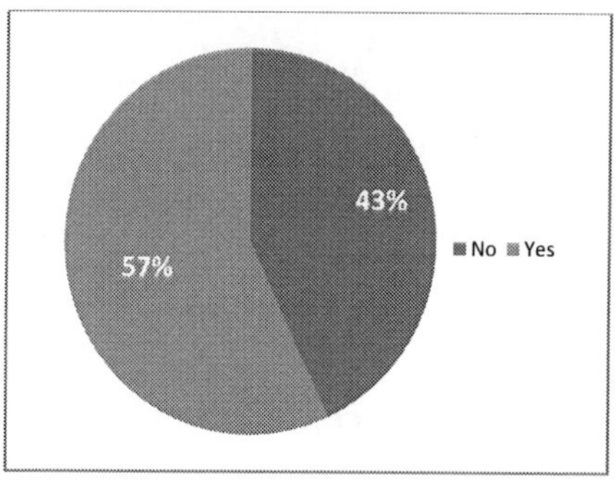

Figure 8: Millets Availability in Fair Price Shop

Data: *x* out of *n*, null probability 0.5
X-squared = 3.2654, df = 1, p-value = 0.03538
alternative hypothesis: true p is greater than 0.5

H0: Proportion of people who agree to the availability of millet in fair price shop is 0.5.

H1: Proportion of people who agree to the availability of millet in fair price shop is more than 0.5.

42.59% of people have no fair price shop at their place, whereas 57.41% people say 'yes', which Intrepretest that they have a fair price shop in their area. There can be an increase in the fair price shop from region to region.

11. Consumption of Millets Leads to a Healthy Life

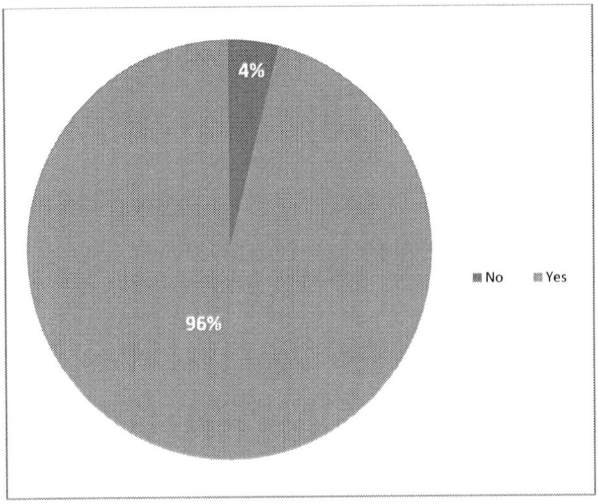

Figure 9: Millet Contribution in Healthy Lifestyle

Data: x out of n, null probability 0.5
X-squared = 148.3, $df = 1$, p-value < 2.2e–16
alternative hypothesis: true p is greater than 0.5

H0: Proportion of people that feel consumption of millet leads to healthy life is 0.5.

H1: Proportion of people that feel consumption of millet leads to healthy life is more than 0.5.

96.30% of people believe that millets do sustain a healthy lifestyle and play an significant role in our day to day life and among total participants, 3.70% people believe that it does not contribute to the lifestyle of people.

12. Millets Being Beneficial for Postmenopausal Women

H0: Proportion of people that feel consumption of millets is beneficial for postmenopausal women is 0.5.

H1: Proportion of people that feel consumption of millets is beneficial for postmenopausal women is more than 0.5.

Postmenopausal women 62.96% believe more millets in diet are less prone to heart ailments, or high blood pressure and high cholesterol whereas 37.03% don't believe.

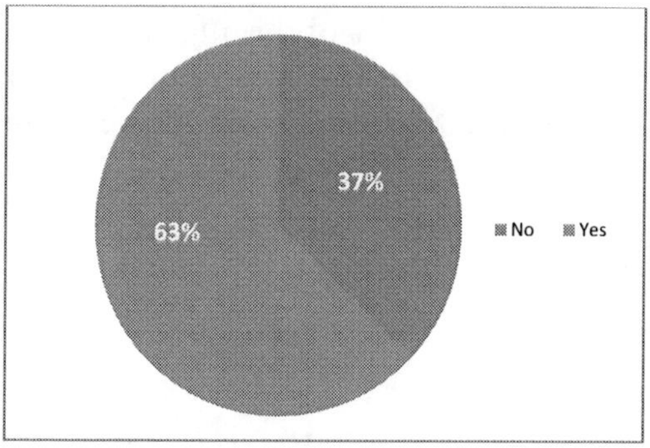

Figure 10: Millet Consumption in Postmenopausal Women

> ***Data:*** *x* out of *n*, null probability 0.5
> X-squared = 10.377, df = 1, *p*-value = 0.0006381
> alternative hypothesis: true p is greater than 0.

13. Millets Properties and its Richness in Protein and Vitamin Content

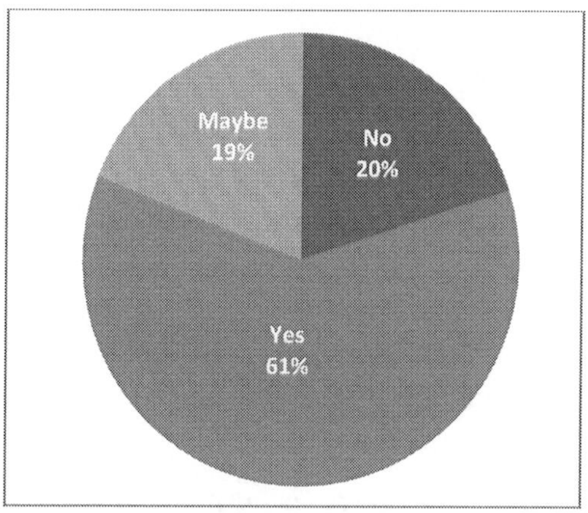

Figure 11: Millet being Gluten Free and Antioxidant Rich

> ***Data:*** *x* out of *n*, null probability 0.5
> X-squared = 7.5617, df = 1, p-value = 0.002981
> alternative hypothesis: true *p* is greater than 0.5

H0: Proportion of people that believe millets are gluten free grains that are rich in antioxidants is 0.5.

H1: Proportion of people that believe millets are gluten free grains that are rich in antioxidants is more than 0.5.

61.11% people replied 'Yes', 20.37% of them replied 'No' whereas remaining 18.52% replied 'maybe' to the fact that millets are gluten free grains that are rich in antioxidants soluble fiber and good source of protein, fiber, key vitamins, and minerals.

14. Millets is Equal to a Healthy Heart

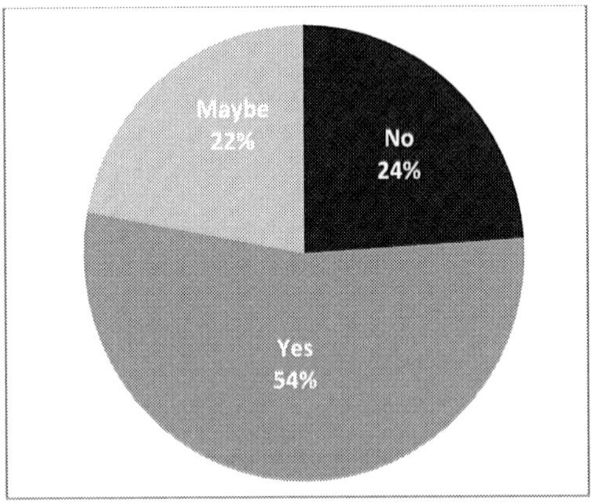

Figure 12: Millet—Good for Heart Health

Data: x out of n, null probability 0.5
X-squared = 0.74691, df = 1, p-value = 0.1937
alternative hypothesis: true p is greater than 0.5

H0: Proportion of people that believe that the consumption of millets leads to a healthy heart is 0.5.

H1: Proportion of people that believe that the consumption of millets leads to a healthy heart is more than 0.5.

22.22% people believe that there is a possibility of millets containing magnesium, which in turn helps to regulate heart rhythm, 24.07% of people believe has a strong belief that millets do not contain magnesium and helps in sustaining the heart rhythm, 54.70% of people do think that there is a good possibility of millets containing magnesium which in turn also plays an essential role in heart rhythm. It should also be known and consumed because of its properties.

Magnesium is a micronutrient used for bone mineralization, teeth maintenance, building up of proteins, enzyme activities, normal muscular contractions and transmission of nerve impulses. [The story of millets ICAR-Indian Institute of Millets Research, Hyderabad, India, 2018]

15. Opportunities of Entrepreneurship

Figure 13: Scope for Entrepreneurship

Data: x out of n, null probability 0.5
X-squared = 148.3, df = 1, p-value < 2.2e–16
alternative hypothesis: true p is greater than 0.5.

H0: Proportion of people that believe that there's a scope/an opportunity in entrepreneurship world due to millets is 0.5.

H1: Proportion of people noe that believe that there's a scope/an opportunity in entrepreneurship world due to millets is more than 0.5.

A huge difference of opinion is observed from the above stat, a huge percentage gap can be seethe out of total participants only 1.85% believe that there is no possibility of future job opportunities and possible entrepreneurship in the area in millets market, apart from that 98.15% people believe that the future holds a magnificent entrepreneurship opportunities possibilities.

The millet market should be promoted by the government and people, a variety of industries are possible because of millets and there's going to be a huge demand for these superfoods, and entrepreneurship opportunities are some best ways to fulfill those demands.

CONCLUSION AND RECOMMENDATION

It has been observed that among all the types of millets, jowar and bajra only constitute 73% of the consumption and rest in combined. Additionally, the majority of the population feels that the millets are rich sources of protein and several vitamins, hence the consumption of millets is essential for a healthy lifestyle, a healthy heart and also to deal with problems like cardiovascular diseases and postmenopausal women's health. Moreover, millet consumption can play a vital role in eradicating malnutrition. Majority of the people prefer the commonly cultivated millet of their respective area for their consumption whereas on the other side of the coin on an average one fifth of the people failed to identify the millet types.

According to the findings, there is a need to formally study about consumers' knowledge, attitudes, and practices related to millets despite growing health consciousness among people, increasing non-communicable diseases in India, and the nutritional properties of millets. Basic information related to awareness of millets and its application should be taught in schools. Frequent awareness programmes should be conducted by stakeholder, Community volunteers and NGO's to maximize community awareness.

REFERENCES

[1] Adithya, Girijavallabhan, Agna Shibu Mathew, Gayathri Rajeev, Sonia Thomas *et al.* (1 January 2022) *International Advance Research Journal in Science, Engineering and Technology,* Perception and consumption pattern of millets among female young adults. https://www.smartfood.org/millet-consumer-survey

[2] Kalaiselvi, A., Razia Fathima, L.A. and Parameswari, M. (February 2016) *Indian Journal of Applied Research* Awareness and Consumption of Millets by Women—A Study on Coimbatore City.

[3] Subramanian Barratry, R., Rajapushpam, R. and Barratry, R. Subramaniya (August 2018). *IOSR Journal of Business and Management* (IOSR-JBM).

[4] Kane-Potaka, Joanna; Anitha, Seetha; Tsusaka. T.W., *et al.* (2021). *Frontiers in Sustainable Food Systems,* www.frontiersin.org

[5] Kane Potaka, J., Anitha, S., Tsusaka, T. *et al.* (2021). Assessing millets and sorghum c ICRISAT: Consumer survey about millets 15,500 face to face interviews in Cities across 7 states in India.

[6] Prasanthi, K. and Sireesha, G. (15 April 2022). *International Journal of Food and Nutritional Security,* Individuals' knowledge, Attitude and Practices on Millets www.ijfans.org

[7] Rizwana, M.; Singh, m Padmalini; Ahalya, N. and Mohanasundaram, T. (2 June 2022) *British Food Journal.* https://www.emerald.com/insight/content/doi/10.1108/BFJ-05-2021-0593/full/html?skipT racking=true

[8] Venkatesh, S.R. (2017). Final technical report *Annex 2.3 Millet usage and its attitude study* Indianet Marketing Services Pvt. Ltd.

[9] Ranjitsing, B., Laxmikant, S. *et al.* [accessed Jan. 18, 2023]. Knowledge Awareness and Practices Regarding Mosquito Borne Diseases among People of Jana, Maharashtra. Available from: https://www.researchgate.net/publication/340273292_Knowledge_Awareness_and_Practices_Regarding_Mosquito_Borne_Diseases_among_People_of_Jana_Maharashtra.

[10] Anita, Seetha; Botha, Rosemary, *et al.* (2021). Can Millets Consumption Help Manage Obesity. doi: https://doi.org/10.3389/fnut.2021.700778

[11] Millets Consumption Probability and Demand in India: An Application of Heckman Sample Selection Model, December 2018.

Growth Anomalies in Average Millets Production in Maharashtra

Rohit Panchal, Taufiq Shaikh, Asmita Mestry and Nitin Labhane

Department of Botany, Bhavans College (Autonomous), Andheri (West), Mumbai, India

ABSTRACT: The cultivation, yield, processing, and production of millets of various sorts are subject to extensive research, Millets are the most efficient and quite nutritious than the other cereals. The states of Rajasthan, Maharashtra, Karnataka, Andhra Pradesh, Tamil Nadu, Kerala, Telangana, and Uttarakhand, show higher average production per unit area. The current analysis is carried out to interpret the difference in the average production of Maharashtra with other states, which have higher yields per unit area under millet cultivation. The review after congregating all the statistical data and reports from the Government and different resources, it is found that Maharashtra's average production per unit area is less. The paper strived to draw attention to the state's limitation of its potential for producing a variety of millets. In defiance of its area and agricultural history, compared to other states, the state lags and is not the leading millet producer or exporter. When compared to other states like Tamil Nadu and Uttarakhand, Maharashtra produces less millets. Maharashtra's underperforming production may be the result of low yield, competition with cash and vegetable crops as well as lack of knowledge can be explored. By proper application of (One district one product) and (One district One Millet) state may achieve a higher yield.

Keywords: Average Production, Millet, Anomalies, Yield.

INTRODUCTION

Millets are major crops in terms of nutrition and mineral values. It has a major market share of over 33% of Maharashtra's economy and is majorly based on agriculture, Millets plays a major role in the total 90% of the production of the state. Millet production in Maharashtra is constricted to 9 districts, i.e. Konkan, Nashik, Aurangabad, Pune, Nagpur, and Latur. Maharashtra is increasing its grains production. [GOI, 2014] In terms of millions of tons yield every year and it is among the major producers of millets all over India. Maharashtra being the largest producer of millets is still lagging in terms of production (tonnes) per unit area (hectares). In the year 2020–2021, ragi, jowar and bajra production was under the area of 1381 million hectares, and this production will uncertainly decrease in the year 2021–22 to 725 million tons. (Perumal and Anbukkani, 2017) The review paper analytically focuses on the growth anomalies in average millets production in Maharashtra on the bases of production (Tonnes) per unit area (hectares). The lack of unexplored potential, strives to draw attention that production of millets is majorly focused in the nine districts of Maharashtra, whereas districts lack in producing millets in constrained 9 districts as well, which leads to an under-explored potential of the production of millets in Maharashtra. States like Odisha, Uttarakhand, and Tamil Nadu, despite having a lesser area are producing more in terms of unit per area. Further to improve the production of millets, Maharashtra can consider schemes like the Odisha millet mission (OMM) or one district one product (ODOP) (NADP, 2015), which will be a beneficial attribute. The focusable part in this term is the area under production is decreasing. In this frame of reference, the present analytical report is to focus on the objective to study the production pattern of millets in Maharashtra in terms of production per unit area. Furthermore, breaking the structure of restricted production of millets in calculated districts and also to emphasize the necessity for implementation of schemes or techniques like ODOP, OMM in state of Maharashtra. (NADP, 2015)

OBJECTIVE

Whether the level of production per unit area is sufficient. Analysis of the unpredictable growth pattern of Maharashtra's average millet production is the study's primary goal.

DATA AND METHODOLOGY

The investigation is based on secondary data collected from various public and government sources. (Samal P., 2018) Data on production per unit area were collected and represented in a form to illustrate whether the production per unit area is up to the mark. The main objective of the study is to analyse the unpredictability in the growth pattern of average millet production of Maharashtra, as well as Comparison between states pattern, techniques to produce the millets, and increased yield in terms of production per unit area.

RESU LT

Table 1: Production Per (Tonnes) Unit Area (Hectare)

	2016–17	*2017–18*	*2018–19*	*2019–20*	*2020–21*
Maharashtra	0.92	0.93	0.76	0.78	0.76
Tamil Nadu	1.49	1.29	1.81	1.50	1.23
Uttarakhand	1.33	1.20	1.38	1.44	1.50
Odisha	1.50	1.46	0.85	1.32	0.96

Source: Ministry of agriculture and welfare.

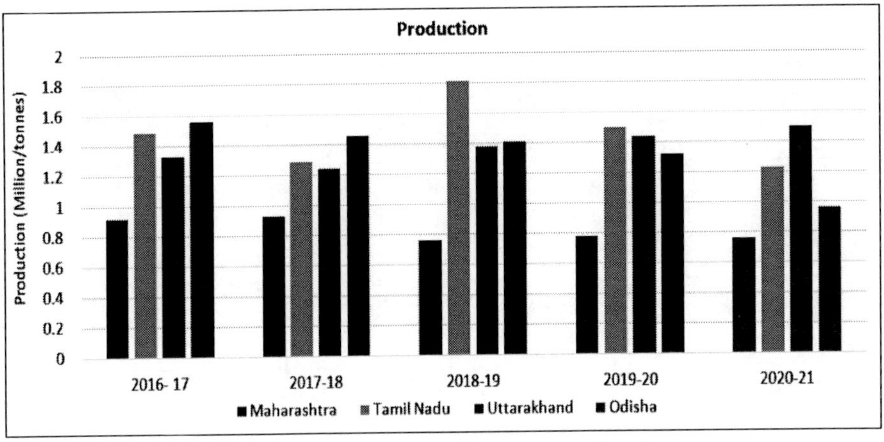

Graph 1: State Wise Production Unit Per Area

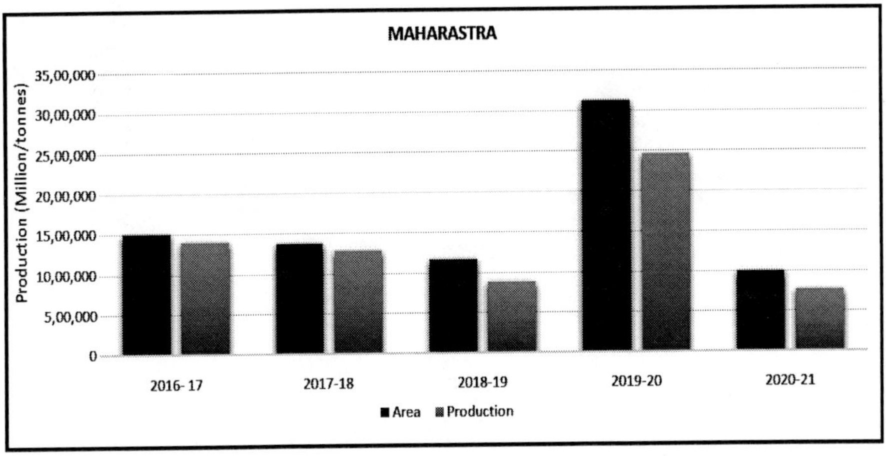

Graph 2: Maharashtra's Production Unit Per Area

Source: Directorate of Economics and Statistics, Economic Survey of Maharashtra [4].

Table 2: Odisha's Production Per Unit Area after OMM

Odisha	*2016–17*	*2017–18*	*2018–19*	*2019–20*	*2020–21*
Production	33330	119530	134880	523600	532300

Source: Department of agriculture and food production (Govt. of Odisha).

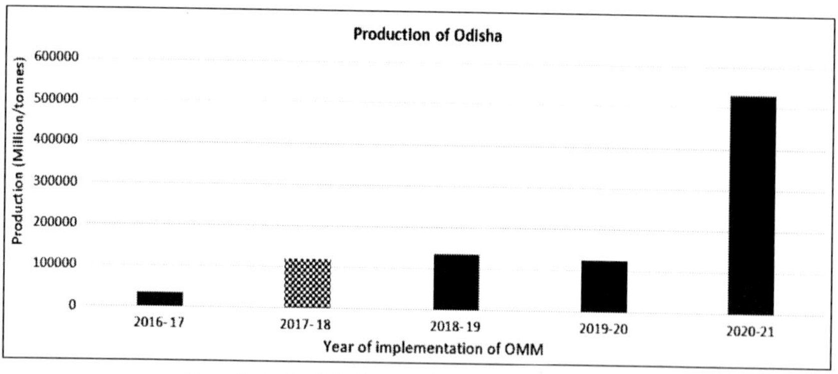

Graph 3: Odisha's Production Per Unit Area

Table 3: Tamil Nadu's Production (tonnes) Per Unit Area (hectare) after ODOP

	2016–17	*2017–18*	*2018–19*	*2019–20*	*2020–21*
Production	1,98,549	3,71,060	2,99,384	4,16,675	9,49,600

Source: Guidelines for implementation of millets mission in Tamil Nadu.

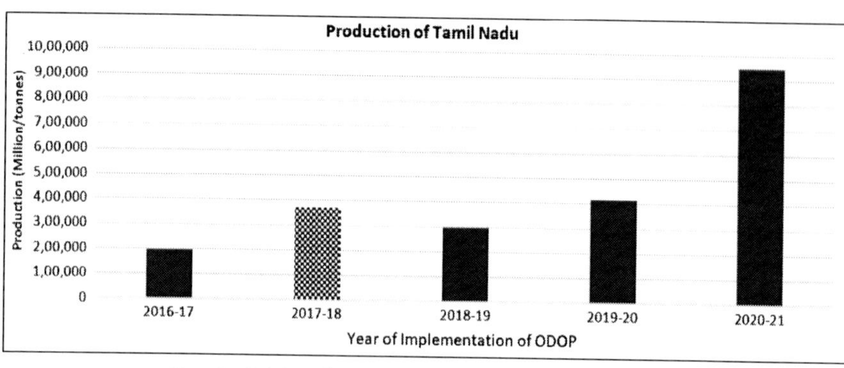

Graph 4: Tamil Nadu's Production Per Unit Area

Formula;

For all the calculation Production per unit area was estimated using,

$P = T_p / A_c$,

Where,

T_p = Total Production

A_c = Area under cultivation

P = Production per unit area

$$\text{Production per unit area (P)} = \frac{\text{Total Production (Tonnes)} \ (T_p)}{\text{Area (Hectares) under cultivation} \ (A_c)}$$

DISCUSSION

Millets are major grown crop in India. The cultivation is mainly in the monsoon session. Presently Rajasthan and Maharashtra are the top most producers of millets. The area under production was studied in the entire period of (2016–21), which showed anomalies in production of millets in Maharashtra. During (2016–21) being within a maximum production limit, the production in the study period was seen decreasing. The statistical data also corroborate that Maharashtra shows growth anomalies in terms of production per unit area. During the subsequent investigation period of 5 years (2016–21); the growth was seen to be decreased by (30%). However, in the recent year there was an increase in the production. But the it was less in the scale of production per unit area. During the year (2016–17), there was 15 (Million/Hectare) of land and production was only 13 (million/tonnes). In 2017–18 area was 13 (million/hectares) and production was 12 million/ tonnes. During the period 2016–18 growth rate was negative as the area decreased in this period for millet cultivation. However, in the next period of investigation i.e. 2018–19, the area decreased subsequently and led to huge depth in the production from 13 (million/Tonnes) to 8 (Million/Tonnes), which showed negative growth from 2016 to 2018. Maharashtra's millets production anomalies have a significant negative influence on the state's ability to keep up with other states production output. The growth, area, and production seen in other states in the same period was significantly positive and growing each year. The growth when was negative in many states, Tamil Nadu, Uttarakhand & Odisha were showing an increasing and developing graph. If the growth excellence is compared with Maharashtra's area under millet cultivation from 2016 to 2021, the state has been at the top in production due to the abundance of the area but shows anomalies in production per unit area than other states. During the investigation period 2016–21 major states like Tamil Nadu, Karnataka, Uttarakhand & Odisha were achieving growth in production in which area even if remains constant between 1 lakh/hectare to 7 lakh/hectares. The production showed a considerable amount of positive and significant growth. (Samal P., 2018)

The growth rate in an area under production, when computed for different states, production is seen to increase in all the states but majorly in Tamil

Nadu & Uttarakhand if the data from 2016–2021 was considered, Tamil Nadu showed absolute increase in the production, the average data from 2016–2021, Tamil Nadu's area under production of millets was 1,32,947 hectare and production showed was 1,98,549 tonnes which is 6% more production then the unit per area or land under use. Similarly, Uttarakhand in the year 2017–18 showed 162 million/hectares of the area under millet cultivation and 216 million/tonnes of production which computes to 7.5% of more production than land under the production of millet. (Ministry of agriculture and welfare, 2021) The production growth i.e. from 2016–21 has recorded a subsequent increase over the years, which are the results of involvement of government support to the farmers resulting in a positive and significant impact in production growth. In states like Tamil Nadu, Uttarakhand due to its agricultural policy, Odisha's millet mission, schemes like one district one millets (ODOM), and guidelines largely played an important role in the production of its crops at an exceptional level. To investigate further in the current year 2020–21, Tamil Nadu, Uttarakhand and Odisha does show an increase in production as well as area. In this state, the major problem of providing support, guidelines, and other amenity were prioritized on a focused level. This facility were appreciated by the farmers to act upon it and adapt modern technique's to increase productivity.

In Odisha's millet mission, the inclusion and distribution of millets as part of policy is a crucial factor in the high production of millets in Odisha. There are three other factors for increased cultivation besides policy intervention; First, the minimum support level for all millets has been set by the government. Second, quality seed supply is available and third, as millet is an essential diet item for Odisha citizens, it is included in the PDS (Yadav, 2022). When compared to other grains, millets are more frequently purchased and consumed. The Odisha millets mission launched in April 2017, subsequently the data shows an increasing figure in terms of production per unit area of millets. The effort has been successful for Odisha during the past five years. The most recent study and statistics Indicated an increase in millet output of up to 14%, or 3333 (ha) in the 2017–18 crop year, which increased over time to 53,230 (ha) in the 2021–2022 crop year. The success of the Odisha millet mission is greatly influenced by the state's quality agricultural practices (OMM) and minimum support responsible shared by the government. (Yadav, 2022)

CONCLUSION

Increasing millet production is not only useful in terms of India's export perspective, but also for Maharashtra from the viewpoint of the employment, food cultivation, and economic importance of the millet. However, the growth of Millets has slowed down or is not up to the limit in terms of production per unit area. This will affect the upcoming years for the farmers as well as the revenue generated from the millets to the state. Tamil Nadu and Odisha missions for millet cultivation have helped the government to develop infrastructure and initiated guidance on the ground level for the farmers to adopt effective farming techniques to increase the production and quality of the crops. Whereas the second perspective is that Maharashtra being called as leading producer of millet, has not opened ends full-fledged for development, awareness about millets, On the bases of the investigation and analytical data the growth of Millets is constrained only in 8 districts of Maharashtra, which is the major drawback for it's less production per unit area, leading to less production each year, The problems related to the observation needs further investigation. This limited production of millet is leading farmers towards profitable crops, Maharashtra being an agricultural state has all the potential for improving the current scenario. The above discussion was an investigation based on the statistical data compiled to bring the focus towards the anomalies of production of millets per unit area in Maharashtra. This discussion brings us to many observations. Tamil Nadu being able to maintain the high level of production per unit area. This is a competing factor. Maharashtra is producing less than a unit per area under production over the years which will lead to a decrease in the cultivation of millets per unit area as compared to other states. Maharashtra in the study period of 2016– 2022 showed a significant decline in production as well as in substantial elimination of land under millet cultivation. (Sreekala, 2022) To overcome the anomalies of millets production in Maharashtra, Bringing additional area of millets in a phased manner will be an appropriate approach as well as eliminating traditional methods which are been followed need a change in technique. (NADP, 2015) Furthermore, the area which is not under cultivation must be brought into cultivation, seed providing, modern millet processing unit and Guidance for post- cultivation and pre-cultivation techniques for millets will help in achieving a positive growth rate in production per unit area of Maharashtra.

REFERENCES

[1] Anamika Yadav (2022). Millet might: Here is how Odisha can succeed in reviving these cereals lost to the Green Revolution, https://www.downtoearth.org.in

[2] Asha Devi, Sukumaran Sreekala, P. Anbukkani, Alka Singh, B. Dayakar Rao and Girish K. Jha (2022), Millet Production and Consumption in India: Where Do We Stand and Where Do We Go, National Academy Science Letters, DOI: 10.1007/s40009-022-01164-0.

[3] Commissionerate of Agriculture (2022). Government of Maharashtra (As published in Economic Survey of Maharashtra, Directorate of Economic and Statistics, Planning Department, GOM).

[4] Directorate of Economics and Statistics (2022). Economic Survey of Maharashtra, Planning Department, Government of Maharashtra, Mumbai.

[5] Government of India (GOI), (2014), tatus paper on coarse cereals, Directorate of Millets Development, Department of Agriculture and Cooperation, Ministry of Agriculture (GOI).

[6] Guidelines for implementation of millets mission in Tamil Nadu (2015), NADP, https://agricoop.nic.in/sites/default/files/Crops.pdf.

[7] Kumara Charyulu D., Moses Shyam D., Bantilan Cynthia, Borikar S.T., Gupta S.K. and Rai K.N. Pearl (2016). Millet Technology Adoption and Impact Study in Maharashtra. Research Report 71, International Crops Research Institute for the Semi-Arid Tropics, pp. 76.

[8] Lessons from Odisha Millets Mission (2021). A Case for Mainstreaming Millets, https://vikalpsangam.org.

[9] Limbore, Nilesh V. and Khillare Shrirang, K. (2015). An Analytical Study of Indian Agriculture Crop, Production and Export with Reference to Wheat Rev. Res., Vol. 4, pp. 1–8.

[10] Ministry of Consumer Affairs, Food & Public Distribution (2019), Global status of Millets: Area and Production region-wise.

[11] Mohammed Arshad Khan, Mohammad Athar Ali, Syed Mohd Minhaj, Vivek, Mohd Afzal Saifi (2022). Agricultural Sector: Status, Challenges and its Role in Indian Economy an Analysis, Agricultural Mechanization in Asia, Vol. 53.

[12] Samal, P., Rout. C., Repalli. S.K. and Jambhulkar, N.N. (2018), State-wise Analysis of Growth in Production and Profitability of Rice in India, *Indian Journal of Economics and Development,* Vol. 14(3), pp. 399–409, DOI: 10.5958/2322-0430.2018.00151.8.

[13] Perumal, Anbukkani (2017). Production and consumption of minor millets in India—A structural break analysis, *Ann. Agric. Res. New Series,* Vol. 38(4), pp. 1–8.

[14] "Production and Export with reference to wheat," (2015), *Rev. Res.*, Vol. 4, No. 6, pp. 1–8, [Online].

[15] Samal, Parshuram Rout, C. Repalli and Jambhulkar, Nitiprasad (2018). State-Wise analysis of growth in production and profitability of rice in India, *Indian Journal of Economics and Development,* Vol. 14, pp. 399, DOI: 10.5958/2322-0430.2018.00151.8.

[16] Shabana Anjum, Madhulika (2018). Growth and instability analysis in Indian agriculture, *International Journal of Multidisciplinary Research and Development,* Vol. 5, pp. 119–125.

[17] Special data dissemination standard division (2016 to 2021), directorate of economic & statistics, ministry of agriculture and welfare (https://aps.dac.gov.in/APY/Public_Report1.aspx).

[18] Tanvi, Deshpande (2017). Policy analytical reports, PRS legislative research, State of Agriculture in India, (https://prsindia.org/files/policy/policy_analytical_reports/State%20of%20Agriculture%20in%20India.pdf)

[19] The Associated Chambers of Commerce and Industry of India (June 2022). Millets the Future Super Food for India.

[20] The Quarterly Newsletter of the Millet Network of India (2019). Vol. 5 (http://www.milletindia.org/ Millets Net in India) (MINI).

[21] Agricultural and processed food products export development authority (APEDA), (2017–21), State Wise Millet Production, Report.

Utilization of Millets

Jigna Kothari and Anagha Sapre
Centre for Vocational and Technical Education, Premlila Vithaldas Polytechnic, SNDT University, Mumbai, India

ABSTRACT: Today, the lifestyle and food habits of India and globally have led to major health issues in humans. Some of the deadly diseases include cardiovascular diseases, stroke, respiratory problems, Tuberculosis, Chronic Obstructive Pulmonary Disease, diabetes, Alzheimer's, Malignant Tumors. The foods consumed, having junk foods included, which are high in sugar, fats, less fiber, deep fried and so on, adds up to reduce immune system to fight these health issues. One of the most nutritious groups of foods is Millets, which will help preserve future generations' health.

Millets are considered poor man's food and neglected, while it is nutritiously rich with high dietary fiber, high proteins, low fat, vitamins like niacin, riboflavin, thiamin, B6, Minerals like Copper, iron, magnesium, manganese, phosphorus, zinc.

Due to the versatility of millets, they can be incorporated in almost all kinds of foods, and can be served to all age groups. India is the world leader of production of millets, so creating the awareness is the most important responsibility that we have to bring about. Suggesting some easy ways to include millets in everyday foods could motivate everyone to use it more frequently.

Hence, this study has been undertaken on millets, major millet crops, easy ways to make quick foods, foods that can be served to different age groups.

Keywords: *Millets, Vitamins, Minerals, Health.*

INTRODUCTION

Millet can be white, green, yellow, or red and is tiny in size and round in form. These are very high in terms of their nutritional content. As for proteins, minerals, and vitamins, each Millet is three to five times nutritionally better than rice and wheat. Millets are rich in B vitamins, mineral rich and gluten-free. They have a low GI, so millets are ideal for wheat allergies/intolerance in humans, diabetics, helps weight loss. Most commonly consumed millets are divided into two:

- **Major Millets:** Sorghum (Jowar), Pearl Millet (Bajra), Finger Millet (Ragi/Mandua),
- **Minor Millets:** Foxtail Millet (Kangani/Kakun), Proso Millet (Cheena), Kodo Millet (Kodo), Barnyard Millet (Sawa/Sanwa/Jhangora), Little Millet (Kutki) and two Pseudo Millets (Buckwheat (Kuttu) and Ameranthus (Chaulai).

Accommodating millets in your daily food routine can be benefitting in many ways:

1. **High in Nutrition:** Millets are gluten and acid free and a good source of nutrients which are excellent for maintaining a healthy life. A balanced diet full of nutritional properties will also help eradicate the risk of heart problems and other such life endangering diseases.
2. **Less Calorie Count:** Millets are manageable, a choice to a healthy weight reduction, as their calorie count is very low. Even for the people who are calorie conscious and trying to eat in a manner, that will help them maintain their energy levels throughout the day without constantly feeling starved.
2. **Improves Immunity:** The weaker the immune system, the more you are prone towards experiencing health issues frequently, which can prove fatal. This is due to the severe deficiency of protein in the system. Adding millets to your diet on a regular basis will help your body gain back all the protein, that your system was short on.
4. **Satiates Hunger:** Many carbohydrates such as rice and wheat are easy to breakdown once consumed, leaving a person hungry within few hours of their meal. This makes people resort to unhealthy eating habits to satiate their hunger. Millets have harder structures and does not breakdown easily in the digestive system, keeping you fuller for a longer time, and far enough away from unhealthy eating habits. The balanced availability of natural dietary fibre inherently makes 'Millets' very special. Out of the total fibre requirement of 38 grams per person a day, eating 'Millets' thrice a day can contribute to about 25 to 30 grams. The remaining 10 grams of fibre can be derived from vegetables and greens. The special arrangement carbohydrate interwoven with the fibre in millets does the trick of regulating glucose into the blood stream.
5. **Reduces Sugar-levels:** Due to its low glycemic index, millets are an ideal food to consume to prevent diabetes from ever occurring in non-diabetic people. It especially aids in keeping type-2 diabetes under control. India, being the world capital for diabetics, an important milestone can be achieved by reducing the increase in diabetic patients, if a large population consumes millets on regular basis.

6. **Aids Heart Problems:** Millets contain essential fats, just the right amount to give our body natural fat. This helps prevent excess fat from depositing over muscles which and avoid high cholesterol, heart strokes, and other heart related disease. Millets are the immediate best alternatives to replace with their unhealthy peers, reducing the risk of cardiovascular diseases by a good measure. Millet are the only grains that contain all the vital nutrients. From proteins to minerals to vitamins, millets have it all. Millets helps you switch to a lifestyle that's healthy and long lasting.

Above are the main benefits, though the list can be much longer. The nutritional comparison with recognized superfoods such as flax, quinoa and chia seeds, showed that a variety of millets have comparable quantities of macro and micronutrients and hence can be considered as superfoods in the nutritional sense. Further, the presence of health promoting flavonoids and phenolic compounds boosts their position in being highly nutritious. In addition, millets also have various non-nutritional advantages that further makes a case for their status as cereal superfoods we will need to nurture for the future.

Table 1

Millets: an approach for sustainable agriculture and healthy world			
↓	↓	↓	↓
Food Security	***Nutritional Security***	***Safety from Diseases***	***Economic Security***
- Sustainable food source for combating hunger in changing world climate - Resistant to climatic stress, pests and diseases	- Rich in micronutrients like calcium, iron, zinc, iodine, etc. - Rich in bioactive compounds - Better amino acid profile	- ***Gluten free:*** a substitute for wheat in celiac diseases - ***Low GI:*** a good food for diabetic persons - Can help to combat cardiovascular diseases, anaemia, calcium deficiency, etc.	- Climate resilient crop - Sustainable income source for farmers - Low investment needed for production - Value addition can lead to economic gains

The special and unique ability of millets to grow in marginal lands with no irrigation facilities could help the world in achieving the targets of food security. With the current challenges of soil erosion and water scarcity in mind and our food security targets set at providing safe and nutritious food to everyone, millets could become the superfood that we need to truly achieve our objectives. So, we can also term Millets as Smart Foods.

How to Use the Millets

Though soaking is optional, yet it is preferred for 4–8 hours. Soaking millet before cooking has so many benefits. Millets can be cooked in boiling water or pressure cooked.

Some of them are:

- Soaking makes it easier to digest, and all nutrients will become bioavailable.
- Soaking gives the millet porridge that creamy texture. Millet will have a hard outer cover, and you can feel that in the cooked millet texture, hence soaking will give creamy porridge.
- Soaking will yield fluffy millet, a preferred fluffy texture if you want to use cooked millet for a salad or pilaf.
- After soaking, rinse the soaked millet 2 to 3 times and completely drain the water from them. Using a colander makes the job easy as millet grain is tiny and difficult to handle with bare hands.

MATERIALS AND METHODS

Dishes can be prepared for all ages. They can be made in different textures – drinkable, soft, medium, crisp. There are some products which can be made with millets only. While some products have to be made with wheat or rice and part replacement with millets, to retain the texture and form of the product. Dosa, Idli, Dhokla (Steamed products), Khichdi, Porridge, Indian Savoury Pancake, Soups, Indian flat breads, sweets, bakery products like cakes, cookies, pizzas, chips, Beverage drinks and many more products can be made either using only millets or partly replace the main ingredient with millets.

Work on many recipes have been done, though the two recipes, I am going to discuss here in details are:

1. Pearl Millet Drink (Bajra Raab) and
2. Khichyu made from Sorghum and Pearl Millet flour.

Though Pearl Millet is healthy and for all ages, special importance is given to Indian Pearl Drink (Bajra Raab) as it is specifically beneficial to lactating mothers, the drink is considered as a Galactagogue. Galactogogues are substances used to induce, maintain, and increase milk production. It is known for its nutritional values like high fibre, high protein and minerals like iron, phosphorous and magnesium, anti-acidic, prevents constipation, helps prevent back pain.

It is infused with dry ginger, carom seeds, gond, ghee, cardamom to increase the nutritional value and taste of the drink. In short, importance of each ingredient added is summarized as below:

- **Ghee** most often prepared and served to strengthen the immune system of a new mother or to someone who is low on energy or has a congested chest, lubricates the joints.
- **Gond** is edible gum sourced from Gum Acacia. – strengthen bones, warm food.
- **Carom seeds** are added to prevent bloating and aid in digestive in the new mothers, internal body healing.
- **Ginger powder** is known to increase breast milk production, supports digestion, alleviate pain, boosts immunity.
- **Poppy seeds** are rich in healthy plant compounds and nutrients like manganese, aid digestion, helps in sound sleep.
- **Cardamom** boosts up the production of milk in lactating women. Not just this but consumption of these spices is known to be good for gut health as well.
- **Almonds** are a good non-dairy source of calcium.

Table 2: Pearl Millet Raab

Ingredients	*Quantity (gms)*
Bajra	120
Ghee	45
Carom Seeds	2
Dry Ginger Powder	0.5
Cardamom	0.5
Gond (roasted and fried in ghee and powdered)	12
Roasted Almonds	7
Dry Roasted Coconut Shavings	15
Poppy Seeds	0.5

- **Dry Coconut** is rich in healthy fats, which can be great to fulfill a nursing mom's increased energy needs, which in turn, encourages the body to produce more breast milk. The healthy fat content in coconut oil can also be a plus for low birth weight babies, as per a study. Coconut Increases Antimicrobial Properties in Breast Milk. Coconut can help Heal the Body Postpartum.

Though the Drink is made on household scale, below is the Standardized recipe, which has also been worked out in a way to make a Premix.

Method of Preparation:

1. Heat a small quantity of ghee and fry the gond. Cool it and powder it. In remaining ghee, sauté nuts, dry coconut shavings and poppy seeds.
2. Heat Ghee, add carom seeds and roast bajra flour.
3. When the above mixes cool, mix them together. Premix is ready.

Directions for Use for the Premix:

1. To the above premix, add 200 ml water and mix well avoiding lumps.
2. Heat 350 ml water and 70 gms jaggery and dissolve it.
3. Add the jaggery water to the bajra ensuring no lumps are formed.
4. Give it a boil till it becomes thick drink.
5. Add cardamom powder and dry ginger powder and remove from flame.
6. Serve hot.

The second most important dish I am discussing here is Pearl Millet – Sorghum (Bajra-Jowar) Khichyu. It has a consistency of porridge. Of course, being healthy for everyone, special importance is due to the texture – soft pasty. It is especially beneficial to those having chewing problems, and hence a Geriatric food too...Being anti-acidic and prevents constipation, helps the elderly people immensely.

Combined with health benefits of Bajra and Jowar together, carom seeds are added to aid digestion. Curds are added for taste, proving an important source of protein and calcium which in turn helps in bone and muscle health.

Though the Khichyu is made on household scales, below is the Standardized recipe, which has also been worked out in a way to make a Premix.

Table 3: Pearl Millet—Sorghum Khichyu

Ingredients	*Quantity (gms)*
Bajra	30
Jowar	30
Ghee	10
Carom Seeds	0.5
Salt	5
Asafoetida	0.25
Green chilli-ginger paste	7
Curds	100
Water	300

Method of Preparation:

1. Heat ghee, add carom seeds, asafoetida, and roast bajra-jowar flour.
2. When the above mix is cooled, add green chilli, ginger powder. Premix is ready.

Directions for Use for the Premix:

1. Boil water to a boil, add curds and the premix.
2. Mix well avoiding lumps and continue cooking till porridge consistency.
3. Garnish with finely chopped coriander leaves and green garlic (optional).
4. Serve hot.

RESULTS AND DISCUSSION

As the above recipes, could be standardized to make premixes, hence they are commercially viable. It is the responsibility of every individual who can understand the benefits of the millets, to spread awareness and build a healthier future generation of people.

CONCLUSION

There is a popular belief that Millets are poor man's foods. Though the millets are diverse and have highly bioavailable nutrients, millets are seen as the poor man's staple as they are cultivated and consumed domestically by the rural poor for decades. Over the decades of efforts and unprecedented situations, there is a shift in perceptions and millets are increasingly sought as healthy and nutritious alternatives to fine cereals. As the hidden hunger led lifestyle diseases are surging irrespective of rural or urban areas, millets are now being highlighted more in the diets of many segments of the population. The strategic promotion will lead to setting the proper awareness in people and place the millets as better choice foods in terms of nutrition and health.

The biggest challenge is that only a few R&D Institutes, NGOs and state governments are involved in awareness creation. So, it is critical that millets are promoted in a way that it is regularly consumed by all.

REFERENCES

[1] https://www.birtheatlove.com/coconut-breastfeeding/#:~:text=Coconut%20is%20rich%20in%20healthy,babies%2C%20as%20per%20a%20study.

[2] https://link.springer.com/article/10.1186/s40066-018-0183-3

[3] https://everydaynourishingfoods.com/how-to-cook-fluffy-millets

[4] White Paper on Millets—A Policy Paper on Mainstreaming Millets for Nutrition Security in India.

[5] Siridhanyalu (Millets)—Foods that Heal.

Nutritional Significance of Mineral Contents in Pearl Millet: A Review

Julekha A. Shaikh

Department of Chemistry, Maharashtra College of Arts, Science and Commerce, 246-A, J.B.B. Marg, Mumbai, India

ABSTRACT: Pearl millet is the world's second most important millet after sorghum. It is commonly known as Bajra and one of the earliest varieties of millet that our ancestors consumed. Bajra is garnering significant attention owing to its abundance in nutrients like protein, fibre, vitamins, mineral elements, leading to a variety of beneficial effects on the body and therefore, nutritionists and dieticians recommend consumption of pearl millet. Literature study reveals 100 g of pearl millet contains ~20 mg calcium, ~6 mg iron, ~330 mg phosphorus, ~400 mg potassium, ~200 mg magnesium, ~10 mg sodium, ~5 mg chlorine and ~120 mg sulphur. In addition, it also contains micro elements such as zinc, cobalt, selenium, manganese and iodine. Pearl millet is the only grain to contain significant levels of chromium, and is only surpassed by barley in terms of magnesium and zinc levels.

Pearl millet contain a variety of bioactive substances that have several health benefits such as antibacterial, antioxidant, antidiabetic, and hypocholesterolemic properties. Currently new pearl millet varieties, that provide women a higher dose of dietary iron are being conventionally cultivated. Bajra is a potent source of magnesium and potassium, which are essential for widening blood vessels and therefore, improving blood circulation. Due to the high fibre content, digestion is slowed, which allows glucose to enter the bloodstream more gradually. This helps to maintain healthy blood sugar levels for a much longer amount of time. The current review addresses the role of mineral elements found in pearl millet.

Keywords: *Pearl Millet, Review, Minerals, Nutrients.*

INTRODUCTION

One of the most significant grains in cereal is millets. Millet is a generic term for tiny grains that belong to the "coarse cereals" group. More than one-third of people worldwide consume millets. Pearl millet, a member of the Poaceae family, is widely cultivated (Figure 1). Other names include bulrush millet, black millet, candle millet, bajra, Indian millet, horse millet, and cattail millet.

Figure 1: Pearl Millet or Bajra

Pearl millet is a wild grass native to West Africa that was domesticated in the Sahara Desert (National Research Council, 1996) that later spread to East Africa and subsequently to India. Pearl millet continues to be the main source of income for millions of people in developing nations. In arid and semi-arid areas of West Africa, East Africa, and Asia, pearl millet is today one of the most significant grain crops grown. Of the world's millet production, India contributes roughly 44% to the production of pearl millet (Foreign Agriculture service, 2022). Up to 75% of the nation's total production of cereals is produced in Rajasthan (36%), Uttar Pradesh (12%), Karnataka (11%), Gujarat (9%) and Haryana (8%) as listed in Table 1.

Table 1: Millet Production Rate of the World

Sr. No.	*Country*	*Pearl Millet Production (%)*
1.	India	44
2.	Niger	11
3.	China	9
4.	Nigeria	7
5.	Mali	6
6.	Sudan	5
7.	Ethiopia	4
8.	Burkina Faso	3
9.	Senegal	3
10.	Chad	2
11.	Rest of the world	11

Minerals serves several significant biological roles such as electrolyte balance and conductivity functions for muscle and nerve excitation, acid-base balance, cofactors in enzymes required for metabolism, crystalline structure of bones and teeth etc. (O'Dell and Sunde, 1997). According to McDowell (2003) and Gropper *et al.* (2009), 17 distinct minerals (Table 2), accounts for 4% of the body's mass, which are necessary for human health. They can be divided into three categories: macro minerals (daily requirement of at least 100 mg), trace minerals (daily requirement of less than 100 mg), and ultra-trace minerals (usually 1 μg/g or less per day) (Offenbacher *et al.,* 1997). The essential macro minerals for human health includes sodium (Na), potassium (K), calcium (Ca), magnesium (Mg), sulphur (S), phosphorus (P), and chlorine (Cl) and micro minerals include elements like iron (Fe), zinc (Zn), copper (Cu), manganese (Mn), iodine (I), fluorine (F), selenium (Se), molybdenum (Mo), cobalt (Co) and boron (B) (Bouis and Welch, 2010). Pearl millet offers geographical as well as therapeutic benefits. In contrast to farming maize or wheat, pearl millet can be grown in areas that have drought, brackish soil, low fertility, low pH, and high temperatures. Pearl millet has been associated to a variety of health benefits due to its chemical composition, which are displayed in Table 2.

Table 2: Essential Minerals with Recommended Daily Intake Level and Key Biological Functions

Sr. No.	*Dietary Mineral*	*Adequate Intake*	*Benefit*
1.	Potassium	4700 mg	Electrolyte essential to sustain membrane action potential and for regulating ATP
2.	Sodium	1500 mg	Electrolyte required to maintain membrane potent and co-regulate ATP with potassium
3.	Calcium	1000 mg	Required for muscular contraction, heart and digestive system health, and bone development
4.	Magnesium	420 mg	Required for ATP processing and bone formation. Reduces the severity of asthmatic patients' respiratory issues. lowers the chance of heart attack
5.	Sulfur	850 mg	Important in tertiary protein structure, including connective tissue and skin, and required by numerous enzymes and antioxidant compounds
6.	Phosphorous	700 mg	Essential for bone growth and development of ATP
7.	Zinc	11 mg	Required for production of immune cells
8.	Iron	8 mg	Help in increasing Hemoglobin
9.	Chromium	200 μg	Required for appropriate sugar metabolism and improve glucose utilization

Pearl millet is more nutritious than common cereals like wheat, rice, and maize in terms of energy value, proteins, fat, and minerals. It contributes significantly to the human diet due to high-quality proteins and high quantities of iron, calcium, lipids, zinc. Due to its richness in high in dietary fibre and micro-nutrients (Malik *et al.,* 2002). The current review aims to explore the mineral content in pearl millet and its favorable implications.

Carbohydrates, Energy, Protein and Fibre: Pearl millet is a good source of carbohydrates, protein and fibre. 100 g of bajra contains 67 g carbohydrate, 361 calories energy, 12 g protein and 1 g fibre. When compared to sorghum, pearl millet has less starch and more protein. Starch accounts for approximately 56 to 65% of the kernel and contains approximately 20 to 22% amylose (Krishnan and Meera, 2018); free sugars account for approximately 2.6 to 2.8% of the grain. Sucrose is the primary sugar found in pearl millets.

According to Nutritive value of Indian foods (NIN, 2003), compared to wheat (346 Kcal/100 g), sorghum (349 Kcal/100 g), and rice (345 Kcal/100 g), pearl millet has a high energy content (361 Kcal/100 g). Pearl millet has a higher food value than other cereals due to its high protein content, excellent amino acid balance, and relatively high vitamin A concentration (Rai *et al.,* 2008). Protein concentration ranges from 9 to 21%, which is greater than rice (6.8 g/100 g), and maize (4.7 g/100 g) but similar to wheat (11.8 g/100 g) (Kaur *et al.,* 2014). It has low levels of tryptophan, lysine, threonine, and sulfur-containing amino acids. Only few grains, such as pearl millet, can alkalinize the stomach, prevent stomach ulcers, or lower the severity of existing ulcers (Taylor *et al.,* 2014). Due to its low prolamin content, the protein is gluten-free. It also retains its alkaline character after cooking, making it an ideal diet for persons who are gluten intolerant. Pearl millet aids in weight loss. Grain consists of high fibre content, resulting in longer time to transit from the stomach to the intestines, which helps to reduce overall calorie intake because it satiates appetite for a prolonged period of time. Additionally, the slow digestion process aids in the maintenance of healthy blood sugar levels (Chandrasekara and Shahidi, 2011) as it releases glucose into the blood at a slower rate, allowing normal blood sugar levels to be maintained for a longer period of time (JinJin *et al.,* 2022). As the insoluble fibre prolongs intestinal transit time and slows bile juice secretion. As a result, the chance of gallstone formation is lowered.

Vitamins: Pearl millet, contain large concentrations of antioxidants and bioactive compounds that can prevent DNA deterioration. Phytonutrients and protective antioxidants such lignin found in pearl millets, helps preventing cardiovascular illnesses. Numerous antioxidants are effective at scavenging different toxins from the body, including those that are located in the kidney and liver, as well as at neutralizing free radicals, which can lead to cancer. Niacin, a B_3 and B_6 vitamin found in millets reduces cholesterol. It contains phytic acid, a phytochemical that is considered to enhance cholesterol metabolism and keep the body's cholesterol levels stable. However, anti-nutrients found in millets such as phytic acid, tannins, and polyphenols tend to limit mineral bioavailability (iron and zinc).

Iron: According to several research, the iron concentration of pearl millet ranges from 3.0 to 18.0 mg/100 g, which is higher than that of other dietary grains. New varieties of pearl millet have recently been conventionally farmed, providing women with a higher dose of dietary iron. Lack of iron affects brain growth and results in all-over weariness resulting in increase in the rate of maternal and newborn death in the obstetric population (Andraca *et al.*, 1997). In a study of children with iron deficiency anemia under three years old, it was discovered that consistent ingestion of pearl millet satisfied their daily iron needs. Iron deficiency anemia is highly common in pregnant women and school-aged children because of their frequent use of rice and wheat, which are relatively poor sources of iron. This encourages including pearl millet on a regular basis in meals. Due to, iron deficiency anemia is a highly common condition among pregnant women and school-aged children.

Zinc: Zinc is an essential nutrient that is required for healthy growth, tissue repair, reproduction, and improved cellular immunity. A strong immune system and healthy growth depend on zinc, which is abundant in pearl millet and provides 12.4 mg per 100 g. It is also important in metabolism regulation, as a component of metallo-protease enzymes that perform catalytic, structural, and regulatory functions, in membrane stabilisation, and in different ionic cell processes (Milner, 1990). Zinc deficiency increases the risk of a variety of infections, limits physical growth, and interferes with pregnancy (Hess *et al.*, 2009).

Calcium and Magnesium: The calcium level of pearl millet varieties ranges from 10 to 80 mg/100 g. Lactating mothers' benefit from eating pearl millet. To improve milk production, lactating moms must eat pearl millet. In addition, pearl millet contains three times as much calcium as milk does. Therefore, pearl millet eating is advantageous for both mothers and their newborns. One of the finest natural treatments for menstruation cramps and the irritation and mood changes that go along with them is pearl millet. Both calcium and magnesium are necessary for bone formation, muscle contraction, the transmission of nerve impulses, blood coagulation, and cell communication (FAO, 2002). Rickets, osteoporosis, hypertension, and stroke are all risks associated with insufficient calcium consumption. Magnesium shortage affects neurologic and neuromuscular function and causes hunger, weak muscles, and lethargy. Pearl millet contains magnesium, which aids in easing asthmatic patients' respiratory issues and reduces the effects of migraines. It controls blood pressure and reduces the risk of heart attacks (Singh, 2018).

Sodium and Potassium: The sodium and potassium level ranges from 10 to 13 and 171.6 to 215.3 mg/100 respectively. Bajra is known for its potassium content, which is essential for people with high blood pressure. It also contains plant lignans, which are transformed into animal lignans in the body. These lignans are also known to help prevent heart disease. Consuming more potassium-rich meals assists in flushing out sodium from body, lowering blood pressure. This also preserves the sodium-potassium ratio in the body. Because of its high potassium concentration, it aids in electrolyte balance, metabolism, and weight loss.

Phosphorus: Phosphorus, which is a key component in the structure of body cells and is abundant in pearl millet and has a value that ranges from 185 to 990 mg/100 g (Datir *et al.,* 2018), It is an important component of several essential chemicals, including adenosine triphosphate (ATP). This element is also an essential component of nucleic acids, which are the genetic code's building blocks. Phosphorus is a lipid-containing substance that is found in cell membranes and nervous system structures.

Micro minerals: Pearl millet is the only grain that has appreciable amount of chromium which is required for appropriate sugar metabolism. Selenium, which is available in 5.4 μg/100 g works as a cofactor required for the action of antioxidant enzymes. Cobalt is a critical component of Vitamin B_{12}. Copper at 1.5 mg/100 g is required for the formation of connective tissues, the release of iron from storage sites, and the maturation of red blood cells. Manganese, which is present at 1.5 mg/100 g, is a cofactor needed by enzymes for the creation of energy and for the synthesis of DNA and RNA. With such abundance of macro and micro nutrients, pearl millet can be used to fight malnutrition brought on by a lack of certain minerals.

CONCLUSION

Pearl millet has the ability to thrive a variety of ecological situations, like drought, brackish soil, low pH and high temperature. Despite growing in a poor nutrient soil, pearl millet is nutrient-rich and contains high amount of protein, fibre, vitamins and minerals. Their revolutionary role in preventing lifestyle diseases and their capacity to fortify the immune system due to their superior nutritional profile make Pearl millet significant.

Chronic micronutrient deficiency affects around 2 billion people globally and constitutes a serious hazard to people's health in today's society (Tulchinsky, 2010). Malnutrition can have an effect on a child's capacity to grow, develop physically and cognitively, perform at work, be productive as an adult

Thus, including pearl millet, which has a high nutritional profile, into the diet will alleviate the problem of a shortage of nutrients and lower the occurrence of malnutrition. The inexpensive and nutrient-rich pearl millet is an excellent source of nutrition. Dieticians and nutritionists are thus working to raise awareness of this millet and encourage its intake by educating all categories about its benefits.

REFERENCES

[1] Andraca, I., Castillo, M. and Walter, T. (1997); Psychomotor Development and Behavior in Iron-deficient Anemic Infants; *Nutrition Reviews,* Vol. 55(4), pp. 125–132.

[2] Bouis, H.E. and Welch, R.M. (2010). Biofortification—A sustainable agricultural strategy for reducing micronutrient malnutrition in the global south; *Crop Science,* Vol. 50(S1), pp. 20–32.

[3] Chandrasekara, A. and Shahidi, F. (2011). Anti-proliferative potential and DNA scission inhibitory activity of phenolics from whole millet grains; *Journal of Functional Foods,* Vol. 3(3), pp. 159–170.

[4] Datir, R.P., Shaikh, A. and Sahare, A.S. (2018). Pearl Millet: boon in mineral deficiency: A review; *Agricultural Reviews,* Vol. 39(4), pp. 327–332.

[5] FAO, Food and Agriculture Organization/World Health Organization Expert Consultation (2002). Magnesium. In: Human Vitamin and Mineral Requirements; Rome and Geneva: Food and Agriculture Organization of the United Nations and World Health Organization.

[6] Foreign Agricultural Service, Millet 2022 World production (2022). U.S. Department of Agriculture (https://ipad.fas.usda.gov/cropexplorer/cropview/Default.aspx)

[7] Gropper, S.S., Smith, J.L. and Groff, J.L. (2009). Advanced Nutrition and Human Metabolism, 5th ed. Wadsworth: Cengage Learning, Australia.

[8] Hess, S.Y., Lönnerdal, B., Hotz, C., Rivera, J.A. and Brown, K.H. (2009). Recent advances in knowledge of zinc nutrition and human health; Food and Nutrition Bulletin, Vol. 30 (1 Suppl), pp. 5–11.

[9] JinJin, P., Umapathy, V.R., Vengadassalapathy, S., Hussain, S., Rajagopal, P., Jayaraman, S., Veeraraghavan, V.P., Palanisamy, C.P. and Gopinath, K. (2022). A Review of the Potential Consequences of Pearl Millet (Pennisetum glaucum) for Diabetes Mellitus and Other Biomedical Applications; *Nutrients,* Vol. 14(14), pp. 2932–2943.

[10] Kaur, K.D., Jha, A., Sabikhi, L. and Singh, A.K. (2014). Significance of coarse cereals in health and nutrition: A review; *Journal of Food Science and Technology,* Vol. 51(8), pp 1429–1441.

[11] Krishnan, R. and Meera, M.S. (2018). Pearl millet minerals: Effect of processing on bioaccessibility; *Journal of Food Science and Technology*, Vol. 55(9), pp. 3362–3372.
[12] Malik, M., Singh, U. and Dahiya, S. (2002). Nutrient composition of pearl millet as influenced by genotypes and cooking methods; *Journal of Food Science and Technology,* Vol. 39(5), pp. 463–468.
[13] McDowell, L.R. (2003). Minerals in Animal and Human Nutrition; 2nd ed. Elsevier, Philadelphia.
[14] Milner, J.A. (1990). Trace minerals in the nutrition of children; *Journal of Pediatrics,* Vol. 117(2), pp. 147–155.
[15] National Research Council (1996). Lost crops of Africa. Volume I: grains; National Academy Press, Washington, DC.
[16] NIN; Nutritive value of Indian Foods (2003). Ed Gopalan and Deosthale; National Institute of Nutrition, Hyderabad.
[17] O'Dell, B.L. and Sunde, R.A. (1997). Handbook of Nutritionally Essential Mineral Elements; Marcel Dekker, Inc. New York.
[18] Offenbacher, E.G., Pi-Sunyer, F.X. and Stoecker, B.J. (1997). Handbook of Nutritionally Essential Mineral Elements. [Chromium. In: O'Dell B.L., Sunde R.A., ed.] Marcel Dekker; New York, pp. 389–441.
[19] Rai, K.N., Gowda, C.L., Reddy, B.V.S. and Sehgal, S. (2008). Adaptation and potential uses of sorghum and pearl millet in alternative and health foods; *Comprehensive Reviews in Food Science and Food Safety,* Vol. 7, pp. 340–352.
[20] Singh, N., Singh, S.P., Kumar, M., Kanhiya, K. and Kumar, A. (2018). Nutri Cereal Pearlmillet: Way Forward; *International Journal of Current Microbiology and Applied Sciences,* Vol. 7(6), pp. 578–581.
[21] Taylor, J., Belton, P., Beta, T., Duodu, K. (2014). Increasing the utilization of sorghum, millets and pseudocereals: developments in the science of their phenolic phytochemicals, biofortification and protein functionality; *Journal of Cereal Science,* Vol. 59, pp. 257–275.
[22] Tulchinsky, T. (2010). Micronutrient deficiency conditions: Global health issues; *Public Health Reviews,* Vol. 32(1), pp. 243–255.

Cultivation, Economy and Consumers: A Review

Manjiri Bhalerao*, Anuradha Ramesh, Ruby Ojha and Jagmeet Madan

S.N.D.T. Women's University, Mumbai, India

ABSTRACT: Millets, a staple Indian crop, contributes to security of food, fodder and fuel. Millet market size is forecasted to grow from USD 9 billion (2018) at 4.5% CAGR till 2025 with value projection of USD 12 billion (milletadvisor.com).

Millets considered 'good for consumers, farmers and the planet' are in alignment with the UN SDG 2 i.e. Zero hunger. Millet consumption is advantageous for nutrient content and health benefits. These 'Nutri-cereals'/'Smart Food' are resilient to climate stress and drought.

The paper has reviewed the data on millet cultivation and production from an economic and consumer perspective along with factors affecting usage, consumer education and promotion of millets and millet products. Surveys on 'Awareness about Products and Consumer Insights' have indicated health and wellness benefits, weight loss and taste as the top three reasons for consumption in urban areas. Higher ratings for RTE products and easy to prepare traditional foods are the prominent reasons for consumer's acceptance of nutri-cereals, implying market opportunities for products categorized as convenience foods and culturally sensitive cuisine.

Social media has been identified as an important influencer for consumers' education about millets and millet products. Role of Government and responsible netizens is crucial in effectively promoting millets along with relevant schemes and policies.

Keywords: *Millet and Climate, Food Security, Consumer and Economy, Nutrient-Rich, Responsible Netizens.*

INTRODUCTION

Millet crop belongs to the family of small grasses, in the Panicoid subfamily and the Paniceae tribe. Finger millet (*Eleusine coracana*) and are in the Chloridoid subfamily (Fuller, D.Q., 2014).

Historical evidence of cultivation and usage of millets for domestic purposes date back to Indus valley civilization (3000 BC). Archeological evidence has suggested excavations from China, India, Europe and different parts of Africa (Fuller, D.Q., 2014).

India suggested and supported the declaration of 2023 as the International Year of Millets by the United Nations General Assembly. The main objectives for the Government of India in the celebration of International Year of Millets, 2023 (IYOM, 2023), is to create awareness of benefits of millets and the acceptance of millets as staple food. The benefits will include the gainful contribution of millet to food security, rise in the production of millets, easy availability of millet crop and its products in markets through efficient distribution and storage and maintaining sustainable practices in production and quality management with aid of all stakeholders (Assocham report; Agriexchange report).

Table 1: Reasons for Millet Cultivation

Sustainability in Agriculture	• Referred as "orphan crops" • Climate resilient • Drought resistant • Xerophilic • Thermophilic • Carbon sequestration crop • Can grow in poor soil, hilly terrain and require less rain
Economy of Investment for Cultivation	• Short crop duration • Less or no fertilizers required • More shelf life than paddy crops • Good export potential • Increased food security for famine areas
Nutrition and Health benefits to Consumers	• High dietary fiber value • Rich in micronutrients like Iron, Zinc etc., Good for Anemia • High Calcium content in Ragi • Protein rich • Low glycemic index, Good for Diabetes • Presence of Phytonutrients • Gluten free • Cancer prevention and Celiac disease • Low level of Triglycerides, Good for Heart health • Probiotic and prebiotic properties, Good for gut health

India is one among the global leaders in production of millets with a share of around 15% of the world total production. The other major millet producers are countries in the African continent, and China. The global Millets market was valued at $9.95 Billion in 2020 and is projected to reach $14.14 Billion in 2028, growing at a CAGR of 4.5% from 2021 to 2028 (Millet Advisor, 2020).

According to Millets Report by Government of India, Millets have the potential to help achieve the sustainable development goals (SDGs)—mainly SDG 2 (Zero Hunger), SDG3 (Good Health and Well-being), SDG 12 (Sustainable Consumption and Production), and SDG 13 (Climate Action).

METHODOLOGY

This paper has reviewed the literature from Secondary sources (Peer reviewed academic studies, reports, conference papers, reviews and policy papers) using online search engines like google scholar, JSTOR and other databases. Key word searches were applied in these search engines using a combination of words, such as "Millets and Economy, Agricultural economy and Millets, Consumerism and Millets, Benefits of Millets" and the like. The search initiated with review of abstract and full text with references available on line or in print. The collected literature was further divided and discussed in three broad categories as:

1. Millet cultivation from economic perspective.
2. Factors affecting millet cultivation and usage.
3. Millets and consumer acceptability.

1. Millet's Cultivation in India from Economic Perspective

The Government of India notified millets as 'Nutri Cereals' which includes major millets—Sorghum, Pearl Millet, Finger Millet; Minor Millets—Foxtail Millet, Proso Millet, Kodo Millet, Barnyard Millet, Little Millet and two Pseudo-Millets/pseudo-cereals (Buckwheat and Amaranthus).

Millets are cultivated in 21 states in India. The area of cultivation is estimated to be 13 million hectares. The crop production is between 15 to 16 million tons. With the major states contributing being Karnataka, Andhra Pradesh, Tamil Nadu, Kerala, Telangana, Uttarakhand, Jharkhand, Madhya Pradesh, and Haryana. The contribution of millets for the Gross Domestic Product is nearly 50%.

(a) *Millets and Climatic Conditions*

Table 2: Millets: Common Names, Origin, Climatic Conditions

Scientific and Common Names	*Origin*	*Climatic Conditions*
Sorghum bicolor Sorghum, great millet, guinea corn, kafir corn, aura, mtama, jowar, cholam, kaoliang, milo, milo-maize	Northeast region of Africa (Ethiopia-Sudan border)	Drought tolerant, recovery mechanism from stresses, highly adapted to wide range of soils, altitudes and temperatures, responsive to high input management
Pennisetum glaucum Pearl millet, cumbu, spiked millet, bajra, bulrush millet, candle millet, dark millet	Tropical West Africa	Highly resilient to heat and drought, grow in poor soils, responsive to high input management
Setaria italica German millet, Siberian millet, Foxtail millet, Hungarian millet, Thenai, Italian millet,	Eastern Asia (China)	Adapted to low rainfall, high altitude (Up to 2000 m amsl short duration, tolerant to low fertility and drought
Panicum sumatrense Little millet, Samai	Southeast Asia	short duration, Adapted to low rainfall and poor soils-famine food; withstand water logging to some extent, Up to 2000 m amsl
Paspalum scrobiculatum Kodo millet, Varagu	India	Long duration, hardy, needs little rainfall, comes up in very poor soils, grow in shallow and deep soil, response to improved management
Panicum miliaceum Proso millet, brown corn, broom-corn millet, common millet, Russian millet, Panivaragu, hog millet,	Central and Eastern Asia	Short duration, adopted to low rainfall and high altitude area, tolerant to heat and drought
Echinochloa crus-galli, Echinochloa colona Japanese barnyard millet, Kudhiraivali, Barnyard millet, sawa millet,	Japan	Very short duration (Fastest growing), voluminous fodder, not limited by moisture, high altitude adapted (Up to 2700 m amsl)
Eleusine coracana Finger millet, African millet, koracan, ragi, wimbi, bulo, telebun, Ragi	Uganda or neighboring region	Moderately resistant to heat, drought and humidity, adapted to wide altitude range (Up to 2100 m amsl), rich source of calcium.

Source: https://spc.tn.gov.in/app_studies/NUTRIENT_COMPOSITION.pdf, https://www.assocham.org/uploads/files/Report_Millets%202022%20(Print%20Version)%20(1).pdf

Literature has stated that millet cultivation was equal to the rice and wheat cultivation. It was post green revolution that a decline was observed in the area of cultivation for millet crops. This has been attributed to lack of input subsidies, low remuneration for farmers, subsidized supply of major cereals through public distribution system and change in consumer preferences.

(b) *Millets and Economy*

The table below indicates findings of the studies reviewed for Millets and Economy.

Table 3: Studies Reviewed for Millets and Economy

Title of study, Authors and year	*Findings*
Millet Scenario in India Gowri U., Shivakumar, (2020)	Data collected from the Ministry of Economics and Statistics (seasonal and crop report from 1950–51 to 2018–19) and analyzed by using compound growth rate. There was a negative growth with a decrease in production till 2005 and a positive growth post 2005.
A short review on millets: A potential nutri cereals Chaudhary S., *et al.*, (2020)	Importance of Millets in relation to nutritional security and climate change has been discussed. The benefits to small farmers for millet cultivation—save water, less fertilizers and can grow in poor soil
Growth and instability analysis of pearl millet cultivation in India Shah P., *et al.* (2020)	State-wise analysis of production of pearl millet has revealed that cultivation area has declined everywhere except in Uttar Pradesh. Instability index was used to measure the production for the years from 1997–1998 to 2016–2017.
Millets for Food Security in the Context of Climate Change: A Review Saxena R., *et al.* (2018)	Authors have reported that in India, there has been a decline in the area allotted for cultivation of millets but during the study period the area-wise production of finger millet and pearl millet has substantially increased with increased crop yield.
Performance of Major Millet Crops in Tamil Nadu: An Economic Analysis N.Narmadha, A. Kandeepan, (2017)	The paper analyzed growth rates, instability, and decomposition analysis of area, production, and productivity of major millets during the period from 1970–91 and 1991 –2015 in Tamil Nadu. Negative growth was reported for millets except for maize.

(c) *Factors Affecting the Economy of Millets*

1. *Pricing:* The pricing for millet crop is considered as volatile determined by supply and demand for the grains.
2. *Cost of Cultivation:* The cost of cultivation for the crop includes variable costs like labour, bullock power, machine power, seeds, farmyard manure, fertilizer, plant protection chemicals, irrigation charges and interest on working capital. The fixed costs consist of depreciation, land revenue, rental value of land and interest on fixed capital. It is important to reduce the cost of cultivation and improve yield to increase profitability.
3. *Income Earned:* Literature has stated that millets have the potential to earn income for farmers, but farmers are inclined to move to cereal and cash crops to earn more income even though millets are drought resistant and rich in nutrients. A Memorandum of Understanding (MoU) has been signed by Agricultural and Processed Food Products Export Development Authority (APEDA) with Indian Institute of Millets Research (IIMR) to boost farmers' income as well as for value-addition.
4. *Exports:* According to Directorate General of Commercial Intelligence and Statistics (DGCIS data), India increased millet exports by 8.02% in the fiscal year 2021–22, with 159,332.16 metric tons exported compared to 147,501.08 metric tons exported the previous year. India's major millet exporting countries are U.K. Tunisia, Yemen, U.A.E, Nepal, U.S.A., Libya, Oman, Saudi Arabia, and Egypt. Some of the millets exported by India are Buckwheat, Jawar, Ragi, Bajra, and Canary. The major countries for importing millets in are the United Kingdom, Japan, Mexico, Germany, the U.S.A, Brazil, Indonesia, Belgium, Italy, and Netherlands.

(d) *Strategies for the Revival of Millets*

Government of India has revised its policy to aid the production and consumption of Millets. Year 2018 was observed as the "National Year of Millets"—an initiative under Rashtriya Krishi Vikas Yojana. The Millets promotion (INSIMP) aimed to increase the production capacity through provision of input kits and promote Nutritional Security. Government has established units for processing and value-addition to millet products. Government has substantially hiked minimum support price (MSP) to motivate farmers for cultivation of millets. Insurance and institutional finance should be offered to millet farmers.

As per Government of India celebrations for IYOM 2023 and notification in All India Radio, "Indian missions abroad will be involved in the branding and promotion of Indian millets, as well as the identification of international chefs and potential buyers such as department stores, supermarkets, and hypermarkets for the purpose of organizing B2B meetings and direct tie-ups. "APEDA has planned to showcase millets and its value-added product at various global platforms such as Gulfood 2023, Foodex, Seoul Food & Hotel Show, Saudi Agro Food, Fine Food Show in Sydney (Australia), Belgium's Food & Beverages Show, Germany's BioFach and Anuga Food Fair, San Francisco's Winter Fancy Food Show, etc." as per government report.

2. Factors Affecting the Millets Cultivation and Usage

(a) *Climate, Water and Soil Requirement*

Millets are grown in tropical as well as subtropical climate up to an altitude of 2,100 m. It is a heat loving plant and for its germination the minimum temperature required is 8–10°C. A mean temperature range of 26–29°C during the growth is best for proper development and good crop yield. Rainfall range from 500–900 mm is suitable for millets. Kodo Millet grows well in moderate rainfall of 50–60 cm as it has a heavy water requirement.

Soil: Millet has wide adaptability from very poor to very fertile soil and can also tolerate alkalinity. The best suitable soils for millet farming are loamy, alluvial, and sandy soil having good drainage. Kodo millet can be grown in gravelly and stony soil such as in the hilly region (Technical Report, ICAR gov.in)

(b) *Type of Millets Cultivation*

The origin of millet is diverse with varieties coming from both Asia and Africa. From traditional times Millets have been considered as the staple crop of people and grown in semi-arid tropical region

Varieties of millet crop in India are listed: a. Foxtail Millet. b. Finger Millet c. Barnyard Millet, d. Browntop Millet, e. Little Millet, f. Kodo Millet, g. Pearl Millet, h. Proso Millet and i. Sorghum.

(c) *Use of Millets as Staple Crop by Consumers*

A research reported that the usage of millets was on the dishes like roti/chapati/Bhakri (Indian Flat bread), Khichidi (mix millets with other grains), Dosa, Ragi balls and sweet preparations like kheer and halwa. It was reported that awareness among consumers was less and only pearl and finger millet were commonly consumed. The barriers for using millets were reported as lack of knowledge of how to cook, which millet to cook and taste factor which was quoted by family members and children (Technical report, Millet usage and attitude study, Indianet).

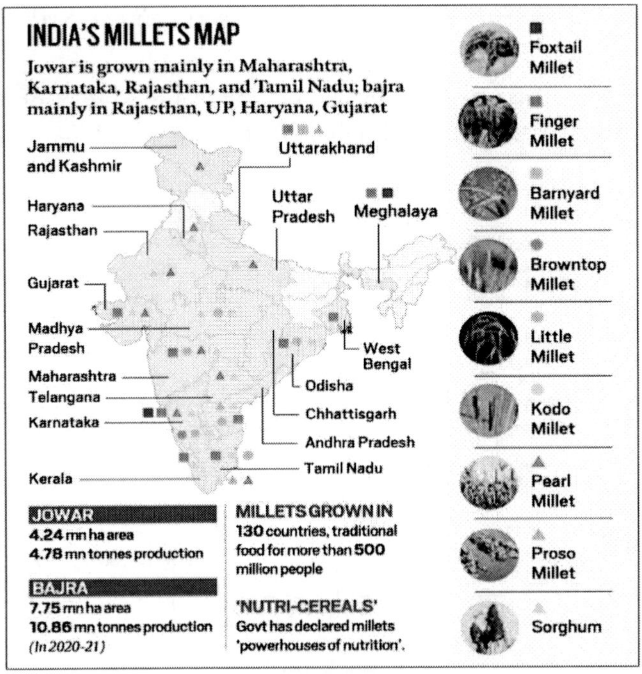

Figure 1: India's Millet Map

(*Source:* https://www.insightsonindia.com/2022/11/26/indias-push-for-millets)

(d) *Awareness about Millets as Staple Consumption*

A survey among 479 participants in Tier 1 and Tier 2 cities in India, pearl millet had a higher recall followed by Jowar and Bajra. Millets were considered to be "good for health", "easy to digest", "good for weight reduction" and "good for people having deficiencies". Lack of knowledge among consumers was reported about variety, availability and usage of millets. Some of the key health benefits mentioned by respondents were easy digestibility, nutrient rich, weight reduction and diabetic food. From existing users of Millets, few (19%) wanted to try a new variety of Millet however among the rare/non consumers of Millets the interest level was higher (50%).

According to retailers, the consumption appears to be done most in Middle/Upper Middle-Class households (68%) and also in low class. The main earner's occupation of Millet consuming households appeared to be from Skilled workers to Professionals. Among businessmen households, consumption appeared to be a little less. Most Millet consuming households (60%) were educated. Main purpose stated for consuming Millet was for occasional preparations and for Healthier living (Technical Report, Indianet).

(e) *Development of Millet Products for Consumers in Small Scale Sectors*

Various food entrepreneurs are trying to promote and commercialize different millet-based products, thereby increasing the consumption of millets and ensuring nutritional security (Sulthana, 2014).

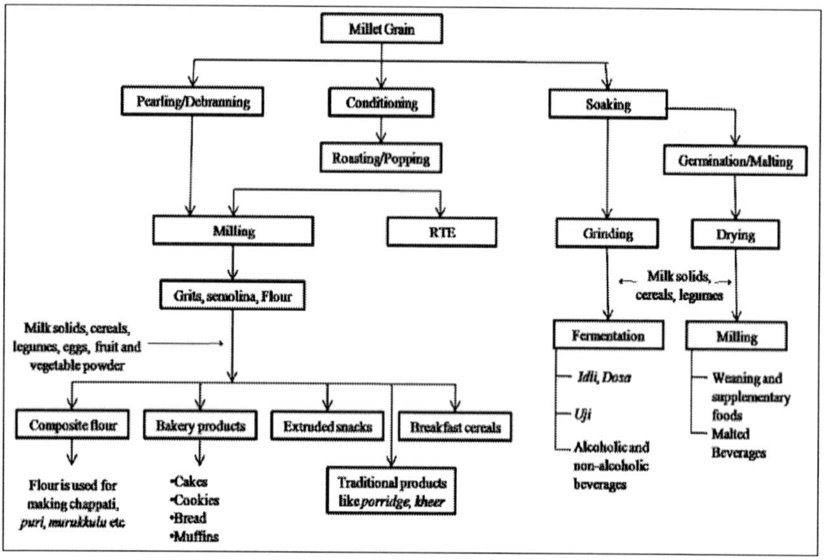

Figure 2: Flowchart for Developing Millet based Foods

(*Source:* Ashwinikumar *et al.*, 2018)

According to a market intelligence report by APEDA 2022, Millets based products such as flour, flakes, cookies, etc. are visible in the consumer market. The high protein content of these grains makes them ideal for the vegetarian/vegan population, largely based in the U.S., Europe and Asia Pacific. Breakfast cereals such as flakes and local recipes have seen a rising trend of consumption by the African and Asian population. Millet based breakfast foods generated revenue over USD 2 billion in 2018 and can increase earning due to want of fibrous and gluten free food consumption. Millet products like porridge for infants, bakery products like cookies, muffins and cakes help to mitigate malnutrition. In Africa, the traditional millet beer is popular in most countries in Asia Pacific, Europe and North American beverage markets. One of the main reasons for the low consumption of millets has been attributed to the weak value chain which includes the unorganized village level primary processing units, distance between the local production unit to processing unit and the inflated cost of production. (Agri exchange report)

(f) *Government Schemes for Millet Cultivation*

Niti Ayog article stated that the Government of India is strengthening the value chain for millet products and making them available to consumers through Minimum Support Price (MSP). The Government has advised for the inclusion of major and minor millets in the public distribution system. The support through various organizations like The Ministry of Agriculture and Farmers Welfare (MoA&FW), via its schemes, supports Self-Help Groups (SHGs)/Farmer Producer Organisations (FPOs) in primary processing at the village level. The Ministry of Small and Medium Enterprises (MSME), via its schemes, supports the secondary processing, provides soft loans, gives accreditation to private entrepreneurs to enable easy access to funds, and promotes adoption of a co-pay model to incentivise decentralized RTE/RTC processing units at sites closer to production. Schemes like National Rural Livelihood Mission (NRLM) and Ministry of Rural Development (MoRD) make a conscious effort to support processing units run by women entrepreneurs/SHGs/cooperatives through National Food Security Mission (NFSM) and PM POSHAN Scheme. Anganwadis and schools under the Integrated Child Development Scheme (ICDS) for provision of nutritious mid-day meal and creating awareness for the use of millet products. APEDA supports farmers for export of Millets and its products. State governments have supported the mission millet—promoting millet entrepreneurship and conducting millet awareness campaigns (Hindustan times, July 28, 2022).

3. Millets and Consumers

(a) *Consumers Perspective Towards Millets Consumption*

From a consumer survey conducted by International Crops Research Institute for the Semi-Arid Tropics (ICRISAT), three factors emerged as driving force for millet consumption—improving health and wellness, weight loss and taste. A study conducted aimed to know the attitude, preference, and consumption pattern of millets among the female young adults between 20 to 28 years of age, reported that 89% of the respondents consumed millets, majority of the respondents (58%) consumed millets as it is healthier and 26% loved the taste, 90% of them recorded millets as nutritious, three percent considered it as expensive and six percent of them considered millets as rural food. Merely 4% of the respondents reported daily consumption of millets and 23% said that they consume millets 3–4 times/week. The millet-based food products consumed among the respondents were roti (25%), snacks (20%), khichadi (16%), kali/porridge (7%), traditional sweets (7%), bread (6%), and dosa/adai (8%). Majority of the respondents knew about multi-millet idli/dosa/adai mix (85±3.4), millet cookies (91±2.3), multi millet noodles/pasta (82±5), little millet vermicelli (88±4.4), millet biscuits (92±2.1), (Adhitya *et al.*, 2022)

1. *Knowledge about millet grain and its nutritional advantages among women*
   A study in Bengaluru revealed that 80.6% of women are aware of millet but only 62.7% of women are consuming millet. (Adhitya *et al.*, 2022)
2. Use of Ready to cook (RTC) and ready to eat (RTE) foods by consumers
   Reasons are fast-paced urban lifestyle, nuclear family structure, rising disposable income and convenience (Takellambam *et al.*, 2016).
3. *Millet product features influencing consumers buying behavior*
   A study results revealed that the majority of the sampled consumers are already acquainted with the consumption of finger millet, sorghum and pearl millet mostly during the breakfast, Dinner and Lunch hours. Frequency of consumption was found to be more as both RTE and RTC forms, (46%), followed by staple grains (29%), only RTC (13%) and RTE (12%). Among all the influencing factors "price holds the first rank followed by taste (second), brand (third), promotional offers (fourth) and packaging (fifth)" (Pravalika *et al.*, 2018).
4. *Major factors that influence the consumption of millets*
   Sirisha and Lakshmi (2018) in their research revealed that the education, income and health level of the family are the major factors that were found to influence the consumption of millets. Rajput *et al.* (2019) conducted a study on consumer acceptance regarding millet-based products among school children for their health benefits. The results reported that the malted drink of kodo millet was accepted by students.

(b) *Types of Millets Products Available for Use by Consumers*

By using either whole millets or processed millets in the form of flour/extruded products, various recipes are being prepared. Since millets are healthy, nutritious and tasty, the millet recipes are chosen for breakfast, snacks, dinner and only few preferred for lunch (Narmadha *et al.*, 2017).

In a study conducted in Telangana, it has been reported that foxtail millet, little millet, pearl millet, kodo millet, finger millet, proso millet and jowar were easily available in the market and were having a varied range of products (Alekhaya *et al.*, 2019).

In a study by Prathyusha *et al.* (2021), an online market survey was done to identify the various millet products prepared under various brand names. The products included flour, noodles and flakes alongwith cookies, energy bars and

Table 4: Research on Millet Products

Author (Year)	*Products Created*	*Findings*
Ballolli *et al.*, (2001)	Nutraceuticals enhanced farm millet treats by consolidating nuts and dry organic products	The farm millet treats were just right with a light and fresh surface, lovely smell and amazing taste.
Varnashree *et al.* (2008)	Idlis by using Ragi and Ragi flour	Fortification and improvement of nutrients without impacting quality of Idli
Vijayakumar *et al.* (2009)	Kodo and Barnyard millets flour with wheat flour enriched chapatti	Improved thickness and taste by the mix of flours with 30% millet flour
Kulkarani *et al.* (2011)	Conventional RTE food items like proso millet sweet pongal, little millet paddu and dosa with chakramuni leaves, little millet idli flavoured with carrot and methi leaves, little millet uppma seasoned with bengal gram leaves and drumstick leaves, farm millet based uppma with drumstick leaves, foxtails millet vada with greens and foxtail millet based besibelebhath.	Consumers and working professionals considered the items to be nutritious, and easy to prepare

laddus. Several local vendors are preparing unbranded products like murukku, chudwa, namak para, sakkar para etc. "Lippia; Trunaturals; Nature's trunk; Saaral; B&B organics; Value life; MilTop" were found to be the popular brands by respondents.

(c) *Education of Consumers about Benefits of Millet Products*

A study Michaelraj and Shanmugam (2008) concluded that consumption and harvesting of millets are falling 2012 onwards. They suggested that developmental effort should be made through minikit demonstration, state level training programs, providing improved seeds, non-monetary inputs and research, start massive education and promotional programs on millets and the government must use its media campaign funds to take up millet promotion.

Diversification of food production must be encouraged both at national and household level with increasing yields as a successful approach for improving household food security (Singh and Raghuvanshi 2012).

IIMR Study has identified a gap among people who were health conscious and who were aware of the health benefits of Millets and suggested the need for promotions of the health benefits in masses. Another reason was that as millets were not consumed at home, education had to be directed towards the decision maker in the household (Pravalika *et al.*, 2018).

The main source of campaign could be through social media as most people have access to a minimum of any one social media platform.

(d) *Challenges Faced in Retailing the Products*

In a study on the awareness of the products, Ready to eat (RTE) and Ready to cook (RTC) forms of millet, choice and consumption form of the products were mostly similar among the respondents and lack of awareness of other varietal forms of millets in the market. Among these forms, only biscuits, cookies, roti mix and rava mixes have gained certain popularity. Awareness has to be made regarding the basket of products and innovations in millet products that are devoid of gluten and in tackling lifestyle diseases (Takhellambam *et al.*, 2016).

Main challenges are:

1. Need for improved methods of production and technology support.
2. Consumer preference towards rice and wheat products.
3. Lack of inclusion of millets in the public distribution system.
4. Need for funding to promote small business ventures for millets production.
5. Absence of strong demand for millet products.

CONCLUSION

Millets are Smart foods and are a powerhouse of major and minor nutrients. The International Year of Millets, 2023 will act as a platform to promote the global production of millet crop and its processed foods and inclusion of millet in the food basket for household consumption. The characteristic of millet being climate resilient and drought resistant would help in global food security and economic benefit. Millets serve dual purpose and can be cultivated as crop and for fodder. Educating consumers on the benefits of Millet and millet products through media will aid productivity increase and remunerative farming. The inclusion of millets in Mid-day meal schemes, Anganwadis, ICDS, state, national and international feeding programs can help in overcoming malnutrition and mitigate micronutrient deficiencies among children.

Millets are sustainable foods with reduced carbon footprint and greenhouse gas emission. The millet cultivation uses less water, fertilizers, pesticides and are easy to store for a longer duration.

It is imperative that the perception of millet being a "poor man's diet" has to be changed. Central and State government, Food industry players, small business entrepreneurs are focusing on the rebranding of millets as nutri cereals.

REFERENCES

[1] Adithya, Girijavallabhan; Agna, Shibu Mathew; Gayathri, Rajeev; Sonia, Thomas and Premagowri, B. (2022). Perception and consumption pattern of millets among female young adults, *IARJSET International Advanced Research Journal in Science,* Engineering and Technology Impact Factor 7.105, Vol. 9, Issue 1, January 2022.

[2] Alekhya, P. and Raj Shravanthi, A. (2019). Buying Behaviour of Consumers towards Millet Based Food Products in Hyderabad District of Telangana, India. *Int. J. Curr. Microbiol. App. Sci.,* 8(10): 223–236.

[3] Ashwani, Kumar; Vidisha, Tomer; Amarjeet, Kaur; Vikas, Kumar and Kritika, Gupta (2018). Millets: A solution to agrarian and nutritional challenges, Agric & Food Secur (2018) 7:31.

[4] Ballolli *et al.,* (2001). Development and quality evaluation of foxtail millet [*Setariaitalica* (L.)] incorporated breads. *J Agric Sci.*; 27(1):52–54.

[5] Fuller, D.Q. (2014). Millets: Origin and Development. In: Encyclopedia of Global Archaeology." Springer, New York, NY.

[6] Kulkarani, L.R., Naik, R.K. and Rokade, C. (2011). Development of Nutri-cereals based Value Added Traditional Products and its Acceptability. NAIP: National Symposium recapturing Nutritious Millets for Health and management of Diseases, UAS, Dharwad. December: 16–17, pp. 44.

[7] Michaelraj Stanly Joseph and Shanmugam (2008). "A study on Millets based Cultivation and Consumption in India", *International Journal of Marketing, Financial Services and Management Research,* Vol. 2, No. 4, pp. 49–50.

[8] Gowri, M. Uma and Shivakumar, K.M. (2020). Millet Scenario in India. Economic Affairs, Vol. 65, No. 3, pp. 363–370, September 2020.

[9] Narmadha, N. and Kandeepan, A. (2017). Performance of Major Millet Crops in Tamil Nadu: An Economic Analysis. *Arthashastra Indian Journal of Economics & Research,* Volume 6, Issue 5, September-October 2017.

[10] Pravallika, D.R., Dayakar Rao, B., Seema, D. Srinivasa Chary and N. Sri Devi (2018), Market Strategies for Promotion of Millets: A Critical Analysis on Assessment of Market Potential of Ready to Eat (RTE) and Ready to Cook (RTC) Millet Based Products in Hyderabad, IIMR, Hyderabad.

[11] Rajput, L.P.S.; Parihar, Pratibha; Dhumketi, Ketki; Naberia, Seema and Tsuji, Koji (2019). Development and Acceptability of Novel Food Products

from Millets for School Children. *International Journal of Current Microbiology and Applied Sciences,* ISSN: 2319-7706 Volume 8 Number 04.

[12] Rachit, Saxena; Sai, Kranthi Vanga; Jin, Wang; Valerie, Orsat and Vijaya, Raghavan Sustainability (2018). Millets for Food Security in the Context of Climate Change: A Review, 10*(7),* 2228.

[13] Ranjita Devi, Bharati V. Chimmad and Jones, Nirmalnath Prkasam (2016). Ready-to-cook millet flakes based on minor millets for modern consumer, J Food Sci Technol., 2016 Feb. 53(2).

[14] Shah Parth, Shrivastava A., Suvagiya. D. and Vani, G.K. (2020). Indian Journal of Economics and Development, Year: (2020) Growth and instability analysis of pearl millet cultivation in India, Volume 16, Issue 4 (511–517), Print ISSN: 2277-5412. Online ISSN: 2322-0430.

[15] Sirisha, S. and Lakshmi, J. (2018). A study on consumption pattern of millets and millet-based products in Guntur city, Andhra Pradesh. P.G. Thesis. Acharya N.G. Ranga University, Rajendranagar, Hyderabad.

[16] Sulthana (2014). Sulthana CP. Consumer Acceptability and Demand for Health Foods. A Study of millet based ANGRAU - Foods in Metro Hyderabad. M.Sc. Thesis. Submitted to ANGRAU, Hyderabad, India 2014.

[17] Sumit Chaudhary, P.S. Negi, Alankar Singh, R.K. Prasad, Pallavi, Ajay Kumar and Rajesh, Kaushal (2020) A short review on millets: A potential nutri cereals. *The Pharma Innovation Journal,* 2020; 9(10): 123–126.

[18] Prathyusha *et al.* (2021) Review on consumer awareness and health benefits about millets. *The Pharma Innovation Journal,* 2021; 10(6): 777–785.

[19] Singh, P. and Raghuvanshi, R.S. (2012). Finger Millet for Food and Nutritional Security, *African Journal of Food Science,* 6, 77–84.

[20] Vijayakumar, P.T., Mohankumar, J.B. and Jaganmohan, R. (2009). Quality Evaluation of Chapatti from Millet Flour Blend Incorporated Composite Flour. *Indian Journal of Nutrition and Dietetics,* 46:144–154.

[21] Varnashree, B.S., Rao, S.G. and Puttaraj, S. (2008). Quality Characteristics of Ragi (Eleucine coracana) based Idli. *Indian Journal of Nutrition and Dietetics,* 45:131–136. Zereyesus YA, Dalton TJ (2017) Rates of return to sorghum and millet research investments: A meta-analysis. PLoS ONE 12(7): e0180414.

[22] Zereyesus, Y.A. and Dalton, T.J. (2017). Rates of return to sorghum and millet research investments: A meta-analysis. PLoS ONE 12(7): e0180414.

Online Resources

1. https://spc.tn.gov.in/app_studies/NUTRIENT_COMPOSITION.pdf
2. https://www.assocham.org/uploads/files/Report_Millets%202022%20(Print %20Version)%20(1).pdf
3. https://newsonair.com/2022/11/11/centre-to-promote-exports-of-millets-and-its-products/
4. https://kvk.icar.gov.in/API/Content/PPupload/k0162_27.pdf

5. https://idl-bnc-idrc.dspacedirect.org/bitstream/handle/10625/57033/57087.pdf
6. https://agriexchange.apeda.gov.in/Weekly_eReport/Millets_Report.pdf)
7. https://www.hindustantimes.com/ht-insight/public-health/increasing-production-and-consumption-of-millets-101658992530337.html
8. https://www.ijcmas.com/8-10-2019/P.%20Alekhya%20and%20A.%20Raj%20Shravanthi.pdf
9. https://www.jetir.org/papers/JETIR2111346.pdf
10. http://gldc.cgiar.org/millet-consumption-survey-provides-largest-baseline-for-india
11. https://www.financialexpress.com/lifestyle/increasing-indias-food-nutritional-and-economic-security-by-reviving-millets/1737642/
12. https://milletadvisor.com/importance-of-millets-in-india/(accessed on January 2nd, 2023)
13. https://impactentrepreneur.com/the-global-impact-economy-of-millets/
14. https://www.weforum.org/agenda/2022/05/ancient-crop-india-water-food-crisis
15. https://www.indianarrative.com/economy-news/will-superfood-millets-become-an-integral-part-of-indias-food-security-52479.html
16. https://www.infokara.com/gallery/38-nov-c112.pdf
17. https://d-nb.info/1256596639/34
18. https://www.ncbi.nlm.nih.gov/pmc/articles/PMC8871339/
19. http://gldc.cgiar.org/millet-consumption-survey-provides-largest-baseline-for-india
20. https://www.iosrjournals.org/iosr-jbm/papers/Vol20-issue8/Version-4/K2008046776.pdf
21. https://idl-bnc-idrc.dspacedirect.org/bitstream/handle/10625/57033/57087.pdf
22. http://jtropag.kau.in/index.php/ojs2/article/view/1058/598
23. http://www.bestpracticesfoundation.org/pdf/A-Marketing-Analysis-Small-Millets_MSSRF.pdf
24. https://ijeponline.org/index.php/journal/article/view/78/102 Challenges to millet consumption- 6th November 2022.
25. https://www.researchgate.net/publication/363364397_Millet_Production_and_Consumption_in_India_Where_Do_We_Stand_and_Where_Do_We_Go
26. https://agricultureandfoodsecurity.biomedcentral.com/articles/10.1186/s40066-020-00282-6
27. https://www.recentscientific.com/sites/default/files/14234-A-2019.pdf
28. https://www.mdpi.com/2072-6643/14/1/225 (accessed on January 6th 2023)

Distribution of Millets for New Mothers under PDS through Public Hospitals

Ashlesha Kamble, Taufiq Shaikh, Sharayu Raje, Asmita Mestry and Nitin Labhane

Department of Botany, Bhavan's College (Autonomous), Andheri (W), Mumbai, India

ABSTRACT: The public distribution system (PDS) ensures that individuals in society who are economically disadvantaged receive requirements. It is expensive for people from economically underprivileged parts to consume a nutrient-rich grain like millet, which is known to be used for nutrition-related silent hunger. In order to enhance nutritious consumption in postpartum women's diets, this research suggests a new policy that would mandate that every government hospital give millet packets to every patient who is discharged for recuperation. Millets can restore nutrition lost during a woman's postpartum discharge, improving the health of both the mother and the child. According to the National Family Health Survey-5 (2019–21), 19.3% of Indian children are underweight. This issue may be resolved by giving women millet when they are lactating so that newborns can obtain their nutritional needs met. During this time, requirements. We wholeheartedly support new regulations or programmes that would enable all postpartum women leaving government hospitals to get a packet of millet as a sign of the hospital's goodwill in addition to their discharge paperwork. As a result, this will be an important step in resolving the malnutrition problem.

Keywords: Millets, Public Distribution System (PDS), Postpartum, Economically Weak, Malnourished.

INTRODUCTION

Although millets significantly contribute to the nation's total agriculture-based economy, their use is still limited. In India, a developing country, malnutrition is a significant issue. Every human needs a balanced range of nutrients for the body's systems to work correctly. Insufficient nutrition is the primary cause of malnutrition, whereas millet is renowned for providing adequate nutrients. A large portion of the population is economically underprivileged; hence the PDS assists individuals in obtaining food grains through rationing shops. The Public Distribution System (PDS) was created to distribute food grains economically while managing shortages. PDS has grown in importance as a component of the government's strategy for managing the nation's food economy. Every state has a PDS Program that consists of wheat, rice, and oil, but this is insufficient to provide people with the necessary nutrition, which can be provided by supplying millet. Since PDS is a government body, millets can be delivered through government programs and distributed from government hospitals.

The breastfeeding lactation period is crucial for every woman and her child. It is a time when a mother and her newborn baby both require adequate nutrition for their well-being. The primary goal of providing millet at public hospitals is to assist new mothers in obtaining enough nutrition for both themselves and their children. Due to the large population that consumes food grain from rationing shops and the excessive throng, postpartum women cannot wait in line for long periods at rationing shops. Low-income families generally give birth in public hospitals. Due to their poor income, they cannot purchase food like millet. It would be a fantastic government initiative to supply millet to postpartum ladies from hospitals. Therefore, millet should be given as a gift to low-income moms who give birth in public hospitals so that they might receive nutritional benefits. As a result, the nutritional values/health index and the malnutrition rate in India will improve.

OBJECTIVE

- To serve millet as a balanced, nutritional diet from public hospitals under PDS Scheme
- To analyze the understanding of postpartum women regarding nutrition following the pregnancy.
- To ascertain whether the mother and the infant are affected by low nutrition.

RESEARCH METHODS

This study, which is a descriptive research study, uses the survey approach. Data for the survey was sourced from the Charkop Maternity Hospital in Charkop Sector 3, Kandivali (west), Mumbai, India. The group includes all of the postpartum mothers from the public hospital. Postpartum ladies responded positively to a survey asking if they would like to get millet directly from the hospital administration. Additionally, they were given packets of millet, including pearl millet, foxtail millet, barnyard millet, and finger millet.

Sr. No.	*Questions*	*Answers*
1.	Name:	
2.	Age:	
3.	Number of children:	
4.	Income	
5.	Are you able to wait in line at a PDS (Ration shop) after the postpartum?	
6.	Is our scheme beneficial to you?	
7.	Do you know the benefits of millet?	
8.	Would it be beneficial if hospitals supply millet?	
9.	After receiving millet packets from the hospital, how do you feel?	
10.	Suggestion	

Charkop Maternity Hospital (Mumbai) Pearl Millet, Foxtail Millet, Barnyard Millet, Finger Millet

Sheetal Patil (Age - 28)

Manisha Sonavane (Age - 22)

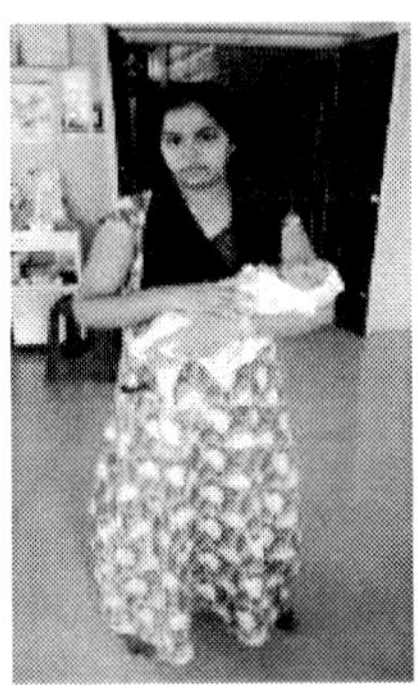

Soni Vishwakarma (Age - 28)

Shabhana Khan (Age - 22)

(With the mothers' approval, all photos were taken.)

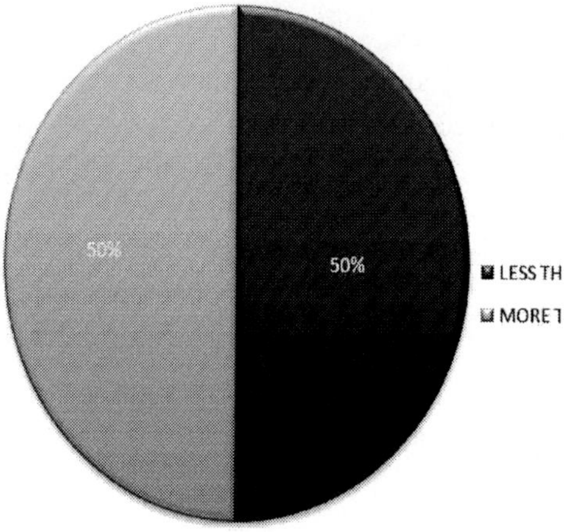

Figure (a): Income of Family

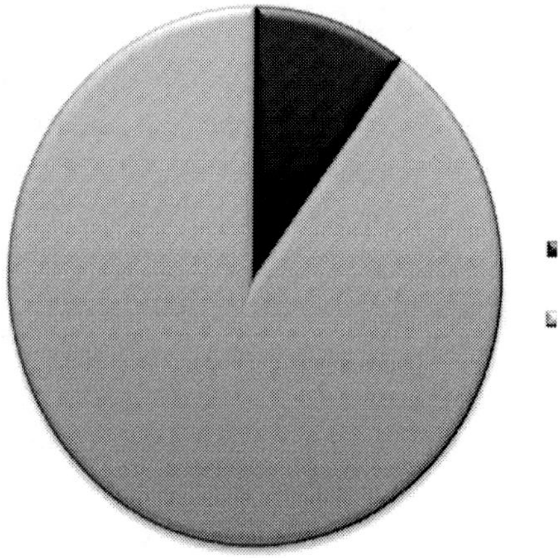

Figure (b): Number of Mothers Ready to Stand in Line after Delivery

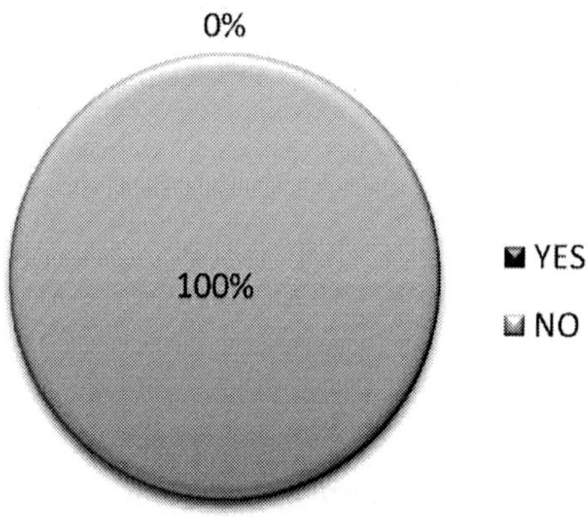

Figure (c): Positive Impact of Scheme

RESULT

All ten of the mothers who were surveyed did so. The majority of mothers claimed that they were unable to wait in line for hours for millet. Even though everyone knows millet's advantages, some women with limited resources cannot afford it. They were relieved to know that someone was helping them out by providing nutrition after receiving a packet of millet as part of the study. After understanding our policies, they were excited and pleased to get millet straight from the hospital as a gesture of gratitude.

CONCLUSION

The outcome of this research through survey and statistical data is to develop a policy whereby all postpartum women from government hospitals should receive some millet as part of the PDS program. The study will provide a clear understanding of millet effects on maternal and newborn health. It will provide postpartum mothers with knowledge and understanding about the kinds of foods they should eat after giving birth. This act will benefit mothers and their newborns by assisting with nourishment. In a study of ten mothers who received millet from public hospitals as a part of the survey, the majority expressed support for the continuation of this program.

ACKNOWLEDGEMENT

I want to express my special gratitude to my teacher Mrs. Asmita Mestry and all my teachers who gave me the golden opportunity to do this project on the topicmillets for new mothers under pds are distributed via public hospital. It helped me do a lot of research, and I learned about many things related to this topic.

I want to express my sincere gratitude to Charkop Maternity Hospital in Mumbai and all the Doctors for their assistance with the survey and for providing me with all the required information. Thanks Mrs. Manisha Pawar for helping me with the survey process.

REFERENCES

[1] Abbi, R., Christian, P., Gujral, S. and Gopaldas, T. (1991). The impact of maternal work status on the nutrition and health status of children. *Food and Nutrition Bulletin* 113, 20–25. doi:10.1177/156482659101300128

[2] Carlson, G.J., Kordas, K. and Murray-Kolb, L.E. (2015). Associations between women's autonomy and child nutritional status: A review of the literature. *Matern. Child Nutr.*; 11:452–482, doi: 10.1111/mcn.12113.

[3] Catherine, J. (2015). Field, The Immunological Components of Human Milk and Their Effect on Immune Development in Infants, *The Journal of Nutrition,* Volume 135, Issue 1, pp. 1–4.

[4] Tejaswini, E., Devi, S. Suchiritha, Vijaya Lakshmi, V., Shirisha, J., Kameswari, S.L. and Sujath, V. (2017). Nutritional status of rural and tribal pregnant woman of Karimnagar district, Telangana, *The Pharma Innovation Journal,* 6(9): 479–481.

[5] Jinali, Goradia and Sarthak, Doshi (2015). Automated Ration Distribution System, December, Procedia Computer Science, 45:528–532, DOI: 10.1016/j.procs.2015.03.096.

[6] Khandelwal, S., Siegel, K.R. and Narayan, K.M. (2013). Nutrition research in India: underweight, stunted, or wasted? Glob Heart. Jun; 8(2):131–7, doi: 10.1016/j.gheart.2013.05.003.

[7] Khandelwal, S., Postgraduate Diploma Program in Public Health Nutrition. 2010. [http://phfi.org/programmes/phn.html.]

[8] Lothian, J.A. (2008). The journey of becoming a mother. *J. Perinat Educ.* Fall; 17(4):43–7. doi: 10.1624/105812408X364071. PMID: 19436533; PMCID: PMC2582416.

[9] National Information Centre (2022, Dec 4). National Food Security Portal. [https://nfsa.gov.in/portal/PDS_page]

[10] Nie, P., Rammohan, A., Gwozdz, W. and Sousa-Poza, A. (2019). Changes in Child Nutrition in India: A Decomposition Approach. *Int J Environ Res Public Health,* May 22;16(10): 1815. doi: 10.3390/ijerph16101815.

[11] Om, Raj Katoch and Aroon, Sharma (2017). Nutritional status and academic performance: A study on school-aged children in rural areas of Jammu & Kashmir, India, *Indian Journal of Social Research,* Vol. 58(5), (Set. –Oct) (659–670), ISSN: 0019-5626, Academic and Law Serials all e ISSN: 2454–3624.

[12] Pathak, P.K. and Singh, A. (2011). Trends in malnutrition among children in India: Growing inequalities across different economic groups. *Soc. Sci. Med.*; 73:576–585, doi: 10.1016/j.socscimed.2011.06.024.

[13] Popkin, B.M., Horton, S., Kim, S., Mahal, A. and Shuigao, J. (2001). Trends in diet, nutritional status, and diet-related noncommunicable diseases in China and India: The economic costs of the nutrition transition. *Nutr Rev. Dec.*; 59(12): 379–390 doi: 10.1111/j.1753-4887.2001.tb06967.x.

[14] Pryer, J., Rogers, S. and Rahman, A. (2004). The epidemiology of good nutritional status among children from a population with a high prevalence of malnutrition. *Public Health Nutrition,* 7(2), 311–317. doi:10.1079/PHN2003530

[15] Robinson, S.M. (2015). Infant nutrition and lifelong health: current perspectives and future challenges. *J. Dev Orig Health Dis.* Oct.; 6(5):384–9, doi: 10.1017/S2040174415001257.

Analysis of Statewise Production of Hybrid Varieties of Pearl Millet and Finger Millet in India

Amita Sarswal, Taufiq Shaikh*,
Nitin Labhane and Sharayu Raje
Department of Botany, Bhavans College (Autonomous),
Andheri (West), Mumbai, India

ABSTRACT: Millets are rich in nutrients. Pearl millet (*Pennisetum glaucum*) and finger millet (*Eleusine coracana*) are both commonly grown cereals that have therapeutic benefits. According to National Agricultural Research (NARS), ICRISAT and various sources, the development of prospective hybrid lines could have a significant influence on India's export rate and GDP value. This review intends to determine the hybrid varieties that are commercially preferred in India. Based on different statistical data and graphical analysis from various government resources and agronomical literature, this review strives to draw attention to the state-wise production and distribution of hybrid varieties of Ragi and Bajra in India. Due to India's diverse Agroclimatic zones, the production of both millets varies significantly. This review further examines the top five varieties of Ragi and Bajra which are cultivated in India based on soil profiling and geographical conditions of various states.

Keywords: Hybrid Varieties, Millets, India, Analysis, Production Rate.

INTRODUCTION

Millets are rich in macronutrients and micronutrients. It possesses an incredible range of therapeutic properties. Being a stress tolerant crop and able to thrive well in harsh environments, makes it one of the most desirable crops to be cultivated. Millets are referred to as 'modern crops' because of their utility in various ways. Belonging to the *Poaceae* (*Gramineae*) family, they are composed of multi-disciplinary constituents. Pearl millet (Bajra), finger millet (Ragi), foxtail (kangni), sorghum (Jowar), Little millet (Moraiyo), Kodo, Barnyard, etc, are amongst most widely cultivated mill*et al*l around the globe. In the dry and semi-arid parts of the world, pearl millet is one of the primary cereals produced, mostly for crop yields, on more than 29 million ha, notably in Asia and Africa. India being the largest producer of pearl millet in Asia, has diverse agro-climatic zones which helps in enhanced growth of different varieties of pearl millet. Throughout 2018–19, pearl millet has occupied 6.93 million ha, yielding an annual crop of 8.61 million metric tonnes at a productivity of 1243 kg per ha (Directorate of Millets Development, 2020). It is regarded as a superfood of the agriculture industry. It is the fourth largest comprehensively cultivated cereal crop in India and is widely cultivated. Ninety percent of India's economic output of pearl millet is generated in the prominent cultivating states of Rajasthan, Maharashtra, Uttar Pradesh, Gujarat, and Haryana (Satyavathi, ICAR-AICRP, 2021). Being rich in macro and micronutrients, it is widely exploited and henceforth the improvement in varieties was initiated in early 1950s by Indian Council of Agricultural Research (Kumara *et al.,* 2014). The hybrid varieties were developed for efficient commercialization of pearl millets. Due to the sustained acceptance of these varieties by Indian farmers, crop efficiency surged from 305 kg/ha between 1951 and 1955. It has resulted in a 400% exponential rise in output. In contrast to the period from 2011 to 2015, the improvement is 7.6% (Figure 1) (Satyavathi *et al.,* 2021). Finger millet (*Eleusine coracana*) commonly known as Ragi or Nachni. With plenty of water, a broad range of temporal adaptability, it is cultivated extensively in Southern parts of India. The hybrid development substantially increased in 1960s in case of finger millet. The enhancement in the outcome of optimum production techniques, high-yielding hybrids of pearl and finger millet and encouragements from farmers paved the way for production of more adaptable varieties. The better growth and disease resistant varieties gained momentum in Indian market (Yadav *et al.,* 2021). This review paper analyses the production of Hybrid varieties of Bajra and Ragi in different states of India. Based on various data accumulated from varied government and research articles, the hybrid varieties cultivated in India, varies with respect to nutrient requirement in soil, diverse agro-climatic zones, soil profiling and irrigation practices. Furthermore, from various sources, the top 5 hybrid varieties of ragi and bajra in India were mentioned on the basis of commercialization and economical prospects.

METHODOLOGY

The secondary data accumulated from various government and public sources have been analyzed to illustrate the substantial production of hybrid varieties of Bajra and ragi in different states of India. This review further emphasizes on the top 5 hybrid vigour of bajra and ragi in India. It's represented in various sources and research articles that hybrid varieties have depicted a progressive outcome with respect to economical growth.

RESULT

The bajra and ragi both possessing different nutritional and temporal requirements, substantially has different productivity rates in Different states.

.

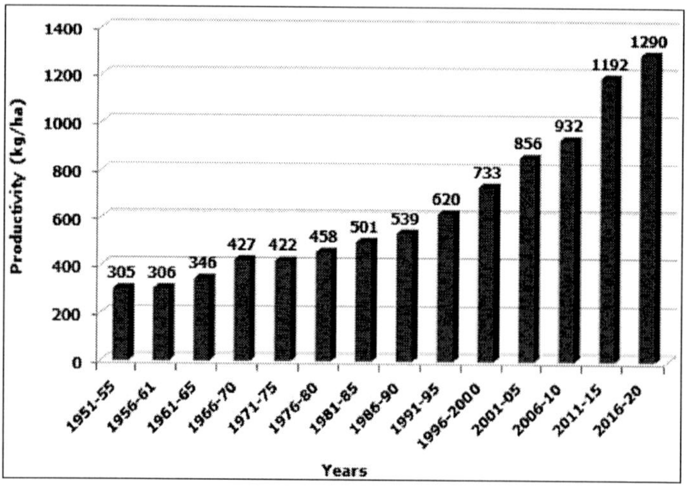

Figure 1: Trends in Pearl Millet Productivity over Years

Source: Genetic Improvement of Pearl Millet in India, 2013.

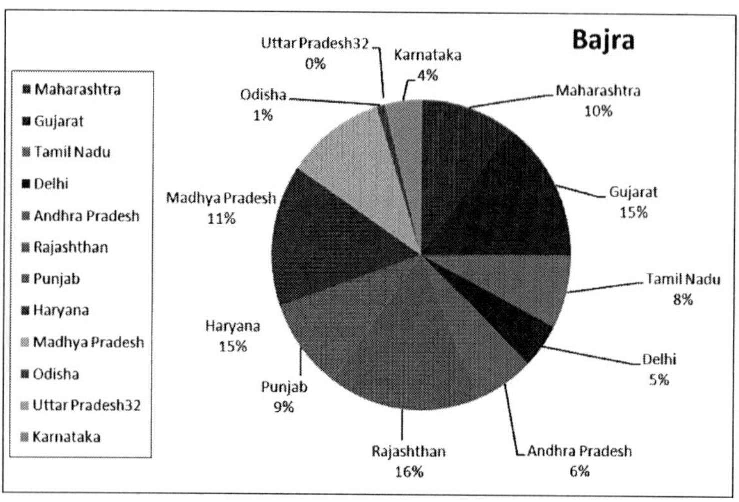

Figure 2: Pearl Millet Production (in %)

(Production rate of hybrid varieties since 1954 to 2014)

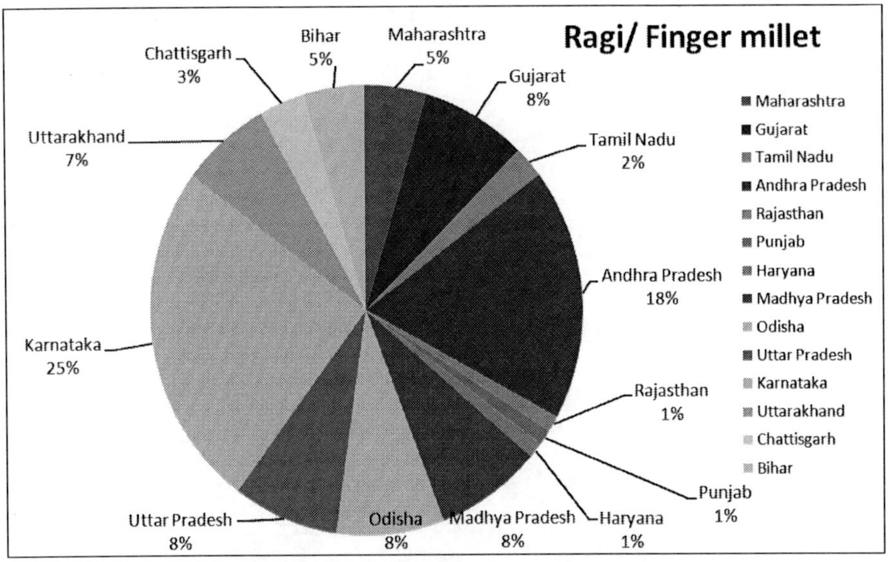

Figure 3: Finger Millet Production (in %)
(Production Rate of Hybrid Varieties since 1960 to 2016)

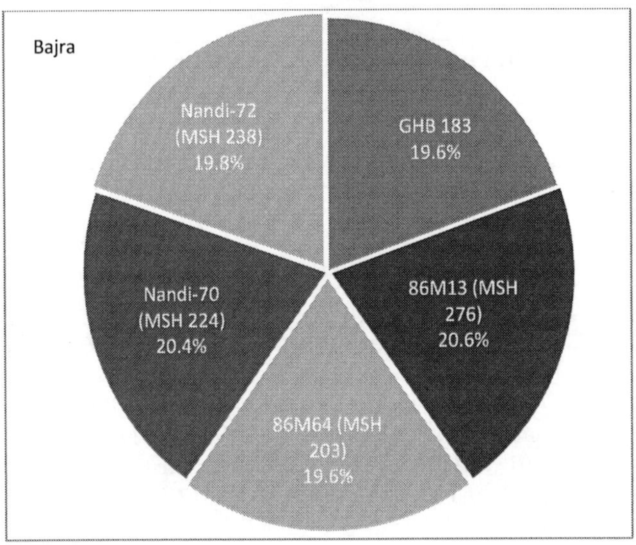

Figure 4: Top 5 Hybrid Varieties of Pearl Millet in Terms of Average Yield (in kg/ha)

The bajra and ragi both possessing different nutritional and temporal requirements, substantially has different productivity rates in Different states. Following table represents the production of hybrid varieties of Bajra introduced from year 1960 to 2014 and ragi varieties introduced from year 1964 to 2014.

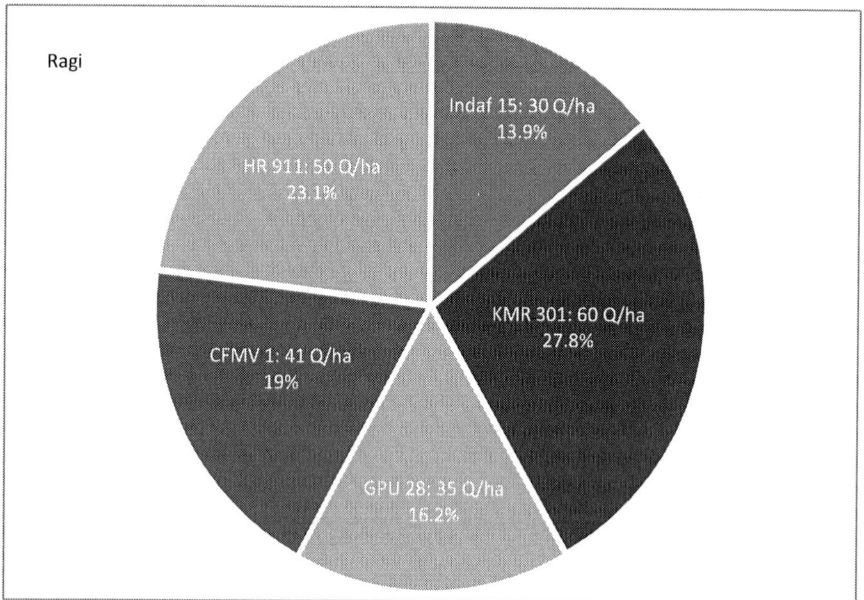

Figure 5: Hybrid Varieties of Ragi/Finger Millet in Terms of Average Yield (in Q/ha)

Table 1: Production of Hybrid Varieties of Bajra and Ragi from 1964–2014

States	*Bajra/Pearl Millet (Pennisetum glaucum)*	*Ragi/Finger Millet (Eleusine coracana)*	*Total No. of both the Hybrid Varieties*
Maharashtra	GHB–526, Saburi, GK–1004, MLBH–504, RHB–58, Nandi–64, Nandi–35, Proagro–9444, Proagro–9555, Pusa–444, B–2301 MLHB–267, Nandi–30, Saburi, GK–1004, PAC–903, Proagro–9443, Nandi–35, GHB–558, B–2095, Pusa–444, GK–1051 (IYoM, 2023), Nandi–72, Kaveri Super Boss, Nandi–70, Pratap, PKV RAJ, 86M86, Shine, PAC–909, 86M64, 86M53, RHRBH 9808, Nandi 64 (Singh *et al.*, 2021).	CFMV 2 (Gira) (Wadhmare *et al.*, 2022), KOPN 235, VL 376, Dapoli 1 (Prabhakar *et al.*, 2017)	Bajra–33 Ragi–4

States	*Bajra/Pearl Millet (Pennisetum glaucum)*	*Ragi/Finger Millet (Eleusine coracana)*	*Total No. of both the Hybrid Varieties*
Gujarat	Guj-Hybrid, Nandi–8, Nandi–32, GHB–526, Pusa–444, Proagro–9444, Pusa–605, Proagro–9555, Nandi–62, Nandi–52, Nandi–62, RHB–58, JKBH–26, GBH–316, 7876, Pusa–605, Nandi–32, Pusa–415, 7688, RHB–121, HHB 67, HHB–146, GHB–577, GHB–538, GHB–719, Nandi–62, HHB–197, GHB–757, GHB–732, GHB–558, Nandi–52, GHB–744, JKBH–676, RHB–154, HHB–216, RHB–173, HHB–223, GHB–526 (IYoM, 2023), KBH–108, Nandi–72, GHB–905, MP MH17, HHB–233, Kaveri Super Boss, Bio–70, Bio–448, MP–7792, Nandi–61, Nandi–65 (Singh *et al.*, 2021).	CFMV 2 (Gira) (Wadhmare *et al.*, 2022), GNN–6, VL–376, GN–5 VL–352, GPU–45, GN–3, Gujarat nagli 3 (Prabhakar *et al.*, 2017)	Bajra–48 Ragi–7
Tamil Nadu	RHB–58, X–6, X–7, PAC–903, COH (Cu)–8, GHB–558, Pusa–444, B–2095, GK–1004, K–1051, Proagro–9443, Proagro–9444, Proagro–9555, GK–1051, B–2301 (IYoM, 2023), Nandi–72, Nandi–35, Kaveri Super Boss, Nandi–70, Pratap, CO–9, 86M86, PAC–909, 86M64, 86M53, GK–1051 (Singh *et al.*, 2021).	CFMV 1 (Indravathi) (Wadhmare *et al.*, 2022), VL 376 (Prabhakar *et al.*, 2017)	Bajra–25 Ragi–2
Delhi	B 2095, JKBH 676, HHB 223, Nandi 61, Nandi 65, RHB 173, 86M66, MP 7792, MP 7872, Bio 448, Kaveri Super Boss, MP MH17, 86M89, GHB 905, RBH 18 (Singh *et al.*, 2021).	–	Bajra–15
Andhra Pradesh	Proagro–9443, GHB–558, B–2301, Pusa–444, Nandi–360, PAC–903, MLHB–504, Proagro–9443, Nandi–35,	CFMV 1 (Indravathi), CFMV 2 (Gira) (Wadhmare *et al.*, 2022), VR	Bajra–21 Ragi–16

States	Bajra/Pearl Millet (Pennisetum glaucum)	Ragi/Finger Millet (Eleusine coracana)	Total No. of both the Hybrid Varieties
	B–2095, GK–1051, GK–1004, Nandi–35, RHB–58, GK–1004 (IYoM, 2023), Kaveri Super Boss, Pratap, 86M86, Shine, 86M64, PAC 909, 86M53 (Singh *et al.*, 2021).	936, VL 376, PPR 2700 (Vakula), Srichaitanya, VL 352, Bharathi, Chilika, B 9–1, Champavathi, PR 230, Saptagiri, Suraj, Padmavathi, VL 149 (Prabhakar *et al.*, 2017)	
Rajasthan	Pusa–444, Nandi–32, GHB–558, RHB–173, HHB–223, HHB–146, MLHB–267, GBH–316, 7886, Pusa–605, Pusa–415, HHB–94, 7688, RHB–58, Proagro–9443, RHB–121, Proagro–9445, GHB–577, JKBH–26, HHB–117, GHB–538, HHB–67, GHB–757, GHB–732, Nandi–52, GHB–719, HHB–197, GHB–744, PHB–2168, JKBH–676, RHB–154, HHB–216, (IYoM, 2023), B 2095, JKBH 676, Nandi 64, Nandi 61, Nandi 65, RHB 177, HHB 226, 86M86, MP 7892, MP 7872, Bio 70, Bio 448, Nandi 70, Kaveri Super Boss, HHB 234, 86M89, Nandi 72, GHB 905, KHB 108 (Singh *et al.*, 2021).	VL 376 (Prabhakar *et al.*, 2017)	Bajra–53 Ragi–1
Punjab	GHB–744, RHB–58, HHB–146, MLHB–267, GHB–316, Pusa–605, Pusa–415, Proagro–9443, BHB–577, Nandi–52, JKBH–26, GHB–744, PHB–2168, JKBH–676, Pusa–444, Nandi–32, (IYoM, 2023), KBH 108, GHB 905, MP MH17, 86M89, Bio 448, Kaveri Super Boss, MP 7892, MP 7872,	VL 376 (Prabhakar *et al.*, 2017)	Bajra–31 Ragi–1

States	*Bajra/Pearl Millet (Pennisetum glaucum)*	*Ragi/Finger Millet (Eleusine coracana)*	*Total No. of both the Hybrid Varieties*
	86M66, 86M86, RHB 173, Nandi 70, HHB 223, Nandi 61, Nandi 65, JKBH 676 (Singh *et al.*, 2021).		
Haryana	RHB–58, GBH–316, 7886, Pusa–605, Nandi–32, Pusa–415, HHB–94, 7688, Proagro–9445 GHB–558, HHB–146, GHB–577, Pusa–444, MLHB–267, PHB–2168, JKBH–676, JKBH–26, HHB–117, GHB–538, HHB–67, GHB–719, HHB–197, GHB–757, Proagro–9443, RHB–121, GHB–732, Nandi–52, GHB–744, RHB–154, HHB–216, RHB–173, HHB–223, Proagro No.1, Avika Bajra Chari (IYoM, 2023), RHB 177, MP 7892, Kaveri Super Boss, MP 7872, JKBH 176, Nandi 61, Nandi 65, 86M86, KHB 108, GHB 905, 86M89, Bio 448, Bio 70, HHB 234 (Singh *et al.*, 2021).	VL 376 (Prabhakar *et al.*, 2017)	Bajra–48 Ragi–1
Madhya Pradesh	MLHB–285, GHB–557, Nandi–62, JKBH–26, GBH–316, 7886, Pusa–605, Nandi–8, Nandi–32, Pusa–415, 7688, RHB–58, Pusa–444, Proagro–9443, GHB–558, HHB–146, HHB–197, GHB–732, Nandi–52, GHB–744, PHB–2168, RHB–173, RHB–121, Proagro–9445 (IYoM, 2023), MP MH17, 86M86, GHB 905, KHB 108, Kaveri Super Boss, MH 7872, 86M66, HHB 223, Nandi 61, JKBH 676, B 2095 (Singh *et al.*, 2021).	VL 376, GPU 45, Chilika, VL 146, Suraj, KM 13, RAU 8 (Prabhakar *et al.*, 2017)	Bajra–35 Ragi–7

States	*Bajra/Pearl Millet (Pennisetum glaucum)*	*Ragi/Finger Millet (Eleusine coracana)*	*Total No. of both the Hybrid Varieties*
Odisha	RHB–58, Pusa–44, Proagro–9443 (IYoM, 2023).	VL–146, Champavathi, BM 9–1, Chilika, OEB–526, OEB–532, VL– 376 (Prabhakar *et al.*, 2017).	Bajra–3 Ragi–7
Uttar Pradesh	Nandi–52, Nandi–62, RHB–58, 7686, 7688, Proagro–9443, RHB–121, Pusa–444, RHB–173, GHB–558, GHB–577, GHB–744, HHB–146, HHB–197, HHB–223, PHB–2168, Nandi–32, Proagro–9445, GHB–558, (IYoM, 2023), Kaveri Super Boss, MP MH 17, 86M89, GHB 905, KHB 108, Bio 448, MP 7872, MP 7892, 86M86, 86M66, Nandi 65, Nandi 61, JKBH 676, B 2095 (MH 1257) (Singh *et al.*, 2021).	VL 376, RAU 8, VL 124, PES 400. KM 65, VL 146, Champavathi (Prabhakar *et al.*, 2017)	Bajra–32 Ragi–7
Karnataka	GHB–526, GK–1004, Nandi–35, GHB–526 (IYoM, 2023), GK 1051, 86M53, 86M64, Shine, 86M86, 86M66, PAC 909, Pratap, Kaveri Super Boss (Singh *et al.*, 2021).	CFMV 1 (Indravathi) (Wadhmare *et al.*, 2022), VL 376, KMR 340, Indaf 9, Indaf 15, Indaf 8, HR 911, Suraj 952, GPU 28, MR 1, Champavathi, L 5, BM 9–1, GPU 26, GPU 45, GPU 48, KMR 301, KMR 340, GPU 66, VL 146, OEB 523, OEB 532 (Prabhakar *et al.*, 2017)	Bajra–13 Ragi–22

States	*Bajra/Pearl Millet (Pennisetum glaucum)*	*Ragi/Finger Millet (Eleusine coracana)*	*Total No. of both the Hybrid Varieties*
Uttarakhand		VL Mandua348, PRM 1, VL 376, VL 146, Suraj, VL 149 [7].	Ragi–6
Chattisgarh		Indira Ragi 1, OEB 526, OEB 532 (Prabhakar *et al.*, 2017)	Ragi–3
Bihar		RAU 8, VL 379, ORB 532, OEB 526 (Prabhakar *et al.*, 2017)	

DISCUSSION

Production of Bajra and Ragi in Indian States varies significantly due to diverse climatic zones, farming practices, moisture content, soil profiling, etc. The top 5 varieties of Bajra and Ragi are aforementioned below. The hybrid varieties of bajra are categorized on the basis of average grain yield per hectare area as represented in Figure 4:

1. **86M13 (MSH 276):** Kharif crop, grown in Rajasthan, Maharashtra, UP, Tamil Nadu, and Gujarat. Released in 2016. The parentage line includes M096F XM174R. Average yield is 5497 kg/ha (Yadav *et al.,* 2021).
2. **Nandi-70 (MSH 224):** Average yield is 5445 kg/ha. Released in 2012. Crossed betn NMS 31A X NMP 55 (Satyavathi *et al.,* 2021).
3. **Nandi-72 (MSH 238):** Average yield is 5274 kg/ha. Extensively grown in Gujarat, Maharashtra, Rajasthan, Tamil Nadu. Released in 2013.
4. **86M64 (MSH 203):** Released in 2011. Widely cultivated in Gujarat, Maharashtra, Tamil Nadu, Rajasthan. Average yield is 5244kg/ha (Satyavathi *et al.,* 2021).
5. **GHB 183:** Grown in arid regions of Gujarat. Kharif crop. Released in 1997. The average grain yield is 5239 kg/ha. Breeding line includes 81 A X J998. Released by AICMIP MRS Jamnagar. Shows presence of tuft of hair on nodal rings. 83 days maturation time.

Hybrid varieties of Finger/Ragi production on basis of Q/ha as represented in Figure 5.

1. **KMR 301:** Average yield is around 60 Q/ha. Present in southern most pat of Karnataka. Thrives well in arid regions. Released in 2009 at Bengaluru (Madhusudan *et al.,* 2015).
2. **HR 911:** Released in 1986 in Bangalore. Average yield is around 50 quintal/Ha. Grown in rain fed and well irrigated soil (Madhusudan *et al.,* 2015).
3. **CFMV 1:** Also known as Indravathi. It is popularly grown in Karnataka, Andhra Pradesh and Tamil Nadu. The grain yield is lower than VL 376. Yield around 41 quintal per hectare grain. It is widely exported from India (Wadmare *et al.,* 2022).
4. **GPU 28:** Yield is 35 quintal/ha. Resistant to Blast. Specifically cultivated in irrigated and rain fed zones. Karnataka and Andhra pradesh are the two states which extensively produces this variety.
5. **Indaf 15:** This is widely grown in Karnataka. Yield ranging between 30 Q/ha. Released in 1991 at Bengaluru. All the other Indaf varieties such as Indaf 8, Indaf 9, have also shown improved version of pre-existing lineage (Wadmare *et al.,* 2022).

CONCLUSION

Due to the multiple agroecological circumstances, hybrid variety production varies per state. The agriculture sector, farming techniques, and economic growth have all been substantially improved through the utilization of hybrid varieties. The production significantly varies and thus the hybrid varieties of bajra and ragi has been enlisted in this review paper. The secondary data accumulated from various sources has been analysed in this review paper to determine the top five hybrid varieties of bajra—86M13 (MSH 276), Nandi 70 (MSH 224), Nandi 72 (MSH 238), 86M64 (MSH 203) and GHB 183 on the basis of average yield per hectare, while the top five hybrid varieties of ragi/finger millet are: KMR 301, HR 911, CFMV 1, GPU 28 and Indaf 15 based on the average yield q/ha.

REFERENCES

[1] Directorate of Millets Development (2020). 'Department of Agriculture, Co-operation & Farmers' Welfare Ministry of Agriculture and Farmers' Welfare, Government of India'.

[2] International Year of Millets (2023). Ministry of Agriculture & Farmers Welfare Government of India. https://www.nutricereals.dac.gov.in/pearlmillet.aspx

[3] Charyulu, D., Bantilan, Cynthia; Rajalaxmi, A., Rai, K.N., Yadav, O.P., Gupta, S.K., Singh, Naveen and Davala, Moses (2015). Development and Diffusion of Pearl Millet Improved Cultivars in India: Impact on Growth and Yield Stability, pp. 76, doi -10.13140/RG.2.1.3064.7520.

[4] Madhusudan, K., Ravishankar, C.R., Raveendra, H.R., Prakash, P., Channa-byregowda and Seetharam, A. (2015). Genetic Enhancement of finger millet in Karnataka—An overview, AICSMIP, GKVK, Bengaluru.

[5] Prabhakar, P.C.G., Boraiah, B., Sujata Bhat, N.C., Kiran T.V. and Manjunath, H.A. (2017). Improved Production Technologies for Finger Millet. Project Coordinating Unit ICAR-AICRP on Small Millets GKVK, Bengaluru-560, 6, 10–12.

[6] Satyavathi, C.T., Ambawat, S., Khandelwal, V. and Srivastava, R.K. (2021). Pearl Millet: A Climate-Resilient Nutricereal for Mitigating Hidden Hunger and Provide Nutritional Security. *Frontiers in Plant Science*, 12, 659938.

[7] Satyavathi, C.T. (2021). ICAR-AICRP Project on Pearl Millet, Jodhpur, Rajasthan. http://www.aicpmip.res.in/pcr2021.pdf

[8] Singh, S.P., Satyavathi, C.T., Sankar, S.M., Chawla, H.S., Shrotria, P.K. and Jeena, A.S. (2014). Hybrid breeding in pearl millet: Past and present status. New paradigms in heterosis breeding: conventional and molecular approaches.

[9] Wadmare, V.B., Pawar, V.S. and Shinde, I.N. (2022). A review of finger millet (*Eleusine coracana*) (L.) its varieties, nutrient composition, cooking qualities and health benefits. *J Pharm Innov*, 11, 3223–3227.

[10] Yadav, O.P., Gupta, S.K., Govindaraj, M., Sharma, R., Varshney, R.K., Srivastava, R.K. and Mahala, R.S. (2021). Genetic gains in pearl millet in India: insights into historic breeding strategies and future perspective. *Frontiers in Plant Science*, 396, 10.3389.

Millets for All

Aditi Abhyankar
Ramnarain Ruia Autonomous College, Matunga, Mumbai, India

ABSTRACT: Millets have been an integral part of Indian household meals for centuries. In recent years there has been much awareness regarding the rich nutritive contents of millets such as pearl millets, finger millets and many other small millets. Also, for health reasons urban Indian households have begun to consciously include these Nutri-rich cereals in their daily meals.

Some of the important millet crops grown in India are Sorghum (Great millet), Bajra (Pearl millet), Ragi (Finger millet) and many other small millets such as foxtail and barnyard millets. These were often referred to as coarse cereals but realizing the nutrient richness of these grains they are now gazetted as "Nutri-cereals" by the Government of India.

The trends in the millet production, area under cultivation of millets over the past few decades and their growth potential and supply channels needs be analysed carefully. Growing awareness regarding health benefits associated with millets consumption has to extend a major boost to the demand and supply of millets. The processed food items varieties of millets also need to grow rapidly to match the demand, in the next few years. These factors will bring the success to the Millet Mission adopted by India.

This paper takes a closer look at different types of millets, their nutrient contents, their benefits in terms of healthy diet, the widespread need for bringing more and more area under millet cultivation in all Indian states and better reach out of millets to larger sections of population of India.

A brief literature review representing recent proliferation of scholarly research articles, policy documents and writings in periodicals, reflecting a rising awareness regarding millet production and consumption and some major findings of a household survey on millets consumption undertaken by the author are presented in this paper.

Keywords: *Nutri-rich Millets, Health Benefits, Consumption of Millets, Production and Supply of Millets.*

INTRODUCTION

The year 2023 has been declared as the international millet year. Traditionally, Millets have been an integral part of the diet in Indian homes. There are many different types of Millets grown and consumed in India. They have been making a resurgence in recent years as more and more consumers have begun to adopt healthier lifestyle and try to include these high nutrition content whole cereal grains in their diet.

Millets are small-seeded cereals that retain excellent nutritional qualities; superior to certain highly consumed cereals such as rice and wheat. Millets have high content of proteins and minerals such as calcium, iron, magnesium, phosphorous and potassium, that can help reducing the incidence of diseases of diabetes, hypertension, and different heart diseases. Therefore, growing awareness regarding health benefits associated with millets consumption will boost both the demand and supply of them. Apart from the full grain, their full marketable potential such as processed food items need to grow rapidly in the next few decades.

There are many different types of Millets grown all over India. Some of them are more common types of Millets, whereas some are not as popular and widely consumed. Some of the important millet crops grown in India are Sorghum (Great millet), Bajra (Pearl millet), Ragi (Finger millet) and small millets viz., Korra (Foxtail millet), Little millet, Kodo millet, Proso millet and Barnyard millet. These were often referred to as coarse cereals but realizing the nutrient richness of the grains they are now gazetted as "Nutri-cereals" by the Government of India.

The small-seeded grasses of millets are hardy and grow well in dry zones as rain-fed crops, under marginal conditions of soil fertility and moisture. Millets are also unique due to their short growing season. According to FAO (Food & Agricultural Organisation of UN), the world production of millets is 89.17 million metric tonnes from an area of 74 million hectares.

India is the largest producer of millets in the world with a share of around 15% of the world total production. In India, millets are cultivated majorly in 21 states over an area of 12.53 million hectares, producing 15.53 million tonnes with a yield of 1237 kg/ha. There is a major impetus on millet production in the states of Karnataka, Andhra Pradesh, Tamil Nadu, Kerala, Telangana, Uttarakhand, Jharkhand, Madhya Pradesh, and Haryana.

Apart from being used for direct consumption millets can be used in many other industries such as food processing industries, animal feed segment and distilleries.

LITERATURE REVIEW

Millets have always been grown widely in different Indian states by farmers and also have been an integral part of Indian meals in most of the Indian states. Only limited data is published annually about the millets production, area under cultivation, procurement, minimum support prices and millets exports by the Government of India.

The awareness regarding health benefits of millets, importance of millets as climate resilient crops is being emphasised globally in the recent years.

Although some authors have talked about production and consumption of millets in general, others have analysed in more details, the benefits of millets in terms of various aspects such as their production, consumption, distribution, exports, nutrient contents and their health benefits, the literature review presented in this paper is organised in merely an alphabetical order.

A. Amarender Reddy, S.S. Raju & others (2017), examined the market structure and value chain of pearl millet grain and fodder in India. The authors have mentioned about a decline in demand of pearl millet for human consumption, with an increase in demand for non-food uses like cattle and poultry feed, raw material for starch and breweries industry. Their main finding is that, as the production of pearl millet is scattered and thin, there is a lot of scope for market aggregators to increase scale economies to reduce market costs to supply in bulk to food and industrial uses. They observed that there would be some demand for human consumption of high-quality grain, but most of the future demand for pearl millet will come from cattle and poultry feed industry, breweries and starch industry.

Asha Devi, A. Perumal & G. Jha (2022) have comprehensively explored the status of production and consumption of millets and related aspects that can guide policymakers and researchers in framing strategies to regain the lost glory of millets. The authors feel that there should be a boost in both production and consumption for the benefits of people and the planet. Creating awareness about millets' nutritional and environmental benefits and developing products suiting consumers' demands, including ready-to-eat foods, will foster consumption.

The authors expect and predict that with the current production and consumption trends in India, there will be a deficit in the supply of jowar and ragi in the near future. According to them this gap can be bridged by improving productivity.

The authors have remarked that the millets are mostly grown as rainfed crops with poor management. Hence popularization and cultivation of high-yielding varieties, provision of irrigation at least during the critical stage of the crop, and good cropping practices will give a boost to the productivity and production.

Benjamin Thomas Ipe and K.J.S. Satyasai (2022), have discussed threefold benefits of Nutri-Cereals for healthy economy, as risk mitigation against droughts for farmers, as nutritive food for consumers and combating climate change for the planet. The authors have pointed out that all-India area under production of nutri-cereals has been declining continuously since the early 1970s, mainly due to the green revolution which focussed mainly on rice and wheat production.

The authors have pointed out that the demand as well as supply side measures will pave the way for nutri-cereals to help bring sustainable evergreen revolution in resource poor region/area of the country.

Heba I. Mohamed1, Eman M. Fawzi1, Abdul Basit and others (2022), have discussed the high levels of production of Sorgham, which is ranked fifth after wheat, rice, maize, and barley, and about its drought resistance, and heat tolerance.

According to the authors some natural components in sorghum are useful in the development of healthy and functional foods. Sorghum farming for both biofuel production and human consumption is gaining popularity. They have mentioned that the pigmented sorghum grain is high in antioxidants such as polyphenols, primarily tannins, which have a variety of health benefits, including antiproliferative properties linked to the prevention of certain cancers, antioxidant activities linked to the prevention of diseases linked to oxidative stress, and anti-inflammatory effects, as well as improving glucose metabolism.

But their application in food business is limited as these chemicals cannot be assimilated. Sorghum is sometimes regarded as a low nutritional grain due to the presence of anti-nutritional components such as strong tannins, which form complexes with proteins and iron, limiting their digestibility. The authors have highlighted the utilization of sorghum as a source of bioactive chemicals and the value they bestow on human health due to the general biological potential it possesses.

Joanna Kane-Potaka & others (2021), to understand public knowledge and practices of consuming millets in urban areas, conducted a survey of around 15,000 individuals from seven major cities of India using a structured questionnaire. The authors found that the largest group among early adopters of millets were people with health problems (28%), followed by those wanting to lose weight (15%) and those selecting millets for its taste (14%).

No statistically significant relationship was found between state-wise per capita production and the frequency of consumption of millets in the urban areas. The authors have recommended developing delicious products to satisfy the taste, providing knowledge on nutritional and health facts on millets and improving the accessibility of millets in urban markets.

Millet Usage and Attitude Study (2018), a market research study on consumption of minor millets and millets in India among consumers and retailers were commissioned by McGill University which was a part of a bigger study for understanding Millet consumption, habits, practices followed in India.

The main objectives of the study were, to find out the awareness levels of the different varieties of Millet, the consumption pattern of the different varieties of Millet- Frequency, volume, price, purchasing habits & practices and to study the attitude and preferences of consumers towards different varieties of Millets.

The survey revealed that around 87% of the respondents interviewed recalled some variety of Millet or the other when asked which ones they were aware of. Top of Mind recall was the highest for Pearl Millet (Bajra) 32% followed at a distance by Sorghum (Jowar) 20% and Finger Millet (Ragi) 15%.

Word of Mouth was found to be the only main source of awareness for Millets—either it was the Family members or Friend's/Neighbors who passed on information on the same. Other sources of awareness were negligible. Millets were considered to be "good for health", "easy to digest", "good for weight reduction" and so on. Only Pearl and Finger Millet were considered for consumption "once a week". For the other varieties of Millets it was "less often than once a month".

The Pearl Millet was the most stocked Millet variety followed by Sorghum and Finger Millet by the retailers. Other types of millets were stocked by only a few retailers. Overall proportion of Millet sales out of 100% grain sales was between 1–10% compared to Rice 40% and Wheat another 40%.

Ministry of Agriculture & Farmers Welfare (2023), the Government of India, the action plan of International Year of Millets -2023 focuses on strategies to enhance production and productivity, consumption, export, strengthening value chain, branding, creating awareness for health benefits etc about the millets. A month-wise action plans have been prepared by

Central Ministries, State Governments and Indian Embassies to create awareness about millets for increasing production and demand.

Coinciding with the International Year of Millets 2023, during the G20 meetings being held in the current year, millets are being promoted on a large scale through millet meals, millet melas and distribution of millet gift hampers. Further interaction with millet farmers is also taking place on the side-lines of the G20 meetings to spread the awareness about millet crops.

Many Central and State Ministries, Agricultural & Food Authorities, and other different groups and individuals have been using all print, electronic and even social media for the promotion of millets.

Major Varieties of Millets Grown in India

Basically, there are nine types of millets grown in India which may be categorized into major and minor millets depending upon the area under production and size of the grains. Sorghum and Pearl Millet come under major millet and the rest are the minor millets.

These millets are categorized into naked and husked grains. Sorghum, Pearl Millet, and Finger Millet come under naked grains and rest millets like Foxtail Millet (kangni or kakum), Barnyard Millet (sangwa), Little Millet (kutki or shawan), Proso Millet (barri or chena), Kodo Millet, and Browntop Millet (hari kangni) are husked millets.

Sorghum Millet/Jowar Popularly known as Jowari or Jowar in India, Sorghum Millet is rich in nutrition. Jowar is mostly used to prepare flatbreads that are called Bhakri. Parathas, Rotis, and Dosas. Other dishes that can be made with this grain are Jowar Upma, Jowar Pulao and Jowar Salad. Jowar is rich source of Protein and dietary fibre. It is also Gluten free and contains minerals such as Magnesium. Sorghum Millets benefits the heart by helping lower cholesterol levels and is also known to boost the immunity of the human body.

Pearl Millet/Bajra is a popular type of Millet in India and known by the name Bajra. Flatbreads prepared with Bajra flour are very popular in the states of Maharashtra, Rajasthan, Punjab, and many other states. Pearl Millets are called so due to the shape of the grain. Like all Millets, Bajra is also naturally Gluten free.

Bajra or Pearl Millet benefits the body by helping build and maintain strength since it is rich in Protein. It is also a rich source of Iron and Calcium which are key minerals for good health, and bone strength and density. Bajra is also rich in dietary fibre which aids in healthy digestion and alleviates many digestions related ailments. Alternative uses of pearl millet grain in the coming years, may come from cattle and poultry feed industry, breweries, and starch industry. Pearl millets are often consumed in roti form particularly in winter seasons in Gujarat and Maharashtra. This fulfilling millet helps keep a check on cholesterol and is also recommended for diabetics.

Finger Millet/Ragi is more popularly known by the names Ragi or Nachni. Ragi is consumed in the form of Ragi Flour which is used to prepare several delicacies including flatbreads such as Rotis, Dosas, and Chillas. Ragi Flour is also used to prepare wholesome breakfast recipes such as ragi idly, ragi porridge.

Finger Millet is naturally Gluten-free and beneficial to consumers who have Gluten sensitivity. Ragi flour is also rich in dietary fibre which helps in digestion and helps alleviate many digestions related problems. It is a source of Protein, Iron, Calcium, and antioxidants which are highly beneficial for the overall well-being of the human body.

Foxtail Millet is another popular type of Millet grown in India. The foxtail Millet is known by many different names across various regions and cultures of the country. In Hindi, Foxtail Millet is known as *Kangni; Kang* in Gujarati; Thina in Malayalam, and Navanakki in Kannada. Unlike Ragi, Foxtail Millet is consumed in both forms, as a whole grain and in the form of flour. Foxtail Millet is often used as a healthier substitute for rice in many recipes such as Pulao, Khichdi, Lemon Rice. Dosas, Porridge and Upma are also prepared with Foxtail Millets. Foxtail Millet are high on protein content as compared to other types of cereals and millets. It is also a tremendous source of Iron and Calcium, and rich in antioxidants.

In addition to these there are some fewer known millets such as Amaranth, also known as Ramdana, is protein-rich grain recommended by experts and it can be used in several recipes to boost immunity and decrease cholesterol.

Buckwheat Millet aka ***Kuttu, a*** popular grain during Navratri, is known to help manage blood pressure and aid weight loss too. From Puris to Cheelas, there is no end to the recipes you can make with Buckwheat.

Barnyard Millet locally known as Sanwa is a gluten-free source of both insoluble and soluble fibres. It can be included in the diet in the form of *porridge, Khichdi and even Pulao.*

Foxtail Millet, also known as Kangni, is a grain well-known for promoting good cardiac health and maintaining good hair and skin. It can be used in *dosas, cheelas* and many more. Recipes.

Kodo Millet is rich in fibre and iron and helps prevent constipation and control blood sugar. The humble grain can be used to make Chapatis, Idli etc.

Different varieties of millets, their health benefits, recipes, and nutritive content per 100 grams are shown in Table 1.

Table 1: Nutrient Contents and Health Benefits of Millets

Millets	*Health Benefits*	*Recipes*	*Nutrition (per 100 gm)*
Sorgham (Jowar)	1. High contents of dietary fibres 2. Good source of vitamins and minerals, 3. Helps in providing strong bones, 4. Lowers blood cholesterol, 5. Rich source of antioxidants, 6. Helps in weight control and improving mood 7. Gelps in reducing tumour growth	Roti, Bhakri, Chilla, Dosa,	Protein 10.4 gm, Energy 349 (KCAL) Fibre 9.6 gm, Carbohydrates 72.6 gm, Fats 1.9 gm
Pearl Millets (Bajra)	1. Reduces cholesterol 2. Promotes bone health 3. Promotes heart health 4. Beneficial in treating stomach ulcers, 5. Aids in weight loss	Khichdi Roti, Upma, Idli, Khakhra, Paratha	Protein 10.96 gm, Energy 348 (KCAL), Folate 36.11 mg, Iron 6.42 mg, Fibre 11.49 gm, Carbohydrate 67,78 gm, Vitamin B3 0.86 mg, Calcium 27.35 mg

Millets	Health Benefits	Recipes	Nutrition (per 100 gm)
Finger Millets (Ragi)	1. Reduces blood glucose level 2. Anti-microbial properties 3. Promotes bone health 4. Revives skin and hair health 5. Repair injured muscle tissues	Roti, Dosa, Cheela, Cookies, Cakes, Porridges, Upma	Protein 7.16 gm, Energy 320.75 (KCAL), Folate 34.66 mg, Iron 4.62 mg, Fibre 11.18 gm, Carbohydrate 66, 82 gm, Vitamin B3 1.34 mg, Calcium 364 mg
Barniyard Millets (sanva in Hindi or vari in Marathi)	1. Rich source of fibre, soluble and insoluble, 2. Gluten free, 3. Anti-mutagenic, 4. Anti-inflammatory 5. Antioxidant	Porridge, Idli Dosas, Roti, Upma, Pulao, Kheer Khichdi,	Protein 10.13 gm, Energy 46.31 (KCAL), Fibre 7.72 gm, Carbohydrate 65.55 gm, Vitamin B3 4.20 mg

Production of Millets in India

Millet production and total area under millet cultivation was quite high in India before the widespread impact of the Green Revolution of 1960s. Consecutive years of droughts during 1960s and low produce of foodgrains had prompted the then government to launch drastic measures to enhance the food grain production in India. Therefore, to achieve self-sufficiency in foodgrain production the Green Revolution was launched as a major agricultural policy reform in India in 1960s.

The production of major cereals of rice, wheat, jowar and bajra and the total area under their cultivation during 1950–51 to 2020–21 (for selected years) has been shown in Table 2.

A massive increase in cereals production and large rise in the area under their cultivation has been achieved in India during last seventy years. However, a sharp rise in production of rice and wheat and a sharp decline in the production of jowar and bajra during 1950–51 to 2020–21 (selected years) is seen as shown in Table 3.

Table 2: Produce and Yield of Cereals: Five Yearly Trends (1950–51 to 2020–21)

Year	*Production of Cereals (million tonnes)*				*Area Under Cultivation (million hectares)*			
Year	*Rice*	*Wheat*	*Jowar*	*Bajra*	*Rice*	*Wheat*	*Jowar*	*Bajra*
1950–51	20.58	6.46	5.50	2.60	30.81	9.75	15.57	9.02
1955–56	27.56	8.76	8.08	3.43	31.52	12.37	17.36	11.34
1960–61	34.58	11.00	9.81	3.28	34.13	12.93	18.41	11.47
1065–66	30.59	10.40	7.58	3.75	35.47	12.57	17.68	11.97
1970–71	42.22	23.83	8.11	8.03	37.59	18.24	17.37	12.91
1975–76	48.74	28.84	9.50	5.74	39.48	20.45	16.09	11.57
1980–81	56.63	36.31	10.43	5.34	40.15	22.28	15.81	11.66
1985–86	63.83	47.05	10.20	3.66	41.14	23.00	16.10	10.65
1990–91	74.29	55.14	11.68	6.89	42.69	24.17	14.36	10.48
1995–96	76.98	62.10	9.33	5.38	42.84	25.01	11.33	9.32
2000–01	84.98	69.68	7.53	6.76	44.71	25.73	9.86	9.83
2005–06	91.79	69.35	7.63	7.68	43.66	26.48	8.67	9.58
2010–11	95.98	86.87	7.00	10.37	42.86	29.07	7.38	9.61
2015–16	104.41	92.29	4.24	8.07	43.50	30.42	6.08	7.13
2020–21	112.27	109.52	4.78	10.86	43.99	31.61	4.24	7.57

Source: Agricultural Statistics at a Glance, Govt. of India 2021.

After the green revolution was launched and its benefits began to be perceptible, there was a massive rise in the production of rice and wheat. The yield per hectare of rice and wheat also rose substantially.

The produce of rice more than doubled from 20.58 million tonnes in 1950–51 to 42.22 million tonnes in 1970–71 and again doubled itself by 2000–01 up to 84.98 mil tonnes (Table 3).

Table 3: Production of Cereals for Selected Years (Million Tonnes)

Year/Cereals	*Rice*	*Wheat*	*Jowar*	*Bajra*
1950–51	20.58	6.46	5.5	2.6
1970–71	42.22	23.83	8.11	8.03
1990–91	74.29	55.14	11.68	6.89
2000–01	84.98	69.58	7.53	6.76
2020–21	112.27	109.52	4.78	10.86

Source: Agricultural Statistics at a Glance, Govt. of India 2021.

The production of wheat increased almost six times during 1950–51 and 1970–71 from 6.46 to 23.83 million tonnes. And later it further increased almost three times up to 69.58 million tonnes by 2000–01 (Table 3).

However, during the same time period, the produce of jowar and bajra began to decline. Its yield per hectare also fell greatly.

The production of jowar marginally increased from 5.5 to 8.1 million tonnes between 1950–51 and 1970–71. It dropped down 7.53 million tonnes by 2000–01.

During the same period the production of bajra increased from 2.6 to 8.03 million tonnes by 1970–71 and dropped down to 6.76 million tonnes by 2000–01.

The following Figure 1 shows the enormous rise in the produce of rice and wheat and a near stagnant situation and in fact, some decline in the produce of jowar and bajra, measured in terms of million tonnes, during the seven decades of 1950–51 to 2020–21.

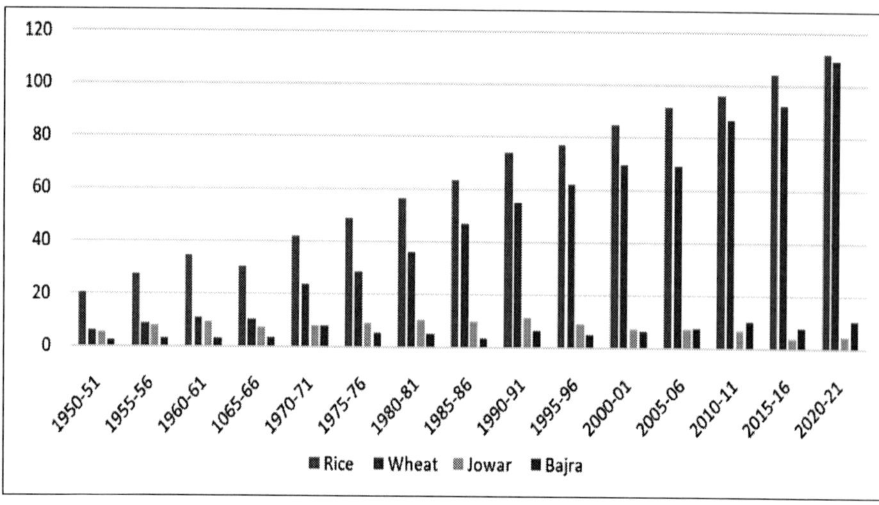

Figure 1: Production of Cereals, 1950–51 to 2020–21 (million tonnes)

Source: Agricultural Statistics at a Glance, Govt. of India 2021 (Graph is drawn by author).

Area under Cultivation of Major Cereals

The total area under the cultivation of rice and whet increased many folds during the post Green Revolution period. The area under cultivation of rice increased from 31 to 44 million hectares and that of wheat from 10 to 31 million hectares over a period of 1950–51 to 2000–01.

The area under the production of jowar in fact declined from 15 to 4 million hectares and that of bajra which has been very low has only marginally declined from 9 to 7.5 million hectares.

During the seven decades of 1950–51 to 2020–21, there was a massive increase in the area under cultivation of rice and wheat, especially during the post Green Revolution years. However, during the same period the area under cultivation of jowar and bajra declined to a great extent.

The following Figure 2 shows the area under cultivation of cereals of rice, wheat, jowar and bajra in terms of millions of hectares.

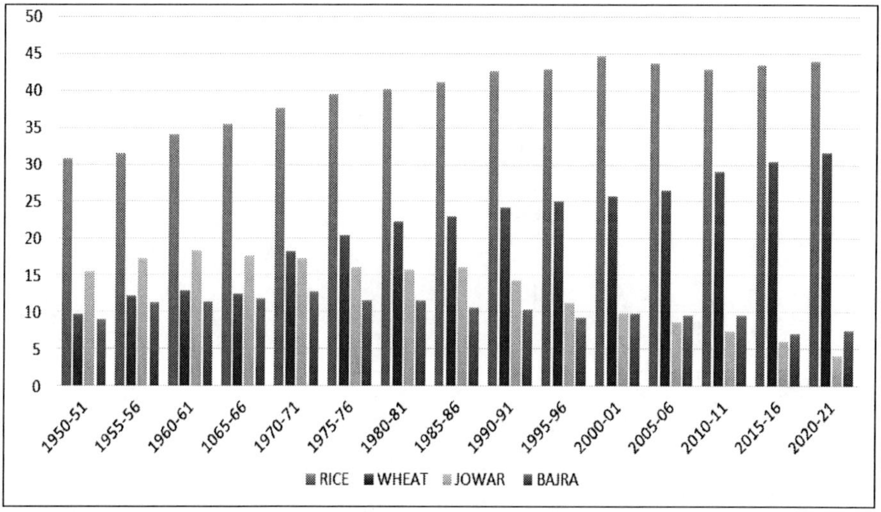

Figure 2: Area under Cereal Cultivation, 1950–51 to 2020–21 (in Million)

Source: Agricultural Statistics at a Glance, Govt. of India 2021 (Graph drawn by author).

Recent Trends in Millets Production in India

Most of the millet is produced in Africa, followed by Asia. India is the largest producer of millet, followed by Niger and China. Other major millet-producing countries include Burkina Faso (West Africa), Mali, and Senegal (West Africa). While millet is not a major food crop in the developed world, it plays a vital role in the diets of many people in developing countries.

The production of Nutri-cereals or millets in India during past ten years has shown year to year fluctuations and no significant rise in the total output. The Produce of jowar was 59 lakh tonnes in 2011–12 which dropped to 43.8 lakh tonnes in 2015–16 and further to 34. 8 in 2018–19 and has again increased to 47.8 lakh tonnes.

The Produce of Bajra was 102.8 lakh tonnes in 2011–12 which dropped down to 80.7 lakh tonnes in 2015–16 and recorded a high of 108.6 lakh tonnes in 2020–21. Ragi crop production remained stagnant between 2011–12 and 2020–21 at around same level of 19.5 lakh tonnes.

The total production of all millet was 186.4 lakh tonnes in 2011–12 which declined to 142.2 lakh tonnes in 2015–16 and again reached the level of 180 lakh tonnes in 2020–21.

The per hectare yield of Jowar in kg was 1119 in 2011–12 which has increased to 1204 kg in 2020–21. During the same period between 2011–12 to 2020–21 the yield of Bajra has grown from 1079 to 1436 kg per hectare. Yield of Ragi has remained constant at around 1700 kg per ectare which is although much higher than jowar and bajra.

The per hectare output of other small millets has increased from 533 to 844 kg per hectare during the same period.

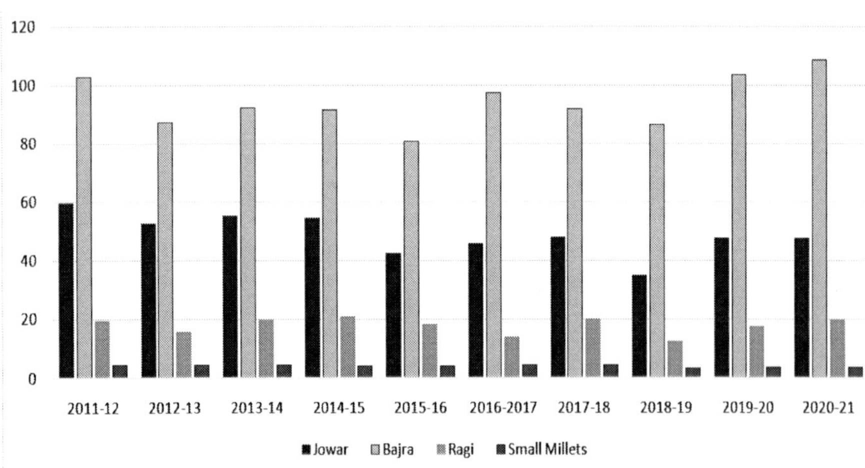

Figure 3: Production of Nutri-cereals, 2010–11 to 2020–21 (Lakh tonnes)

Source: Agricultural Statistics at a Glance, Govt. of India 2021 (Graph drawn by author).

Incentives for Millets Production in India

Some positive developments have been witnessed in the recent years with regard to raising of MSP (Minimum Support Price) and providing incentives to the exports of millets and millet products.

Since 2017–18, the MSP (Minimum Support Price) of Jowar, Bajra and Ragi has steadily been increasing. Ragi saw the maximum increase in MSP (78%) increasing from Rs. 1900 per quintal in 2017–18 to Rs. 3377 per quintal in

2021–22. During the same period, the MSPs of Jowar and Bajra have also increased by 61% and 58%, respectively. This is a key supply side measure and may encourage farmers to bring larger areas under millet cultivation and to adopt best technologies and farm practices.

The exports of Nutri-cereals although have picked up since 2018, and although India is the largest producer of Nutri-cereals in the world but is only the fifth largest exporter of the same. Export can act as a very important tool for demand creation and India needs to tap the global millet export market. Private & Public partnership and Corporates need to come forward for the brand development of ready to eat and ready to cook products of millets.

APEDA (Agricultural and Processed Food Products Export Development Authority) under the ministry of commerce has been promoting millets and processed products' exports. The varieties of millets exported from India in 2021–22 have been shown in the Figure 4.

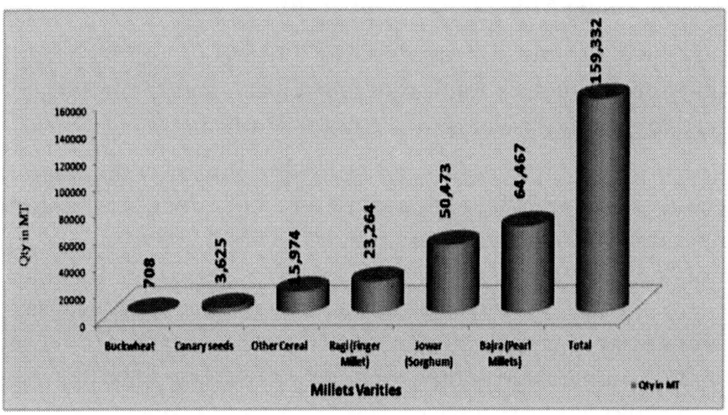

Figure 4: Indian Millet Exports, 2021–22 (Million Tonnes)

Source: Agricultural and Processed Food Products Export Development Authority (APEDA).

Consumption of Millets

In India, as per the NABARD (National Bank for Agricultural and Rural Development) estimates, the per capita consumption of millets fell drastically from 32.9 kg to 4.2 kg from 1962 to 2010. The demand-side factors responsible were: (i) increasing urbanization and per capita incomes that changed consumer tastes and preferences, (ii) poor social status and inconvenience to prepare millet dishes, (iii) lower shelf life of milled grains, and, (iv) primacy of rice and wheat in the PDS.

The supply-side factors included: (i) lack of industrial demand for value-added millet products discourages farmers from cultivating millets, (ii) low profitability, (iii) the Green Revolution has favoured the production of rice and wheat, including output price incentives and input subsidies, (iv) lack of access to quality seeds, (v) inadequate infrastructure such as processing technologies, and unique milling equipment to address the total value chain, (vi) weak value chain, including inadequate primary processing at the village level.

The Survey on Consumption of Millets in Urban Areas

The survey on Consumption pattern of Millets aimed at finding out the main cereals consumed apart from rice and wheat, factors influencing consumption of millet—habit, taste, health benefits, season, price, food preparation process and so on; frequency of consuming millet and its different varieties; who all in the household (family members) eats millet; varieties of regular and special preparations made with millet, reasons for eating millet and If started consuming Millet recently the reasons for it.

The study tried to find out the Purchase Process in terms of frequency of purchase, volume purchased at a time, price paid, and the factors considered while purchasing millets such as quality, gradation etc.

A detailed structured questionnaire was used to interview the respondents in a household survey on Millets Consumption. The survey was undertaken on mobile phone and/or through direct interview and conversation. A random sample of hundred respondents of a diverse socio-economic strata, doing various professions, of level of education ranging between Std X-Post Graduation, of age between 25–65, from Greater Mumbai, Thane and Pune region of Maharashtra were interviewed, out of which seventy-five respondents were women and rest were men.

Out of the total hundred respondents, sixty said that they consume jowar and/or ragi roti twice or thrice a week. Almost all these sixty respondents said that they consumed bajra roti in winters. Apart from roti they would consume ragi idly, dosa, upma, dhokla, porridge and many other innovative recipes. Around 30 respondents said that they consume jowar and ragi roti sometimes. Around ten respondents mentioned that their families do not have a tradition of consuming millets in any form, but they have started buying ready to eat millet products from the market recently.

More than fifty respondents said that they have been consuming millets from childhood at home, now more consciously, for health reasons, and that they consume jowar and ragi more regularly in their meals. It was observed that the incidence of diseases such as hypertension, diabetes and different heart diseases have increased awareness regarding consuming healthy diet and particularly, millets along with rice and wheat. The respondents were asked to describe the millet recipe they cooked and most of the respondents enthusiastically responded and described the detailed process of cooking *bhakri, idly, dhokla* and so on.

The respondents often purchased full grain millets or millet flour monthly in quantities of 1 to 2 kg jowar, bajra and ragi and it was mostly purchased from a nearby retail grocery shop.

Apart from consumption of millets at home, which has been quite common in India, some restaurants, public meetings, food fairs have been offering bajra roti, ragi dosa, idly, noodles and many innovative recipes as a new addition to their menu.

The three boxes presented above depict responses of some selected respondents.

Box 1 shows the recipe of millet dosa/uttapam narrated by a respondent who consumes all types of millets daily in the evening meals and also makes newer varieties of food snacks using jowar, bajra and ragi.

Box 2 is a response of another respondent who showed great awareness about high nutritive value of millets, their health benefits, marketing of new millet food items, the price trends of millets, the distribution channels and so on.

Box 3 is a response of an old north-Indian Mumbai taxi driver of age around 65 who talked about very old tradition of including millets in food in Indian meals and said that in recent years, it is grown and consumed in low rainfall drought prone areas of India. He mentioned that consuming ragi roti has helped in controlling his diabetes.

As the survey covered majority of female respondents, however, more detailed information about millets consumption, purchase quantity, recipes were obtained with much ease.

As far as findings of the survey are concerned, it was observed that overall awareness about millets as Nutri-rich cereals is on the rise, Many individuals and families are consciously trying to include millets in their regular meals. Ragi kheer and porridge is very commonly consumed and given and recommended as a source of iron and calcium to the infants in most of the families and their relatives and neighbours. Innovative uses of millets and new recipes of jowar khichdi, ragi dosa/idly are being tried out and are becoming very popular. This will give more options apart from usual wheat and rice preparations as wholesome meals.

Box 1: Recipes

Ragi Dosa or Uttapam

Ingredients: Whole ragi 2 cups, brown rice 1/3 cup, urad dal 1 cup, brown poha ½ cup, methi seeds 1 teaspoon and salt to taste.

Preparation: Soak ragi, urad dal, brown rice and methi in water overnight. Grind everything together into a fine batter/paste and let it ferment for 7–8 hours.

Soak brown poha for half an hour. Add salt and add water for the right consistency for making dosas. Also, add vegetables of your choice and keep the batter thick for making uttapam.

Heat a griddle/pan and sprinkle few drops of oil. Pour a ladle full of batter in the centre of pan and spread it evenly in a round shape dosa. Cook till it turns golden brown.

Serve hot with chutney and sambhar.

As a response to a question pertaining to mention of a recipe you regularly make, this recipe was mentioned by a respondent.

Box 2: Responses

A respondent who herself very regularly consumes millets in daily meals was quite aware of the elitist inclusion of millet recipes in the diet and the souring prices of millets in certain select departmental stores in the urban hubs.

She also mentioned how millets could be made more popular by using them in making of some tasty snacks such as burgers, noodles, cookies, cakes, chips etc.

Box 3: Healthy Diet

A taxi driver of Mumbai of age around 65, although called millets an akal (drought) dana, said he was suffering from diabetes and started eating ragi roti to control it.

He said, ragi is grown and consumed in areas of low rainfall. He called it makra, which according to him is eaten from the days of raja Harishchandra.

He mentioned of various recipes of millets such as kachori, patties, green peas stuff millets (ragi) parathas.

Particularly, the respondents of this survey, and those with health issues have been consuming jowar, bajra, ragi rotis very regularly.

Promotion of Millets and Processed Millets Products

Millets can be easily processed further by way of malting, popping and fermentation and can be used in making several food products. Its chemical composition and the properties of its flour make it suitable for developing several value-added fried snacks, fermented breakfast food, bakery products and sports and diabetic foods.

The promotion of products made of millets with proper packing and marketing strategies has the potential to strengthen social and economic security even for entrepreneurs. Its promotion in Government hospitals, in the Mid-Day Meal Programme and Government offices would improve consumption and also encourage the development of new food formulations that use millets.

The genetic diversity of millets lends itself to many diverse and innovative applications of millets in areas such as therapeutics and pharmaceuticals.

Millets support the biodiversity of crops, which is the need of the day for achieving food security. Millets offer rural families multiple forms of security in terms of food nutrition, fuel, fodder for animals and to some extent as a source of fuel. As millets can sustain higher heat regimes, drought and excess moisture, it is also projected as a suitable crop for climate change. Minimal care is required for the growth of millets. The total production cost is also relatively low when compared to other crops. As millets are grown in eco-friendly organic conditions, fetch better prices, especially as the demand for organically grown grain is on a rise. Awareness about the 1grain and its many benefits can spur marketing efforts and benefit self-help groups and commodity interest groups in the rural areas of India.

Apart from India, in many regions of the globe millets can be grown as the required soil & water, climate & temperature, and the overall length of cropping season of millet cultivation is very short compared to other crops. This will ensure food security to the population of the whole world.

CONCLUSION

The millets are to be consumed by all as they are there when others are not, that is, they will feed the rising populations of the globe with nutrition even withstanding the climate change challenge facing their production. Millets are a drought-tolerant crop that can be grown in dry, arid climates where other crops would fail.

Millets can contribute to a healthy diet as they provide antioxidants, minerals, and protein. As whole grains, each millet variety also offers different types and amounts of fibre, which play a vital role in regulating bowel function, blood sugar and lipids. Millets are gluten-free with a low-glycaemic index making them a great food option for those with celiac disease or gluten intolerance, high-blood sugar or diabetes. Inclusion of millets in our diets can provide us with nutritious and healthy alternatives to the usual refined grains.

The 'Millet Mission' or 'Millets for all' can be a new way of life, globally. The production, and consumption of millets is beneficial as they possess high nutritional value, offer economic security to human beings, represent fodder security for animals and can provide solutions to the challenge of meeting food security globally.

REFERENCES

[1] Reddy, A. Amarender, Raju, S.S. and others (2017). 'Analysis of pearl millet market structure and value chain in India', *Emerald Insight at: www.emera ldinsight.com/2044-0839.htm*

[2] Agricultural Statistics at a Glance (2021). Government of India, *Ministry of Agriculture & Farmers Welfare Department of Agriculture & Farmers Welfare Directorate of Economics & Statistics.*

[3] Asha, Devi; Perumal, A. and Jha, G. (2022). 'Millet Production and Consumption in India: Where Do We Stand and Where Do We Go? *National Academy Science Letters, September* 2022.

[4] Ipe, Benjamin Thomas and Satyasai, K.J.S. (2022). 'Nutri-Cereals for a Healthy Economy', Rural Pulse, NABARD, Issue XXXIX.

[5] Grunert, K.G. (2005). "Food quality and safety: consumer perception and demand", *European Review of Agricultural Economics,* Vol. 32, No. 3, pp. 369–391.

[6] Heba, I. Mohamed, Eman, M. Fawzi, Abdul Basit and others (2022). 'Sorghum: Nutritional Factors, Bioactive Compounds, Pharmaceutical and Application in Food Systems: A Review', *Phyton, International Journal of Experimental Botony.*

[7] Joanna Kane-Potaka and others (2021). 'Assessing Millets & Sorgham Consumption Behaviour in Urban India: A Large Scale Survey', *Smart Food for Healthy, Sustainable and Resilient Food System,* Vol. 5, 2021.

[8] Millet Usage and Attitude Study (2018), India Net Marketing Services Pvt. Ltd. MCGILL University, https://idl-bnc-idrc.dspacedirect.org/bitstream/handle/10625/57033/57087.pdf

[9] Pocket Book of Agricultural Statistics (2017). Government of India Ministry of Agriculture & Farmers Welfare Department of Agriculture, *Cooperation & Farmers Welfare Directorate of Economics & Statistics New Delhi.*

[10] Rai, K.N., Gowda, C.L.L., Reddy, B.V.S. and Sehgal, S. (2008). "Adaptation and potential uses of sorghum and pearl millet in alternative and health foods", *Comprehensive Reviews in Food Science and Food Security,* Vol. 7, No. 4, pp. 340–352.

[11] The Associated Chambers of Commerce and Industry India (2022). *Knowledge Paper on Millets, Future Super Food of India, New Delhi,* June 2022.

Millets and Marginality: A Study Encompassing Gender, Tribes and Caste

Kirti Y. Nakhare

Department of Business Communication, S.I.W.S. College, Mumbai, India

ABSTRACT: Millets have the capacity to brave extreme temperatures, while requiring limited quantities of water as compared to cash crops. The 'Green Revolution' which was ironically chemically fuelled, began in the 1960s, stressed on cultivation of cash crops. It claimed to create a 'sujalam, sufalam' Bharat, till the alarm bells of global warming went buzzing. Millets largely became marginalised as they were not the ones earning profit and were not on the Green Revolution-Cash Crop repertoire. As a result, farmers turned to cultivating Wheat and Paddy bidding adieu to a variety of millets.

Millets, although marginalised provided succour to women farm hands during the Covid 19 pandemic. Women's Collectives organised women farm hands and helped them cultivate millets on barren land. Systematic training programmes providing guidance helped these women take charge of their lives.

Millets were consumed regularly by the tribals and the 'non-creamy' layer of the society, mainly because of the hardy quality of millets and their capacity to grow within a short span of time, without much care and fear of pest attack. The tribals practiced multi-cropping and such methods which kept the fertility of the land intact. Several small millet varieties that are now nearly extinct were consumed by them.

This paper attempts at studying the role played by women's collectives in the lives of women farm hands in cultivating millets leading to their empowerment. The revival of millet farming by tribals, especially the small millets and the place enjoyed by millets in Dalit autobiographies.

Keywords: *Green Revolution, Women Collectives, Traditional Crops, Food Security, Millet Farming.*

INTRODUCTION

Millets and Marginality: A Study Encompassing Gender, Tribe and Caste

Myths are our reality. Millets find a mention in our epics this can be deduced on the basis of the famous Kannada poet Kanakadasa's poem 'Rama Dhanya Charite' which focuses on lord Rama's victorious return from Ayodhya, and his stop over at ashrama of Gautama rishi. It was at the ashrama that Sri Rama and his entourage was served dishes made from nine grains.Hanuman was questioned about which grain was tasty to answer this question each dish had to be tasted closely. Finger millet was chosen as the best amongst the ones presented. Rice took offence at the judgement as it considered itself as the essence of all dharma. How could an inferior grain like ragi be considered superior as compared to wheat and rice. Rice considered itself superior to ragi and was upset that a 'kula-heena', a 'low born' was elevated to a similar status. Rice was for the rich, whereas ragi was for the poor and humble.

Ragi maintained her dignity by refusing to answer for the indecent conduct in a gathering which didn't deserve a reply. Lord Rama was at his wits end, hence he imprisoned both ragi and rice. He made them undergo a test of endurance. After six months, the other gods and Indra were present at the court. Ragi had passed the endurance test by remaining fresh and rice had gone stale. Ragi was declared the winner by lord Indra. The debate between rice and ragi is a reflection of the disparity that exists in society. The one between the 'creamy' and 'non-creamy' layer. Food, thus becomes an arena of politics, and is not merely a simple means of satiating hunger.Finger-millet (ragi)emerged successful in Kanakadasa's poem and it has deep emotional ties between people residing in various regions of Andhra and Tamil Nadu and in Southern Karnataka. It has connections with people of these regions and castes. However, most upper castes abstain from eating ragi because of its colour. It is ironic, that even ragi eaters prefer eating rice on special occasions.

Based on the above tale and the conclusions drawn on its basis, it can be expressed in a nutshell that,

'Diets are not only the outcome of nutritional demands in specific ecological conditions but also evolve in specific cultural contexts. Eating practices develop in tandem with cosmological perceptions, moral attitudes, social norms, ideologies, and power-structures' (Markowitz and Avieli, 2022).

Food consumption is not an apolitical arena. As aptly expressed by Heide and Muhleisen *et al.* 'Food is now fore-grounded in local, national and global policies, in the ethics involved in resource distribution patterns and eating practices and their broader connections to social justice, sustainability and climate change (Doring, Heide, and Muhleisen 2003; Sack 2000,138). In connection to climate change and sustainability, it would be interesting to note that millets are hardy crops with some, like Proso millet, needing just seventy days to be ready for harvest. The short cultivation times is what made this the staple grain of nomadic communities across the Central Asia, spreading far as these tribes moved from place to place. (A brief history of millets—The Millet Foundation) As far as social inequality is concerned Dalit autobiographies depict the rationale behind consumption of millets. The 'low maintenance' and highly resilient crop was the preferred choice of tribes across the world.

This research paper will undertake a study involving millets and various perspective including the utility of millets with regard to gender, the role it played in the life of various tribes and the mention of millets in Dalit autobiographies.

It would be pertinent to refer to Bronner (1981, 122 qtd in Chaudhuri 2019, 10) while referring to Dalit autobiographies, "Since eating is intrinsically "biform" i.e. it is both a social activity and also a very private and singular one, thus, humiliation can often be traced through food, because it exists in the domains of both the private and the public." Food is very often viewed as a necessity than a luxury in Dalit autobiographies. Food is not central and it is seen as a means of subsistence. Excerpts from Dalit Autobiographies that mention bread made out of millets or any other preparation involving millets are enumerated below.

Kamble proclaims that the taste of bhakri with tea in the morning is better than any expensive biscuit that she has eaten till today (Kamble 1986, 51 qtd in Chaudhuri 2019, 12).

Urmila Pawar has a different point of view. She reminisces that their mother motivated them with tales of resilience associated with food. She remembers during tough times, the food that was prepared was almost "impossible to swallow" but their mother would say to them, "only people who can eat such food can do good in life". She would narrate the story of the Dalit boy who was so obedient that he would imagine curry in his empty bowl and eat it with his dry bhakri (flatbread) and hearing this, they would all scamper to finish their food silently (Pawar 2003, 94–95 qtd in Chaudhuri 2019, 10).

Bama on the other hand, writes of her pain at having to eat wild greens with thickened ragi (finger millet) dough or sometimes, just boiled drumstick leaves (Bama 2000, 70–71 qtd in Chaudhuri 2019, 11).

In almost every, Dalit autobiography certain foods are thought to be medicinal, some even miraculously so. As a narrative of pride, emphasis on the wholesomeness of certain traditional Dalit food is an important instrument to construct a self within the modern discourse of healthy eating. C.S. Lakshmi writes of how Perumal Murugan's students had made horsegram kuzhambu (a tamarind broth), sambar (lentil soup) and finger millet dosais (crepes) for her as they had heard she was diabetic and they believed this food would help with her condition (Murugan 2018, xi qtd in Chaudhuri 2019, 12).

Just as ragi was not considered 'fit' for celebration, the place of puran-poli could not be taken by the bhakri during celebrations. Daya Pawar notes his father's extreme generosity, to the point wherein he refused to serve his guests bhakri and insisted on serving them puran-poli, which they could barely afford (Pawar 1978, 58 qtd in Chaudhuri, Amiya 2019, 14).

Baby Kamble elaborates on the offerings that were collected and given to all the residents of the chawdi (neighbourhood). During rede jatra or the buffalo fair, the sacrificial buffalo was cooked with chilli powder and jowar (sorghum) to make panji (meat soup) and this was distributed to all the men of the community, and when they had had their fill, to the women and children (Kamble 1986, 35 qtd in Chaudhuri 2019, 14).

Thus, it can be observed that millets were compulsorily a part of the Dalit culinary repertoire. More out of affordability and availability and also at times attributing medicinal qualities to the food for achieving solace and creating some rationale. These and several other reasons made it a staple of Dalit cuisine.

Gender and the Role Played by Women's Collectives

Moving on to gender and the role that was played by millets in bringing women together can be elaborated in this section of the article. At the international conference on millets, organised by Ruia Autonomous college, a session titled Kissan Goshti was moderated by Dr Sunil R. Karad which focused on the socio-cultural, economic and historical aspects of millets. During this session, women farmers from Maharashtra shared 'traditional knowledge' about farming and experiences while farming millets. For novices in the audience, it proved to be an eye-opener as this session introduced one to the fact that even jowar had various varieties like wheat. The management of fields and the economic independence and confidence gained due to farming millets was evident amongst the robust women of the soil, who urged the audience to consider getting their daughters married to farmers, who too could be considered as 'eligible and worthy' bachelors.

Similarly, women's collectives across the country play an important role in women empowerment. DHAN (Development of Humane Action) for instance is a collective of fifteen villages of Odisha's Koraput district which offers millet items to the urban population in a ready-to-eat-form. Regional actors were also roped in their support, for promotion. SABALA a women's collective, now going strong in a district in south India. The women in this district readily accepted to be involved with millet farming and its revival as they were anaemic and faced health issues related to malnutrition. Today, the above-mentioned women's collective has grown strong with a large number of female farmers who cultivate millets and intercrop it for commercial as well as for personal use. Women in the district who felt disempowered because of their caste, class and gender, after being introduced to the nutritive value of millets, have taken charge of their families in their own hands and have gained recognition.

Umbrella organizations such as the 'Millet Network of India' extends support to women's collectives. It helps women face problems and provide succour to revive millet farming. Another platform, 'All India Millet Sisters' has a membership boasting of a huge number of participating women, from the country. Such groups work hard for millets to be included as part of all public food schemes. (Rodrigues, 2020)

The Sangtin Kisan Mazdoor Sangathan encourages women to grow millets and it has been conducting awareness campaigns, demonstrations in farms and cooking sessions for villagers that involve the 'nutricereals'.Barley and millets (foodcultures.org). Thus, with regard to gender, millets play an important role in empowering women through various women's collectives.

Millets and Tribes

Several tribes in Odisha and the Gond tribals in Bundelkhand have grown millets across ages. However, the tribals have also experienced a 'loss' of little millets. The tribals revered land and treated it as sacred. The Dongria Kondh tribes have been involved in reviving heirloom millets varieties. A seventy-year-old senior from the same tribe mentions reviving it, "We have intensified our search for lost indigenous varieties." After ousting the British Mining group, Vedant in 2013, the community felt to reoccupy the hills and bring it under traditional farming. The reason for the decline in cultivating millets was providing subsidy on rice, to the otherwise millet eating people, the Dongria Kondhs, besides support by the governments policy to promote cash crops like turmeric, pine apple and ginger has cut into over 50% of the millet land over the past sixteen years.

The Dongria Kondhs are on a mission of gathering information about millets among clan meetings and at market meetings about preservation of abandoned millet varieties. Kodo millet variety, high on fibre and energy content and is a panacea for diabetics, besides two varieties of sorghum in Jangojodi village and another foxtail millet variety in Sagadi village are amongst the ones revived. Dasara Kadraka is hopeful of adding many more to this repertoire. (Jena, 2017)

The director of the Indian Institute of Millets, Vilas Tonapi, expressed, "The promotion of certain millets like ragi in Odisha, jowar in Karnataka and kodo and kutki in Madhya Pradesh need not be at the cost of small millets."(India's millets policy: Is it heading in the right direction? Niyogi, 2020).

Vilas Tonapi further mentions, "… There is a need to go back to our culture of multi mixed and intercropping as it provides insurance against monsoonal failure and also enables sustainable livelihoods." (Ibid Niyogi, 2020)

Sanyasi Rao, a programme manager at Wassan, a Telangana-based organisation which works on rainfed agriculture, draws attention to the problem with supply of millet seeds. He further points at the inability in providing seeds of little millets of the nearly extinct variety. Due to this constraint farmers use only seeds that are available. Seed subsidy on 'released' or 'certified' crop is the reason for entering into a monoculture and risking agrobiodiversity.

Debjeet, who works with Kondh tribals in Odisha, suggests that millets have always been on the tribal cropping repertoire. Tribals practice of mixed cropping and growing different varieties of millets along with other crops like pigeon pea, cowpea, turmeric and sesame is not a novel idea, he further draws attention to the spiritual and ecological connection with cropping which needs to be revived.

"All these crops have different durations. Some are deep rooted whereas others are shallow rooted. It is a complex arrangement. Each crop also has a ritual attached to it. Festivals bring communities together and reinforce the sense of belonging to a collective space. When we delink spiritual, cultural and ecological dimensions and bring productivity into the framework, it becomes a linear arrangement." (Ibid Niyogi, 2020)

Millet hotspots in Madhya Pradesh, Odisha, the Garhwal Himalayas and in Tamil Nadu should be focused on for 'diverse-centric' agriculture. Israel

Oliver King, director in-charge regional centre, MS Swaminathan Research Foundation further expresses,"Diversity is the hope in today's world, as it is not only about food, but also about soil, bees and pollination." (Ibid Niyogi, 2020)

In conclusion, millets are sturdy crops, and are an answer to the environmental woes. They can be considered to be a boon as they are not only the answer to environmental problems but also provide a reliable solution to lifestyle diseases like diabetes and obesity. Women empowerment is brought about, as seen through women's collectives and their united synergy. Planting traditional crops like millets, would not only help revive traditional methods of farming, like intercropping, but if done in a responsible manner, would help to bring about cultural awareness and help in synchronising life once again with the weather, festivals and the overall wholesome culture and community living will be synchronised, bringing about sustainability.

REFERENCES

[1] A brief history of millets—The Millet Foundation Accessed on 29 December, 2022.

[2] Bama (2000). Karukku. Translated by Lakshmi Holmstrom. Macmillan India Limited. In Cooking_with_Caste_An_ethnography_of_foo.pdf. Accessed on 30 December, 2022.

[3] Bronner, Simon (1981). The Paradox of Pride and Loathing, and Other Problems. Western Folklore, Vol. 40, No. 1, Foodways and Eating Habits: Directions for Research (Jan., 1981), pp. 115–124 in Cooking_with_Caste_An_ethnography_of_foo.pdf. Accessed on 30 December, 2022.

[4] Chaudhuri, Amiya. Cooking with Caste: An ethnography of food and the modern, Urban Dalit identity. Paper submitted under Opportunities for Undergraduate Research (OUR) 2018–19, 22nd April, 2019. Cooking_with_Caste_An_ethnography_of_foo.pdf. Accessed on 30 December, 2022.

[5] Doring, T.M. Heide and Muhleisen, eds. (2003). *Eating Culture: The Poetics and Politics of Food.* Heidelberg: Universtatsverlag Winter. Sack, D. 2000. Whitebread Protestants: Food and Religion in American Culture. New York: St. Martin's. In Food, Culture & Society, 2022, Vol. 25, No. 4. 640-646. 'Eating religiously: food and faith in the 21st century' Markowitz Fran & Avieli, Nir. https://doi.org/10.1080/15528014.2022.2116195. Accessed on 7 January, 2023.

[6] Gowda, Chandan. Millets: A different Journey- Bangalore Mirror-Updated: Apr 26, 2017, 07.13 PM IST-Millets: A different journey—Bangalore Mirror (indiatimes.com) Accessed on 29 December, 2022.

[7] Jena, Manipadma (2017). The Journey of the Humble Millet from Tribal Farms to Dinner Tables (thewire.in) Accessed on 29 December, 2022.

[8] Kamble, Baby (1986). The Prisons We Broke. Translated by Maya Pandit. Orient BlackSwan. In Cooking_with_Caste_An_ethnography_of_foo.pdf. Accessed on 30 December, 2022.

[9] Markowitz Fran, Avieli, Nir. Food, Culture and Society (2022). VOL.25, N0.4.640-646. 'Eating religiously: food and faith in the 21st century' https://doi.org/10.1080/15528014.2022.2116195. Accessed on 7 January, 2023.

[10] Murugan, Perumal, ed. (2018). Black Coffee in a Coconut Shell. SAGE Publications and YodaPress. In Cooking_with_Caste_An_ethnography_of_foo.pdf. Accessed on 30 December, 2022.

[11] Pawar, Daya (1978). Baluta. Translated by Jerry Pinto. Speaking Tiger. In Cooking_with_Caste_An_ethnography_of_foo.pdf. Accessed on 30 December, 2022.

[12] Pawar, Urmila (2003). The Weave of My Life: A Dalit Woman's Memoirs. Translated by Maya Pandit. Stree. In Cooking_with_Caste_An_ethnography_of_foo.pdf. Accessed on 30 December, 2022.

[13] Niyogi, Deepanwita Gita, India's millets policy: is it headed in the right direction? Published on 27 July, 2020.India's millets policy: is it headed in the right direction? (mongabay.com) Accessed on 29 December, 2022.

[14] Rodrigues, Annie Pinto. Indian Women Turn to Ancient Grains to Feed their Families and their Futures - YES! Magazine (yesmagazine.org) November 26, 2020. Accessed on 29 December, 2022.

THEME-2

Millets Sustainable Solution to Food Security

Development, Organoleptic and Nutrient Evaluation of Pearl Millet (*Bajra*) Food Preparations for the Dietary Management of Type 2 Diabetes

Suman Bala

Department of Home Science, Kurukshetra University, Kurukshetra

ABSTARCT: Today, diabetes is a serious and growing challenge for public health and ranks among the top ten causes of mortality. The global prevalence of diabetes is projected to rise from 537 million in 2021 to 783 million in 2045. Improvements in lifestyle and nutritional management are mostly responsible for therapeutic approaches for diabetes prevention in high-risk patients. Pearl millet is a highly nutritious grain, nutritionally comparable and even superior in calories, protein, vitamins, and minerals to many other big cereals, and phenolic substances present in it have anti-diabetic properties. Keeping in view the nutritional significance of pearl millet, food preparations made from it were developed and organoleptically evaluated. The pearl millet recipes namely bajra roti, khichri and bajra dalia were prepared and organoleptically evaluated by a nine-point hedonic scale. All the developed recipes of pearl millet were organoleptically acceptable and should be included in the dietary management of diabetes. Daily consumption of meals prepared from pearl millet is associated with a decreased risk of diabetes. There is a need to spread awareness of pearl millet recipes for diabetic patients, even though their use is restricted to lower-income groups of people.

Keywords: *Pearl Millet, Diabetes, Nutritional Management.*

INTRODUCTION

Currently, diabetes is one of the top 10 causes of death, posing a severe and growing threat to public health. By 2045, there are expected to be 783 million people worldwide who now have diabetes, up from 537 million in 2021 (Ramchandran *et al.,* 2012; Wild *et al.*, 2004; Anjana *et al.*, 2017). The most effective therapeutic strategies for type 2 diabetes prevention in high-risk patients are those, that improve lifestyle and dietary control. Diet plays a decisive role in controlling and maintaining blood glucose levels in type 2 diabetes as it is directly linked with the metabolism of major nutrients such as carbohydrates, protein and fat (Thondre *et al.*, 2009; Hussain, 2018.) Recently, millets have gained popularity for their high nutritional contents and potential to prevent lifestyle diseases (Geeta *et al.*, 2020). Millets have gained importance due to their high nutritional value, established health advantages, adaptability to a variety of environmental conditions, and sustainability in low-input agriculture and organic farming. Small-grain cereal food crops known as millets are particularly nutrient-dense and cultivated on soils with poor fertility or marginal soils with minimal inputs like pesticides and fertilizer. These crops make a significant contribution to the nation's food and nutritional security. Millet crops are known as "nutri-cereals" since they are largely composed of the nutrients the body needs to continue functioning.

The sixth-largest cereal in terms of production and the millet with the highest potential for drought tolerance is pearl millet, also referred to as bajra. After rice and wheat, pearl millet is the third most extensively grown crop in India. It is rich in essential nutrients like protein, fiber, phosphorous, magnesium, and iron. Pearl millet offers a lot of health benefits because of its high mineral and protein content (Nambiar *et al.*, 2011). It has also been evidenced to efficaciously help diabetic patients, particularly those who have type 2 diabetes, in keeping their blood sugar levels consistent over an extended period of time. Additionally, it contains significant levels of carotenoids, phenols, and phenolic acids, all of which have antioxidant potential. Due to their high fibre content and alpha amylase inhibitory capabilities, which are known to lessen starch digestion and absorption, diets based on pearl millet have demonstrated decreased glycemic reactions. As a nutritional choice, pearl millet has recently come under special emphasis in the fight against diabetes. (Nambiar *et al.*, 2011). Foods made from pearl millet are considered preventive medicine for type 2 diabetes because they are good sources of protein, fiber, and other nutrients. They also aid in better nutrition. Consuming pearl millet thus benefits the management and prevention of type 2 diabetes. Therefore, keeping the nutritional importance of this mighty millet, the current study was executed with the aim to develop commonly consumed pearl millet food preparations for type 2 diabetics and to evaluate their acceptability organoleptically.

MATERIALS AND METHODS

Raw Materials: Pearl millet andits flour were purchased from the local market of Kurukshetra district of Haryana, India.

Processing of the Sample: Pearl millet grains were cleaned, washed with water, and then strained for making *khichri* and *dalia.* The flour of pearl millet was strained and stored in an airtight container.

Organoleptic Evaluation: For assessing the sensory attributes such as color, appearance, aroma, texture, taste, and overall acceptability of developed food preparations, namely *bajra roti, khichri,* and *bajra dalia,* organoleptic evaluation was done. A panel of ten members of trained judges including type 2 diabetic persons was selected to assess the overall acceptability of developed food preparations. A nine-point hedonic scale was used for the organoleptic scoring of bajra food preparations.

Statistical Analysis: The data was subjected to statistical analysis using Statistical Package for Social Science (SPSS) version 16 IBM. Mean±SD was used to obtain the differences in organoleptic scores of each food preparations.

RESULTS AND DISCUSSION

Development of Pearl Millet Food Preparations

***Bajra roti*:** Bajra roti is a healthy food choice in place of roti or chapati made from wheat flour. Bajra roti has become a nutritious substitute for diabetic population due to its excellent nutrient profile despite the slightly higher calorie count than wheat. It is cooked on tava and roast over open flame till brown spots appear. It is strongly suggested by many health experts that regular intake of bajra roti in the diabetic's diet plan can help in controlling the blood glucose levels.

Bajra khichri: *Khichri* as the name suggests, is a nutritious, healthy and delicious recipe simply made from rice and pulse also enjoyed at any-time. Pearl millet was added in *khichri* to improve the nutritional profile. *Khichri* made from *bajra* is rich in various nutrients such as protein, iron, potassium, phosphorus, magnesium and fiber which reduces degenerative diseases and helpful in aiding digestive processes.

Bajra dalia: *Dalia is a* nutritious porridge made with combination of seasonal vegetables for all the age groups. For improving the nutrient profile of *dalia,* pearl millet was added and prepared in pressure cooker.

Organoleptic Evaluation

Mean sensory scores of developed food preparations was presented in Table 1 and all the pearl millet food preparations namely *roti, khichri* and *dalia* were found to be organoleptically acceptable. The mean scores for the sensory

Table 1: Sensory Evaluation of Pearl Millet Food Preparations

Food Preparations	*Appearance*	*Color*	*Texture*	*Aroma*	*Taste*	*Overall Acceptability*
Bajra roti	8.05±0.48	8.21±0.12	7.94±0.08	8.07±0.71	8.00±0.62	8.21±0.71
Bajra khichri	8.45±0.79	7.92±0.80	8.15±0.82	8.23±0.84	8.37±0.81	8.42±0.81
Bajra dalia	8.41±0.64	7.08±0.81	8.05±0.45	8.14±0.41	8.65±0.80	8.35±0.82

Values are mean ±SD.

attribute of appearance were highest reported in *khichri* (8.45±0.79) then *dalia* (8.41±0.64) and *roti* (8.05±0.48) whereas mean scores in regard to color of *dalia* (7.08±0.81) and *khichri* (7.92±0.80) scored lesser in comparison to *roti* (8.21±0.12). The sensory scores in respect to taste attribute *roti* scored least (8.00±0.62) and *dalia* scored maximum (8.65±0.80). The scores given to texture and aroma sensory attributes was noticed highest in *khichri* i.e. 8.15±0.82 and 8.23±0.84 and also lowest in *bajra roti* (7.94±0.08 and 8.07±0.71) respectively. The highest mean overall acceptability score was observed within *bajra khichri* (8.42±0.81) followed by *bajra dalia* (8.35±0.82) and *bajra roti* (8.21±0.71).

Nutrient Evaluation

Table 2: Nutrient Evaluation of Pearl Millet Food Preparations (one serving)

Food Preparations	*Cooked Weight (gm)*	*Calories (Kcal)*	*Protein (gm)*	*Carbohydrates (gm)*	*Fat (gm)*	*Fiber (gm)*	*Iron (mg)*
bajra roti	40	144	8.8	58.4	3.3	6.8	2.4
bajra khichri	50	235	12.1	72.2	4.5	5.8	3.5
bajra dalia	60	161.2	7.5	85.5	6.8	7.4	2.8

Nutrient content of pearl millet food preparations is presented in Table 2. The calorific values of food products prepared from pearl millet was varied from 235 kcal of *bajra khichri* to 144 kcal of *bajra roti.* Protein content was highest in khichri (12.1 g) followed by roti (8.8 g) and dalia (7.5 g). Carbohydrates varied from 85.5 g of *dalia* to 58.7 g of *roti.* The highest fiber content was noticed in bajra dalia i.e. 7.4 g, 6.8 g in roti and 5.8 g in dalia. *Bajra khichri* comprised maximum iron content 3.5 mg in comparison to dalia (2.8 mg) and bajra roti (2.4 mg).

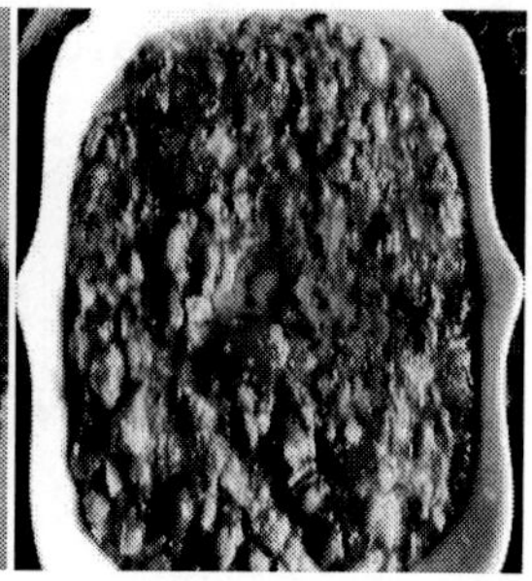

Bajra Roti *Bajra Khichri* *Bajra Dalia*

CONCLUSION

All the developed pearl millet food preparations were organoleptically acceptable and the mean scores for overall acceptability was highest for bajra *khichri* followed by *dalia* and *roti.* As good sources of protein, fiber, and other nutrients, foods made from pearl millet are regarded as preventive therapy for type 2 diabetes and also improved nutritional status of diabetic people. Therefore, eating pearl millet based food preparations assists in managing and preventing type 2 diabetes.

REFERENCES

[1] Anjana, R.M., Deepa, M. and Pradeepa, R. (2017). Prevalence of diabetes and prediabetes in 15 states of India: Results from the ICMR–INDIAB population-based cross-sectional study. *Lancet Diabetes Endocrinol,* 5: 585–596.

[2] Geetha, K., Yankanchi, G.M. and Hulamani, S. (2020); Glycemic index of millet-based food mix and its effect on pre diabetic subjects. *J. Food Sci Technol,* 57, pp. 2732–2738.

[3] Hussain A. 2018, Diabetes in Asia: Special challenges and solutions. *J. Diabetol.,* 9: 69–72.

[4] Nambiar, V.S., Dhaduk, J.J., Sareen, N., Shahu, T. and Desai, R. (2011). Potential functional implications of pearl millet (Pennisetum glaucum) in health and disease. *J. App Pharma Sci.,* (I), pp. 62–67.

[5] Ramachandran, A., Snehalatha, C., Shetty, A.S. and Nanditha, A. (2012), Trends in prevalence of diabetes in Asian countries. *World Journal of Diabetes,* 3(6):110–117.

[6] Thondre, P.S., Jeya, C. and Henry, K. (2009), High-molecular-weight barley β-glucan in chapatis (unleavened Indian flatbread) lowers glycemic index. *Nutrition Research,* 29(7): 480–486.

[7] Wild, S., Roglic, G., Green, A., Sicree, R. and King, H. (2004). Global Prevalence of Diabetes Estimates for the year 2000 and projections for 2030. *Diabetes Care*; 27(5): 1047–1053.

A Functional Food Composition to Combat Pre-Diabetes: A Thrust on Preventive Health

Shalaka Kshirsagar, Anuradha Majumdar and Nikhil Sambhare

Department of Pharmacology, Bombay College of Pharmacy, Mumbai, Maharashtra-400098, India

ABSTRACT: Pre-diabetes is associated with moderately elevated blood glucose levels with fasting plasma glucose of 100 to 125 mg/dl. Poor management of pre-diabetes leads to various complications which increases the risk of developing Type 2 diabetes, prediabetic nephropathy, and cardiovascular diseases. Therefore, it is necessary to address the prediabetic stage to prevent its progression into Type 2 diabetes. The study was aimed at developing a Functional Food (FF) composition and evaluating its benefits in pre-clinical model of prediabetes. A functional food was prudently prepared with the right blend of carbohydrates, dietary fibres, proteins, prebiotics, herbo minerals and oligo elements so as to maintain a low glycemic index. This blend consisted of Buckwheat, PQQ (Pyrroloquinoline Quinone Disodium salt), Kala Jamun Powder, Moringa Powder, Vitamin E, Barley, Stevia and Essence. It acted by improving glucose tolerance, decreasing catabolism, improving carbohydrate homeostasis, decreasing the appetite, alleviating the oxidative stress and thereby lowering the lipoprotein indices. The study revealed beneficiary effects of the developed functional food in prediabetes associated hyperglycemia, impaired glucose tolerance, polyphagia, lipid homeostasis, weight management in sucrose (35%) induced prediabetes like condition in rats. Therefore, the developed functional food may serve as promising intervention in terms of food intake viz. life style change in combating pre-diabetes and its complications.

Keywords: Pre-diabetes, Functional Food, Glucose Tolerance, Antidiabetic, Antioxidant.

INTRODUCTION

Pre-diabetes is a condition characterized by slightly elevated blood glucose levels, regarded as indicative that a person is at risk of progressing to Type 2 diabetes (Abdul *et al.*, 2009). Most individuals pass through a phase of prediabetes before developing full blown diabetes. Pre-diabetes increases the risk of developing type 2 diabetes, heart disease, and stroke, which produces symptoms such as reduction in weight, polyuria, blurry vision, tiredness than usual (Khetan *et al.,* 2017). Diet and lifestyle plays an important role in prevention and management of pre-diabetes. So the combination of drugs with dietary interventions might be more beneficial. Nowadays demand for dietary interventions (functional foods, nutraceuticals and health supplements) is increasing prominently because they are contemplated as safe, readily available, usually considered as natural substances (Sathya *et al.,* 2013). Pre-diabetes increases the risk of developing type 2 diabetes and diabetes is catered with many complications so the single-drug therapy might not always be sufficient, therefore diabetic patients are usually subjected to polypharmacy which might show greater side effects, drug interactions or adverse reactions and also leads to the socio-economic burden. Thus, functional foods which are safe and showing beneficial effects with standard drugs are more utilitarian (Boonloh K. *et al.,* 2015). The developed functional food includes carbohydrates, dietary fibres, proteins, prebiotics, herbo minerals and oligo elements. The antidiabetic effect of functional food was attributed to D-chiro inositol as a source of insulin mediator (Lee *et al.,* 2016). Prebiotic source refer to a group of non-digestible carbohydrates that alter the composition and activity of gut microbiota (Dehghan P. *et al.,* 2016). Water-soluble redox cofactor in the given supplement protected oxidative stress and decreased the levels of free radicals as well as lipid peroxidation (Devasani *et al.,* 2019). The vitamin constituent of the developed food is a potent oil-soluble antioxidant, which increased antioxidant activity and glucose tolerance (Pavithra *et al.,* 2018).

MATERIALS AND METHODS

Metformin was kindly gifted by Herman Finochem Ltd., Mumbai. Following products were purchased from listed sources.

Sr. No.	*Drugs/Product*	*Source*
1.	Metformin	Herman Finochem Ltd., Mumbai
2.	Prroloquinoline quinine disodium salt (PQQ)	Wuxi Cima Science Co., Ltd., China
3.	Buckwheat Groats	Objectwitty Agrotech Pvt Ltd., Gurgaon
4.	Moringa Powder	Earth Expo Company, Bhavnagar
5.	Jambu Beej Powder (Kala Jamun Powder)	Isha Agro Developers Pvt. Ltd., Pune
6.	Barley	Natureland Organic Foods Pvt. Ltd., Rajasthan
7.	Stevia	Ankur Drugs and Pharma Ltd., Solan
8.	Spirulina	Spirulina Production India, Sangli
9.	Rice protein hydrolysate powder 50%	Akola Chemicals, Palghar
10.	Vitamin E powder 50%	Shreenath Chemicals, Palghar
11.	Jaggery Powder	Krushidut Agro Farm, Sangli
12.	Essence	Aromatic Essence

Male Sprague Dawley rats (160 ± 20 g) used for main study obtained from National Institute of Biosciences, Pune.

METHODS

Constituents of functional food were selected by literature search. Compositions of constituents were decided according to following guidelines:

1. https://www.fda.gov/media/110030/download (PQQ GRAS notice)
2. https://fssai.gov.in/cms/health-supplements.php (FSSAI neutraceuticals guideline)
3. https://www.fda.gov/media/98081/download (rice protein GRAS notice)
4. https://www.who.int/health-topics/diabetes#tab=tab_1

Table 1: Role of Each Component Used in Functional Food I

COMPONENT	ROLE
Buckwheat	Antidiabetic
PQQ	Growth promotional activity, Mitochondrial Biogenesis
Kala Jamun Powder	Antidiabetic
Moringa Powder	Antidiabetic
Vitamin E	Antioxidant
Barley	Prebiotic
Stevia	Sweetening Purpose
Essence	As Flavour

Experimental Model of Pre-diabetes Using Sucrose (35%) in the Rats

Total 48 animals were divided in to 6 groups. The sucrose solution (35%) was administered to positive control group only for 9 weeks. The energy provided by sucrose was 3870 kcal/kg and by the normal rat chaw was 3020 kcal/kg hence the total energy provided was 6890 kcal/kg. At the end of the 9 weeks blood glucose was measured by Morephen Glucometer, animals those having blood glucose more than 140 mg/dl were considered as pre-diabetic and were randomised and into the following groups.

1. Vehicle Control
2. Positive Control
3. Standard Treatment
4. Functional Food I
5. Functional Food II
6. Combination (Met 500 mg/kg + Functional Food I).

The rats from the group number 3,4,5 & 6 were followed by treatment with Standard Treatment, Functional Food I, Functional Food II and Combination (Met 500 mg/kg + Functional Food -I) respectively for the four weeks. The energy provided by the normal diet was 3020 kcal/kg.

Following Anthropometric, Dietary, Antioxidants, Biological parameters were evaluated in study:

A. *Anthropometric Parameters*

1. ***Body Weight:*** Body weight (Gms) of each animal was measured once in a week throughout the experimental period using digital weighing balance.
2. ***Body Mass Index:*** BMI was calculated using the following formula. BMI = Body Weight (g)/Body Length (cm^2) Where, body length of a rat refers to the length from tip of the nose to the anus on the dorsal surface.

B. *Dietary Measurements*

1. ***Total Feed Intake:*** Three rats were housed in a polypropylene cage. While calculating the feed and water intake an assumption was made that each of a rat amongst three consumed equal amount of feed and water. Total feed intake was calculated using the following formula, Total feed intake = standard rat chow consumed (g).
2. ***Water Intake:*** Three rats were housed in a polypropylene cage. While calculating the water intake an assumption was made that each of a rat amongst three consumed equal amount of water. Total water intake was calculated in ml.

C. *Biological Estimation*

1. Fasting Blood Glucose Levels
2. Oral Glucose Tolerance Test

OGTT was performed before initiation of treatment and after termination of treatment on 15th day of the experimental period. After an overnight fast (12 h), Rats were orally dosed with a D-glucose solution (2.0 g/kg b.w.) and blood glucose concentrations were subsequently measured at 0 (just prior to oral glucose dosing), 30, 60, 120 and 180 min after the oral dosing of glucose by withdrawing blood through tail vein.

(a) Before treatment (On day 63)

(b) After treatment (On day 92).

D. *Biochemical Estimation in Serum*

1. Total Protein estimation
2. Serum cholesterol
3. Serum triglyceride
4. Serum HDL-C
5. Serum LDL-C
6. Serum VLDL-C
7. Serum non-HDL cholesterol
8. Hexokinase activity.

E. *Antioxidant Assays on Liver Tissue*

1. Catalase Activity
2. Superoxide Dismutase
3. Reduced Glutathione.

Statistical Analysis

Statistical analyses were performed using GraphPad Prism 8 software (GraphPad Software, San Diego, CA, USA). The differences among the experimental groups were analyzed with one-way analysis of variance, followed by Tukey's multiple comparison test as a post-hoc analysis. For OGTT data, two –way analysis of variance (with time and treatment as variables) followed by Tukey's multiple comparison test as a post-hoc analysis was used. All the values were expressed as mean ± SEM. $P < 0.05$ was considered statistically significant.

1. ****, ***, ** and * denote statistical significance at ($p < 0.0001$), ($p < 0.001$), ($p < 0.05$) respectively in comparison to positive control group.
2. ####, ###, ## and # denote statistical significance at ($p < 0.0001$), ($p < 0.001$), ($p < 0.05$) respectively in comparison to vehicle control group.
3. $$$$, $$$, $$ and $ denote statistical significance at ($p < 0.0001$), ($p < 0.001$), ($p < 0.05$) respectively in comparison to standard treatment group.

RESULTS

Composition of Functional Food

Nutritional Value for Functional Food 1: Carbohydrate (27%), Proteins (18.29%), Fat (2.28%), Dietary fiber (7.89%). The energy provided by Functional food 1 3320.3 kcal/kg.

Functional Food I were prepared as per FDA-GRAS notices and FSSAI guidelines and individual constituents are as follow (1000 gm).

Table 2: Composition of Functional Food 1

Sr. No.	*Constituents*	*Quantity*
1.	Buckwheat	1000 gm
2.	PQQ	50 mg
3.	Kala Jamun Powder	25 gm
4.	Moringa Powder	25 gm
5.	Vitamin E	8 mg

Nutritional Value for Functional Food 2

Table 3: Composition of Functional Food 2

Component	**Role**
Buckwheat	Antidiabetic
PQQ	Growth promotional activity, Mitochondrial Biogenesis
Kala Jamun Powder	Antidiabetic
Moringa Powder	Antidiabetic
Vitamin E	Antioxidant
Barley	Prebiotic
Spirulina	Protein Source
Jaggery Powder	Sweetening Purpose

Following Anthropometrical, Dietary, Biological, Biochemical parameters were evaluated in main study.

A. Anthropometrical Parameters

1. ***Body Weight:*** Administration of Sucrose (35%) to the rats except vehicle control group resulted in significant rise in body weight and failure in surmounting the same. Body weights were recorded on day 92 (termination of the study). Subsequent to feeding of the rate functional food composition I and II and intervention with metformin where the rate were fed with composition I (on day 92) resulted in decrease in the average body weights when compared with the positive control group. Positive control group showed significant increase in weight on day 92.

Table 4: Body Weight

Experimental Groups	*Body Weight in gm*
Control	365.333 ± 3.40
Positive control	402.106 ± 5.20
Standard Treatment	359.333 ± 4.16
Functional food-I	335.6667 ± 4.30
Functional food-II	352.6667 ± 6.57
Met 500 mg/kg + Functional Food I	325 ± 6.61

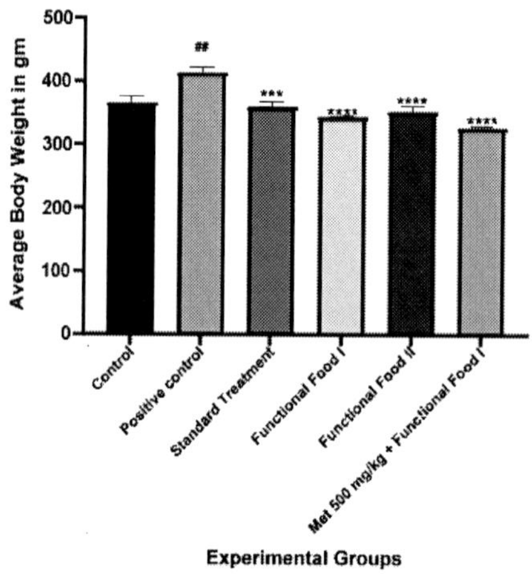

Figure 1: Average Body Weight

2. ***Body Mass Index:*** Composition of 35% sucrose resulted in significant increase in BMI when compared with normal control rats' Functional food 1 and Combination (Met 500 mg/kg + Functional Food 1) significantly prevented 35% sucrose induced increase in anthropometrical parameters.

Table 5: Body Mass Index

Experimental Groups	*Average Body Mass Index in (gm/cm^2)*
Control	0.755921 ± 0.06
Positive control	0.838997 ± 0.026
Standard treatment	0.600532 ± 0.032
Functional food-I	0.536922 ± 0.033
Functional food-II	0.584353 ± 0.08
Met 500 mg/kg + Functional food-I	0.510993 ± 0.024

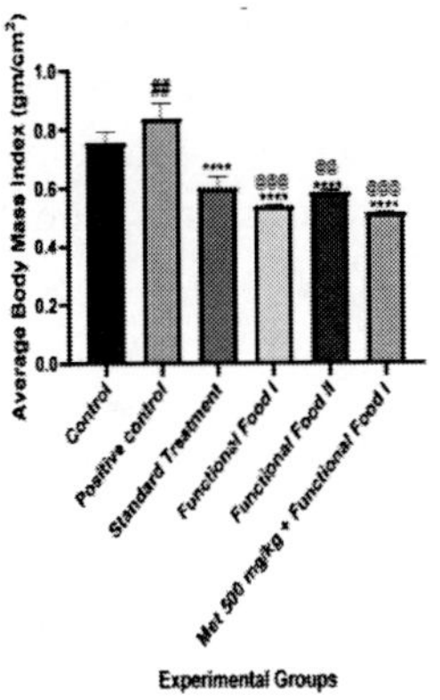

Figure 2: Average BMI

B. Dietary Parameters

1. ***Feed Intake:*** Feed intake was calculated considering consumed rat chow in grams. Administration of sucrose (35%) to the rats except vehicle control resulted in significant increase in the feed intake. Subsequent to feeding of the rate functional food composition I and II and intervention with metformin where the rate was fed with composition I (on day 92) resulted in decrease in feed intake when compared with the positive control group.

Table 6: Feed Intake

Experimental groups	Average Food Intake in gm
Control	26.55333±0.600
Positive control	50.95±0.291
Standard Treatment	37.95±0.400
Functional food-I	42.37556±0.716
Functional food-II	38.10556±0.447
Met 500mg/kg + Functional Food I	39.33556±0.627

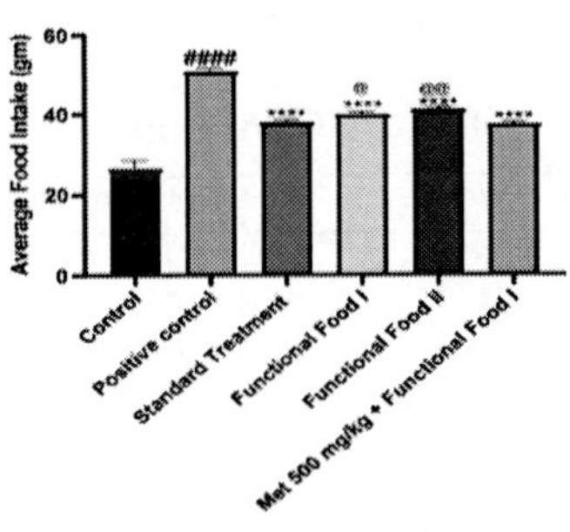

Figure 3: Average Food Intake

2. ***Water Intake:*** Administration of sucrose (35%) to the rats except vehicle control resulted in significant increase in the increase water intake. Subsequent to feeding of the rate functional food composition I and II and intervention with metformin where the rate were fed with composition I (on day 92) resulted in decrease in water intake when compared with the positive control group.

Table 7: Water Intake

Experimental groups	Average Water Intake in ml
Control	36.89111±0.044
Positive control	46.38667±0.067
Standard Treatment	35.28334±0.077
Functional food-I	32.79556±0.01
Functional food-II	33.22222±0.08
Met 500mg/kg + Functional Food I	31.71334±0.044

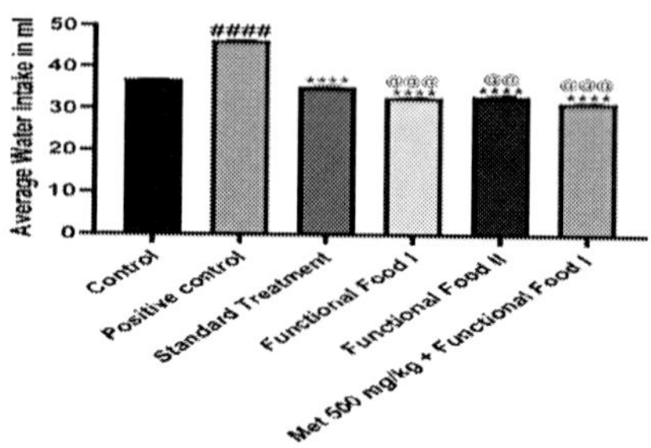

Figure 4: Average Water Intake

3. ***Calorie Intake:*** Calorie intake was calculated considering calories provided through rat chow and Functional Food 1 & 2. Administration of sucrose (35%) to the rats of Positive Control resulted in significant increase in the calorie intake as compared with Standard Treatment, Functional Food I, Functional Food II and Combination (Met 500 mg/kg + Functional Food I).

Table 8: Calorie intake

Experimental groups	Average Calorie Intake in kcal/gm
Control	80.19106±0.044
Positive control	351.0455±0.067
Standard Treatment	114.609±0.077
Functional food-I	140.6996±0.01
Functional food-II	135.3866±0.08
Met 500mg/kg + Functional Food I	130.6059±0.044

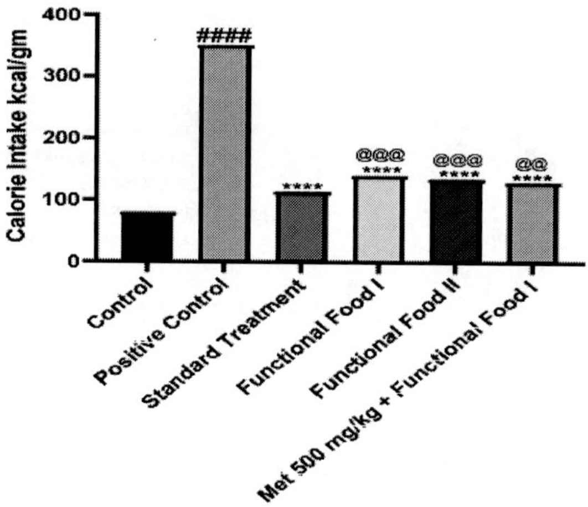

Figure 5: Calorie Intake

C. Biological Estimation

Oral Glucose Tolerance Test (OGTT)

(a) ***Oral Glucose Tolerance Test (Before treatment) (On Day 63):*** After the glucose load, all group of rats except vehicle control group showed significant decrease in glucose tolerance when compared to that of vehicle control. This indicated an impaired glucose tolerance, which is indication of diabetes. The AUC glucose (0–180 min) was the greatest in the all groups which is also indicative of insulin resistance. Treatment was initiated post confirmation of glucose intolerance.

Table 9: OGTT before Treatment

Experimental groups	**Blood glucose (mg/dL)**				
	0 min	**30 min**	**60 min**	**120 min**	**180 min**
Vehicle control	93.375± 2.54	90.625± 2.27	93.375± 1.94	88.875± 1.62	89.75± 2.38
Positive control	130.875±8.30	150.875±4.16	140.375±3.14	138±2.60	135.375±7.08
Metformin (500 mg/kg)	129.125±8.56	157.25±9.85	138.125±4.43	129.375±3.29	128.75±3.21
Functional Food- I	126.375±4.15	161±5.12	141.25±3.97	132.875±2.06	129.5±2.72
Functional Food- II	128.75±2.57	160.25±4.06	150.25±5.89	138.25±4.37	129.875±2.08
Met 500mg/kg + FunctionalFood- I	129.625±4.40	165±11.08	149.375±5.16	138.25±31.03	128.75±9.16

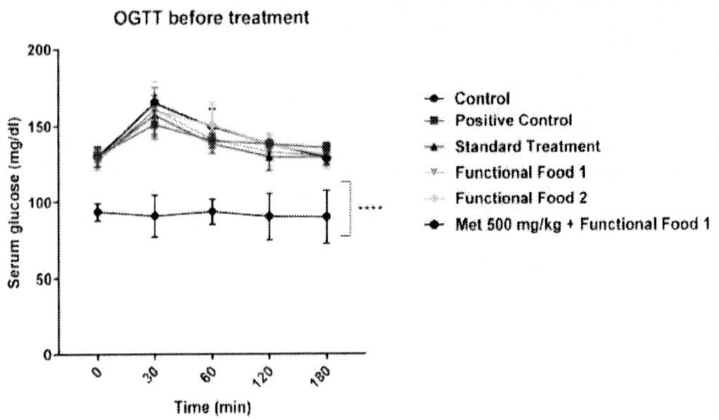

Figure 6: OGTT before Treatment

(b) ***OGTT (After Treatment) (On Day 92):*** On day 92, after the glucose load, positive control group rat showed significant decrease in glucose tolerance when compared to that of vehicle control. After 180 min of post glucose load, the blood glucose concentration of positive control group rat was 166.5 ± 17.977. This indicates an impaired glucose tolerance, which is a harbinger to pre-diabetes. Treatment with Functional Food-I, Functional Food-II, metformin 500 mg/kg per se significantly prevented sucrose (35%) induced impaired glucose tolerance, but this was more pronounced in combination treatment groups (metformin 500 mg/kg + Functional Food-I) which shown marked improvement in the glucose tolerance.

Table 10: OGTT after Treatment

Experimental groups	Blood glucose (mg/dL)				
	0 min	30 min	60 min	120 min	180 min
Vehicle control	93.375±6.54	99.75±9.56	99.75±3.67	103.375±5.67	96.625±7.30
Positive control	130.875±7.89	150.875±8.87	140.375±9.63	138±5.23	135.375±8.63
Metformin (500 mg/kg)	110.375±5.36	120.25±8.36	115.125±12.3	113.75±9.63	105.75±8.98
Functional Food- I	114.875±5.69	119.75±9.33	120.875±2.36	120.625±8.96	112.75±10.03
Functional Food- II	112.5±4.65	120.875±9.63	117.25±6.36	117.75±6.36	110.625±7.32
Met 500mg/kg + Functional Food- I	111.75±6.39	125.75±4.65	121.125±8.96	119.5±9.65	108.875±10.25

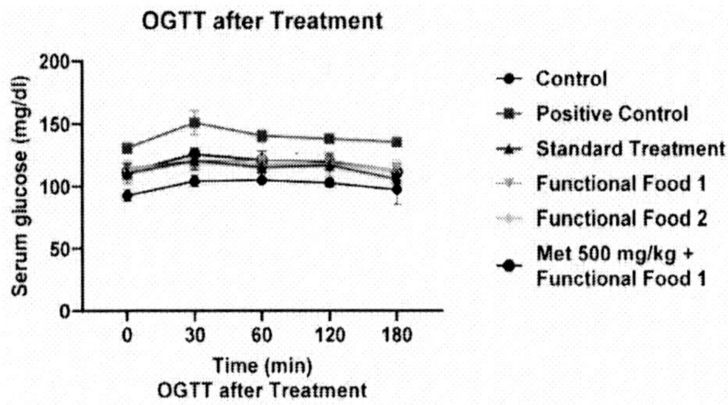

Figure 7: OGTT after Treatment

D. Biological Estimation in Serum

1. ***Total Protein Estimation:*** A significantly lower serum total protein levels were found to be observed in positive control group rats. Subsequent treatment with metformin 500 mg/kg, functional food-I (*ad libitum*), functional food-II (*ad libitum*), and combination (metformin 500 mg/kg + functional food *ad libitum*) resulted in rising of serum proteins levels.

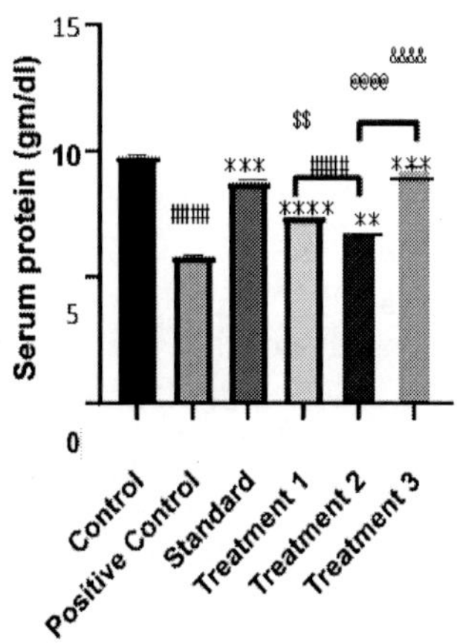

Figure 9: Total Proteins

2. ***Serum Cholesterol (TC):*** A significantly higher serum cholesterol levels were found to be observed in positive control group rats. Subsequent treatment with metformin 500 mg/kg, functional food-I (*ad libitum*), functional food-II (*ad libitum)*, and combination (metformin 500 mg/kg + functional food *ad libitum*) resulted in lowering serum cholesterol.

Table 11: Serum Cholesterol

Experimental groups	Serum Cholesterol (mg/dl)
Control	82.04±0.435
Positive control	147.08±0.398
Standard Treatment	88.20±0.283
Functional food-I	107.72±0.400
Functional food-II	121.71±0.270
Combination (Met 500mg/kg + Functional Food I)	98.84±0.341

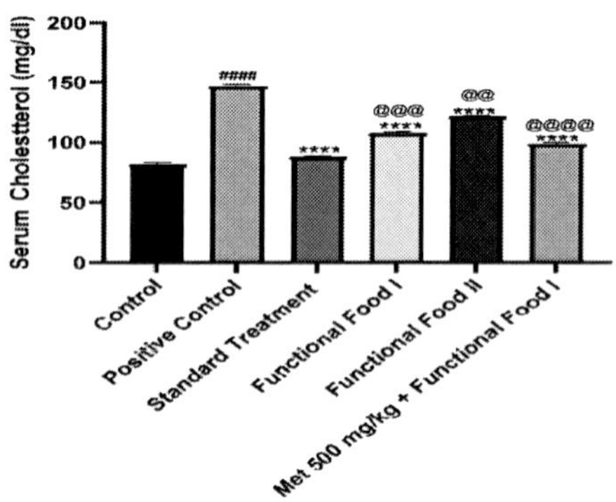

Figure 10: Serum Cholesterol

3. ***Serum Triglyceride:*** A significantly higher serum cholesterol levels were found to be observed in positive control group rats. Subsequent treatment with metformin 500 mg/kg, functional food-I (ad libitum), functional food-II (ad libitum), and combination (metformin 500 mg/kg + functional food ad libitum) resulted in lowering of serum cholesterol levels.

Table 12: Serum Triglyceride

Experimental groups	Serum Triglyceride (mg/dl)
Control	87.87±0.802
Positive control	130.89±0.396
Standard Treatment	99.37±0.315
Functional food-I	105.98±0.713
Functional food-II	111.97±0.476
Combination (Met 500mg/kg + Functional Food I)	103.53±0.474

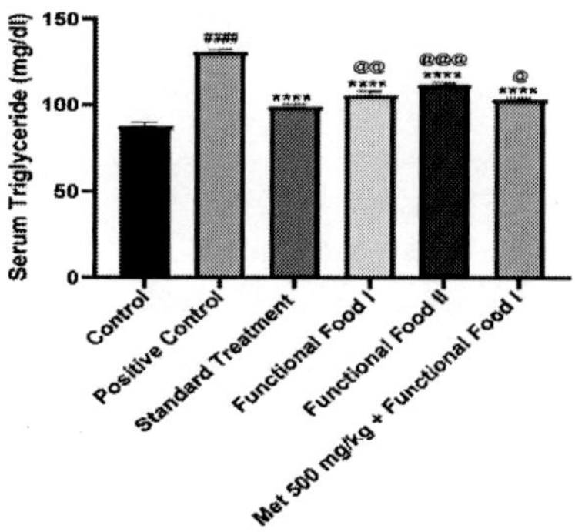

Figure 11: Serum Triglyceride

4. ***Hexokinase Activity:*** A significantly higher serum hexokinase levels were found to be observed in positive control group rats. Subsequent treatment with metformin 500 mg/kg, functional food-I (*ad libitum*), functional food-II (*ad libitum)*, and combination (metformin 500 mg/kg + functional food *ad libitum*) resulted in lowering of serum glucose levels.

Table 13: Hexokinase Activity

Experimental Groups	Serum Hexokinase (mg/dL)
Control	90.73±0.146
Positive control	186.67±0.540
Standard Treatment	126.23±0.136
Functional food-1	143.06±0.121
Functional food-2	153.02±0.134
Met 500mg/kg + Functional Food 1	131.85±0.152

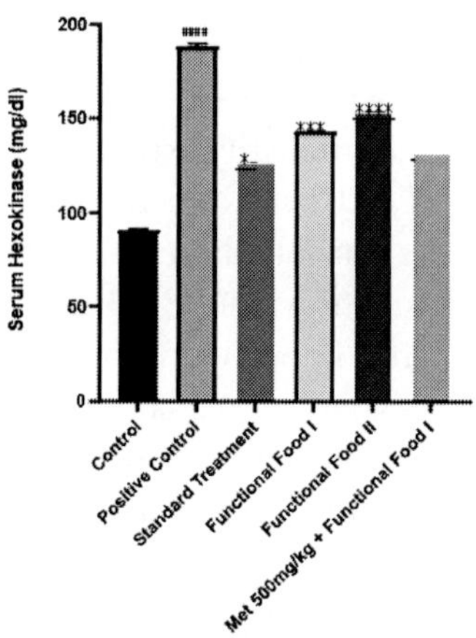

Figure 12: Hexokinase Activity

E. Antioxidant Assays on Liver Tissue

Catalase Activity

The liver Catalase activities significantly dropped in positive control group rats. Subsequent treatment with metformin 500 mg/kg, functional food-I (*ad libitum*), functional food-II (*ad libitum)*, and combination (metformin 500 mg/kg + functional food *ad libitum*) resulted in enhancement of liver Catalase activities.

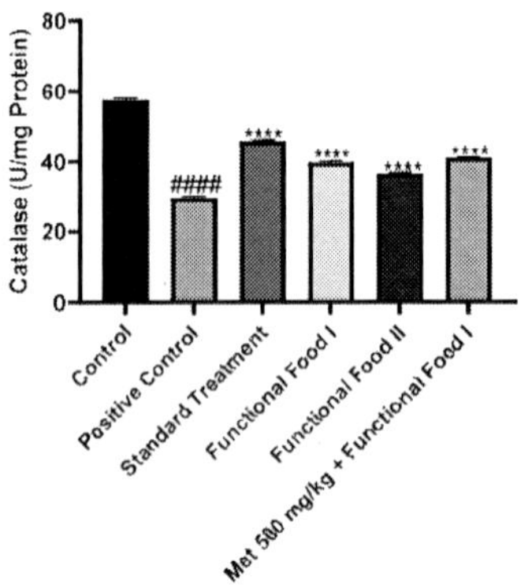

Figure 13: Catalase Activity

DISCUSSION

Pre-diabetes is a disorder with both short- and long-term undesirable complications which leads to diabetes and diabetic complications with various pathogenic conditions including dyslipidemia, vascular dysfunction, oxidative stress, sub-clinical inflammation, and altered signaling pathways. Ineffectiveness of the current medical treatments in management of long-term pre-diabetes confirms that other complementary approaches are required; the use of functional foods and bioactive compounds is one of these new approaches (Kurosaki *et al.,* 2013). Functional foods and their bioactive compounds could attenuate carbohydrate metabolism and hyperglycemia, improve pancreatic β-cell function and insulin secretion as well as insulin resistance, regulate lipid and lipoprotein metabolism and adipose tissue metabolism, modulate oxidative/antioxidative balance and inflammatory processes, improve weight management and prevent micro and macro vascular complications (Groover *et al.,* 2017), (Mirmiran P *et al.,* 2014).

The present work was undertaken to investigate the possible benefit of developed functional food per se and in combination of reference drug metformin treatment in pre-diabetes and its related complications. Sucrose

35% induced prediabetes model is used to investigate decreased insulin secretion and utilization by body tissues, which occurs due to partial destruction of the pancreatic ß cells (Konopka *et al.,* 2016). Glucose homeostasis of the body id regulated by lipid metabolic enzyme and glucose uptake cycle of body tissues (Fan S *et al.,* 2020). Suggestive evidence is also becoming increasingly available that indicates that a diet based on low glycemic index foods may contribute to diabetes prevention (Kurosaki E *et al.,* 2013). Oligo elements (i.e., magnesium) present in significant amounts in whole-grain foods may have an insulin-sensitizing effect (Salunkhe *et al.,* 2018).

Fasting blood glucose levels were observed decreasing after 29 days treatment groups but marked decline was seen in combination groups, VI. Combination (Met 500 mg/kg+ Functional Food-I). Interpretation of oral glucose tolerance test (OGTT) data suggest that there is impaired peripheral glucose uptake, which is a hallmark of pre-diabetes and diabetes. Significant change were observed in before and after treatment OGTT levels as compared to Positive Control and Vehicle Control Group. Significant decrease in Total Protein, HDL-cholesterol was observed in Positive Control Group as compared to Treatment groups. Significant increase in Serum Cholesterol, Serum Triglyceride, VLDL-cholesterol, LDL-cholesterol, Hexokinase activity levels in Positive Control Group as compared to Treatment Groups. There was also significant decrease in Catalase Activity in Positive Control Group as compared with Treatment Groups.

CONCLUSION

We observed that developed functional food compositions in combination with metformin had a more pronounced potential in combating the detrimental anthropometrical and biochemical changes that were triggered in the rodent model of pre-diabetes. Also, showed propitious protective effect against progression of related cardiovascular complications of pre-diabetes. Hence, we propose that developed functional food may serve as ancillary treatment with metformin in combating pre-diabetes and its complications. It may differ or prevent transformation of pre-diabetes to diabetes.

REFERENCES

[1] Abdul-Ghani, M.A., DeFronzo, R.A., Pathophysiology of prediabetes. *Curr Diab Rep.* 2009; 9(3):193–199. doi:10.1007/s11892-009-0032-7

[2] Kazeem, M.I., Davies, T.C., Anti-diabetic functional foods as sources of insulin secreting, insulin sensitizing and insulin mimetic agents. *J. Funct Foods*, 2016; 20: 122–138. doi:10.1016/J.JFF.2015.10.013

[3] Lee, D.G., Jang, I.S., Yang, K.E., *et al.* Effect of rutin from tartary buckwheat sprout on serum glucose-lowering in animal model of type 2 diabetes. *Acta Pharm*, 2016, 66(2): 297–302, doi: 10.1515/acph-2016-0021.
[4] Kumar, N. and Kar, A., Pyrroloquinoline quinone (PQQ) has potential to ameliorate streptozotocininduced diabetes mellitus and oxidative stress in mice: A histopathological and biochemical study. Chem Biol Interact. 2015; 240: 278–290. doi:10.1016/j.cbi.2015.08.027
[5] Devasani, K., Kaul, R. and Majumdar, A., Supplementation of pyrroloquinoline quinone with atorvastatin augments mitochondrial biogenesis and attenuates low grade inflammation in obese rats. *Eur J. Pharmacol*, 2020; 881:173273. doi:10.1016/j.ejphar.2020.173273
[6] Dehghan, P., Farhangi, M.A., Tavakoli, F., Aliasgarzadeh, A. and Akbari, A.M. Impact of prebiotic supplementation on T-cell subsets and their related cytokines, anthropometric features and blood pressure in patients with type 2 diabetes mellitus: A randomized placebo-controlled Trial. *Complement Ther Med.*, 2016; 24: 96–102. doi:10.1016/j.ctim.2015.12.010
[7] Boonloh, K., Kukongviriyapan, V., Kongyingyoes, B., Kukongviriyapan, U., Thawornchinsombut, S. and Pannangpetch, P. Rice bran protein hydrolysates improve insulin resistance and decrease pro-inflammatory cytokine gene expression in rats fed a high carbohydrate-high fat diet. *Nutrients*, 2015; 7(8):6313–6329. doi:10.3390/nu7085292
[8] Sathya, A. and Siddhuraju, P., Protective effect of bark and empty pod extracts from Acacia auriculiformis against paracetamol intoxicated liver injury and alloxan induced type II diabetes. *Food Chem Toxicol*, 2013; 56:162–170. doi:10.1016/j.fct.2013.02.031.
[9] Moreira, L.M., Behling, B. del S., Rodrigues, R. da S., Costa, J.A.V. and Soares, L.A. de S. Spirulina as a protein source in the nutritional recovery of Wistar Rats. *Brazilian Arch Biol Technol.*, 2013; 56(3): 447–456. doi:10.1590/S1516-89132013000300013
[10] Al-Malki, A.L. and El Rabey, H.A., The antidiabetic effect of low doses of moringa oleifera lam. Seeds on streptozotocin induced diabetes and diabetic nephropathy in male rats. *Biomed Res Int.*, 2015; 2015. doi:10.1155/2015/381040
[11] Vlachos, D., Malisova, S., Lindberg, F.A. and Karaniki, G. Glycemic Index (GI) or Glycemic Load (GL) and Dietary Interventions for Optimizing Postprandial Hyperglycemia in Patients with T2 Diabetes: A Review. *Nutrients*, 2020; 12(6). doi:10.3390/NU12061561
[12] Ojo, O., Ojo, O.O., Adebowale, F., Wang, X.H. The effect of dietary glycaemic index on glycaemia in patients with type 2 diabetes: A systematic review and meta-analysis of randomized controlled trials. *Nutrients,* 2018; 10(3). doi:10.3390/NU10030373
[13] Sarwar, N., Gao, P., Kondapally Seshasai, S.R., *et al.* Diabetes mellitus, fasting blood glucose concentration, and risk of vascular disease: A collaborative meta-

analysis of 102 prospective studies. *Lancet*. 2010; 375 (9733): 2215–2222. doi:10.1016/S0140-6736(10)60484-9

[14] Chowdhury, K.K., Legare, D.J. and Lautt, W.W. Exercise enhancement of hepatic insulin-sensitising substance-mediated glucose uptake in diet-induced prediabetic rats. *Br J Nutr*. 2013; 109(5): 844–852. doi:10.1017/S0007114512002267

[15] Zhu, F., Buckwheat starch: Structures, properties, and applications. Trends Food Sci Technol. 2016; 49: 121–135. doi:10.1016/J.TIFS.2015.12.002

[16] Zhu, F., Buckwheat proteins and peptides: Biological functions and food applications. *Trends Food Sci. Technol.*, 2021; 110: 155–167, doi:10.1016/J.TIFS.2021.01.081

[17] Lee, C.C., Hsu, W.H., Shen, S.R., Cheng, Y.H. and Wu, S.C. Fagopyrum tataricum (Buckwheat) improved high-glucose-induced insulin resistance in mouse hepatocytes and diabetes in fructose-rich diet-induced mice. *Exp Diabetes Res.*, 2012; 2012. doi:10.1155/2012/375673

[18] Ikeda, K., Ishida, Y., Ikeda, S., Asami, Y. and Lin, R. Tartary, but not common, buckwheat inhibits αglucosidase activity: its nutritional implications. *Fagopyrum*, 2017; 34: 13–18. doi: 10.3986/fag0002

[19] Surono, I.S., Wardana, A.A., Waspodo, P., Saksono, B., Verhoeven, J. and Venema, K. Effect of functional food ingredients on gut microbiota in a rodent diabetes model. *Nutr Metab.*, 2020; 17(1):1–16. doi:10.1186/s12986-020-00496-2

[20] Dorcely, B., Katz, K., Jagannathan, R., *et al.* Novel biomarkers for prediabetes, diabetes, and associated complications. Diabetes, *Metab Syndr Obes Targets Ther.*, 2017; 10: 345–361, doi:10.2147/DMSO.S100074

[21] Montonen, J., Knekt, P., Järvinen, R. and Reunanen, A. Dietary antioxidant intake and risk of type 2 diabetes. *Diabetes Care.*, 2004; 27(2):362–366. doi:10.2337/DIACARE.27.2.362

[22] Kurosaki, E., Ogasawara, H. Ipragliflozin and other sodium–glucose cotransporter-2 (SGLT2) inhibitors in the treatment of type 2 diabetes: Preclinical and clinical data. *Pharmacol Ther.* 2013; 139(1):51–59. doi:10.1016/J.PHARMTHERA.2013.04.00

[23] Konopka, A.R., Esponda, R.R., Robinson, M.M., *et al.* Hyperglucagonemia Mitigates the Effect of Metformin on Glucose Production in Prediabetes. *Cell Rep.*, 2016; 15(7):1394–1400. doi:10.1016/J.CELREP.2016.04.024.

[24] Salunkhe, V.A., Veluthakal, R. and Kahn, S.E. and Thurmond, D.C. Novel approaches to restore beta cell function in prediabetes and type 2 diabetes. *Diabetologia*, 2018; 61(9): 1895–1901. doi:10.1007/S00125-018-4658-3/FIGURES/2.

[25] Riccardi, G., Capaldo, B. and Vaccaro, O., Functional Foods for Diabetes and Obesity. *Nutr Metab Bases Cardiovasc Dis.* Published online January 12, 2011:138–146. doi:10.1002/9781444318456.CH18.

MILLET—Nutritional Profiles and Health Benefits

Pushplata Kurmi
Department of Agriculture, Mangalayatan University, Jabalpur (M.P.)

ABSTRACT: MILLET are good source of energy. They provide protein, fatty acid, minerals, dietary fiber and polyphenols. Climate change, water scarcity, higher food prices, growing populations, and other socioeconomic consequences are predicted to pose a serious challenge to agriculture as well as the global food security in the twenty-first century. This implies especially, for the world's poorest people of arid and sub arid areas. Scientists and nutritionists were challenged by these impacts to look into the production, processing and consumption of alternate food sources in order to eliminate hunger and poverty. The most important source of food in the world is cereal grains, that also play a major role in the diet of world people. In the semiarid Tropics of Africa and Asia, millets are one of the most prominent drought-resistant crops and is a key source of carbohydrates and protein for the native community. Furthermore, millet grain is currently attracting greater interest from food scientists, technologists and dieticians due to its significant contribution to national food security and potential health advantages.

Keywords: *Millet, Nutrients, Health Benefits.*

INTRODUCTION

The grass family includes the small-grained, warm-weather cereal known as MILLET. They have a comparable nutrient profile to other main cereal and are very resistant to drought and other adverse weather conditions. One of the most significant cereal grains is millets. More than one-third of people on the planet eat millets. In terms of the amount of agriculture produced worldwide, it is the sixth cereal crop. In rural places, millets are frequently consumed as food. They have been cultivated for thousands of years and are used all over the world. In the mediaeval times, the Romans and Gauls preferred millets to wheat for their porridge. About 21 Indian states produce millets, including Karnataka, Tamil Nadu, Andhra Pradesh, Kerala, Hyderabad, Telangana, Uttarakhand, Jharkhand, and Madhya Pradesh. Karnataka state accounts for 58% of the world's millets production.

Millets plants can grow anywhere between 0.5 and 4 metres tall. The millet grain can be virtually white, pale yellow, brown, grey, slate blue, or purple. It also varies greatly.

Production and Consumption

Tables 1 and 2 showed that millets are coarse crop growing where the most of grain world fail. Millets yield decrease from 68.4 million hectare in 1974, 43.4 million hectare in 1980 and 37.6 million in 1997 to 98. The average of millets in India is about 7.9 quintal per hectare the enhanced yield of millets was around 82% in finger millet 95% in little millet, 83% in kodo millet, 43% in Foxtail millet, 76% in proso millets and 82% in barnyard millet due to use of new recommended Technology. Millets have very important role in Indian foods according to NSSO unit the consumption pattern of small millets and finger millets was examined. The small millets consumption pattern was highest in Assam 18.82 kg/ha/m and 18.6 kg/ha States. Madhya Pradesh has highest area of a small millets are 32.4%after that 19.5% Uttrakhand 8%, Maharastra 7.8%, Gujarat 5.3% and 3.9% Tamil Nadu. The productivity has highest as 1174 kg/ha in Uttrakhand followed by 1067 kg/ha Tamil Nadu and 1056 kg/ha in Gujarat.

Table 1: Area, Production and Yield of Small Millets and Finger Millets

Particulars	*1965–66*	*1975–76*	*1985–86*	*1995–96*	*2005–06*	*2013–14*
Area (Lakh/ha) 53.35	45.64	46.72	31.55	16.62	10.64	6.82
Production (Lakh tones) 20.7	15.55	19.24	12.17	7.79	4.72	4.29
Yield (q/ha) 388	341	412	386	469	443	633

Source: DES, MOA, GoI.

Table 2: State Wise Area Production and Yield of Small Millets

State	*Area (000 ha)*	*Production*	*Yield (Kg/ha)*
Andhra pradesh	30.00(3.88)	26.00(6.07)	866.67
Arunachal pradesh	22.09(2.86)	21.54	975.03
Gujarat	49.00(6.34)	48.60	991.84
Jharkhand	24.95(3.23)	11.95	478.96
Karnataka	24.21(3.13)	11.85	489.61
Madhya pradesh	246.22(31.85)	28.72	360.09
All India	773.13(100)	428.20(100)	553.85

TYPES OF MILLETS

- **Pearl Millet (*Pennisetum glaucum*):** Pearl millet originated in Central tropical Africa and is widely distributed in the drier tropics and India. It was introduced into the Western state in the 1850's and become established as minor forage in the Southeast and Gulf coast States. The plant was probably domesticated as a food crop some 4000 to 5000 years ago along the southern margins of the central Highlands of the sahara.
- **Finger Millet (*Eleusine coracana*):** finger millet is a serial grass grown mostly for its grain finger millet is a robust tufted during annual grass up to 170 cm high (FAO, 2012, De Wet, 2006, Quattrocchi, 2006). The inflorescence is a panicle with 4–19 finger-like spikes that resembles a first when mature, hence the name finger millet.
- **Proso Millet (*Panicum miliaceum*):** Proso millet is a annual grass, growing form each year is origin goes back in history at least as far as 2000 B.C. when it is reported to have been grown in the central reason of Europe this plant is especially well suited to dry climate such as Central Russia, the Middle East Northern India, Africa Manchuria and the great planes area of North America, Proso millet was first introduced to Canada in the 17th century and was used in a limited way as a forage crop in the early 1900's.
- **Foxtail Millet (*Setaria italica*):** Foxtail millet is regarded as a native of China, it is one of the world's oldest cultivated crops foxtail millet ranks second in the total world production of millet and continue to have an important place in the world agriculture providing approximately 6 million tons of food too million of people mainly on poor or marginal soils in southern Europe and in temperate, Sub tropical and tropical Asia.
- **Little Millet (*Panicum sumartrence*):** little millet was domesticated in India. It grown throughout India to a limited extent up to altitude of 2100 m, but is of little importance elsewhere. The seeds of little millets are smaller than those of common millet these species of serial is similar in habit to the proso millet except that it is smaller.

NUTRITIONAL COMPOSITION OF MILLETS

Some nutritional values of millets are similar to that of wheat and rice. millets are staple food for many African countries, however they are low in macro nutrients such as protein and fat but rich in vitamins and minerals millets are a good source of magnesium which reduce the saverty of asthma frequency of migraines, lower high blood pressure and reduce the risk of heart attack these nutrients play important role in human nutrition the grains are also a good source of diet for growing children and expectant mothers. They are a good source of phytochemicals such as polyphenols tannins and fatty Acid which helps to lower cholesterol and reduce cancer risk, high blood pressure, heart disease and diabetes. Other potential health benefits and medical function of millets are increasing in time span of gastric emptying and provides roughage to gastro intestine. Millets are also known as an alkaline farming food.

Table 3: Nutritive Value of Millets, Nutritive Value of Indian Foods

Content	*Foxtail Millet*	*Kodo Millet*	*Barnyard Millet*	*Pearl Millet*
Moisture	11.2	12.8	11.9	12.4
Protein	11.50–12.3	9.8	6.2	11.6–11.8
Fat	2.38–4.3	1.3	2.2	4.8–5
Mineral	0.47–3.3	2.6	4.4	2.2–2
Dietary fiber	2.5–8.5	2.47	1.98	11.3
Carbohydrate	60.9–75.2	65.9–66.6	65.5	67–67
Energy	331	309	307	361–362
Calcium	13	27	11	38

BIOFORTIFICATION IN MILLETS

Biofortification in Millets can be achieved through Two Strategies

- By enhancing the accumulation of nutrients in milled grains.
- By reducing the anti nutrients to increase the bioavailability of minerals.

Staple crop bio fortification is a practical proven, and demand lead response to hidden hunger particularly among the 100 of millions of small holder farming families who eat mostly what they grow themselves cannot afford nutritionally divers diets, and are also not easily reached by food fortification on supplementary initiatives.

"Biofortification" or "biological fortification" refers to nutritionally enhanced food crop with increased bioavailability to the human population that are developed and grown using modern biotechnology techniques conventional plant breeding, and agronomic practices. The United National Food and Agriculture Organization has estimated that around 792.5 million people across the world are malnourished out of which 780 million people live in developing million people live in developing countries 1) Apart from this around two billion people across the world suffer form and other type of Hunger known as "hidden hunger" which is caused by an indicuate intake of essential micronutrients in the daily diet. Fortification of grain foods was found to be an effective strategy that can be used to overcome nutrient deficiencies micro nutrients deficiency especially of Vitamins a, iron, iodine and a zinc are widely prevalent in both developing as well as some developed countries. However iron deficiency is a major public health problem in developing countries it affects up to 50% of infants children and women of child bearing age in poorer population of Africa, Asia and Latin America. it has been established that fortified pearl millet flour seems to be a satisfactory candidate for fortification with Zink and so can be exploited to address zinc deficiency.

Why We Need Biofortification

The world population was continuously increasing suffer from a lack of food so that fighting hunger continuous to be a challenge for humanity.

HEALTH BENEFITS OF MILLETS

Table 3 are shows health benefits of different types of millets.

Millets—Diabetes

Millets have ability to reducing glucose by enzymatic millets have ability to reducing glucose by Enzymatic hydrolysis of complex carbohydrate in hyperglycemia. The aldose reductase Enzyme helps in preventing accumulation of Sorbitol and decrease the risk of diabetes. Millets also help in controlling sugar level in Blood and delay wound healing process (Rajesh Karan N.S. *et al.*, 2004). The whole Grain foods are consumed to be effective for the prevention and management of diabetes Mellitus, and showed that millet intake Population epidemiologically lower of Diabetes (American Diabetes Association 2005; Shobana and others 2009; Kim and Others 2011). Some studies find out on Humans (male & female) the effect of millet Grains and their effect on diabetes. Results Showed that persons who has taken millet in Diet, found to decrease the level of blood Glucose (Pradhan *et al.*, 2010).

Millets—Cancer

Millets are rich in antinutrients such as Phenolic, tannins and phytates which help in Reducing the risk of cancer. It contains Phenolics which effective for the prevention of The cancer initiation and progression (Chandrasekaran, A. *et al.*, 2011). Millets Contains linoleic acid which helps in preventing Tumor (Nobihou *et al.*, 2007). Sorgam has Anticarcinogenic properties and antimutagenic Property due to presence of tannins and Polyphenols' (Grimmer *et al.*, 1992). Millet Grains are rich in phenolic acids, tannins, and Phytate that act as "antinutrients" (Thompson 1993). A recent study has showed that Phenolics of millets may be effective in the Prevention of cancer initiation and progression in vitro (Chandrasekara and Shahidi 2011).

Millets—Celiac Disease

Celiac disease is a genetically problem Associated with consumption of gluten one of the component of cereal grain mainly in Wheat. Millets are gluten free which help to Decrease irritation of other cereal grains (Saleh *et al.*, 2013). Millets are gluten-free foods and Beverages that can be suitable for celiac Disease person (Taylor and others 2006; Taylor and Emmambux 2008; Chandrasekara and Shahidi 2011, 2011).

Millets—Heart Disease

Millets are good source of magnesium which Helps in reducing heart attack. Millets are rich in phyto-chemicals which help in lowerin.

Types of Millets	*Health Benefits*
Finger millets	Inhibit cataract eye lens lower plasma Glucose level, anti microbial activity against bacillus cereus and Aspergillos, flavas.
Foxtail millet	Anti-hyperglycemic activity
Proso millet	Improved HDL, lower triglycerides prevent cardiovascular disease
Kodo millet	Inhibit glycation and cross linking of collagen of aging.
Pearl millet	Inhibiting of the growth of the phytopathogenic fungi.
Barnyard millet	Improved the level of HDL.
Little millet	Inhibiotary effects on lipid peroxidation.

Source: Verma (2012), Fereidoon shahidi (2013).

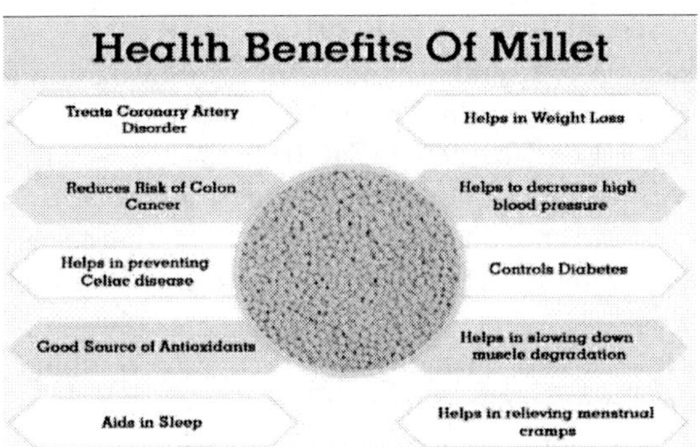

waw.hybrata.com

The important nutrients present in millets include resistant starch, oligosaccharides, lipids, antioxidants such as phenolic acids avenanthramides, flavonoids, lignans phytosterols which are believed to be responsible for many health benefits.

CONCLUSION

Now it is established fact that the whole world is facing many health challenge because of fiber-less foods it is also clear to 1000s of patience that all the lifestyle disease can be made to disappear just by eating millet for breakfast, lunch and dinner and removing refined foods like rice wheat, refined floor persist meets refined oil packet and ready to consume kind of foods and milk. Millets have multiple health benefits to include these incident price like seed in our regularly diet most of the city civilized people have not have an hard about merits and much less understand the benefits of millet nutrition. The aim of this study is to help the people to recognize the importance of food and to introduce the millets as a nutritious foods, fulfilling the nutritional needs of global population and to find ways to consume the millets nutritionally effectively and to reduce the problem of male nutrition and other health problems.

REFERENCES

[1] Vijaya Kumari, J.M., Begum, S., Pandy, A., Shivaleela, H. and Meena, Kumari (2003). Nutritional composition and sensory profile of baked products from finger millet). In: Recent, 45: 97–102.

[2] Blandino, A., Al-Aseeri, M.E., Pandiella, S.S., Cantero, D. and Webb, C. (2003). Cereal-based fermented foods and beverages. *Food Research International,* 36(6), 527–543. http://dx.doi.org/10.1016/ S0963-9969(03)00009-7.

[3] Chandra, Anjali. Singh, Ajeet Kumar and Matos, Bharat (2018). Processing and Value Addition of Finger Millet to Achieve Nutritional and Financial Security. *International Journal of Current Microbiology and Applied Sciences,* 2319–7706, Special, Issue 7, 2901–2910.

[4] Chandrasekara, A., Shahidi, F. (2011b). Antioxidant phenolics of millet control lipid peroxidation in human ldl cholesterol and food systems. *J. Am Oil ChemSoci,* DOI: 10.1007/s11746-011–1918–5. Available from Springer [http://www.springerlink.com]. Posted August 20, 2012.

[5] Chandrasekara, A. and Shahidi, F. (2011c). Antiproliferative potential and DNA scission inhibitory activity of phenolics from whole millet grains. *J. Funct Foods,* 3:159–70.

[6] Chandrasekara, A. and Shahidi, F. (2012). Bioaccessibility and antioxidant potential of millet grain phenolics as affected by simulated in vitro digestion and microbial fermentation. *J. Funct Foods,* 4: 226–37.

[7] Chauhan, S. and Malleshi, N.G. (2007). Finger millet polyphenols: Optimization of extraction and the effect of pH on their stability. *Food Chem.,* 105(2): 862– 870.

[8] Chowdhury, S. and Punia, D. 1997. Nutrient and antinutrient composition of pearl millet grains as affected by milling and baking. *Nahrung,* 41(2): 105–7.

[9] Coulibaly, A., Kouakou, B. and Chen, J. Phytic Acid in Cereal Grains: Structure, Healthy or Harmful Ways to Reduce Phytic Acid in Cereal Grains and their Effects on Nutritional Quality. *Am. J. Plant Nutr.* Fert. technol2011;1:–22.

[10] https: //ejfa.me/index.php/journal/article/view/981/712.

[11] Manach, C., Mazur, A. and Scalbert, A. (2005). Polyphenols and prevention ocardiovascular diseases. *Current Opinion Lipidology.* 16: 77–84.

[12] Millets: Future of Food & Farming-millet network of India-deccan development of India FIAN-INDIA.

[13] Reddy, O.S.K. (2017). Smart Millet and Human Health, Green Universe Environmental Services Society.

[14] Palanisamy Bruntha Devi, Rajendran Vijayabharathi, Sathyaseelan Sathyabama, Nagappa Gurusiddappa Malleshi and Venkatesan Brindha Priyadarisini (2014). Health benefits of finger millet (Eleusine coracana L.) polyphenols and dietary fiber: a review, *Journal of Food Science and Technology,* 51(6):1021–1040.

[15] Rajasekaran, N.S., Nithya, M., Rose, C. and Chandra, T.S. (2004). The Effect of Finger millet feeding on the early responses during the process of wound healing in Diabetic rats Biochimica et Biophysica. Acta, 1689: 190–201.

[16] Saleh, A.S.M., Zhang, Q., Chen, J. and Shen (2013). Millet frains: Nutritional quality, processing and potential health benefits. *Comprehensive reviews in Food Science and Food Safety,* 12: 281–295.

[17] Shobana, S. and Malleshi, N.G. (2007). Preparation and functional properties of decorticated finger millet (Eleusine coracana). *Journal of Food Engineering,* 79: 529–538.

[18] Shweta, Malik (2015). PEARL Millet-Nutritional Value and Medicinal Uses (Food & Nutrition) Dept. of Home Science, B.P.S. Women's University Khanpur Kalan (Hry) www.ijariie.com, Vol. 1, Issue 3.

[19] Singh, K.P., Mishra, A. and Mishra, H.N. (2012). Fuzzy Analysis of Sensory Attributes of Bread prepared from millet based composite flours. *LWT-Food Science Technology,* 48: 276–82.

[20] Deshpande, S.S., Mohapatra, D., Tripathi. M.K. and Sadvatha, R.H. (2015). Kodo Millet-Nutritional Value and Utilization in Indian Foods, ICAR-Central Institute of Agricultural Engineering, *Journal of Grain Processing and Storage,* Vol. 2: 16–23.

[21] Upadhyaya, H.D., Gowda, C.L.L. and Reddy, V.G. (2007), Morphological diversity in finger millet germplasm introduced from Southern and Eastern Africa, *The Journal of Semi-Arid Tropical Agricultural Research.* 3: 1–3.

[22] Veenu, Verma and Patel, S. (2012), Nutritional security and value added products from finger millets (ragi), *Journal of Applicable Chemistry,* 1 (4): 485–489.

[23] Veena, B. (2003). Nutritional, functional and utilization studies on barnyard millet. M. Science Thesis, University of Agricultural Sciences, Dharwad (Karnataka), India.

[24] Pragya Mishra, H.G. Prakash, Suman Devi, Semasonkar, Shewera Yadav and H.S. Singh Nutritional quality of millets and their value added products with the potential health benefits.

[25] Malleshi, N.G. and Hadimani, N.A. (1993). Nutritional and technological Characteristics of small millets and preparation of value-added products from Them. In: Riley KW, Gupta SC, Seetharam A, Mushonga JN, editors. Advances in small millets. New Delhi: Oxford and IBH Publishing Co Pvt. Ltd., pp. 271–87.

[26] Mamiro, P.R.S., Van Camp, J., Mwikya, S.M. and Huyghebaert, A. (2001). In vitro Extractability of calcium, iron, and zinc in finger millet and kidney beans during processing. *J. Food Sci.,* 66:1271–5.

[27] Manach, C., Mazur, A. and Scalbert, A. 2005. Polyphenols and prevention of Cardiovascular diseases. *Current Opin Lipidol,* 16:77–84.

[28] Miller, G. 2001. Whole grain, fiber and antioxidants. In: Spiller GA, editor. Handbook of dietary fiber in human nutrition. *Boca Raton, FL: CRC Press. P.,* 453–60.

[29] Mohamed, E.A, Ahmed, I.A.M. and Babiker, E.E. 2011. Preservation of millet flour By refrigeration: changes in total protein and amino acids composition During storage. *World Acad Sci Eng Technol,* 52:11–4.

[30] Mohamed, E.A., Ahmed, I.A.M., Yagoub, A.E.A. and Babiker, E.E. (2010). Effects of Radiation process on total protein and amino acids composition of raw and Processed pearl millet flour during storage. *Intl J. Food Sci Technol,* 45:906–12.

[31] Mohamed, T.K., Issoufou, A. and Zhou, H. (2012). Antioxidant activity of Fractionated foxtail millet protein hydrolysate. *Intl Food Res J.,* 19(1):207–13.

[32] Mohamed, T.K., Zhu, K., Issoufou, A., Fatmata, T. and Zhou, H. 2009. Functionality, n vitro digestibility and physicochemical properties of two varieties of Defatted foxtail millet protein concentrates. *International J. Molecular Sci.,* 10:522–438.

[33] Monnier, V.M. (1990). Nonenzymatic glycosylation, the Maillard reaction and The aging process. *J. Gerontol,* 45:105–11.

[34] Mugocha, P.T., Taylor, J.R.N. and Bester, B.H. (2000). Fermentation of a compositeFinger millet-dairy beverage. *World J. Microbiol Biotechnol,* 16:341–4.

[35] Nantanga, K.K.M., Seetharaman, K., de Kock, H.L. and Taylor, J.R.N. (2008). Thermal Treatments to partially pre-cook and improve the shelf-life of whole pearl illet flour. *J. Sci. Food Agric.,* 88: 1892–9.

[36] Obilana, A.B. (2003). Overview: importance of millets in Africa. In: Belton PS, Taylor JRN, editors. Proceeding of the Workshop on the Proteins of Sorghum and Millets: Enhancing Nutritional and Functional Properties for Africa. 2–4 April 2003. Pretoria, South Africa. Available from http://www.afripro.org.uk. Accessed July, 2004.

[37] Oghbaei, M. and Prakash, J. (2012). Bioaccessible nutrients and bioactive Components from fortified products prepared using finger millet (Eleusine Coracana). *J. Sci. Food Agric,* 92(11), 2020.

Evaluating Nutritional Potential of the Millets and Oats

Gaurav Salaskar, Nihal Kamble, Kajal Sharma and Sharayu Raje

Department of Botany, Bhavan's College (Autonomous), Andheri (W), Mumbai, India

ABSTRACT: India has been producing millets in great quantities for the past few years. While the oats are cereal grain that are grown all over the world and are popularly known as healthy nutrition source. As compared to oats millets are richer in Copper, Manganese, Phosphorus, Magnesium, Vitamin A, Vitamin B1, Vitamin B5, and Vitamin B3. They are also rich source of fibres and proteins. Millets are particularly high in dietary fibre, antioxidant, phytochemical and polyphenols all of which have a broad range of positive effects on human health.

The cost price of millets is very less as compared to oats. Several regions of the world which are dry and arid are more suitable for the cultivation of millets as compared to oats. Millets can easily grow in locations where low nutrition is a concern. These characteristics make millets most probable candidate to replace oats in third world countries where malnutrition is taking a toll on human life.

Keywords: *Millets, Oats, Nutrition, Production, Environment.*

INTRODUCTION

The word 'millet' derived from old French word 'mille' which means 'Thousand' as well as the Greek and Latin words 'millium' and 'meline' according to several prehistoric assertions about its origin (21). Since ancient times, millets have been acknowledged as group of cereal grains like Finger millet *(Eleusine coracana)*, Foxtail Millet (*Setaria italica*), Pearl millet (*Chenchrus americanus*), etc. It is also one of the oldest known foods by human civilization because it has been used as food source for more than a thousand years. Similarly, oats *(Avena sativa)* are types of cereal grains. From nutritional standpoint. the nutritional value of 100 gm of Oats is 379 Calories, 6% Fat, 13% Protein, and 68% Carbohydrates, (17). Millets, on the other hand, have a nutritional value of 378 Calories per 100g 4% Fats, 11%

Protein, and 73% Carbohydrates (14). Facts from the global scenario show that there are 87 countries that produce millets and 83 countries producing oats. Following are the production values in 2020 from the leading millet producers: India (12490000 tonnes), Niger (3508903 tonnes), China (2300000 tonnes), Nigeria (2000000 tonnes) [FAOSTAT 2020]. Whereas, production values in 2020 for the chief oat producers are as follows: Canada (4575800 tonnes), Russia (4132096 tonnes), Poland (1627140 tonnes), and Spain (1377690 tonnes) [FAOSTAT 2020]. Millets can resist harsh environmental condition and they can be cultivated in regions with low nutrient concern (15). Unlike millets, oats are not drought-resistant crops (18). The main aim of this study is to compare the nutritional value of millets and oats. Based on this, we also provide an overview of their production, expenditure and environmental suitability.

MATERIAL AND METHODOLOGY

The research is based on secondary data collected from various public sources, government sites, PubMed and Google scholar search engines. The topic was chosen based on a literature review that sought to fill a knowledge gap about the nutritional value of millets compared to oats. The evaluation of any beneficial effects on human health was also the main goal of this review (6).

DISCUSSION

Nutritional Value

Compared to other major cereal grains, millets have similar or even better nutritional value also millets are a suitable healthy food because of their high fibre content, low glycaemic index, richness in bioactive compounds and they are gluten-free (9). Millets contain 60–70% carbohydrates, 1.5–5% fat, 7–11% proteins, 2–7% crude, 50 to 70% of starch and a large quantity of vitamins and minerals (1). Millets contain a significant amount of protein (7). Dietary fiber in Finger millets is more (11.5gm) than that of oats (10.6 gm). Similarly, Calcium content of Finger millets is 344 gm which is far greater than oats that contain 54 gm of calcium. Finger millets are rich in micronutrients like Molybdenum, Chromium, Sulphur and vitamins like Carotene, Folic acid and Choline, whereas these nutrients are present in minimal quantity in oats (8). The nutritional content of oats comprises of 60% starch (2), 11–15% Protein (11), 5–9% Lipids (4). They also contain the following trace minerals—Calcium 0.54%, Iron 0.047% and vitamins—Thiamine 0.002%, Riboflavin 0.001%, Niacin 0.032% (3).

Table 1

Sr. No.	*Nutrients*	*Millets Amount*	*Units*	*Oats*	*Units*
1.	Energy	119	Kcal	68	Kcal
2.	Protein	3.51	G	2.37	G
3.	Total Lipid (Fat)	1	G	1.36	G
4.	Ash	0.41	g	0.56	G
5.	Carbohydrates	23.7	g	11.7	G
6.	Total Dietary Fiber	1.3	g	1.7	G
7.	Calcium (Ca)	3	mg	80	Mg
8.	Iron (Fe)	0.63	mg	5.96	Mg
9.	Magnesium (Mg)	44	mg	26	Mg
10.	Phosphorus (P)	100	mg	77	Mg
11.	Potassium (K)	62	mg	61	Mg
12.	Sodium (Na)	2	mg	49	Mg
13.	Zinc (Zn)	0.91	mg	0.62	Mg
14.	Copper (Cu)	0.161	mg	0.066	Mg

(The nutritional value of millets and oats (amount is specified per 100 g)

Source: USDA's Food Central (19, 20).

The above (Table 1) demonstrates that this comparison is based on nutritional content by which one or other cereals crop grains is richer. Relative to oats, millets are higher in energy, protein, carbohydrates, and minerals like magnesium, phosphorous, potassium, zinc, and copper. Simila to millets, oats have higher levels of lipids, ash, dietary fibres, and minerals like calcium, iron, and sodium.

Production and Expenditure

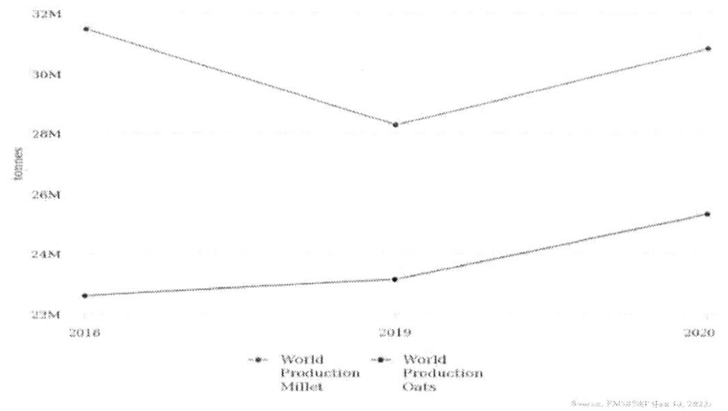

Figure 1: Global Trends in Production of Millets and Oats (FAOSTAT)

Figure 1 shows the global-level production of millets in comparison with oats from 2018 to 2020. The data indicates that the highest production was observed in millets (2018–31, 471, 221.16 Mt, 2019–28, 273, 641Mt, 2020–30, 825, 05.54 Mt) in comparison with oats (2018–22, 611, 221.16, 2019–23, 140, 981.44 Mt, 2020–25, 319, 509.06 Mt). (FA0STAT2018, 2019, 2020).

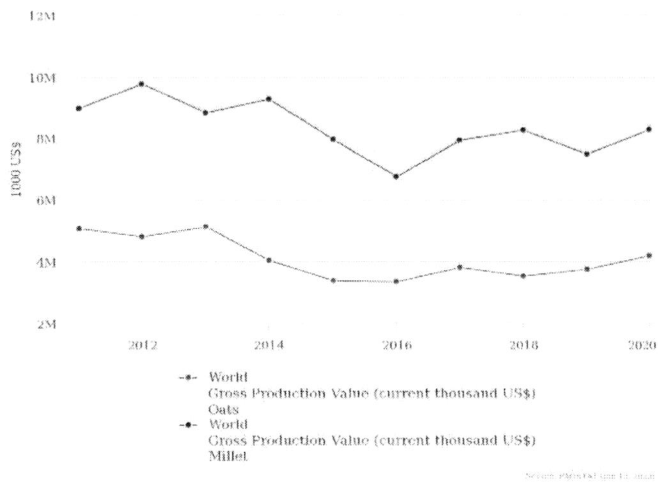

Figure 2: Global Trends in Production Value of Millets and Oats (FAOSTAT)

The time series data (Figure 2) shows the total world gross production for millets and oats in terms of (1000 US$) from year 2011 to 2020. The gross production of millets was as follows: 2011 (8,966,202 1000 US$), 2012 (9,754,604 1000 US$), 2013 (8,818,541 1000 US$), 2014 (9,260,939 1000 US$), 2015 (7,959,192 1000 US$), 2016 (6,733,145 1000 US$), 2017 (7,934,671 1000 US$), 2018 (8,258,322 1000 US$), 2019 (7,480,301 1000 US$), 2020 (8,284,271 1000 US$). Oats production values are as follows: 2011 (5,065,992 1000 US$), 2012 (4,802,541 1000 US$), 2013 (5,120,933 1000 US$), 2014 (4,033,236 1000 US$), 2015 (3,374,932 1000 US$), 2016 (3,341,476 1000 US$), 2017 (3,801,371 1000 US$), 2018 (3,520,833 1000 US$), 2019 (3,742,805 1000 US$), 2020 (4,190,579 1000 US$). The overall data indicates that the gross world production for millets has been more as compared to oats over the span of 2011–2020. (FA0STAT 2011, 2012, 2013, 2014, 2015, 2016, 2017, 2018, 2019, 2020)

The data of (Figure 3) shows the total world import value between millets and oats from year 2011 to 2020. The millets import values are as follows: 2011 (151,255 1000 US$), 2012 (146,513 1000 US$), 2013 (170,990 1000 US$),

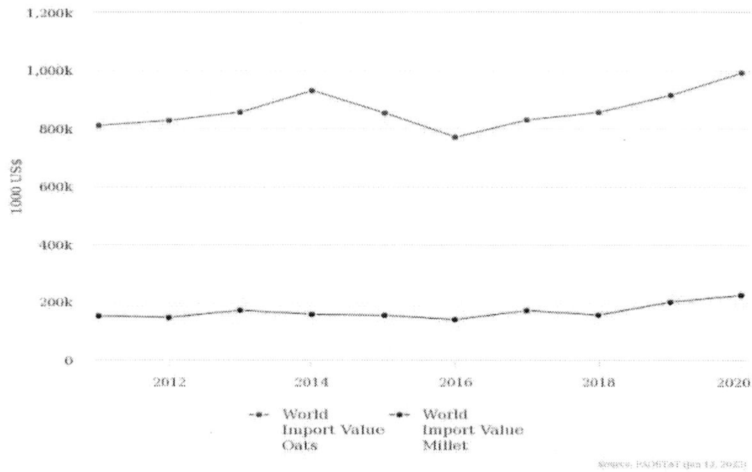

Figure 3: Global Trends in Import Value of Millets and Oats (FAOSTAT)

2014 (157,062 1000 US$), 2015 (153,569 1000 US$), 2016 (139,222 1000 US$), 2017 (170,818 1000 US$), 2018 (155,199 1000 US$), 2019 (200,417 1000 US$), 2020 (223,973 1000 US$).

Import values of oats are as follows: 2011 (809,724 1000 US$), 2012 (827,153 1000 US$), 2013 (855,784 1000 US$), 2014 (930,144 1000 US$), 2015 (853,284 1000 US$), 2016 (769,569 1000 US$), 2017 (829,754 1000 US$), 2018 (855,994 1000 US$), 2019 (914,337 1000 US$), 2020 (992,177 1000 US$). The data indicates that the import values of oats for the span of 2011 to 2020 is high compared to that of millets. (FA0STAT 2011, 2012, 2013, 2014, 2015, 2016, 2017, 2018, 2019, 2020)

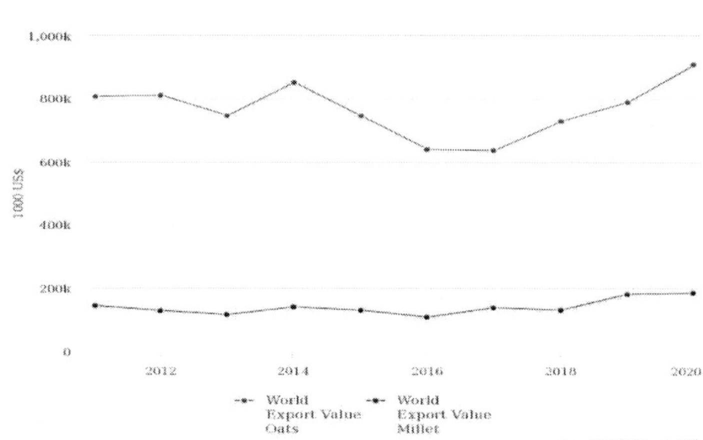

Figure 4: Global Trends in Export Value of Millets and Oats (FAOSTAT)

The data of (Figure 4) show the total world export value of millets and oats from year 2011 to 2020.

The millets export values are as follows: 2011(145,625 1000 US$), 2012 (129,609 1000 US$), 2013 (117,224 1000 US$), 2014 (141,494 1000 US$), 2015 (131,104 1000 US$), 2016 (109,812 1000 US$), 2017 (139,565 1000 US$), 2018 (131,985 1000 US$), 2019 (182,850 1000 US$), 2020 (786,906 1000 US$).

The oats export values are as follows: 2011 (807,039 1000 US$), 2012 (810,857 1000 US$), 2013 (746,441 1000 US$), 2014 (851,592 1000 US$), 2015 (745,812 1000 US$), 2016 (640,206 1000 US$), 2017 (637,108 1000 US$), 2018 (728,731 1000 US$), 2019 (789,205 1000 US$), 2020 (907,805 1000 US$).

As per this data, the export values of oats are high compared to that of millets for the span of 2011–2020. (FA0STAT 2011, 2012, 2013, 2014, 2015, 2016, 2017, 2018, 2019, 2020)

Environmental Suitability

Millets have ability to survive in drought-prone areas and can withstand higher temperatures (12). Millets thrive in shallow, low-fertile soil with PH values ranging from 4.5 to 8.0 and have a minimum water requirement (5). Millets are more environment friendly and can aid in reducing atmospheric carbon dioxide and phasing out climate uncertainties (9). Oats are typically planted in cool, damp regions since they are too delicate to grow in hot, dry conditions. Oats are sown in the autumn in warmer regions as they cannot sustain summer heat and drought (13). Compared to other small-grain crops, oats need more moisture for their growth (16). Thus, millets are a good option for the future of agriculture because of their resilience to environmental stress (10).

CONCLUSION

The conclusion drawn from the comparison of millets and oats is that millets are a strong competitor to oats in many areas, including nutritional content, production, expenditure, and environmental suitability. Although millets are rich in macronutrients and micronutrients, oats also offer the best nutritional value. From global production standpoint, millets are more productive cereal grain crop than oats. Production of oats is declining due to environmental impact on farming which reflects in their high import and export costs. As compared to oats, millets are resilient to harsh environmental conditions, making them suitable to grow in hot regions affected by draughts. All these properties make millets an effective alternative for nutrition requirements, especially in third world countries impacted by famine and malnutrition.

REFERENCES

[1] Anjali Eduru, Ankush Kamboj, P. Manisha Reddy, Basudev Pal, Vikas. Nutritional and health benefits of millets, present status and future prospects: A review. *Pharma Innovation,* 2021; 10(5): 859–868.

[2] Berski, W., Ptaszek, A., Ptaszek, P., Ziobro, R., Kowalski, G., Grzesik, M. and Achremowicz, B. Pasting and rheological properties of oat starch and its derivatives. *Carbohydr Polym,* 2011; 83: 665–671.

[3] Chavan, J.K. and Kadam, S.S. Nutritional improvement of cereals by fermentation. *Crit Rev Food Sci Nutr.,* 1989; 28: 349–400.

[4] Flander, L., Salmenkallio-Marttila, M., Suortti, T. and Autio, K. Optimization of ingredients and baking process for improved wholemeal oat bread quality. *LWT - Food Sci Technol.* 2007; 40:860–870.

[5] Gangaiah. Agronomy—Kharif Crops Finger Millet (2008). http://nsdl.niscair.res.in/jspui/bitstream/123456789/527/1/Millets%20(Sorghum,%20Pearl%20Millet,%20Finger%20Millet)%20-%20%20Formatted.pdf. Accessed 23 Mar 2018.

[6] Gowda, N.A.N., Siliveru, K., Prasad, P.V.V., Bhatt, Y., Netravati, B.P. and Gurikar, C. Modern Processing of Indian Millets: A Perspective on Changes in Nutritional Properties. Foods. 2022; 11(4): 499. https://doi.org/10.3390/foods11040499

[7] Hassan, Z.M., Sebola, N.A. and Mabelebele, M. The nutritional use of millet grain for food and feed: a review. *Agric & Food Secur,* 10, 16 (2021). https://doi.org/10.1186/s40066-020-00282-6

[8] Shaji, K.P. (2017). A Comparative Kap Study on Ragi, A Forgotten Wonder Grain with Oats among South Indian Families. *Current Research in Diabetes & Obesity Journal,* 3. 10.19080/CRDOJ.2017.03.555601

[9] Kumar, A., Tomer, V., Kaur, A. *et al.* Millets: A solution to agrarian and nutritional challenges. *Agric & Food Secur* 7, 31 (2018). https://doi.org/10.1186/s40066-018-0183-3

[10] Meena, Rajendra, Joshi, Dinesh, Bisht, Jaideep and Kan*t*, Lakshmi. (2021). Global Scenario of Millets Cultivation. 10.1007/978-981–16-0676-2_2.

[11] Robert, L.S., Nozzolillo, C. and Altosaar, I. Characterization of oat (Avena sativa L.) residual proteins. *Cereal Chem.* 1985; 62: 276–279.

[12] Satyavathi, C.T., Ambawat, S., Khandelwal, V. and Srivastava, R.K. (2021). Pearl Millet: A Climate-Resilient Nutricereal for Mitigating Hidden Hunger and Provide Nutritional Security. *Frontiers in Plant Science,* 12, 659938. https://doi.org/10.3389/fpls.2021.659938

[13] Stevens, E., Armstrong, K., Bezar, H., Griffin, W. and Hampton, John (2004). Fodder oats: an overview. Fodder Oats: A World Overview.

[14] The Nutrition Source, T.H. Chan School of Public Health, Harvard University. 2020. Retrieved 14 August 2020.

[15] Tiwari, Himanshu, Naresh, R.K, Kumar, Lalit, Kataria, S., Tewari, Sumit, Saini, Ankit, Kumar Yadav, Rakesh and Asati, Ruchi (2022). Millets for Food and Nutritional Security for Small and Marginal Farmers of North West India in the Context of Climate Change: A Review. *International Journal of Plant & Soil Science.* 10.9734/IJPSS/2022/v34i232594.

[16] Valenzuela, Hector & Smith, Jody. (2002). Common Oats (Green Manure Crops).

[17] https://healthjade.com/millet

[18] https://sarep.ucdavis.edu/covercrop/oat

[19] https://fdc.nal.usda.gov/fdc-app.html#/food-details/168871/nutrients

[20] https://fdc.nal.usda.gov/fdc-app.html#/food-details/171662/nutrients

[21] https://www.etymonline.com/word/millet

Feeding the Gender: A Study of Genderism in the Crop and Food Industry

Prachi Gupta

Ramnarain Ruia Autonomous College, Matunga, Mumbai, India

ABSTRACT: This paper studies the significance of representation of alternative sexual identities in crop and food industry. As a consequence of genderism, millets and other subsistence crops are gendered as "women's crops" while cash crops like cotton are gendered as men's crops. The dichotomous agrarian space ostracises the spectrum of identities and mainstreams agricultural activities for men and women.

From farm to fork, genderism is reflected in the commercial interests of the food industry. Advertisements employ gender biases and perpetuate stereotyped marketing by gendering food. By viewing this through the queer lens, the paper demonstrates importance of food accessibility and the mainstreaming agriculture for social inclusion and economic development of non-binaries.

The paper studies advertising patterns of commercials, marketing techniques and theories to illustrate genderism and explore how these binaries exist in our daily surroundings. It also intends to devise a theoretical tool for evaluating and neutralising such notions of heteronormativity.

Keywords: Genderism, Binarism, Millets, Heteronormativity.

INTRODUCTION

Genderism as defined in the SAGE Encyclopaedia for Trans Studies is the cultural belief or ideology that gender nonconformity or incongruence is negative. Genderism earlier was used as a gender-based alternative to sexism. However, the current use of the term refers to anti-trans-prejudice emerged in the 2000s.

The Food and Agriculture Organisation of United Nations highlights the traditional gender norms in the agrarian communities where cash and export crops are frequently gendered as "men's crops" and subsistence crops as "women's crops."

The International Crops Research Institute for the Semi-arid tropics conducted a household survey in 2016; where groundnuts and sunflower were seen as women's crops and cotton as men's crop. The paper intends to show how this persistent dualism of gender in agriculture makes farming a mainstream activity.

Feminisation to Queering

From edifying "Sons of the Soil" to making space for women in soil by reducing gender gap and discrimination, we are witnessing 'feminisation' of agriculture. From government's policies which didn't recognise women as farmers, to the present government which aims to promote millet production, train and equip women farmers for agro-processing and marketing; there has been a slow change. Through organisations like The Millets Sisters Network, female farmers are being economically empowered and sustaining the economy.

Whereas other research papers highlight the gendered patterns of cropping, food industry, re-orienting gender roles and problematic heteropatriarchal relations, this paper focuses on the exclusion of third-gender. Huli Timmapura, a small village in Karnataka, harbingers social inclusion with a group of transgender women sowing seeds and uprooting the genderism. They are cultivating vegetables like potatoes, corn and hyacinth beans for the last four years. Despite this, we have gendered terminologies which categorise bean as women's crop. Even the disparity in the division of farm labour only recognises men's and women's farm. The dichotomous agrarian space ostracises the alternative sexual identities and mainstreams agricultural activities for men and women. By 'queering' our approach, we can understand the ubiquity of heterosexual representation seen in terminologies, media, and literature.

Dichotomous Study and Stereotyping

From farm to fork, genderism is reflected in the commercial interests of the food industry. Silvia Benso says, "Food is culture, and everything—especially gender—is affected by culture." Advertisements employ gender biases and perpetuate stereotyped marketing by gendering food as masculine and feminine. For example, a yoghurt brand was endorsed using words like "powerful" with visuals of "bull" on the packaging. Kinder Joy ad also depicts two colours—pink for "girls" and blue for "boys". The restaurants and cafes also utilize interiors that are gender targeted. Not surprisingly, the transgender is saved from this stereotyping because the heterosexual matrix knows only the duality of men and women to stereotype.

As noted in the journal of Gender and Development, food availability does not ensure equality in access to food and the distribution of food and nutrition depends on the traditions and norms set by communities. Carole Counihan in her book, *Food and Gender: Identity and Power,* examines the significance of food-centred activities to gender relations, how food production, consumption and distribution contribute to men's and women's power in a social structure. Here, while this literature explores how food symbolically connotes maleness and femaleness there is no space for interaction between the two genders delineating rigid heterosexual boundaries and failing to recognise the social value of alternative sexual identities in the food hierarchy.

It restricts the possibility of shifting identities and nonconformity to conventional gender expressions as John Gosling notes in his paper: Gender fluidity reflected in contemporary society. Judith Butler in his essay, Performative Acts and Gender Constitution: An Essay in Phenomenology and Feminist Theory, points out that cultivation of bodies into discrete sexes with 'natural' appearances heterosexual dispositions is a way by which the system of compulsory heterosexuality is reproduced and concealed.

Recognising Non-binaries

The conventional gender and sexual relations are so prevalent that they are neither noticed nor questioned. Acknowledging the significance and centrality of alternative sexualities in food and agriculture is a pre requisite for changing the focus of gender studies and reconstructing the cisgenderism in crop and food industry.

Existing papers on gender differences in food choices study the consumption preference of male and female consumer groups, nutritional attitudes, gender inequality and food insecurity in which the relation of gender in nutrition largely revolve around men and women. There are no studies and research on queer and transgender farmers. There is a lack, rather absence, of any qualitative data on queer people. Ensuring food justice is not possible without understanding the specific dietary needs and food preferences of the queer. We can feed the gender only if we heed the gender in our studies, policies, commercials, and reform the gendered notions and constructs.

Every aspect of food industry; production, consumption, distribution patterns and its marketing depict binary representation. The aim of this paper is to highlight this gap and suggest neutralising the genderism in the crop and food industry. In order to dismantle the binary representation of male and female, the first tool is to integrate the spectrum in literature while studying the sociological, economic, and cultural structures of the society. The experiences of rural and urban queer in accessing resources, queer farmers, resistance movements and the discrimination and oppression faced by them needs to be documented.

MATERIALS AND METHODS

Universe and Sample

The sample of this study constitutes a total of 42 respondents from the Mumbai suburban region. The survey was done among people in the age group of 16 to 60 years of age. It consisted mostly students, working men and home-makers among others.

For this study, qualitative and quantitative research methods were used. The research is exploratory and analytical in nature as it studied literature, surveys and data to examine the genderism in crop and food industry. To illustrate the exclusion of the broad spectrum of identities, a study of advertising patterns, marketing and evaluation of gendered terminologies was done.

For the survey, two commercials—Amul India and PepsiCo were shown and questions around inclusivity, gender roles, discrimination and stereotyping were asked. Both commercials depicted the cultural and normative gender binary in farming and household setting.

RESULTS AND DISCUSSIONS

The commercial by Amul showed working woman and a man helping his wife in kitchen against the conventional gender expectations. There was gender role inversion. However, it was not gender inclusive.

Watch this ad by Amul India. Do you think it is inclusive and recognises various identities in our society?

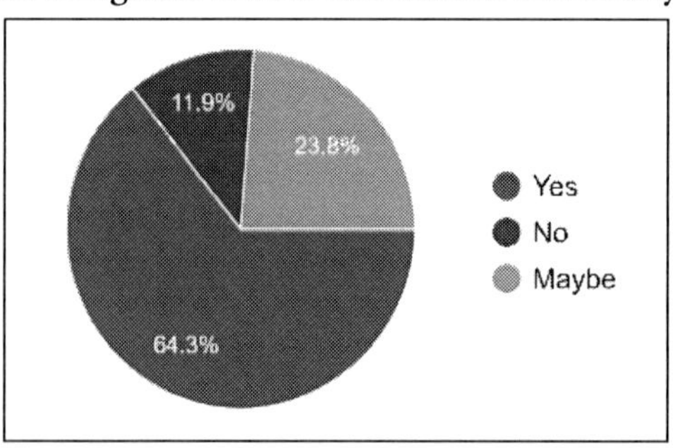

Does this ad depict stereotypical gender roles and discrimination?

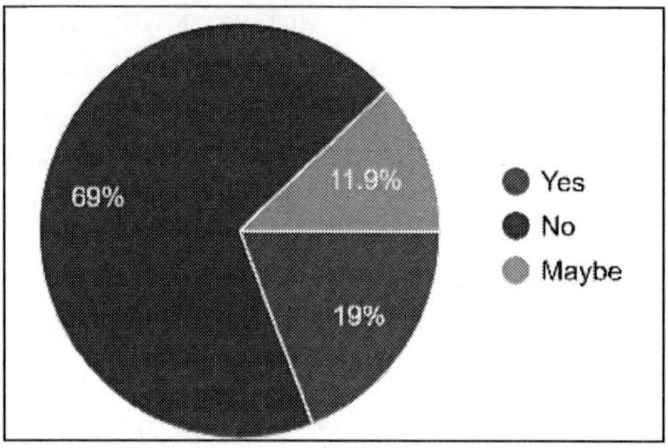

64% of the respondents believed it to be an inclusive one as it did not portray gender inequalities. This implies that heteronormativity is deep-rooted such that the concepts of gender disparity and gender equality revolve around men and women for most people.

What was the division of labour as reflected in the above ad according to you?

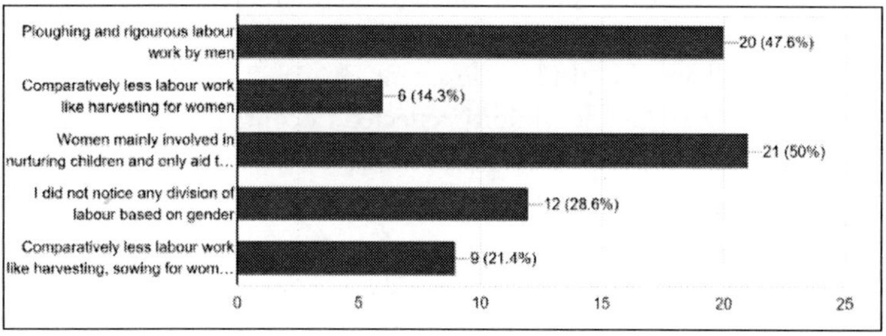

The PepsiCo commercial showed a clear division of farm labour based on traditional gender roles. It is interesting to note that 28% of the respondents did not notice any gender-based division. This reflects the firm establishment of gender roles of men and women. While delineating these disparities, the dualism of men and women is seen.

CONCLUSION

The exclusion of alternative sexual identities in the food and crop industry leads to problems of recognition, funding, encouragement, and support to them in the sector. Their representation is important to understand the specific challenges that affect the transgender and other non-binaries in farming and agricultural activities. The lack of cognisance seen in advertisements and commercials reflects the indifference of society at large to this genderism.

The government policies should be gender sensitive, promote agri-entrepreneurship for the transgender and work towards inclusive farming. Food gap can be eliminated and justice can be done by understanding these hierarchies. Gender dynamics on the farm, food production and the entire food chain calls for an awareness of its impact on queer, making sexuality and gender studies central to every aspect of the crop and food industry.

The sexual and gender minorities in the agrarian space highlights the perpetuating heteropatriarchal roots in the farm and food industry. We need a queer lens and understanding of the food production, distribution and consumption systems in order to achieve sustainability and growth.

REFERENCES

[1] Butler, J. (1998). Performative Acts and Gender Constitution: An Essay in Phenomenology and Feminist Theory Theatre Journal, Vol. 40, No. 4.

[2] Brandth, B. (1995). Rural masculinity in transition: Gender images in tractor advertisements. *Journal of Rural Studies.*

[3] Counihan, C. (1998). Food and Gender: Identity and Power.

[4] Doss, C.R. (2002). Men's Crops? Women's Crops? The Gender Patterns of Cropping in Ghana. World Development, 30, 1987–2000.

[5] Gosling, J. (2018). Gender fluidity reflected in contemporary society.

[6] The SAGE Encyclopaedia for Trans Studies.

[7] The Routledge Dictionary of Literary Terms.

[8] Smyth, J.D., A. Swendener, and E. Kazyak. 2018. Women's work? The relationship between farmwork and gender self-perception. *Rural Sociology.*

[9] Whatmore, S. (1991). Life cycle or patriarchy? Gender divisions in family farming. *Journal of Rural Studies 7.*

[10] Lekshmi, P. (2019). Trans People in This Karnataka Village Are Farming Their Way Out of Exploitation. https://www.thebetterindia.com/180036/karnataka-farming-organic-inspiring-transgender-discrimination-india

[11] PepsiCo India supports Indian farmers through agricultural training, YouTube video, 00.50. PepsiCo India, September 11, 2019. https://youtu.be/IVccuDdlqLs

[12] Amul Taste of India-Gentleman, YouTube Video, 1.00. Amul TV, September 1, 2022. https://youtu.be/QOwPFzH0kD0

[13] Amul Taste of India-Woman, YouTube Video, 1.00. Amul TV, August 30, 2022. https://youtu.be/AXjlkLcqJC8

Nutritional Evaluation and Potency of Millets in Curing Arrhythmias in *Daphnia Magna*

Ashwathi Prakash Ambala*, Sneha Salgaonkar and Prashant Masali

*Student, Department of Biochemistry, Ramnarain Ruia Autonomous College, Mumbai University, Mumbai, India

ABSTRACT: Millets have been widely used in the world as a high value nutritional food. In India, there are a lot of varieties of millets but out of the many, the most common are Jowar (*Sorghum bicolor*), Bajra (*Pennisetum glaucum*) and Ragi (*Eluisine coracana*). Millets are easy to grow and require less care. They are high in fibers, nutrients, and minerals. Added to that, millets are rich in carbohydrates and proteins which make them a necessary part of the diet. Due to the increasing challenges of dry lands and deepening water levels, there is a low possibility of increasing the production of major staple cereals. Cultivation of millets can solve this problem. However, the use of millets as a medicine is not researched upon a lot even though it can have a potential in curing certain diseases. Nowadays, heart diseases are the most common among today's generation and therefore the research on millets in the field of medicine needs to be studied. This project gives a nutritional check on the three millets and determines their toxicity as well as their ability to cure tachycardia and bradycardia using *Daphnia magna* as a model organism. Three varieties of millets and their mix are considered, where they are analyzed for their nutritional value according to the standard methods. The toxic effects of millets is studied using *Daphnia magna*. The millets are found to be non-toxic till a particular period of time. The millet extracts are seen to have a change in the effect of tachycardia and bradycardia induced in these organisms.

Keywords: *Millets, Nutrition, Daphnia Magna, Tachycardia, Bradycardia.*

INTRODUCTION

Millets along with maize and sorghum, are part of the Poaceae family of plants. They occur naturally in arid areas and semi-arid areas such as central Africa and Asia. The rainfall requirement of certain millets like pearl millet and proso millet (Panicum miliaceum) is as low as 20 cm, which is several fold lower than the rice, which requires an average rainfall of

120–140 cm. After sowing, most of the millets mature in 60 to 90 days which saves a lot of water. Important millet crops grown in India are Sorghum (Great millet), Bajra (Pearl millet), Ragi, Korra, Little, farmyard and Kodo (Selvamani, 2019). The most abundant varieties of millets are pearl millet (*Pennisetum glaucum*), which make up to approximately 40% of the worldwide production, followed by foxtail millet (*Setaria italica*), proso (or white) millet (*Panicum miliaceum*), and finger millet (*Eleusine coracana*) (Sharma, & Niranjan, 2018). Gupta *et al.* (2017) concluded that finger millet is a solution to food security and hidden hunger and therefore termed it as the "certain" crop for an "uncertain" future.

Millets are of high nutritional value (Rao *et al.*, 2017; Chauhan *et al.*, 2018; Nithiyanantham *et al.*, 2019). Besides nutrients, millets are considered a 'super food' with phytochemicals as an impressive range of secondary metabolites (Niharika *et al.*, 2020). Rich in proteins, carbohydrates, vitamins, and minerals, they also contain a good amount of antinutrients Amadou *et al.* (2014), Muthamilarasan *et al.,* (2016). Pragya, *et al.* (2012) researched on the use of chemical composition and the nutritional value of finger millet and the various methods of millet processing to increase the bioavailability of certain minerals.

World is in the clinch of several health disorders and chronic diseases. This can be due to a nutrient imbalanced diet. The urban lifestyle choices and related food habits have given rise to a number of diseases such as diabetes, obesity and cardiovascular problems such as heart attack, coronary artery disease, arrhythmias etc. Kim *et al.* (2016) conducted a meta-analysis of prospective cohort studies and concluded that there might be a relationship between the dietary fibre intake and mortality from cardiovascular disease and all cancers. Bazzano *et al.* (2003) and Threapleton *et al.* (2013) concluded from their systematic review that dietary fibre intake and risk of cardiovascular disease are interrelated. Crowe *et al.* (2012) published their work where they concluded that a higher consumption of dietary fibre is associated with a lower risk of fatal IHD with no clear difference in the association with Ischemic Heart Disease (IHD) for fibre from cereals, fruits or vegetables. Research on diets high in fibre—specifically from cereal or vegetable sources conveyed that they are significantly associated with lower risk of Congenital Heart Defects (CHD) and Cardiovascular Disease (CVD). (Khaw *et al.*, 1987; Streppel *et al.*, 2008; Wu *et al.*, 2015) published their findings which supported the hypothesis that high dietary fiber intake is protective for ischemic heart disease mortality.

The toxicity to millet extracts is studied using the model organism *Daphnia magna* (Baylor 1942; Buikema *et al.*, 1980; Campbell *et al.*, 2004; Persoone *et al.*, 2009; Bownik 2020). Many external agents have been used to check their effects on the heart rate of *Daphnia magna.* Kaas *et al.* (2009) explained the effects of melatonin and ethanol on the heart rate of *Daphnia magna.* Studies were conducted in the same manner with ethanol, nicotine and caffeine on *Daphnia magna* (Corotto *et al.*, 2010; Cloar *et al.*, 2018).

In this study, sucrose and ethanol are used to induce tachycardia and bradycardia respectively in the organism and the potency of millet extracts in reversing the effects is studied across various defined time intervals.

MATERIALS AND METHODS

Sample: The millets considered in this study were of three major types:

1. *Sorghum bicolor*—Sorghum or Jowar
2. *Pennisetum glaucum*—Pearl millet or Bajra
3. *Eleusine coracana*— Finger millet or Ragi.

Sorghum bicolor

Pennisetum glaucum

Eleusine coracana

METHODS

Procurement of the Samples

All the 3 samples were procured from a local shop in Mulund, Maharashtra. They were cleaned, dried, powdered and sieved thoroughly. The prepared sample was stored in Ziplock covers for further use.

Preparation of Millet Extracts

Acid water extracts were prepared—1 gm powder dispersed in 20 ml acidified water, it was made to stand for 24 hrs, filtered and filtrate was used for further analysis.

Proximate Analysis

The carbohydrate, protein, minerals, ash and moisture content was analyzed according to standard AOAC procedures. The tests were carried out in triplicates.

Phytochemical Analysis

The samples were analyzed for their phytochemical constituents such as flavonoids, tannins, alkaloids, phenols, etc.

Testing for Toxicity and Arrhythmic Reversal of Millets on Daphnia magna:

Test Animals and Test Setup

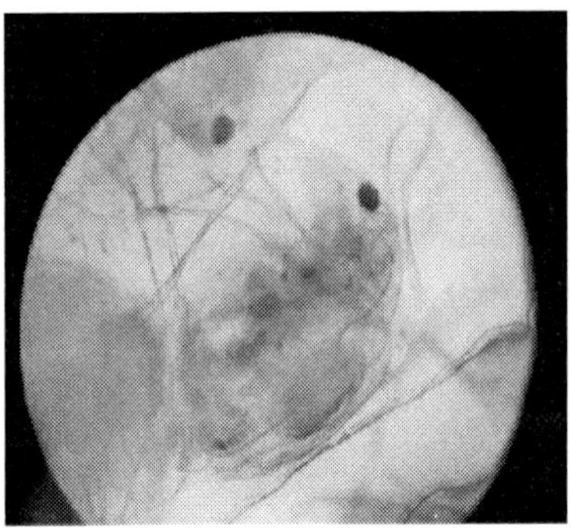

Figure 1

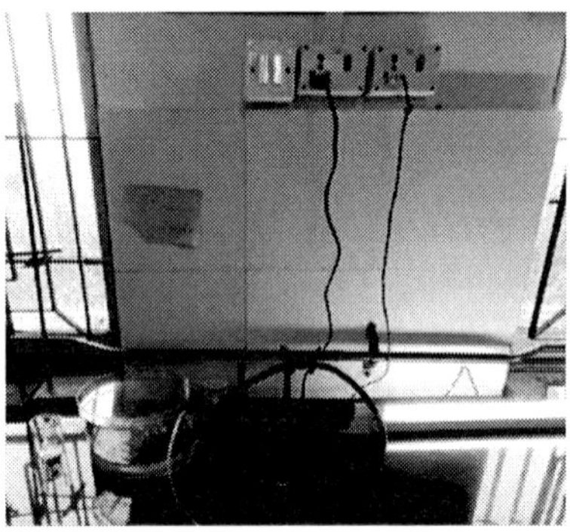

Figure 2

- *Daphnia* was procured from an aquarium in Mumbai. They were maintained under laboratory conditions in the medium in which they were shipped. Natural light was provided and the culture was kept undisturbed, aerated and at room temperature.
- The tests were conducted only after the *Daphnia* were acclimatized to the laboratory conditions (approximately 24 hours).

- The culture was kept in a tank which was aerated and the tank consisted of around 500–800 ml of the culture.
- The cultures lasted around 4 to 5 days.

METHOD

On a slide with a small amount of cotton mesh, two drops of the culture were added. The normal heart beat of the organism was noted. Tachycardic conditions were induced using 2–3 drops of sucrose solution and the change in the heartbeats were noted at defined time intervals. Once the organism showed a regular tachycardic pattern, each millet extract (2–3 drops) was added and the difference in the heartbeat of the organism was recorded. In order to ensure the precision in the values obtained, the test was conducted in triplicates. The same procedure was done for bradycardia.

RESULTS AND DISCUSSION

Qualitative Analysis

Test	*Jowar*	*Bajra*	*Ragi*	*Mix (1:1:1)*
Carbohydrates	✓	✓	✓	✓
Proteins and Amino acids	✓	✓	✓	✓
Starch	✓	✓	✓	✓
Reducing sugars	×	×	×	×
Flavonoids	×	×	×	×
Resins	✓	✓	✓	✓
Quinones	×	×	×	×
Phenols	✓	✓	✓	✓
Alkaloids	✓	✓	✓	✓
Tannins	✓	×	×	✓
Saponins	✓	✓	✓	✓
Glycosides	×	×	×	×
Terpenoids	✓	×	×	✓

Key: ✓: Present ×: Absent

With respect to the nutritional basis, millets showed a good amount of proteins, carbohydrates and minerals such as calcium and phosphorus. In addition to that, they are also a good source of phenols and other phyto-chemicals.

Quantitative Analysis

Test	*Jowar*	*Bajra*	*Ragi*	*Mixture*
Moisture	7.1%	3.2%	10.35%	10.3%
Ash	4.68%	1.13%	1.65%	1.29%
Carbohydrates	80%	60%	20%	60%
Proteins	7%	10%	8%	11%
Calcium	300.6 mg	300.6 mg	460.9 mg	360.7 mg
Phosphorous	0.085 mg	0.06 mg	0.068 mg	0.063 mg
Iron	0.025 mg	0.045 mg	0.02 mg	0.05 mg
Phenols	2.2 mg	1.3 mg	2.8 mg	3.0 mg

- Among the millet varieties, Jowar was found to contain maximum amount of ash content (4.68%), carbohydrates (80%) and phosphorus (0.085%). Bajra contained a good amount of proteins (10%), carbohydrates (60%) and iron (0.045 mg). Ragi contains the most amount of moisture (10.35%) and calcium (460.9%). The mixture of the three millets was found to contain a good amount of moisture (10.3%), proteins (11%), iron (0.05%) and phenols (3 mg).

Effect of Toxicity in Millets

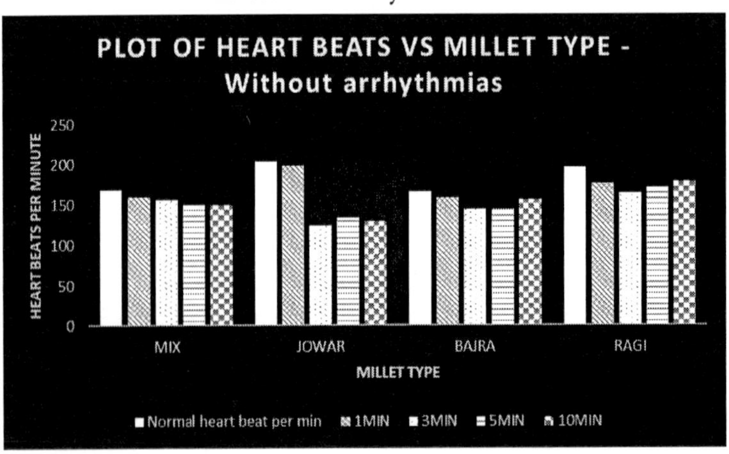

Tap Water Extract of Millets and Sucrose Used to Induce Tachycardia

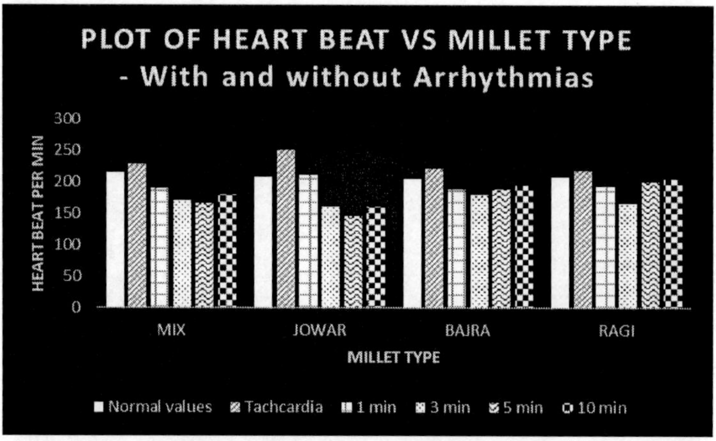

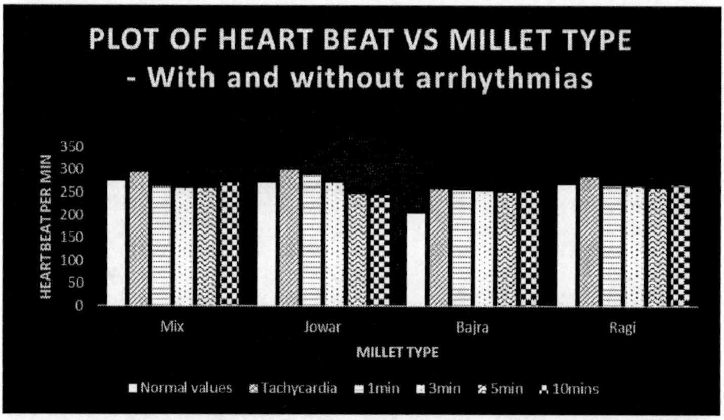

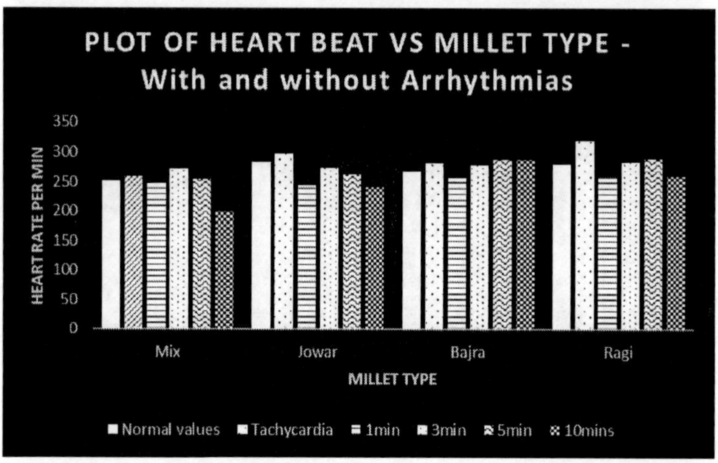

Tap Water Extract of Millets and Ethanol Used to Induce Bradycardia

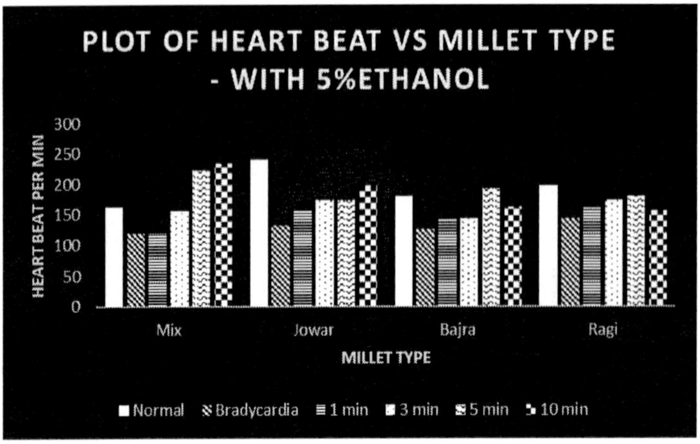

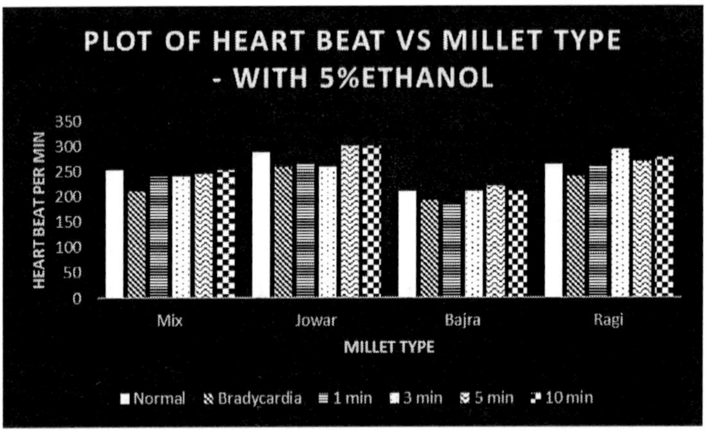

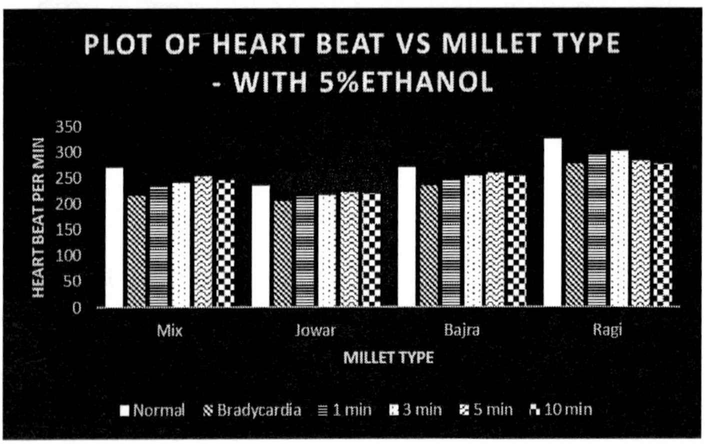

- Millets do not show toxicity at low levels.
- There were some changes observed in the heartbeat of *Daphnia* when millet extracts were used.
- With respect to tachycardia, the millet extracts were able to bring down the heart beat towards the normal range. The time required for the reversal was also seen to be minimum. Among the millet extracts, jowar and ragi showed better results.
- For bradycardia, the extracts showed an increase to the normal levels of heartbeat within some minutes. Among the millet extracts, mixture and bajra showed better results.
- This change in the heartbeat can be due to the presence of phytochemicals or maybe due to the amount of various minerals such as calcium which is said to be associated with cardiac properties.
- Daphnia, an insect used as the model organism, shows a change in the heartbeat easily. These studies can be further extended to higher animals which can give a better idea regarding the potency of millets in curing tachycardia.

CONCLUSION

From the above findings, we can conclude that millets are an important source of nutrition. They are high in essential nutrients and phytochemicals which need to be included as a part of the diet. Millets also help in reversing the heartbeat in *Daphnia magna* and hence help in normalizing the values to certain extent.

ACKNOWLEDGEMENTS

I would like to express my heartfelt gratitude to Dr. Anushree Lokur-Principal, Ramnarain Ruia Autonomous College, Dr. Kamini Donde, Research Dean. I would also like to extend my gratitude to all my teachers in the Department for their significant guidance. I would also like to thank Mr. Mali—Support staff, Department of Zoology, Ramnarain Ruia Autonomous College for his support and help in procuring the samples. Special thanks to the support staff of the Biochemistry Department, Ramnarain Ruia Autonomous College for their support throughout the entire project. Finally I would like to acknowledge my Family and friends for their continuous support and encouragement in pursuing this project.

REFERENCES

[1] Amadou, I., Gounga, M.E. and Le, G.W. (2013). Millets: Nutritional composition, some health benefits and processing-A review. *Emirates Journal of Food and Agriculture*, 501–508.

[2] Amadou, I., Gounga, M.E., Shi, Y.H. and Le, G.W. (2014). Fermentation and heat-moisture treatment induced changes on the physicochemical properties of foxtail millet (Setaria italica) flour. *Food and Bioproducts Processing*, 92(1), 38–45.

[3] Auwal, M.S., Saka, S., Mairiga, I.A., Sanda, K.A., Shuaibu, A. and Ibrahim, A. (2014). Preliminary phytochemical and elemental analysis of aqueous and fractionated pod extracts of Acacia nilotica (Thorn mimosa). *In Veterinary research forum: an international quarterly journal* (Vol. 5, No. 2, p. 95). Faculty of Veterinary Medicine, Urmia University, Urmia, Iran.

[4] Baylor, E.R. (1942). Cardiac pharmacology of the cladoceran, Daphnia. *The Biological Bulletin,* 83(2), 165–172.

[5] Bazzano, L.A., He, J., Ogden, L.G., Loria, C M. and Whelton, P.K. (2003). Dietary fiber intake and reduced risk of coronary heart disease in US men and women: the National Health and Nutrition Examination Survey I Epidemiologic Follow-up Study. *Archives of internal medicine,* 163(16), 1897–1904.

[6] Bownik, A. (2020). Physiological endpoints in daphnid acute toxicity tests. *Science of the Total Environment,* 700, 134400.

[7] Buikema, A.L., Geiger, J.G. and Lee, U D. (1980). Daphnia toxicity tests (pp. 46–69). *ASTM International.*

[8] Campbell, A.K., Wann, K.T. and Matthews, S. B. (2004). Lactose causes heart arrhythmia in the water flea Daphnia pulex. *Comparative Biochemistry and Physiology Part B: Biochemistry and Molecular Biology*, 139(2), 225–234.

[9] Ch, H., Patro, T.S.S.K., Anuradha, N., Triveni, U. and Jogarao, P. (2020). Estimation of nutritive composition of seven small millets. *Journal of Pharmacognosy and Phytochemistry*, 9(3), 1871–1874.

[10] Chauhan, M., Sonawane, S.K. and Arya, S.S. (2018). Nutritional and nutraceutical properties of millets: A review. *Clinical Journal of Nutrition and Dietetics*, 1(1), 1–10.

[11] Cloar, G., Dinh, T., Hoffen, A., Humphrey, J. and Vesy, M. (2018). Ethanol decreases heart rate of Daphnia compared to water. *Journal of Undergraduate Biology Laboratory Investigations,* 1(2).

[12] Corotto, F., Ceballos, D., Lee, A. and Vinson, L. (2010). Making the most of the Daphnia heart rate lab: optimizing the use of ethanol, nicotine & caffeine. *The American Biology Teacher*, 72(3), 176–179.

[13] Crowe, F.L., Key, T.J., Appleby, P. N., Overvad, K., Schmidt, E.B., Egeberg, R., ... & Riboli, E. (2012). Dietary fibre intake and ischaemic heart disease mortality: the European Prospective Investigation into Cancer and Nutrition-Heart study. *European journal of clinical nutrition*, 66(8), 950–956.

[14] Dayakar Rao, B., Bhaskarachary, K., Arlene Christina, G. D., Sudha Devi, G., Vilas, A. T. and Tonapi, A. (2017). Nutritional and health benefits of millets. *ICAR_Indian Institute of Millets Research (IIMR): Hyderabad, Indian*, 112.

[15] Devi, P. B., Vijayabharathi, R., Sathyabama, S., Malleshi, N. G. and Priyadarisini, V. B. (2014). Health benefits of finger millet (Eleusine coracana L.) polyphenols and dietary fiber: a review. *Journal of food science and technology,* 51(6), 1021–1040.

[16] Ebert, D. (2005). Ecology, epidemiology, and evolution of parasitism in Daphnia. *National Library of Medicine.*

[17] Gaikwad, N.S., Panat, A.V., Deshpande, M.S., Ramya, K., Khalid, P.U. and Augustine, P. (2012). Effect of shilajit on the heart of Daphnia: A preliminary study. *Journal of Ayurveda and integrative medicine.*

[18] Ganguly, S. (2015). Effect of bajra flour (Pearl millet) on some quality and sensory attributes of chicken nuggets. *Asian Journal of Animal Science,* 10(2), 107–114.

[19] Gupta, S.M., Arora, S., Mirza, N., Pande, A., Lata, C., Puranik, S., ... and Kumar, A. (2017). Finger millet: a "certain" crop for an "uncertain" future and a solution to food insecurity and hidden hunger under stressful environments. Frontiers in plant science, 8, 643.

[20] Kaas, B., Krishnarao, K., Marion, E., Stuckey, L. and Kohn, R. (2009). Effects of melatonin and ethanol on the heart rate of Daphnia magna. Impulse: the premier journal for undergraduate publications in the neurosciences, 1.

[21] Khaw, K.T. and Barrett-Connor, E. (1987). Dietary fiber and reduced iscremic heart disease mortality rates it men and women: A 12-year prospective study. *American Journal of Epidemiology,* 126(6), 1093–1102.

[22] Kim, Y. and Je, Y. (2016). Dietary fibre intake and mortality from cardiovascular disease and all cancers: A meta-analysis of prospective cohort studies. *Archives of cardiovascular diseases,* 109(1), 39–54.

[23] Kumar, A., Tomer, V., Kaur, A., Kumar, V. and Gupta, K. (2018). Millets: a solution to agrarian and nutritional challenges. *Agriculture & food security,* 7(1), 1–15.

[24] Kumar, A., Tripathi, M.K., Joshi, D. and Kumar, V. (Eds.). (2021). Millets and Millet Technology. Springer Singapore.

[25] Lewis, P.A. and Weber, C.I. (1985). A study of the reliability of Daphnia acute toxicity tests (pp. 73–86). *ASTM International.*

[26] Malik, S. (2015). Pearl millet-nutritional value and medicinal uses. *International Journal of Advance Research and Innovative Ideas in Education*, 1(3), 414–418.

[27] Muthamilarasan, M., Dhaka, A., Yadav, R. and Prasad, M. (2016). Exploration of millet models for developing nutrient rich graminaceous crops. *Plant Science,* 242, 89–97.

[28] Niharika, B., Jaipuriar, D.S., Ranjan, R. and Vaishnav, V. (2020). *A Study to Explore the Biochemical Properties of Locally Grown Millets.*

[29] Nithiyanantham, S., Kalaiselvi, P., Mahomoodally, M.F., Zengin, G., Abirami, A. and Srinivasan, G. (2019). Nutritional and functional roles of millets—A review. *Journal of food biochemistry*, 43(7), e12859.

[30] Paul, R.J., Colmorgen, M., Pirow, R., Chen, Y.H. and Tsai, M.C. (1998). Systemic and metabolic responses in Daphnia magna to anoxia. *Comparative Biochemistry and Physiology Part A: Molecular & Integrative Physiology*, 120(3), 519–530.

[31] Persoone, G., Baudo, R., Cotman, M., Blaise, C., Thompson, K.C., Moreira-Santos, M., ... and Han, T. (2009). Review on the acute Daphnia magna toxicity test–Evaluation of the sensitivity and the precision of assays performed with organisms from laboratory cultures or hatched from dormant eggs. *Knowledge and Management of Aquatic Ecosystems*, (393), 01.

[32] Pragya, S. and Rita, S.R. (2012). Finger millet for food and nutritional security. *African Journal of Food Science*, 6(4), 77–84.

[33] Ramar, P. (2018). Investigate on the leaves of Cinchona officinalis L—it's distinct in vivo reversal of tachycardia induced by synthetic cathinones on microcrustacean Daphnia magna arrhythmic heart.

[34] Rao, B.N. (2003). Bioactive phytochemicals in Indian foods and their potential in health promotion and disease prevention. *Asia Pacific Journal of clinical nutrition*, 12(1).

[35] Rao, P.P., Birthal, P.S., Reddy, B.V., Rai, K.N. and Ramesh, S. (2006). Diagnostics of sorghum and pearl millet grains-based nutrition in India. *International sorghum and millets newsletter*, 47, 93–96.

[36] Saleh, A.S., Zhang, Q., Chen, J. and Shen, Q. (2013). Millet grains: nutritional quality, processing, and potential health benefits. *Comprehensive reviews in food science and food safety*, 12(3), 281–295.

[37] Selvamani, R. (2019). The Role of Millet Production and Processing in India. *Think India Journal*, 22(10), 3486–3492.

[38] Sharma, N. and Niranjan, K. (2018). Foxtail millet: Properties, processing, health benefits, and uses. *Food reviews international*, 34(4), 329–363.

[39] Shobana, S., Krishnaswamy, K., Sudha, V., Malleshi, N.G., Anjana, R.M., Palaniappan, L. and Mohan, V. (2013). Finger millet (Ragi, Eleusine coracana L.): a review of its nutritional properties, processing, and plausible health benefits. *Advances in food and nutrition research*, 69, 1–39.

[40] Sholichah, A.R. (2017, November). Phytochemical screening and antioxidant activity of ethanolic extract and ethyl acetate fraction from basil leaf (Ocimum basilicum L.) by DPPH radical scavenging method. *In IOP Conference Series: Materials Science and Engineering* (Vol. 259, No. 1, p. 012008). IOP Publishing.

[41] Siong, T.E., Choo, K.S. and Shahid, S.M. (1989). Determination of calcium in foods by the atomic absorption spectrophotometric and titrimetric methods. *Pertanika*, 12(3), 303–311.

[42] Spicer, J.I. (2001). Development of cardiac function in crustaceans: patterns and processes. *American zoologist*, 41(5), 1068–1077.

[43] Streppel, M.T., Ocké, M.C., Boshuizen, H.C., Kok, F.J. and Kromhout, D. (2008). Dietary fiber intake in relation to coronary heart disease and all-cause mortality over 40 y: The Zutphen Study. *The American journal of clinical nutrition*, 88(4), 1119–1125.

[44] Threapleton, D.E., Greenwood, D.C., Evans, C.E., Cleghorn, C.L., Nykjaer, C., Woodhead, C., ... and Burley, V.J. (2013). Dietary fibre intake and risk of cardiovascular disease: systematic review and meta-analysis. *Bmj,* 347.

[45] Truswell, A.S. (2002). Cereal grains and coronary heart disease. *European journal of clinical nutrition*, 56(1), 1–14.

[46] Wei, S., Cheng, D., Yu, H., Wang, X., Song, S. and Wang, C. (2018). Millet-enriched diets attenuate high salt-induced hypertension and myocardial damage in male rats. *Journal of Functional Foods,* 44, 304–312.

[47] Whiteoak, A.M. and Penson, P.E. (2017). 220 Daphnia magna as a model for quantifying chaos in cardiac arrhythmia. *Heart*, 103(Suppl 5), A143–A143.

[48] Wilson, J., Collette, J., Miller, A., Hoang, A. and Dubose, T. (2019). The opposite effects of caffeine and ethanol on Daphnia magna's heart rate. *Journal of Undergraduate Biology Laboratory Investigations,* 2(2).

[49] Wu, Y., Qian, Y., Pan, Y., Li, P., Yang, J., Ye, X. and Xu, G. (2015). Association between dietary fiber intake and risk of coronary heart disease: A meta-analysis. *Clinical nutrition*, 34(4), 603–611.

Anti-Nutritional Factors in Kodo Millet and Comparison of their Reduction Strategies

Amita Dhadphale[1], Manali Nimse[2], Akshay Bhogle[3], Saira Mulla[2], Pradnya Prabhu[3] and Kamini Donde[1*]

[1]Department of Chemistry, Ramnarain Ruia Autonomous College of Arts and Science, Mumbai, India

[2]Department of Chemistry, N.G. Aacharya D.K. Marathe College of Arts, Science and Commerce, Mumbai, India

[3]Department of Chemistry, K.J. Somaiya College of Science and Commerce, Mumbai, India

ABSTRACT: For centuries, varieties of Kodo millet have been dietary staple for humans in India, Vietnam, Philippines, Thailand, Indonesia, Pakistan and West Africa. It produces good yields in almost any type of soil and is resistant to drought stress. The nutrient composition of Kodo millet i.e., total protein, amino acid composition, minerals and vitamin compositions etc. has been widely studied and has proven comparable, or even better with other millets. Despite of nutritional qualities, certain anti-nutritional factors such as tannins, polyphenolic substances and phytates are also present in Kodo millet, each of which might directly or indirectly affect the digestibility of the nutrients. Presence of these factors leads to chelation of dietary minerals in the gastro- intestinal tract, thereby reducing their bio-accessibility and bioavailability. To improve the nutritional value of millets, reduction of antinutritional factors is necessary. Numerous processing techniques are reported in the literature for instance dehulling, milling, malting, blanching, parboiling, acid and heat treatment to reduce anti-nutritional content of millets. But very less literature is available on the anti-nutritional factors present in the Kodo millet and processes to reduce them. Therefore, the focus of this study is to compare the various processing techniques for the reduction of anti-nutritional factors and their impact on nutritional value present in the Kodo millet. The study will prove helpful to develop commercial processes for post-harvest handling and safer usage of kodo millet.

Keywords: *Kodo Millet, Anti-Nutritional Factor, Phytates, Oxalic Acid, Processing Techniques.*

INTRODUCTION

Millet is a generic name for tiny grains, commonly known as 'coarse cereals'. Millets have agricultural importance as they possess tolerance to extreme weather, produce good yields in almost any type of soil and are resistant to drought stress. For centuries, millets have been a dietary staple for humans and are mostly grown in African and Asian countries. Foxtail millet, pearl millet, little millet and finger millet are the most exploited and commonly used millets. While proso millet, kodo millet, brown top and barnyard millet are examples of underused minor millets (Jaybhaye *et al.*, 2014). Millets are known colloquially as "poor people's produce" or "poor man's meal" because traditionally they were consumed by poor people of the community as an energy dense and healthy food especially during drought and scarcity (Banakar, 2005).

Kodo millet (*Paspalum scrobiculatum*) is one of the oldest underutilized crops. For centuries varieties of Kodo millet has been a dietary staple for humans in India, Vietnam, Philippines, Thailand, Indonesia, Pakistan and West Africa (Deshpande *et al.*, 2015). As public understanding grows, dietitians, researchers, and food manufacturers have become more aware of the health and nutritional advantages of this underused millet. It becomes a plausible policy to create various meals from kodo millet to improve the nutritional well-being of individuals and develop new variances in food basket to overcome issues related to public health. The nutrient composition of Kodo millet i.e., total protein, amino acid composition, minerals and vitamin compositions have been widely studied and have proven comparable, or even better with other millets (Sharma *et al.*, 2017). It is a rich source of essential amino acids, like lysine, valine, threonine and sulphur containing amino acids (Ravindran, 1992). Kodo millet grain contains vitamin B6, vitamin B3, folic acid, minerals such as calcium, magnesium, potassium, and zinc in high quantities as well as 8.3 per cent protein in which major protein is glutelin (Sudharshana *et al.*, 1988).

In addition to nutritional qualities, certain anti-nutritional factors such as tannins, polyphenolic substances, oxalates and phytates are also present in millets, each of which might directly or indirectly affect the digestibility of the nutrients (Lorenz, 1983). Presence of these factors leads to chelation of dietary minerals in the gastro-intestinal tract, thereby reducing their bio-accessibility and bioavailability (Nour *et al.*, 2014). Moreover, existence of antinutrients in the pericarp and endosperm areas may cause the development of unpleasant grey colour and taste to the finished product (Rathi *et al.*, 2004). Development of off odours and taste in the flour and its products was mainly attributed with the occurrence of lipase activity in pericarp, aleurone layer and germ of grains (Galliard, 1999; Yadav *et al.*, 2012). In conjunction with their primary impact on nutrient absorption, several antinutrients, such as oxalate or phytic acid, may be harmful over a certain threshold (Novak and Haslberger, 2000). Oxalate is one of the key antinutrient present in millets which causes risk of kidney stone after prolonged use as a dietary staple (Pathan *et al.*, 2014). Kodo millet contains higher quantities of oxalate as compared to other millets (Agnihotri *et al.*, 2008). As a consequence, elimination of these factors becomes crucial.

To improve the nutrition value of millets, reduction of antinutritional factors is necessary. Numerous processing techniques are reported in the literature for instance dehulling, milling, malting, blanching, parboiling, acid and heat treatment to reduce anti-nutritional content of millets (Rani *et al.*, 2018). These techniques are known to improve bioavailability of nutrients and digestibility of millets. Very less literature is available on the anti-nutritional factors present in the Kodo millet and processes to reduce them.

Therefore, the focus of this study is to compare the various processing techniques for the reduction of anti-nutritional factors and their impact on nutritional value present in the Kodo millet. This will prove helpful in the development of commercial processes for post-harvest handling and safer usage of kodo millet.

MATERIAL AND METHODS

Kodo millet, purchased at the local market of Mumbai, Maharashtra, was cleaned and washed before using for processing and proximate analysis in triplicates.

All other chemicals used were of analytical grade and obtained from Fisher Scientific and Sigma- Aldrich.

Processing of Kodo Millet

Milling

Milling of Kodo millet grains was performed by using agate mortar and pestle to obtain coarse flour. In order to achieve further size reduction, milling of one part of coarse flour was performed on laboratory model of mixer-grinder (Make: Havells) to obtain medium milled flour. A part of medium milled flour was further sieved through 44 mesh sieve to get fine flour. The coarse flour contained broken fragments of whole millet grains due to impact of pestle. The medium milled flour consisted mostly of bran and large fragments of millet endosperm while the fine flour contained endosperm and finely ground bran. Sample of coarse flour, medium milled flour and fine flour were collected for analysis.

Roasting

Roasting of Kodo millet grains was performed on hot metal pan with periodic stirring on low flame till colour changes to light brown. Samples were collected for three different durations (10, 20 and 30 min.) and milled using laboratory model of mixer-grinder (Make: Havells) to generate millet samples of roasting followed by milling for further analysis.

Similar treatment was given to fine flour of Kodo millet obtained from milling process. The fine flour was subjected to roasting on hot metal pan with periodic stirring on low flame till colour changes to light brown. Samples were collected for three different durations (10, 20 and 30 min.) to generate millet samples of milling followed by roasting for further analysis.

Blanching

The millet grains were loosely tied in muslin cloth and dipped in boiling water at 98°C for 10 or 20 sec, then dipped in water at room temperature. Samples were collected for three different durations (5, 10 and 15 min.) and were ground to flour using laboratory model of mixer-grinder (Make: Havells) for further analysis.

Fermentation

The fine flour of Kodo millet obtained from milling process was dispersed into distilled water (1:1.5, w/v). The dispersion was incubated at 37°C for 12 hrs. Samples were collected for three different durations (12, 24 and 36 hrs.) for further analysis. Incubation beyond 36 hrs led to mould formation in the fermented batter.

Germination/Malting

Grains were steeped in water at a ratio of (1:2 w/v) for 24 h at room temperature. Water was drained off. The soaked grains were wrapped in moist cotton cloth and subjected to warm humid environment to effect germination. Samples were collected for three different durations (24, 48 and 72 hrs). The sprouted grains of Kodo millet were ground to flour using laboratory model of mixer-grinder (Make: Havells) for further analysis.

Proximate Analysis

Thousand Grain Weight

Three groups of 1000 undamaged grains were chosen at random. Using an electronic weighing balance, the weight of chosen grains was recorded in triplicate and the average was computed. The average weight was expressed in g/1000 grains.

Thousand Grain Volume

The volume of millet grains was measured by using 10 ml measuring cylinder.

Bulk Density

The following formula was used to compute bulk density from the weight and volume of Kodo millet and average of 5 replications was calculated.

$$\text{Bulk density} = \frac{\text{Weight of sample (g)}}{\text{Volume of sample (ml)}}$$

Moisture Content

Moisture content of the samples was analysed on Metrohm Karl Fischer auto titrator (model: KF Ti-touch). Samples were soaked in methanol for 2 hrs at room temperature for extraction of moisture and then filtered using Whatman filter paper. The filtrate was titrated by Karl Fischer titration method in the Metrohm auto titrator till a stable reading of% moisture content was observed.

Loss on Drying (LOD)

Loss on drying was analysed on a moisture balance of 2 gram scale. Sample was exposed to heat in moisture balance at 100°C till a constant reading was observed. Average of 3 replications was calculated for each sample.

pH

Samples were soaked in water for 1 hr. at room temperature for extraction of total acidity and then filtered using Whatman filter paper. The pH of the filtrate was measured on research grade pH meter (make: Hanna, Model: HI5221).

Total Alkaloids Presence

5 gram samples were soaked in a mixture of 50 ml 10% acetic acid and ethanol for 4 hrs and then filtered using Whatman filter paper. Filtrate was

concentrated by evaporation to its 1/4th volume. Concentrated ammonium hydroxide was added dropwise to get the precipitate and filtered using a pre-weighed Whatman filter paper. Precipitate collected on filter paper was washed with 1% ammonium hydroxide. Filter paper was dried in oven at 60°C for 30 min and weighed till a constant reading was observed.

Analysis of Anti-Nutrients

Quantification of Phytates

4 grams millet sample was soaked in 100 ml of 2% hydrochloric acid for 5 hrs and then filtered using Whatman filter paper. 5 ml of 0.3% ammonium thiocyanate solution was added to 25 ml of filtrate and titrated against ferric chloride solution till yellow to brown colour persisted for 5 minutes.

Quantification of Oxalates

High performance anion-exchange chromatography followed by pulsed Conductivity detection was chosen as suitable technique to analyse total oxalate content present in Kodo millet samples. IC system was comprised with Metrohm 940 Professional IC (vario ONE/HPG) equipped and Conductivity detector and Metrosep A Supp 5 column (15 cm × 4 mm). The chromate-graphic conditions were optimized experimentally. In brief, a reverse phase C18 column was employed (Metrosep A Supp 5 column (15 cm × 4 mm)). The mobile phase consisted of a mixture of 3.2 mM Sodium carbonate and 1.0 mM Sodium bicarbonate in Water and was run at a flow rate of 0.7 mL/min.

RESULTS AND DISCUSSION

The present study investigated the effect of various processing techniques on reduction of two anti-nutrients, phytates and oxalates, present in the Kodo millet in order to improve its nutritional value, and to minimize the health hazard associated with the prolonged use.

Proximate Analysis

Physicochemical properties of control ample of Kodo millet were studied.

Kodo millet grains were found to have thousand grain weight of 3.92 g and a thousand grain volume of 3.4 ml. Kodo millet sample demonstrated a bulk density of 1.15 g/ml. The functional properties like moisture content, loss on drying, pH was found to be 3.34%, 9.15% and 7.0 respectively. Presence of alkaloids and phenolics was confirmed qualitatively by chemical analysis. Anti-nutrients i.e. total phytates and total oxalates was found to be 2.25% and 0.071% respectively. Total oxalate content in the control sample was high as compared to the values reported in literature. Table 1 summarizes the findings of proximate analysis of Kodo millet control sample.

Table 1: Proximate Analysis of Control Sample of Kodo Millet

Sr. No.	*Parameter*	*Value*
1.	Thousand Grain weight (g)	3.92 g
2.	Thousand grain volume	3.4 ml
3.	Bulk density	1.15 g/ml
4.	Moisture content	3.34%
5.	Loss on drying	9.15%
6.	pH	7.0
7.	Total alkaloids	Present
8.	Total phytate content	2.25%
9.	Total oxalate content	0.071%

Effect of Milling

Milling was performed to separate endosperm and bran to maximum extent and for size reduction of endosperm to facilitate the formation of fine flour. The milling is difficult as its small kernel has firmly embedded germ and hard bran and endosperm. Hence milling was performed on the laboratory mixer grinder. Table 2 summarizes effect of milling on three samples i.e., coarse flour, medium milled flour, and fine flour. Moisture content and loss on drying was found to be increasing with increase in the fineness of flour. This is due to separation of bran and size reduction of hard endosperm which facilitates exposure of moisture and other volatile components present in endosperm. Milling showed no effect on the change in pH as it remained neutral and constant. Total oxalate content did not show any significant change upon milling whereas total phytate content showed slight decrease with increase in fineness of flour. Milling facilitated 13.33% reduction in total phytate content of fine flour due to almost absence of bran in the fine flour of kodo millet.

Effect of Roasting

Effect of roasting was studied with the help of two types of experimental sets in combination with milling process. In the first set millet grains were roasted and then subjected to milling whereas in the second set millet grains were milled to obtain fine flour and then the flour was subjected to roasting. Table 2 summarizes the results of samples collected at three different time intervals of two sets. Both sets showed the reduction in moisture content and loss on drying with increase in exposure time to heat; but milling flowed by grinding shows less% loss on drying as compared to grinding followed by roasting. Both sets showed no effect on the change in pH as it remained neutral and constant. Both sets showed reduction in total oxalate and total phytate content with increase in exposure time to heat. Reduction of total oxalate contents in 30 min time interval sample of both sets was found to be 18.31% and 22.54% respectively. Similarly, Reduction of total phytate contents in 30 min time interval sample of both sets was found to be 6.22%.

Table 2: Effect of Milling and Roasting on the Kodo Millet in Comparison with Control Sample

Sample interval	*Control Sample*	*Changes in the of Kodo Millet after Milling*			*Changes in the of Kodo Millet after Roasting Followed by Milling*			*Changes in the of Kodo Millet after Milling Followed by Roasting*		
		Coarse	*Medium*	*Fine*	*10 min*	*20 min*	*30 min*	*10 min*	*20 min*	*30 min*
% Moisture	3.43	3.51	5.75	6.85	2.84	1.63	0.94	2.38	1.49	0.73
% LOD	9.15	8.42	9.64	10.23	6.12	3.36	2.77	5.26	2.88	1.77
pH	7	7	7	7.03	6.87	6.89	6.92	6.77	6.74	6.74
% Oxalates	0.071	0.073	0.73	0.071	0.067	0.06	0.055	0.069	0.062	0.058
% Phytates	2.25	2.24	1.98	1.95	2.22	2.16	2.11	2.23	2.19	2.11

Effect of Blanching

One of the most effective techniques for enhancing the shelf life of pearl millet flour is blanching, which slow down the enzymatic activity without having any significant effect on its nutritional composition. The effects of blanching process on Kodo millet are summarized in Table 3. Moisture content and loss on drying was found to be increasing with increase in the blanching time. No significant change in pH was observed in samples after blanching process. Samples showed reduction in total oxalate and total phytate content with increase in blanching time. Reduction of total oxalate contents in 15 min time interval sample was found to be 30.99%. Similarly, Reduction of total phytate contents in 15 min time interval sample was found to be 18.22%.

Table 3: Effect of Blanching, Fermentation, and Germination on the Kodo Millet in Comparison with Control Sample

	Control Samples	*Changes in the of Kodo Millet after Blanching*			*Changes in the of Kodo Millet after Fermentation*			*Changes in the of Kodo Millet after Germination*		
	0 hrs	*5 min*	*10 min*	*15 min*	*12 hrs*	*24 hrs*	*36 hrs*	*24 hrs*	*48 hrs*	*72 hrs*
% Moisture	3.43	19.56	26.47	29.5	30.44	30.45	30.95	29.5	27	26.8
% LOD	9.15	20.1	27.15	29.8	31.27	31.2	31.18	29.8	27.79	27.4
pH	7	6.98	6.93	6.93	6.6	6.56	6.56	6.5	5.01	5.01
% Oxalates	0.071	0.061	0.052	0.049	0.056	0.037	0.035	0.052	0.044	0.046
% Phytates	2.25	1.93	1.88	1.84	1.85	1.48	1.46	1.87	1.66	1.63

Effect of Fermentation

Effect of fermentation on Kodo millet flour is presented in Table 3. The samples of fermentation process were collected at three-time intervals i.e., 12, 24 and 36 hrs. Any significant change in moisture content and loss on drying was not found with respect to fermentation time. Decrease in pH was observed with increase in fermentation time shifting it to acidic range. pH of 36 hrs time interval sample was found to be 6.23 and no difference was observed in the pH of 24 hrs and 36 hrs time interval samples. Significant decrease in total oxalate and total phytate content with increase in fermentation time was observed. Reduction of total oxalate contents in 36 hrs time interval sample was found to be 50.70%. Similarly, Reduction of total phytate contents in 36 hrs time interval sample was found to be 35.11%.

Effect of Germination

Table 3 summarizes the effects of germination on composition of anti-nutrients present in kodo millet. The samples of germination process were collected at three-time intervals i.e., 24, 48 and 72 hrs. Reduction in moisture content and loss on drying was observed with increase germination time as moisture present in the endosperm was utilized by germ part of millet during the germination process. Significant decrease in the pH was observed with increase in germination time. pH of 72 hrs time interval sample was found to be 5.01 indicating the increased acidity during germination process. It was observed that increase in germination time causes significant reduction in total oxalate and total phytate content. Reduction of total oxalate contents in 36 hrs time interval sample was found to be 35.21%. Similarly, Reduction of total phytate contents in 36 hrs time interval sample was found to be 27.56%.

Graphical comparison of effects of above-mentioned processes on moisture content, loss on drying and pH is shown in Figure 1 and effect on % reduction in anti-nutrients is represented in Figure 2.

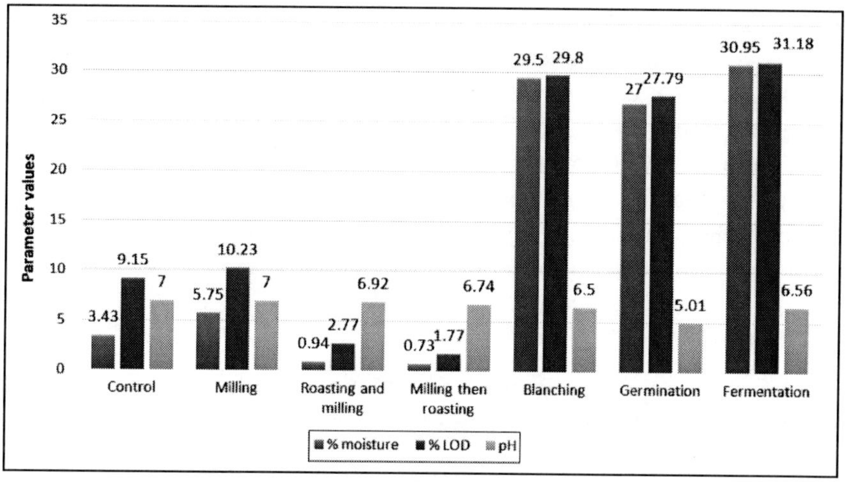

Figure 1: Graphical Comparison of Effects of Processes on Moisture Content, Loss on Drying and pH of Kodo Millet

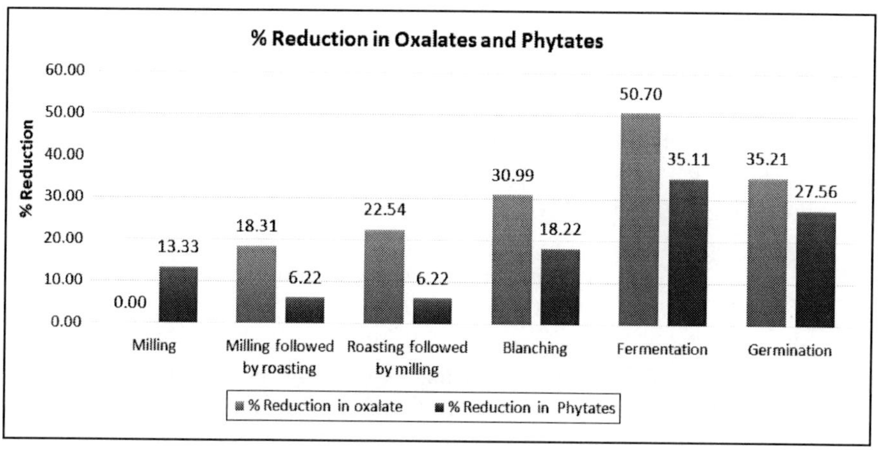

Figure 2: Graphical Comparison of Effects of Processes on % Reduction in Anti-Nutrients

CONCLUSION

In the present study, various processing techniques for the reduction of two anti-nutritional factors i.e., phytates and oxalates present in Kodo millet were studied. Processing techniques like milling, roasting, blanching, fermentation and germination were found to have impact on various parameters of Kodo millet. Blanching and fermentation showed the highest amount of moisture and % loss on drying which enhances the digestibility and bioavailability of nutritional factors of Kodo millet. Fermentation and germination processes showed significant reduction in anti-nutritional factors. Fermentation process exhibited the highest reduction in total phytate and total oxalate content. Reduction of total oxalate and total phytate contents in 36 hrs time interval sample was found to be 50.70% and 35.11% respectively. The observed reduction in anti-nutritional factors may be considered as a good indicator to employ fermentation treatment for the safer usage of Kodo millet; however, further studies should be conducted on effects of fermentation on the nutritional factors and optimization of fermentation process.

REFERENCES

[1] Agnihotri, R. and Shrivastava, S.K. (2008). Chemical and bio-chemical studies on some hybrid millet seeds; *Asian Journal of Chemistry;* Vol. 20, No. 4, 2657–2662.

[2] Banakar R. (2005). Formulation, nutritional quality and growth promoting characteristics of supplementary food. M.H. Sc. *Thesis, Univ. Agric. Sci.*, Dharwad (India).

[3] Deshpande, S.S., Mohapatra, D., Tripathi. M.K. and Sadvatha R.H. (2015). Kodo millet-Nutritional Value and Utilization in Indian Foods, *Journal of Grain Processing and Storage,* Vol. 2.

[4] Galliard, T. (1999). Rancidity in cereal products, Allen JC, Hamilton RJ (eds.) Rancidity in foods. *Aspen Publishers, Gaithersburg,* pp. 140–156.

[5] Jaybhaye, R.V., Pardeshi, I.L., Vengaiah, P.C. and Srivastav, P.P. (2014); Processing and technology for millet-based food products: A review. *J. Ready Eat,* 1(2):32–48.

[6] Lorenz, K. (1983). Tannins and Phytate contents in proso millet (*Panicum miliaceum*); *Cereal chem;* 60(6):424–426.

[7] Nour, A., Sokrab, A.M., Ahmed, I. and Babiker, E. (2014). Supplementation and cooking of pearl millet: Changes in anti-nutrients, and total minerals content and extractability, *Innovative Romanian Food Biotechnology;* Vol. 15, pp. 9–22.

[8] Novak, W.K. and Haslberger A.G. (2000). Substantial equivalence of antinutrients and inherent plant toxins in genetically modifed novel foods; *Food and Chemical Toxicology,* Vol. 38, pp. 473–483.

[9] Pathan, S.H., Tumbare, A.D. and Kamble, A.B. (2014). Effect of agronomic management on oxalate and silica content in pearl millet (*Pennisetum glaucum*) × napier (*Pennisetum purpureum*) hybrid; *Indian Journal of Agronomy*; 59(3), 415–420.

[10] Rani, S., Singh, R., Sehrawat, R., Pal, B.K. and Upadhyay, A. (2018). Pearl millet processing: A review; *Nutrition & Food Science,* 48(3).

[11] Rathi, A., Kawatra, A. and Sehgal, S. (2004). Influence of depigmentation of pearl millet (*Pennisetum glaucum* L.) on sensory attributes, nutrient composition, in vitro protein and starch digestibility of pasta; *Food Chemistry,* Vol. 85, No. 2, pp. 275–280.

[12] Ravindran, G. (1992). Seed proteins of millets: amino acid composition, proteinase inhibitors and in vitro digestibility, *Food Chemistry,* 44(1): 13–17.

[13] Sharma, S. Sharma, N. Handa, S. Pathania, S. (2017); Evaluation of health potential of nutritionally enriched Kodo millet grown in Himachal Pradesh, India; *Food Chemistry;* 214:162–168.

[14] Sudharshana, L., Monteiro, P.V. and Ramachandra, G. (1988). Studies on the proteins of Kodo millet (*Paspalum scrobiculatum*), *Journal of the Science of Food and Agriculture,* 42(4): 315–323.

[15] Yadav, D.N., Anand, T., Kaur, J. and Singh, A.K. (2012). Improved storage stability of pearl millet flour through microwave treatment, Agricultural Research; Vol. 1, No. 4, pp. 399–404.

Millets: A Key to Nutritional Security

Shama Zaidi
Aries Agro Ltd.

ABSTRACT: Millets are a group of small grained grasses called as nutri-cereals or dry-land cereals. They have a glorious history of being among the earliest crops grown by humans and can be traced back to the Stone Age. Millets are a treasure trove of nutrients and are a good source of Ca, Mg, K, Fe and other micronutrients, minerals and fibre. They are largely cultivated in arid and semi-arid areas to provide food as well as fodder and can be mix-cultivated (polyculture) with pulses and vegetables. Being cultivated in rain fed regions, their average crop yields are 3 to 4 fold lower than the potential due to the agro-ecology. Efficient nutrient management is the key to enhance millet yields in rain fed regions. Along with the application of NPK, use of compost, vermicompost and biofertilizers like *Azospirillum* and PSB enhanced crop growth and quality parameters. It can also improve the physical, chemical and biological properties of the soil. Foliar application of essential micronutrients like Zn and Fe; either in sulphate or chelated forms can lead to effective absorption and translocation of other macro and micro nutrients at the critical growth stages of the crop.

Increased awareness regarding the health benefits of millets has led to an increase in its consumption. To promote cultivation and consumption of millets and millets-based products, the Government of India has adopted and implemented certain measures like increasing the MSP, creation of supply chains and public distribution systems, encouraging Public Private Partnership (PPP) projects, entrepreneurships and startups, setting up of post-harvest processing units, etc. Millets, being a staple of many regions worldwide seem to be the most promising sustainable approach to tackle nutritional security around the world. Millets can not only serve as income crop but can also prove to be the preferred crop to improve the health of the community as a whole.

Keywords: *Nutri-cereals, Nutrient Management, Foliar Application.*

INTRODUCTION

Millets are a group of small grained grasses, differing from cereals mainly in the size of the seed. They are also known nutri-cereals or dry-land cereals. In India, different types of millets are grown such as pearl millet, sorghum, finger millet, foxtail millet, proso millet, barnyard millet, kodo millet, and little millet (Table 1). Based on the grain size, millets have been classified as major millets which include sorghum and pearl millet which have a sizeable area under cultivation in India whereas the rest of the millets are categorised as minor millets (Louhar *et al.*, 2020).

There are almost 6000 varieties of millets grown throughout the world as a major source of energy and protein. Millets have a glorious history of being among the earliest crops grown by humans. Millets can be traced back to the Stone Age and in fact, there are reports of many of them being found in the archaeological sites of ancient civilizations like Mohenjo Daro and Harappa.

Table1: Millet Cultivation in India (Louhar *et al.*, 2020)

Crop	*Scientific Name*	*Place of Origin*	*Major Growing States in India*
Pearl millet	*Pennisetum glaucum*	Africa	Rajasthan, Maharashtra, Uttar Pradesh, Haryana and Gujarat
Sorghum	*Sorghum bicolor*	North Eastern Africa	Maharashtra, Karnataka, Madhya Pradesh, Andhra Pradesh, Gujarat, Rajasthan, Uttar Pradesh and Tamil Nadu
Ragi (Finger millet)	*Eleusine coracana*	East Africa	Madhya Pradesh, Maharashtra, Orissa, Uttarakhand, Uttar Pradesh and Tamil Nadu
Foxtail/Italian millet/Kauni	*Setaria italica*	Central Asia-India	Karnataka, Rajasthan, Madhya Pradesh and Chattisgarh
Proso or common millet (Cheena)	*Panicum miliaceum*	Central Asia-India	Bihar, North-east India and Maharashtra
Indian barnyard millet (Sawan)	*Echinochloa frumentacea*	India	Karnataka, Madhya Pradesh, Uttarakhand, Uttar Pradesh and North-east India
Kodo millet	*Paspalum scrobiculatum*	India	Maharashtra, Tamil Nadu and Chattisgarh
Little millet/Kutki	*Panicum sumatrense*	India	Karnataka, Madhya Pradesh, Andhra Pradesh, Tamil Nadu, Jharkhand, Odisha, Maharashtra and Chattisgarh

In fact, it was this food group that was a staple food in many diets across the world like the Indian, Korean and the Chinese. India is a major producer of Millets, accounting for 80% of Asia's production and 20% of global production. India's average yield of Millets (1239 kg/hectare) is also higher than Global average yield of 1229 kg/ha (https://pib.gov.in). (provide the reference is APA format) Major Millets crops grown in India and their percentage share of production are Pearl Millet (Bajra) – 61% share, Jowar (Sorghum) – 27%, and Finger Millet (Mandua/Ragi) – 10%.

Millets—Nutrition

Millets are called nutri-cereals as they are a powerhouse of nutrients like minerals and B-complex vitamins like iron, folate, calcium, zinc, magnesium, phosphorus, copper, and antioxidants. The grains of these crops have better nutritional qualities compared to other cereals. Millets are good source of Ca, Mg, K, Fe and other micronutrients (Figure 1B). They are also a good source

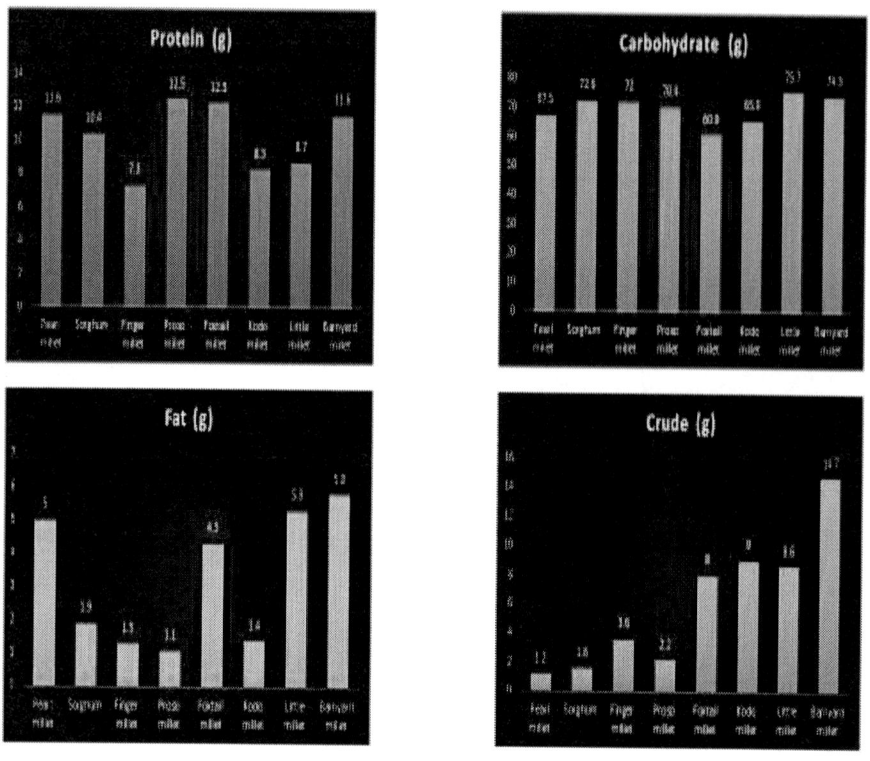

Figure 1A: Amount of Nutrients in Millets per 100 gms

Source: National Academy of Agricultural Sciences, New Delhi (2018).

of dietary fibre, free of gluten and have a low glycemic index. Fibre also regulates the speed of digestion and helps absorption of nutrients (Figure 1A). Millets contains about 10 times more Ca and 2–10 times higher Fe than wheat or rice (Bala *et al.*, 2010). The grain protein of these crops is rich in essential amino acids especially sulphur-containing amino acids and phytochemicals such as phytic acid compared to other major cereals. Besides fibre, millets are also rich in health-promoting phytochemicals like polyphenols, lignans, phytosterols, phyto-oestrogens, phytocyanins. These function as antioxidants, immune modulators, detoxifying agents etc. and hence protect against age-related degenerative diseases like cardiovascular diseases (CVD), diabetes, cancer etc. (Rao *et al.*, 2011).

Among the millet groups generally found in India, Proso millet contains the highest amount of protein 12.5 g/100 g and least amount of fat 1.1 g/ 100 g whereas Barnyard millet contains the highest amount of carbohydrate 74.3 g/100 g, fat 5.8 g/100 g, crude fibre 14.7 g/100 g and Iron 17.47 mg/100 g. Fox tail millet contains the highest amount of phosphorus 290 mg/100 g and zinc 60.6 mg/100 g as compared to the other millets in India. The highest amount of calcium 344 mg/100 g is present in Finger millet (Figure 1A and 1B).

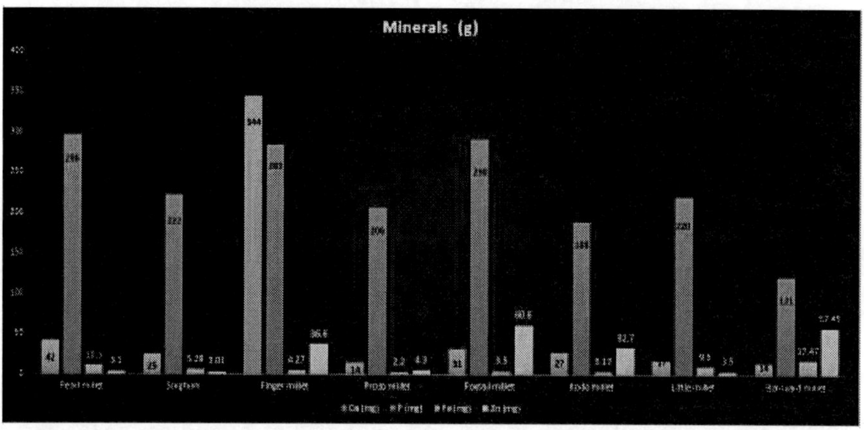

Figure 1B: Amount of Minerals (Ca, P, Fe and Zn) in Millets per 100 gms

Source: National Academy of Agricultural Sciences, New Delhi (2018).

According to the traditional medicine, millets help in improving appetite and aid in digestion, help in blood deficiency, functions as a galactogogue and helps the body to relax by improving sleep. Apart from these benefits, millets are also a source of Phosphorus which plays a vital role in energy production and storage and a good source of magnesium which enhances nutrient delivery and plays a role in insulin sensitivity. Lignans in millets are thought to play a role in certain types of cancers like breast cancer and cardiovascular diseases.

Millets—Cultivation in India

Jowar (sorghum), bajra (pearl millet), and ragi (finger millet) are the three main millet crops that are currently being grown in India. India also produces a wide variety of indigenous, biogenetically diverse "small millets" like kodo, kutki, chenna, and sanwa. Rajasthan, Andhra Pradesh, Telangana, Karnataka, Tamil Nadu, Maharashtra, Gujarat, and Haryana. are the major producers of millet in the country.

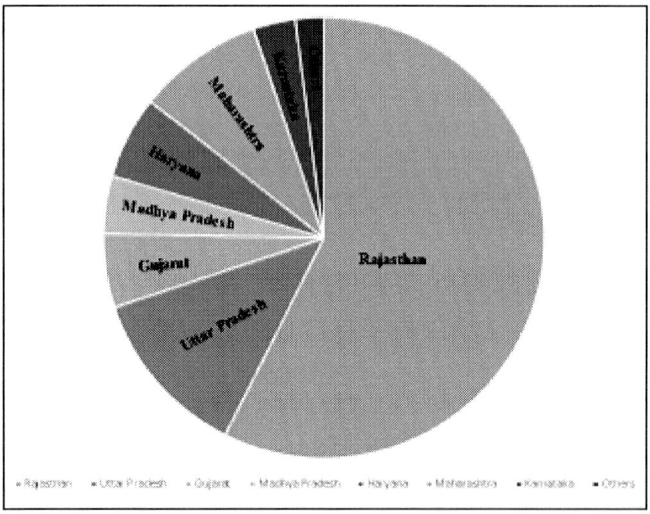

Figure 2: Area under Pearl Millet Production, India

Source: Government of India, 2018. (Agricultural statistics at a glance, Directorate of Economics and Statistics, 2018) and AICRPSM, 2020, ICAR-All India Coordinated Research Project on Small Millets, Bengaluru.

Millets are largely cultivated in arid and semi-arid areas and are well referred to as 'Crops of antiquity', largely due to their drought tolerance capacity and pest resistance (Devi *et al.,* 2014). The other advantages of cultivating millets include crop sturdiness, short to medium duration, low labour requirement and minimal purchased inputs. Millets being C4 crops are compliant to climate change. They can also be grown in mountainous, low-fertility, dry, rain-fed, and tribal areas and can be a long-term source of income for farmers due to low investment requirements. Millets sequestrate carbon and thereby reduce the burden of greenhouse gas.

Millet cultivation is the mainstay of rain-fed farming on which 60% of Indian farmers depend. They provide food as well as fodder and can be mix-cultivated (polyculture) with pulses and vegetables. Despite these attributes, millets are losing their pride of place in production and consumption in India. In recent years, there has been some effort towards reviving millets. India has the 3rd largest area under cultivation of small millets in the world. Since 1950 to 2018, the area under pearl millet declined from 9.02 mha to 7.38 mha, in sorghum it came down from 15.57 mha to 4.96 mha, whereas in small millets the area under cultivation came down from 7.56 mha to 1.86 mha (GOI 2018; AICRPSM 2020) (Figure 2). Contrarily, the total production of pearl millet has increased from 2.60 mt to 9.13 mt. Whereas, in sorghum and minor millets, total production has seen a decrease from 5.50 mt to 4.95 mt and 3.78 mt to 2.29 mt, respectively. Rajasthan is the leading state in both area and production for pearl millet (Fig.2), whereas Maharashtra has the highest area and production for sorghum (Figure 3) (GOI 2018).

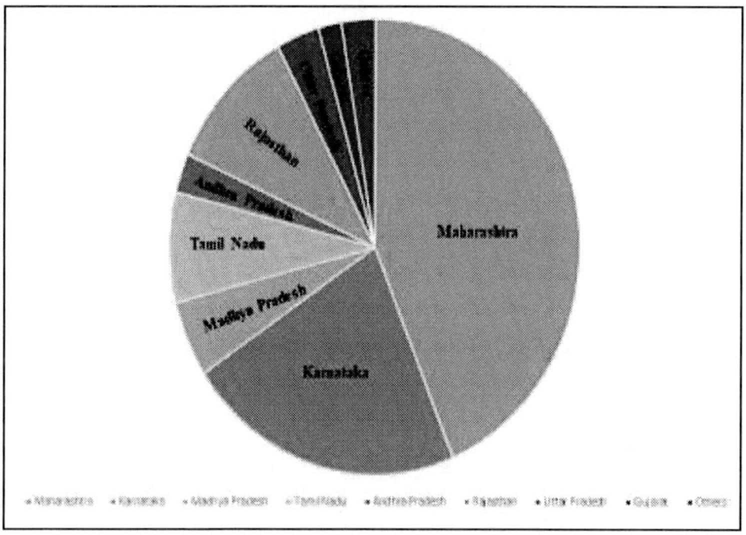

Figure 3: Area under Sorghum Production, India

Source: Government of India, 2018. (Agricultural statistics at a glance, Directorate of Economics and Statistics, 2018) and AICRPSM, 2020, ICAR-All India Coordinated Research Project on Small Millets, Bengaluru

As most of the millets are being cultivated in arid and semi-arid ecologies, without any assured supply of irrigation water, therefore environmental and weather-related factors are the biggest limitation in the millet production in India (Bana *et al.*, 2013). Inadequate and uneven distribution of rainfall, late onset and early cessation of monsoon rains, prolonged dry spells during the crop growing season and excessive evaporation are some paramount factors which affects millet productivity adversely (Bana *et al.,* 2013; Dass *et al.,* 2014).

One of the biggest reasons of poor adoption of millets by the cultivators is the low level of productivity of these crops (Bana *et al.,* 2013) and consequently poor economic returns. Nutrient and moisture stresses are considered as the greatest constraints in millet productivity (Bana 2014; Bana *et al.,* 2018; Mubeena *et al.,* 2019). Another reason for the decline of millets cultivation is that almost 50% area under millets has been diverted largely to soybean, maize, cotton, sugarcane and sunflower. This has been due to factors like low remuneration as compared to other food crops, lack of input subsidies and price incentives, subsidised supply of fine cereals through Public Distribution System (PDS), and change in consumer preference.

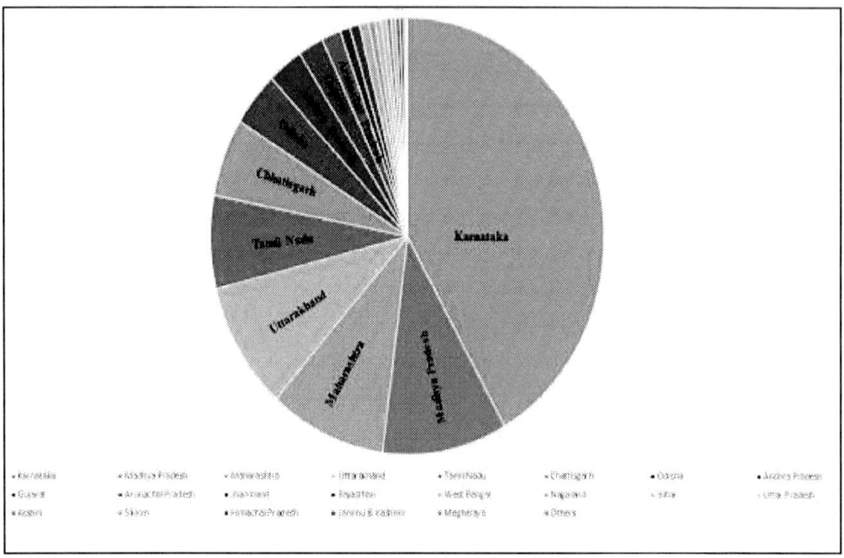

Figure 4: Area under Minor Millets Production, India

Source: Government of India, 2018 (Agricultural statistics at a glance, Directorate of Economics and Statistics, 2018) and AICRPSM, 2020, ICAR-All India Coordinated Research Project on Small Millets, Bengaluru.

Millets—Nutrient Management

Millets are mainly being cultivated (91%) in rainfed regions and the average crop yield levels of these agro-ecologies are 3 to 4 folds lower than the potential. Therefore, to make millet cultivation remunerative, it is important to achieve the higher productivity of these crops. Efficient nutrient management approaches are among key strategies to realize higher millet yields in rainfed regions (Bamboriya *et al.*, 2017; Bana *et al.*, 2016).

Nutrient requirement depends on various factors like soil type, cultivar, source of nutrients etc. Generally, as compared to irrigated crops, almost half doses of fertilizers are applied for rainfed crops (Rana and Bana 2012). As per the nutritional requirements, to obtain a good yield, in Sorghum, a NPK application of 80–100:40–50: 40:N:P_2O_5:K_2O kg ha^{-1} is recommended. Nitrogen application is done in 2–3 splits to match the crop N requirement as it is prone to volatalization, leaching and denitrification. Hence half the nitrogen dose is applied at the time of planting below the seed and the remaining half is done 35 DAS as side dressing (Cothren *et al.*, 2000). However, full dose of P and K are applied at the time of planting (Louhar *et al.*, 2020).

For pearl millet the recommended fertiliser application is 60–100:40–50:40–50:: N:P_2O_5:K_2O kg ha^{-1}. Similar to Sorghum, in Pearl millet also, nitrogen doses are split into two; half dose at the time of planting with full dose pf P and K and the remaining half 25–30 days after sowing.

In finger millet, the general recommendation is 40-60:20:20::N:P_2O_5:K_2O kg ha^{-1} where again N is applied in two splits, first as basal placement along with full doses of P and K, and second split is applied as side dressing at late tillering stage.

Along with the application of NPK, use of compost, vermicompost and biofertilisers like *Azospirillum* and PSB are used to enhance crop growth and quality parameters like grain yield, dry matter accumulation, test weight, increased photosynthetic activity, ear head length and increase in the number of tillers. It also improves the physical, chemical and biological properties of the soil. *Azospirillum* and PSB also affect the plant growth through the production of growth hormones like IAA, GA and Cytokinin (Louhar *et al.*, 2020).The combined use of organic and inorganic sources has good effect on physiological process of plant metabolism and growth, leading to higher yield. The better availability of nitrogen due to mineralization of organics thereby influence the shoot and root growth favouring absorption of other nutrients also (Basavaraju and Purushotham 2009). Foliar application of nutrients like Zn and Fe along with the recommended doses of NPK also impacts the yield. This could be due to the role of Fe in photosynthesis and the physiological role of Zn in phytohormone metabolism in plants (Rurinda *et al.*, 2014). Foliar application of essential micronutrients like Zn and Fe; either in sulphate or chelated forms can lead to effective absorption and translocation of other macro and micro nutrients at the critical growth stages of the crop.

Millets—A Solution to Nutritional Security

In the context of Nutritional security, sustainable crop substitutes are being explored to meet the challenges of world hunger (cereal demand) and at the same time improve income of the farmers. Millets are a sustainable food source for combating hunger in a changing global climate because they are resistant to climatic stress, pests, and diseases and hence can play a major role in combating nutritional security. Increased awareness regarding the health benefits of millets has led to an increase in its consumption. The role of millets in designing the modern foods like multigrain and gluten-free cereal products is well known (Saleh *et al.,* 2013).

As per 2016 Global Nutrition report, 44% population of the countries with available data i.e. 129 countries approximately face very serious levels of to the estimates of United nutritional issues like under nutrition, adult overweight and obesity. According Nations Food and Agriculture Organization, about 795 million people (10.9% of world population in 2015) were reported undernourished. While on the other hand more than 1.9 billion (39% of world's population) adults ≥18 years of age were overweight and further 13% were reported to be obese (36, 37). Concerning India's statistics, about 194.6 million people, i.e. 15.2% of total population of India, are undernourished. Protein energy malnutrition (PEM) was found to be a serious issue resulting in 4,69,000 deaths and around 84,000 deaths due to deficiency of other vital nutrients such as iron, iodine and vitamin A. Obesity is another major health concern in India with the prevalence rate of 11% in men and 15% in women. Millet cultivation secures sixth position in terms of world agricultural production of cereal grains and are still a staple food in many regions of world. These are rich source of many vital nutrients and hence, promise an additional advantage for combating nutrient deficiencies in the third world countries (Kumar *et al.,* 2018).

Millets have been used as food and feed since ancient times and has been a staple food of various countries around the world, particularly Africa and Asia. These are largely consumed as bakery products like bread, cookies, flour, porridge beverages, etc, Studies have shown that plant based diets play a role in combating non-communicable and lifestyle diseases like obesity, diabetes and cardiovascular ailments. Thus is due to the presence of protective phytochemicals and nutrients. In line with this, millets are a rich source of nutrients and protective phytochemicals and hence have gained importance as functional foods.

However, the effect of addition of millets on the sensory acceptability of commercial foods has not been widely studied. Various proportions of millet blends with other food groups like pulses, legumes and cereals have been reported to have an increase in the acceptability of the product (Deshpande and Poshadri, 2011; Devi *et al.*, 2013; Okpala *et al.*, 2013). Alongwith the sensory aspects, presence of certain anti-nutritional factors like phytic acid, phenols and tannins hamper their widespread use. However, processing methods like roasting, malting, germination, etc reduce the anti-nutritional content.

International Year of Millets–2023

At the initiative of Prime Minister Shri Narendra Modi, the Government of India had proposed to the United Nations to declare 2023 as International Year of Millets (IYoM). The proposal of India was supported by 72 countries and United Nations General Assembly (UNGA) declared 2023 as International Year of Millets (IYOM) in March, 2021. The action plan of IYOM-2023 focuses on strategies to enhance production, consumption, export, branding etc. Government plans to hold several programmes in the country and abroad for popularizing Millets and other nutri-cereals. Besides, all Ministries/Departments of the Government of India, along with the State Governments, will promote the nutri-cereals in coordination with the Department of Agriculture and Farmers Welfare (DA&FW) and Department of Agricultural Research and Education (DARE).

The Government has evolved the 'Seven Sutras' (themes) in the run-up to the IYOM, which will be rolled out by the concerned Ministry/Departments—Enhancement of Production/Productivity (DA&FW/DARE), Nutrition & Health benefits (Ministry of Health/FSSAI), Value Addition, Processing & Recipe Development (Ministries of Food Processing Industries & Tourism), Entrepreneurship/Startup/Collective Development (Commerce and DA& FW), Awareness creation including Branding Labelling & Promotion (All Ministries), International Outreach (Commerce & Ministry of External Affairs) and Policy Interventions for Mainstreaming (Dept of Food & Public Distribution and DA&FW).

The Government has taken several steps for promotion of Millets. To create domestic and global demand and to provide nutritional food to the people,

The National Year of Millets was celebrated in 2018. In view of the nutritional value of the millets, the Government also notified Millets as nutri-cereals in April, 2018 and Millets were included under the POSHAN Mission Abhiyan. Over 500 Startups are working in Millet value chain while the Indian Institute on Millet Research, has incubated 250 Startups under RKVY-RAFTAAR. More than Rs. 6.2 crores has been disbursed to over 66 Startups while about 25 Startups have been approved for further funding (pib.gov.in)

Millets—Government Initiatives

To promote cultivation and consumption of millets and millets-based products, the Government of India has adopted and implemented certain measures.

MSP

The government increased the MSP of millets, providing farmers with a significant price incentive. Major millets such as sorghum, pearl millet and finger millet are covered under Minimum Support Price (MSP) with increase of 42, 55 and 67% in respective MSP. The government has included millets in the public distribution system to ensure a steady market for the produce.

Creation of Value Chains

The government has included millets in the public distribution system to ensure a steady market for the produce. The initiation of the Public Private Partnership (PPP) project, mooted by the Directorate of Sorghum Research (DSR), Hyderabad was done to promote demand based millet cultivation, procurement and processing for continuous supply-chain management. It also focussed on development of new commercial food products and upscaling their production. It also encouraged new startups in this sector.

The PPP also included consortium partners like National Institute of Nutrition (NIN), Sate agriculture universities (SAU's), private partners such as ITC and linkages with Defence Food Research Laboratory (DFRL), Central Food Technological Research Institute (CFTRI), CIAE, Central Institute of Post-Harvest Engineering and Technology (CIPHET) and Home Science Colleges. The e-*choupal* infrastructure of ITC was leveraged by DSR to network hundreds of farmers in Andhra Pradesh and Maharashtra as stakeholders for technology transfer, and improvement of product-specific millets, and primary processing at farm level to value addition.

The government has made seed kits and other inputs available to farmers helped build value chains through Farmer Producer Organizations and made millets more marketable.

Encouraging Start-Ups

Apart from bringing more land under cultivation of millets, some entrepreneurship-driven manufacturing units were also set-up. The government has help set up about 300 Post-Harvesting units in Andhra Pradesh, Gujarat, Madhya Pradesh, Mahrashtra, Tamil Nadu and UP. These units have increased the supply of raw material for value added products. Marketing was facilitated by ITC and nutritional and health studies were done at NIN. A host of sorghum based products were developed and entrepreneurship promoted. This is a promising approach which can generate income and improve access to products made from nutri-cereals.

Women Employment and Empowerment

Some Home Science Colleges have involved Self Help Groups (SHGs) in preparing traditional millet-based products, teaching quality control measures, packaging and labelling and hand-holding for local marketing. A successful case study is from the University of Dharwad, where women of North Karnataka are making *roti* (salted pancakes) from sorghum, and pearl millet, and supplying to hotels "Kharnavalis" retail and wholesale dealers. A variety of other traditional and contemporary (noodles) millet-based products including small millets have been developed under projects like NAIP and IDRC. SHG women have been trained in their production and help is being given for direct marketing as quality control and market are major challenges. Such a venture can be successful if either a government institution or NGO provides supervision and marketing.

Deccan Development Society (DDS) is conducting a unique experiment of empowering many '*dalit*' rural women by organizing them into '*Sanghams*' (Voluntary village-level association of poor) and involving them in dryland agriculture using green methods of farming. Location is Zahirabad mandal of Medak District of Andhra Pradesh. A variety of minor millets besides SAM, as well as pulses, vegetables, medicinal plants etc. are cultivated on own or leased land. Seeds using scientific methods are being preserved. A millet-based, decentralised Public Distribution System (PDS), and food outlet are being run. DDS has also started a virtual Millet Network (*www.ddsindia.com*) for sharing experiences.

Other Schemes

The Ministry of Women and Child Development has been establishing Nutri-gardens, promoting research on the connections between crop diversity and dietary diversity, and running a behaviour change campaign to increase consumer demand for Nutri-cereals.

Some recent initiatives to rejuvenate millets from production to Consumption, include: "Initiative for Nutritional Security through Intensive Millets Promotion" (INSIMP), under the Rashtriya Krishi Vikas Yojna of Government of India, "Revalorising Small Millets in the Rain-fed regions of South Asia (RESMISA) funded by International Development Research Centre (IDRC) and CIDA (Canadian funds), and DSR-led value chain development approach for commercialisation of millets. Millets are an important component of the National Agriculture Innovation Projects of ICAR, and All India Coordinated Project in Home Science, Other policy initiatives include: price and procurement support for millets, inclusion of millets in the Mid-day meal programme and, promotion of Nutrifarms.

CONCLUSION

Millets, being a staple of many regions worldwide seem to be the most promising sustainable approach to tackle nutritional security around the world. They have a potential to combat hidden hunger and are nutritionally superior to cereals. Apart from being a treasure trove of nutrients, they can be cultivated in arid, semi-arid, mountainous and tribal areas without heavy inputs and can be a good source of income to tribals and farmers. These can not only serve as income crop but also improve the health of the community as a whole. The incorporation of millet-based foods into the various diets that consumers indulge in can help in promoting the cultivation of more millets that would revitalize the farmers. The inclusion of millet-based foods in international, national and state-level feeding programs will help to overcome the existing nutrient deficiencies of protein, calcium and iron in developing countries.

Millets are mainly being cultivated (91%) in rainfed regions and the average crop yield levels of these agro-ecologies are 3 to 4 folds lower than the potential. Efficient nutrient management approaches are among key strategies to realize higher millet yields in rainfed regions. Generally, as compared to irrigated crops, almost half doses of fertilizers are applied for rainfed crops. Nitrogen application is done in 2–3 splits to match the crop N requirement as it is prone to volatalization, leaching and denitrification. However, full dose of P and K are applied at the time of planting. Alongwith the application of NPK, use of compost, vermicompost and biofertilisers like *Azospirillum* and PSB are used to enhance crop growth and quality parameters. The combined use of organic and inorganic sources has good effect on physiological process of plant metabolism and growth, leading to higher yield. Foliar application of essential micronutrients like Zn and Fe; either in sulphate or chelated forms can lead to effective absorption and translocation of other macro and micro nutrients at the critical growth stages of the crop.

Despite the fact that millets are highly nutritious, their consumption is still limited to the conventional and poor population due to lack of awareness towards its nutritional values. There is lack of processing technologies, lack of food subsidies and inconvenience in food preparations which makes millets more obsolete.

With the aim to create awareness and increase production & consumption of millets, United Nations, at the behest of the Government of India, declared 2023 the International Year of Millets. The Government of India decided to celebrate IYOM, 2023 by making it a peoples' movement so that the Indian millets, recipes and value added products are accepted globally.

With the appropriate promotion of millets for their health benefits and benefits as a sustainable crop to cultivate, it can widely increase the consumption of millets. This will further contribute to improving nutrition, food security, and the welfare of farmers.

REFERENCES

[1] Bala, R.S., Swain, S., Sengotuvel, D. and Parida, N.R. (2010). Nutritious millets for enhancing income and improved nutrition: A case study from Tamilnadu and Orissa. *Minor millets in South India.*

[2] Bamboriya, S.D., Bana, R.S., Pooniya, V., Rana, K.S. and Singh, Y.V. (2017). Planting density and nitrogen management effects on productivity, quality and water-use efficiency of rainfed pearlmillet (*Pennisetum glaucum*) under conservation agriculture. *Indian Journal of Agronomy,* 62(3): 363–366.

[3] Bana, R.S., Rana, K.S., Dass, A., Choudhary, A.K., Pooniya, V., Vyas, A.K., Kaur, R., Sepat, S. and Rana, D.S. (2013). A Manual on Dryland Farming and Watershed Management. India. IARI, New Delhi, p. 104.

[4] Bana, R.S. (2014). Agrotechniques for conserving water and sustaining production in rainfed agriculture. *Indian Farming,* **63**(10): 30–35.

[5] Bana, R.S., Pooniya, V., Choudhary, A.K., Rana, K.S. and Tyagi, V.K. (2016). Influence of organic nutrient sources and moisture management on productivity, biofortification and soil health in pearlmillet (*Pennisetum glaucum*)

+ clusterbean (*Cyamopsis tetragonaloba*) intercropping system of semi-arid India. *Indian Journal of Agricultural Sciences,* 86(11): 1418–25.

[6] Bana, R.S., Sepat, S., Rana, K.S., Pooniya, V. and Choudhary, A.K. (2018). Moisture-stress management under limited and assured irrigation regimes in wheat (*Triticum aestivum*): Effects on crop productivity, water use efficiency, grain quality, nutrient acquisition and soil fertility. *Indian Journal of Agricultural Sciences,* 86(10): 1606–12.

[7] Basavaraju, T.B. and Purushotham, S. (2009). Integrated nutrient management in rainfed ragi (*Eleusine coracana* L. Gaertn.), *Mysore Journal of Agricultural Sciences,* 43: 366–368.

[8] Cothren, J.T., Matocha, J.E. and Clark, L.E. (2000). Integrated crop management for sorghum. *Sorghum: Origin, History, Technology, and Production,* pp. 409–441. C W Smith and RA Federiksen (Eds). John Wily and Sons Inc.

[9] Dass, A., Rana, K.S., Choudhary, A.K. and Bana, R.S. (2014). Climate Resilient Dryland Farming and Watershed Management, p 250. Post Graduate School, IARI, New Delhi and ICAR, DARE, Ministry of Agriculture, Govt. of India, New Delhi, India.

[10] Deshpande, H.W. and Poshadri, A. (2011). Physical and sensory characteristics of extruded snacks prepared from Foxtail millet based composite flours. *International Food Research Journal,* 18(2):751–6.

[11] Devi, N.L., Shobha, S., Tang, X., Shaur, S.A., Dogan, H. and Alavi, S. (2013). Development of protein-rich sorghum-based expanded snacks using extrusion technology. *International Journal of Food Properties,* 16(2):263–76.

[12] Devi, P.B., Vijayabharathi, R., Sathyabama, S., Malleshi, N.G. and Priyadarisini, V.B. (2014). Health benefits of finger millet (*Eleusine coracana*) polyphenols and dietary fiber: a review. *Journal of Food Science and Technology,* 51(6): 1021–1040.

[13] Kumar, A., Tomer, V., Kaur, A., Kumar, V. and Gupta, K. (2018). Millets: A solution to agrarian and nutritional challenges. *Agriculture & Food Security,* 7:31.

[14] Louhar, G., Bana, R.S., Kumar, V. and Kumar, K. (2020). Nutrient management technologies of millets for higher productivity and nutritional security. *Indian Journal of Agricultural Sciences,* 90(12): 2243–50.

[15] Mubeena, P., Halepyati, A.S. and Chittapur, B.M. (2019). Effect of date of sowing and nutrient management on nutrient uptake and yield of foxtail millet (*Setaria italica* L.). *International Journal of Bio-Resource and Stress Management,* 10(1): 92–95.

[16] Okpala, L., Okoli, E. and Udensi, E. (2013). Physico-chemical and sensory properties of cookies made from blends of germinated pigeon pea, fermented sorghum, and cocoyam flours. *Food Science and Nutrition,* 1(1):8–14.

[17] Rana, K.S. and Bana, R.S. (2012). Pearl millet. (*In*) *Text Book of Field Crops Production,* Vol. 1, p. 396. ICAR, New Delhi.

[18] Rao, B.R., Nagasampige, M.H. and Ravikiran, M. (2011). Evaluation of nutraceutical properties of selected small millets. *Journal of. Pharmaceutical and. Bioallied. Sciences*, 3:277–9.

[19] Rurinda, J., Mapfumo, P., Van-Wijk, M.T., Mtambanengwe, F., Rufino, M.C. and Chikowo, R. (2014). Comparative assessment of maize, finger millet and sorghum for household food security in the face of increasing climatic risk. *European Journal of Agronomy*,55: 29–41.

[20] Saleh, A.S.M., Zhang, Q., Chen, J. and Shen, Q. (2013). Millet grains: nutritional quality, processing, and potential health benefits. *Comprehensive Reviews in Food Science and Food Safety*, 12:281–95.

Safety Assessment of Consumption of Proso Millet and Milk

(A case study with Ayurveda perspective)

S. Deeksha, J. Deeksha, N. Pragati, B. Siddhesh, Sachin Palekar and Gauri Risbud

Ramnarain Ruia Autonomous College, Department of Bioanalytical Sciences

ABSTRACT: Millets are a group of small, round whole grains that belong to the Poaceae family. Proso (*Panicum miliaceum*) also known as Vari/Chena is a type of millet which has various health benefits. Viruddha Ahara is an important concept in Ayurveda and if consumed it can cause disbalance of dosha leading to many disorders. One such wrong food combination (Viruddha Ahara) has been mentioned in Charaka Samhita i.e. consumption of proso millet with milk which can cause psoriasis. To test this hypothesis; the pH factor of millets and millet-milk combination as well as the protein and lipid profile of millet, milk and millet-milk combination are taken into consideration. The best method for protein extraction was selected based on the protein yield obtained. For protein profiling, SDS-PAGE was performed and a comparative analysis of proteins in milk and milk-millet combination was done to check for the presence of any extra protein formed in the millet-milk combination. Further confirmatory studies for analyzing the protein profile of millet, milk, and millet milk combination have to be done by HPLC. A peak for eicosanoic acid (also known as arachidic acid) was observed in the GC-MS the for millet milk combination.

Keywords: *Millets, Proso Millet, Vari, Chena, Viruddha Ahara, Charaka Samhita, Psoriasis, Protein Profile, Lipid Profile, SDS-PAGE, GC-MS.*

INTRODUCTION

Proso millet (*Panicum miliaceum*), also called as the white millet, hog millet, or Kashif millet is widely grown and consumed in India. It is sold as a health food that has an egg-shaped fertile floret and its hull is hard and shiny which firmly encloses the seeds when it is shed. It comes in a variety of colors, from brownish black, olive brown, light cream, orange-red, or even golden color. It has various health benefits like it strengthens the nervous system, helping in lowering cholesterol, serving as an anti-ageing source, preventing pellagra conditions, being beneficial for coeliac disease, reducing the risk of diabetes and it strengthens bones. Food-food interaction is a serious issue requiring much attention to prevent any possible chronic harmful effects caused by food incompatibility. Viruddha Ahara (incompatible food combination) is an important concept in Ayurveda. According to Charaka Samhita, people who consume Viruddha ahara are prone to an imbalance of dosha which leads to many disorders. Many of these food-food interactions don't have an immediate effect, but it is always better to know their long-term effect. Viruddha Ahara, if taken regularly, can induce inflammation at a molecular level which disturbs the eicosanoid pathway, and creates more Arachidonic acid, leading to increased Prostaglandin 2 and thromboxane [Pandey, J., (2015)]. Several food incompatibilities are mentioned in classical Ayurvedic literature like Charaka and Sushruta Samhita. One such wrong food combination according to Ayurvedic literature is consuming Proso millet with milk which can lead to ekakushta (psoriasis).

The purpose of this research is to generate scientific evidence(s) regarding the viruddha ahara—milk and proso millet combination.

METHODOLOGY

A detailed literature survey was conducted to determine various methods, which are followed throughout this research project. The methods are discussed below.

This study included both cow milk and buffalo milk due to the variety in people's milk preferences.

1. ***pH Monitoring:*** The pH of the millet and millet with milk samples was monitored. Grounded proso millet powder was used for this purpose. 1 g of millet powder was added to 10 ml of water, 10 ml of cow milk, and 10 ml of buffalo milk separately. The samples were boiled and the pH was checked after the samples had cooled down.
2. **Selection of a Protein Extraction Method:** Proteins had to be extracted from millet, milk, and millet milk combination samples. Milk and millet are complex matrices. Therefore, selecting a proper extraction method that gives a good yield of proteins is an important step for establishing consistency in results. Two methods were used:
   (a) *Acetone Precipitation:* Proteins are insoluble in acetone (particularly at low temperatures), whilst many small molecules which could interfere with downstream protein work are soluble. By precipitating proteins in this solvent, buffer contaminants can be eliminated and proteins can be concentrated in a pellet which can be redissolved by other solvents.

Acetone precipitation was done via two techniques. In one technique, samples were directly added to acetone and allowed to precipitate overnight. In the second technique, the proteins were firstly extracted in ice-cold 0.125 M tris buffer (pH 8) containing 0.1% SDS & 0.01% beta-mercaptoethanol using a mortar and pestle. The samples were centrifuged and the supernatant was allowed to precipitate overnight in 2 folds of ice-cold acetone. The precipitate was collected via centrifugation it was air-dried, and reconstituted in 1ml of 0.125 Tris Buffer.

(b) *Methanol/Chloroform Extraction:* The protein is denatured in a homogeneous solution of water, methanol, and chloroform. Further addition of water makes the chloroform precipitate out of this solution as fine droplets, with the protein at the interphase. After centrifugation, the protein forms a thin film between the layers. The hydrophilic contaminants can be eliminated with the aqueous phase, and the lipophilic ones can be eliminated with the chloroform phase. [Vincent, D., Ezernieks, V., Elkins, A., Nguyen, N., Moate, P.J., Cocks, B.G. & Rochfort, S. (2016).]

The best protein yield providing method was selected from the methods listed above by performing the Folin Lowry method.

Lowry Method for Protein Quantitation: It is used to confirm the presence, and quantify the proteins in the extracted samples. The Lowry protein assay is based on the biuret reaction with additional steps and reagents to increase the sensitivity of detection. Copper interacts with four nitrogen atoms of peptides to form a cuprous complex. Lowry adds phosphomolybdic/proteins acid also known as the Folin-Ciocalteu reagent. This reagent interacts with the cuprous ions and the side chains of tyrosine, tryptophan, and cysteine to produce a blue-green color that can be detected between 650 nm and 750 nm.

3. **Sodium Dodecyl-Sulfate Polyacrylamide Gel Electrophoresis (SDS PAGE):** It is a technique for the separation of the charged protein molecules, and it depends upon the relative mobility of charged species. The smaller molecules migrate faster due to less resistance during electrophoresis. The structure and the charge of the proteins also influence the rate of migration. Sodium dodecyl sulfate and polyacrylamide eliminate the influence of the structure and charge of the proteins, and the proteins are separated based on the length of the polypeptide chain.

SDS PAGE was performed for the extracted protein samples to observe if there is the formation of any extra protein (which might be toxic) in the milk millet combination or/and for the absence of any protein in the milk millet combination on comparison with the only millet sample.

4. **Gas Chromatography-Mass Spectrometry (GC-MS):** Gas chromategraphy is a common type of chromatography used in analytical chemistry for separating and analyzing compounds that can be vaporized without decomposition. Typical uses of GC include testing the purity of a particular substance or separating the different components of a mixture.

   GC-MS analysis was performed for fatty acid profiling of millet, milk, and millet milk combination samples. Proso millet powder (1 g), millet-milk combination (0.1 g in 1 ml of milk—buffalo milk and cow milk), and milk (1 ml of both buffalo and cow milk) were added to 9 ml n-Hexane and kept in the shaker for 2 hours. The samples were syringe filtered and subjected to GC-MS analysis.

RESULTS AND DISCUSSION

1. *Monitored pH:* The pH readings of the milk and millet milk combination were found to be 6.4. However, after boiling the samples, the pH changed as follows:

Sr. No.	*Sample*	*pH*
1.	Millet + Water	6.29
2.	Millet + Cow milk	4.88
3.	Millet + Buffalo milk	4.99

   Decreased pH of milk-millet combination samples indicates that the samples have turned acidic. According to Ayurveda, consuming acidic food can aggravate pitta dosha, causing many problems like indigestion, heartburn, chest burn, itching all over the body, etc.

2. ***Selection of a Protein Extraction Method:*** On performing Folin-Lowry for estimating the protein concentration in the protein extracts obtained from three methods, it was found out that extraction of proteins using 0.125 M Tris followed by acetone precipitation gave better yield of proteins compared to other methods. Therefore, this method was finalized for performing protein extraction from the samples.

3. *SDS-PAGE Observations:*

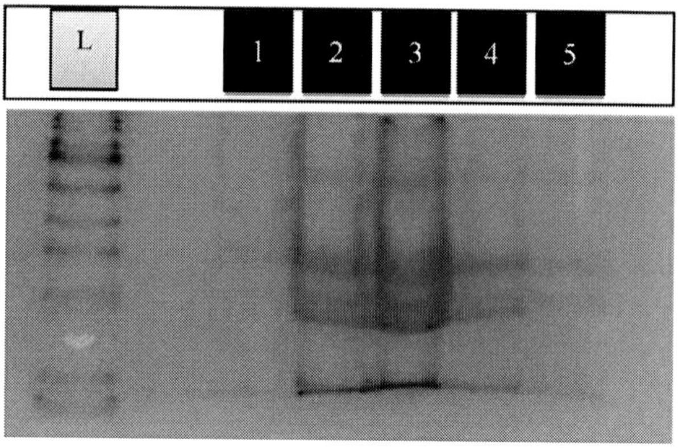

Figure 1: SDS-PAGE

No additional protein bands were observed in the milk-millet combinations. Deletion of band were not observed.

4. *GC-MS Findings:*

Number of (significant) peaks obtained in each sample:

Sr. No.	*Sample Name*	*No. of Peaks*
1.	Only millet	11
2.	Cow milk	15
3.	Buffalo milk	20
4.	Millet + Cow milk	18
5.	Millet + Buffalo milk	13

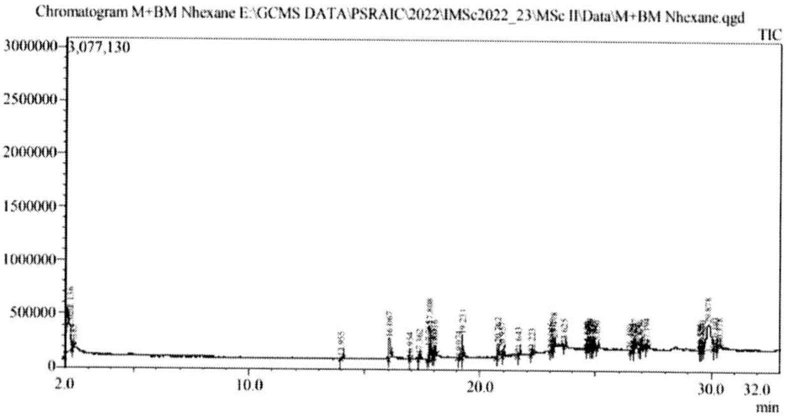

Figure 2: Millet and Cow Milk Combination

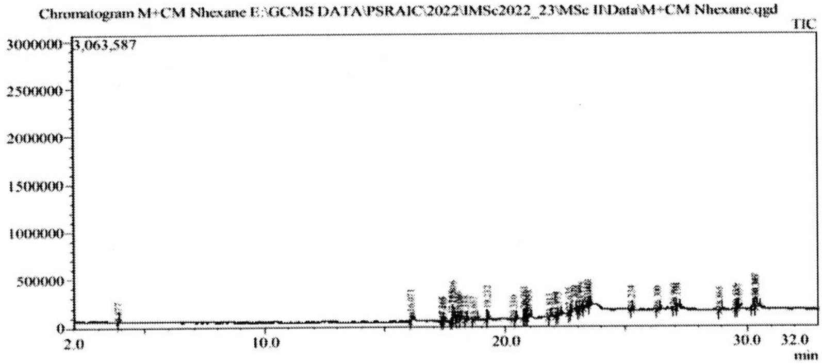

Figure 3: Millet and Buffalo Milk Combination

Table 1: GC-MS Data

Sample	*Mol. Name*	*Rt (Mins)*	*Area (%)*	*Toxic Effects*
Millet and cow milk	1. Eicosanoic acid, 2-[(1-oxo hexadecyl)oxy]-1-[[(1-oxo hexadecyl)oxy]methyl]ethyl	23.41	5.92	Arachidonic acid worsens psoriasis symptoms.
	2. Decanoic acid, 1, 2, 3–propanetriyl ester And 9-Octadecenoic acid, 1, 2, 3-propanetriyl ester, (E, E, E)-	22.85 20.33	5.73 6.08	Known to prevent blood clots but can also lead to skin itching and irritation.
	3. Citronellol epoxide	23.18	4.97	Citronellol is used in perfumes, a frequent cause of allergic contact dermatitis.
Millet and buffalo milk	9-Octadecenoic acid, 1, 2, 3-propanetriyl ester, (E, E, E)-	20.76	2.14	Known to prevent blood clots but can also lead to skin itching and irritation.

CONCLUSION

Preliminary tests were performed to check for the formation of harmful molecules in milk-millet combinations. No extra protein band was observed however, few harmful lipid molecules were found in the millet milk combinations. Hence, we can conclude that toxicity can be due to lipids and not proteins. Confirmatory tests for further analysis of proteins would be done by HPLC.

REFERENCES

[1] https://www.zizira.com/blogs/plants/7-health-benefits-proso-indian-millet

[2] Pandey, J. (2015). Viruddhahara (Antagonist food), National health portal.

[3] http://nhp.gov.in/viruddhahara-antagonistic-food-_mtl

[4] https://www.easyayurveda.com/2019/11/14/millets/amp/

[5] Vincent, D., Ezernieks, V., Elkins, A., Nguyen, N., Moate, P.J., Cocks, B.G. and Rochfort, S. (2016). Milk Bottom-up Proteomics: Method optimization. Frontiers in genetics, 6, 360. https://doi.org/10.3389/fgene.2015.00360.

[6] https://www.sciencedirect.com/topics/biochemistry-genetics-and-molecular-biology/lowry-protein-assay

[7] https://ruo.mbl.co.jp/bio/e/support/method/sds-page.html

THEME-3

Entreprenerurship, Start-Ups, Product Development and Marketing Strategies

Formulation of Value-Added Beauty Products from Jowar

Zeel Maru*, Ritika Pingle, Ashwathi Prakash, Sneha Ganpule, Aditi Patwardhan and Sneha Salgaonkar

*Student, Department of Biochemistry,
Ramnarain Ruia Autonomous College, Mumbai University

ABSTRACT: Millets have been used as a nutritional source of food for ages and are widely researched in all the fields of biology from botany, biochemistry, food science, nutrition, agriculture and others. The ease of growing, maintenance and usage classifies millets as sustainable crops. They form a major staple food in most of the regions around the globe. Additionally, they also possess the characteristics, potential and benefits to be exploited in the field of cosmetics. Thus, our study revolves around the development and utilization of millet in cosmetics. Among the most commonly cultivated millets in India namely, Ragi, bajra and jowar, our study involves the use of Sorghum due to its more appealing and promising properties compared to the other millets. The developed natural cosmetic products in our study comprises a blend of jowar (major ingredient), rose (flavouring agent). The physical properties of the millet based cosmetics was observed to be similar to the market-available cosmetics. Thus, our study paves a new pathway of using millets in cosmetics which can serve as natural alternatives to the market-available cosmetics.

Keywords: *Jowar, Rose, Compact Powder, Scrub, Natural Cosmetics.*

INTRODUCTION

Millets are a very nutrient-dense food and a great source of bioactive compounds. They are an important part of the diet and there has been a lot of research going on its agricultural and nutritional values. One of the major millets in India is Jowar (*Sorghum bicolor*). It is the third important food crop of the country after rice and wheat both in terms of area and production, therefore occupying a madjor prominence in India. It is also useful as animal feed and industrial raw materials. Nearly half the total percentage of sorghum is grown directly for human consumption and the rest is used primarily as fodder, alcohol production and industrial products (Suradkar *et al.*, 2020).

The European Directives 93/35/EEC, defines "cosmetic products", as any substance or preparation intended to be placed in contact with the various external parts of the human body (epidermis, hair system, nails, lips and external genital organs) or with the teeth and the mucous membranes of the oral cavity with a view exclusively or mainly to cleaning them, perfuming them, changing their appearance and/or correcting body odors and/or protecting them or keeping them in good condition (European Commission, 1993). Cosmetics have become an essential aspect of everyone's life, particularly for those in the workforce, as a result, the market is flooded with a variety of cosmetic items, and the industry is growing.

Facial scrubs cleanse skin, remove dirt and dead cells. They are also known as exfoliating cleansers. They help to cleanse, condition and treat the skin (Talpekar *et al.*, 2016). Natural scrubs usually contain small particles of natural ingredients. Compact loose powder is often applied after moisturizer or foundation. The powder helps to fix the make-up base, evens out and mattifies skin tone. Added to that, it evens out skin color and tones down imperfections (Valet *et al.*, 2007). Further there has been an increasing demand for herbal cosmetics due to less side effects and high potential in enriching the body with nutrients and essential minerals (Bouidin, *et al.*, 1999). Indian herbs have been used in large numbers for cosmetic and toiletry formulations (Gediya *et al.*, 2011).

Naturally derived raw materials or from organic origin may be used for formulations (IBD, 2010). Natural and organic cosmetics will be similar qualitatively with respect to the formulations but differ quantitatively. Sometimes, semi-synthetic raw materials may be used, with some exceptions. Essential oils, extracted from virtually every part of the plant, are natural fragrances used in skin care, hair care and sometimes as co-preservatives. One of the commonly known essential oils is the Rose oil—extracted from the petals of *Rosa damascena* and *Rosa centifolia*, family Rosaceae. It has topical applications and is mainly used in the perfumery industry. Multani mitti (Calcium bentonite) is widely used as a substitute for kaolin which helps in the cleaning of oil and dust particles from the skin. The varied advantages of using multani mitti in cosmetic products include diminished pore sizes of skin; removal of blackheads and whiteheads; reduction of acne and blemishes; soothening for sunburns; skin cleanser; improved blood circulation and complexion; and glowing effect to the skin as it contains healthy nutrients (Ghode *et al.*, 2019).

The great impact on the environment regarding the choice of products, packaging, and production processes has paved the way for Sustainable products (Fonseca-Santos, 2015). Eco-friendly products are the ones that were developed absolutely in an environment friendly manner—starting from conception till manufacture. Packaging should also allow the re-use or recycling without much disposal. The use of aluminum, glass or paper can be termed as green packaging (Chick, 1992). Thus, non-green products must be replaced with green products in the market, so they need to be high quality, affordable and useful products (Ottman, 1993).

Furthermore, Beauty and business do seem opposite terms but in reality, have had an important relationship that business historians are still exploring (Peiss, 2000). Aesthetics is a relative term—usability; how important is beauty as a part of experience; what is the value users attach to it; and what "consequences" of beauty are sparse and results are inconsistent. There are three major consequences of beauty: sources of value, appealing to self-referential goals and finally, its attractiveness (Hassenzahl, 2008). Therefore, taking these into consideration, it is important to form products that follow most of these parameters for better results.

Millets possess the characteristics, potential and benefits to be exploited in the field of cosmetics. Therefore, the aim of this project was to use Jowar and Rose petals powder together with other natural products to formulate a loose facial powder and a natural facial scrub. The biochemical properties and phytochemical screening of Jowar and Rose petals powder were analyzed to ascertain their quality in the formulation of cosmetic products. The formulated products were assessed for their microbial, stability and organoleptic properties which were essential to demonstrate their potential use in the cosmetic industry.

MATERIALS AND METHODS

All the raw materials used in this study to formulate the cosmetic products, namely, loose facial powder and facial scrub, were procured from a local shop in Mumbai. They were stored in airtight containers for further analysis.

Phytochemical Analysis

Jowar

Jowar grains were cleaned, dried, powdered and sieved thoroughly. The prepared sample was stored in ziplock covers for further use.

Preparation of Millet Extracts

Acid water extracts were prepared and made to stand for 24 hrs, filtered and filtrate was used for further analysis.

Figure 1: Jowar Sample

Figure 2: Jowar Extracts

Proximate Analysis

The carbohydrate, protein, minerals, ash and moisture content was analyzed according to standard AOAC procedures. The tests were carried out in triplicates.

Phytochemical Analysis

The samples were analyzed for their phytochemical constituents such as Flavonoids by Alkaline reagent test, Quinones by Sulfuric acid test, Phenols by

Folin Ciocalteau reagent test, Alkaloids by Meyer's test, Tannins by Ferric chloride test, Saponins by Foam test, Terpenoids by Salwoski test, etc.

Rose

Preparation of Millet Extracts

Rosa indica (Indian Rose) flower petals were washed thoroughly, dried at 40°C for about 24 hours and crushed to fine powder. The aqueous rose petal powder extract was then subjected to phytochemical screening.

Figure 4: Rose Petal Powder

Formulation of Loose Facial Powder

Materials: Jowar powder, arrowroot powder, rose petal powder, rose essential oil, Multani Mitti, Vitamin E tablets, mineral color.

Table 1: Formulation of Loose Facial Powder

Ingredient	*Category*	*Quantity*
Jowar powder	Brightens skin	40%
Arrowroot powder	Binding agent, moisture absorber	20%
Multani mitti	Cleansing dust and oil	40%
Rose essential oil	Fragrance	As required
Vitamin E	Free radical scavenger	1 Capsule
Mineral color	Color	3.8%

Figure 5: Loose Facial Powder

Methodology

- All the ingredients were weighed in a fixed proportion.
- Jowar powder, arrowroot powder and Multani mitti were first mixed thoroughly.
- Then mineral color and Vitamin E were added along with essential oil dropwise.
- The contents were mixed manually and slowly to avoid lump formation.
- The mixture was stored in an airtight container for further use.

Formulation of Facial Scrub

Materials: Jowar powder, arrowroot powder, Rose powder, mineral color, Jowar granules, Vit E tablets, mineral color.

Table 2: Formulation of Facial Scrub

Ingredient	*Category*	*Quantity (g/ml)*
Jowar powder	Brightens skin and removes dead cells	45%
Arrowroot powder	Binding agent, moisture absorber	9%
Multani Mitti	Cleansing dust and oil	18%
Rose petal powder	Fragrance and color	3%
Jowar granules	Removes dirt and dead cells	18–27%
Rose essential oil	Fragrance	As required
Vitamin E	Free radical scavenger	2–3 Capsules
Mineral color	Color	5%

Figure 6: Facial Scrub

Methodology

- All the ingredients were added at fixed proportions and mixed evenly.
- Granules were added at the end in small proportions.
- The mixture was stored in an airtight container.

Microbial Assessment

Sterile Agar was poured in autoclaved petri plates aseptically and allowed to solidify. Then a small amount of both the products was added on the separate agar plates. These plates were incubated for 72 hours and examined for any microbial growth.

Freeze Thaw Stability

Freeze thaw stability includes the study of the stability of the product at variable temperatures. Here the products i.e. Loose powder and Scrub were kept at 45°C for 12 hours in hot air oven, then removed and kept at room temperature for 12 hours. This same procedure was repeated by varying the temperature to –4°C for 12 hours and then again at room temperature for 12 hours. After this process, the samples were checked for their colour retention, odour, pH and texture.

Organoleptic Studies

A google questionnaire consisting of questions related to the properties of the product was prepared. Samples of both the products were distributed among a group of 32 female volunteers who consented for the study. The questions were targeted to seek the opinion of the respondents regarding the texture, colour, fragrance, skin-feel, side effects (if any) and the overall preference for both the products.

RESULTS AND DISCUSSION

Phytochemical Analysis

The phytochemical analysis of Jowar and Rose indicated that the quality of both these ingredients were good to be used in cosmetics.

Qualitative Analysis

Table 3: Qualitative analysis of Jowar and Rose

Test Parameter	*Jowar*	*Rose*
Carbohydrates	✓	✓
Proteins and Amino acids	✓	✓
Starch	✓	-
Reducing sugars	×	✓
Flavonoids	×	✓
Resins	✓	-
Quinones	×	✓
Phenols	✓	✓
Alkaloids	✓	✓
Tannins	✓	✓
Saponins	✓	×
Glycosides	×	×
Terpenoids	✓	✓

Key: ✓: Present

×: Absent

Quantitative Analysis

Table 4: Quantitative Analysis of Jowar and Rose

Test Parameter	*Jowar (mg%)*	*Rose (mg%)*
Carbohydrates	80	43
Proteins	9	5.88
Calcium	12.02	7.34
Phosphorus	3.38	2.92
Iron	0.80	–
Phenols	11.34	–

Microbial Assessment

The microbial study demonstrated that both the formulated products, namely, loose powder and facial scrub, were free of any microbial growth in them for 72 hrs. Thus, indicating good microbial stability of the formulated products.

Freeze Thaw Stability

The colour, odour and pH of the scrub and loose powder did not change after the freeze thaw stability testing.

Organoleptic Studies

The responses of the female volunteers in the organoleptic studies.

Facial Scrub

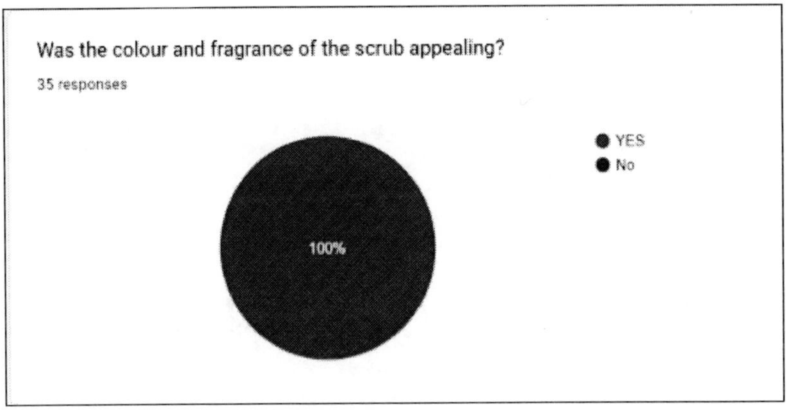

Graph 1

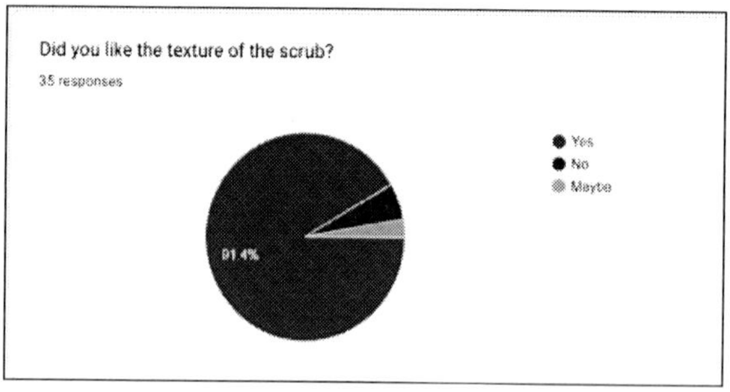

Graph 2

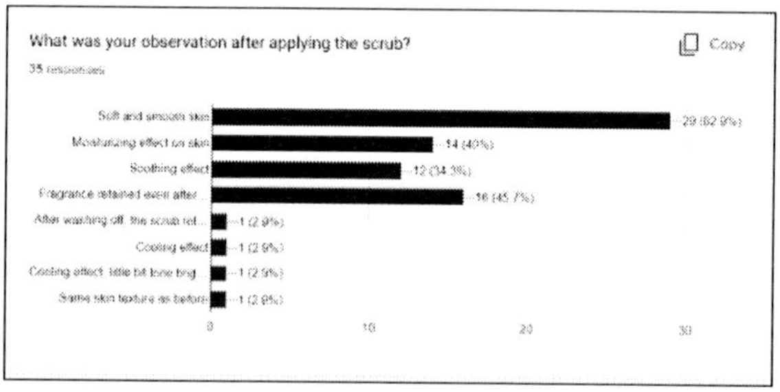

Graph 3

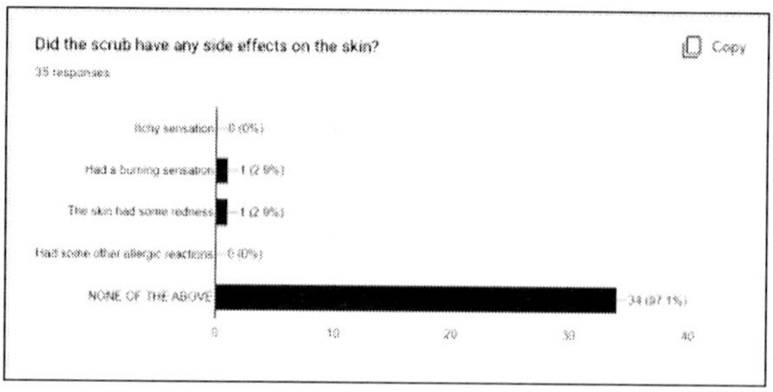

Graph 4

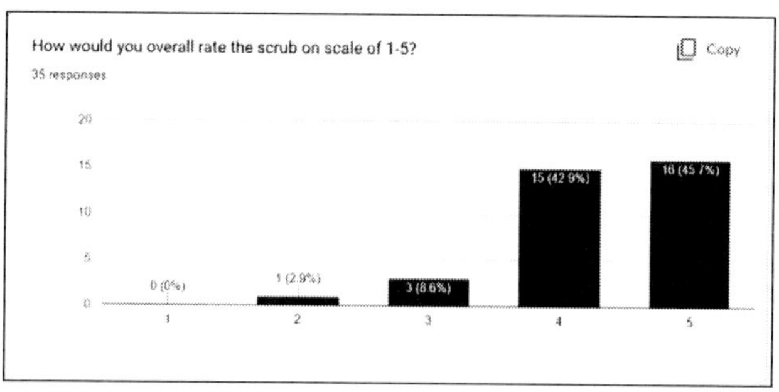

Graph 5

Loose Powder

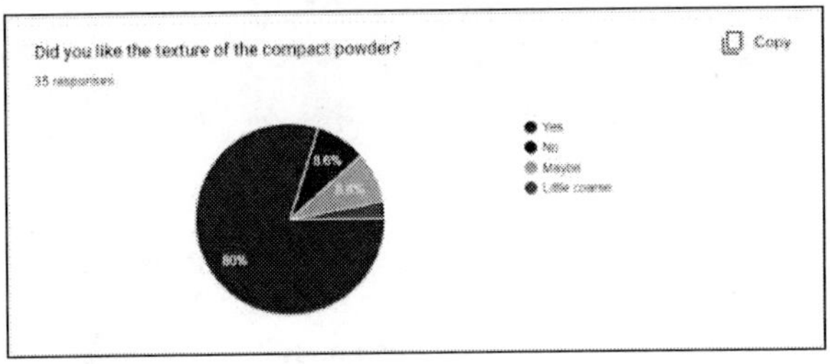

Graph 6

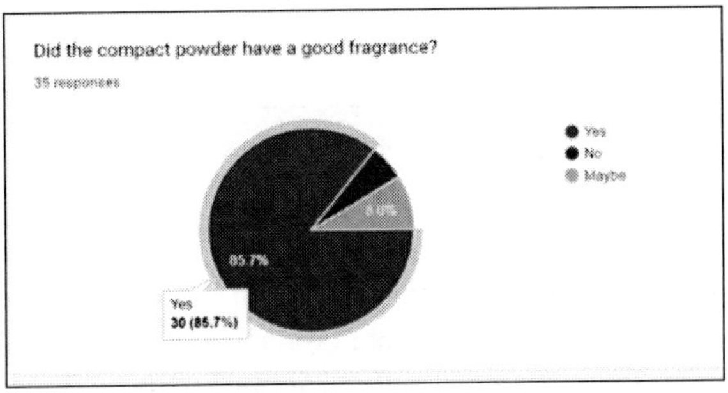

Graph 7

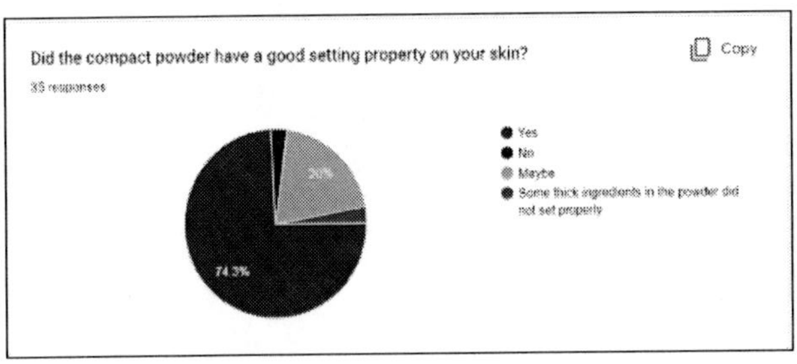

Graph 8

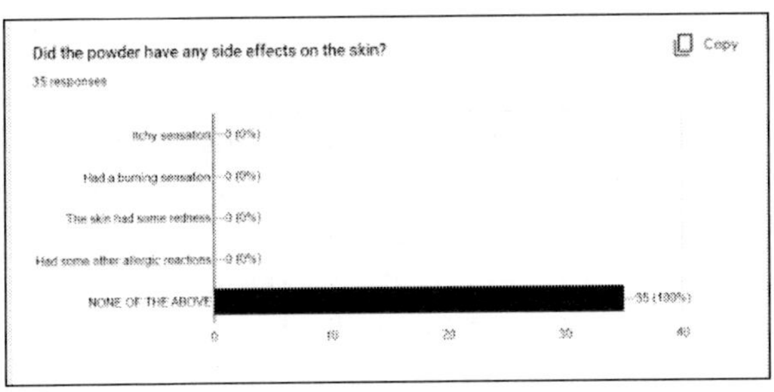

Graph 9

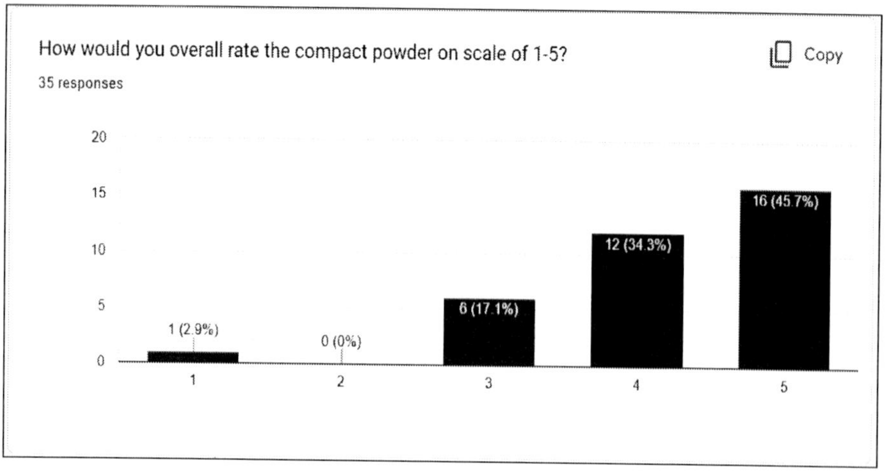

Graph 10

CONCLUSION

- Millets have good properties to be used in cosmetics and can therefore serve as a natural alternative to the market available cosmetics.
- Rose petals impart natural fragrance and colour to the cosmetics. However, its colour is not stable due to deterioration of anthocyanins present in it.
- Value added ingredients helped to enhance the properties of the formulated product.
- The formulated products were found to have minimal side effects and can thus be used more often as natural cosmetics.

ACKNOWLEDGEMENTS

We would like to express our heartfelt gratitude to Dr.Anushree Lokur, Principal, Ramnarain Ruia Autonomous College, Dr. Kamini Donde—Research Dean, Mr. Prashant Masali, Head of Department of Biochemistry, Ms. Priyanka Koli, Assistant Professor, Department of Biochemistry, Ramnarain Ruia Autonomous College, Mumbai University for their extended support and guidance throughout the project. Special thanks to the support staff of the Biochemistry Department, Ramnarain Ruia Autonomous College for their support throughout the entire project. Our humble gratitude to all the junior staff teachers—Ramnivas Ruia College, Mumbai University, for their active participation in the project. Finally I would like to acknowledge my Family and friends for their continuous support and encouragement in pursuing this project.

REFERENCES

[1] Amadou, I., Gounga, M.E. and Le, G.W. (2013). Millets: Nutritional composition, some health benefits and processing-A review. Emirates Journal of Food and Agriculture, 501–508.

[2] Bouidin, A.S., et al. Social science medicine, 1999, 49, 279–289.

[3] European Commission, Directives 93/35/EEC, official journel of European Commission I. Series., 1993, 151.

[4] Fonseca-Santos, B., Corrêa, M.A. and Chorilli, M. (2015). Sustainability, natural and organic cosmetics: Consumer, products, efficacy, toxicological and regulatory considerations. Brazilian Journal of Pharmaceutical Sciences, 51, 17–26.

[5] Gediya, S.K., Mistry, R.B., Patel, U.K., Blessy, M. and Jain, H.N. (2011). Herbal plants: used as a cosmetics. J Nat Prod Plant Resour, 1(1), 24–32.

[6] Ghadage, P.K., Mahamuni, S.S. and Kachare, D.S. (2021). Formulation and evaluation of herbal scrub using tamarind peel. Research Journal of Topical and Cosmetic Sciences, 12(1), 39–42.

[7] Ghode, D.S.P., Chatur, V.M., Ghode, D.P.D., Shaha, N., Prajapati, S. and Thorave, A. (2019). Formulation and Evaluation of Facial Scrub Containing Sunflower Seeds and Other Natural Ingredients. World Journal of Pharmaceutical Research, 8(9), 1772–1781.

[8] Grace, X.F., Vijetha, R.J., Shanmuganathan, S. and Chamundeeswari, D. (2014). Preparation and evaluation of herbal face pack. Adv J Pharm Life Sci Res, 2(3), 1–6.

[9] Hassenzahl, M. (2008). Aesthetics in interactive products: Correlates and consequences of beauty. In Product experience (pp. 287–302). Elsevier.

[10] Peiss, K. (2000). On beauty… and the history of business. Enterprise & Society, 1(3), 485–506.

[11] Rao, P.P., Birthal, P.S., Reddy, B.V., Rai, K.N. and Ramesh, S. (2006). Diagnostics of sorghum and pearl millet grains-based nutrition in India. International sorghum and millets newsletter, 47, 93–96.

[12] Suradkar, S.W. and Waghmare, S.A. Effect on Germination Pattern of Jowar (Sorghum Vulgare) of Some Sulphur and Nitrogen Containing Heterocyclic Compounds.

[13] Talpekar, P. and Borikar, M. (2016). Formulation, development and comparative study of facial scrub using synethetic and natural exfoliant. Research Journal of Topical and Cosmetic Sciences, 7(1), 1–8.

[14] Valet, B., Mayor, M., Fitoussi, F., Capellier, R., Dormoy, M. and Ginestar, J. (2007). Colouring Agents in Decorative and other Cosmetics. Analytical Methods. Analysis of cosmetic products, 141–152.

Development of Nutrient-Dense Cookies for Female Athletes between the Age Group (of 18–23 years) in Mumbai

Samruddhi Shinde, Yashika Shah and Sengupta Rupali

Smt. Manjulaben Gunvantri Shah Department of Post Graduate Studies
Post Graduate Diploma in Sports Science Fitness Nutrition
Dr. BMN College of Home Science (Autonomous)
SNDT Women's University, Matunga, Mumbai

ABSTRACT

Introduction: The female athlete triad is a disorder that women can develop. The disorder affects body functions that are connected to each other. The main cause of the female athlete triad is an energy imbalance.

Aim: To prepare modified nutrient-dense cookies for female athletes in sports.

Objective: 1. To modify a cookie into a nutritious snack. 2. To study the shelf life of the product. 3. To do a sensory evaluation of the product. 4. To incorporate millet for female athletes.

Methodology: This process's first step was standardisation of the amount of ingredients to be added to make the product more nutritious and tasty. Naive panellist was there for sensory evaluation. Then microbial testing was done using pore plate method.

Conclusion: Cookies were crispy and had high level of energy and fats.

Keywords: *Millets, Female Athlete Triad, Millet-rich Cookies, Energy Enhancement, Micronutrient Rich, Sports Athlete.*

INTRODUCTION

In 2020 Sahaja *et al.* reported that Finger millet, also known as ragi, is a cereal popular in South India and many African countries. It is regarded as a weight-loss miracle grain. It contains a lot of fibre and is high in calcium. It is regarded as the most effective treatment for diabetics. It aids in the control of blood sugar levels. Anemia: Ragi is high in iron and aids in the improvement of haemoglobin levels. Insomnia: Ragi contains antioxidants such as tryptophan and amino acids, which help with sleep. This aids in the reduction of anxiety, depression, and insomnia. It is high in protein, vitamins, minerals, fibre, and energy. Manganese, phosphorus, iron, copper, chromium, magnesium, zinc, molybdenum, and selenium are also present. Also, it is gluten free, it is suitable for gluten intolerant people.

Finger millet has the most calcium (344 mg %) and potassium (408 mg %) of any cereal or millet. The cereal has a low fat content (1.3%) and is mostly made up of unsaturated fat. 100 gram of finger millet contain approximately 336 Kcal of energy on average.

MATERIALS AND METHODS

The product was standardized first by standardizing every ingredient on weighing machines. The ingredients ragi, oats, wheat flour, butter, and almonds were used for developing the product. The process started with the roasting of ragi and oats followed by making a syrup of jaggery. All syrup and flour were mixed to make a dough while gradually adding milk. The dough was refrigerated for 30 mins then was baked in an oven for 15–20 mins at 180 degrees. Material used were ragi flour, wheat flour, oats, almonds, Jagger and the ratio was 2:1:1:1:2.

Sensory Evaluation

The sensory evaluation was done by naive panellists. A total number of 20 members had participated in the sensory evaluation. The panellists were given a simple score card. The attributes given to evaluate were, appearance, taste, aroma, colour and texture. The product was approved as the majority of the scores were above average.

Microbial Test

The microbiological testing was performed to determine the number of colonies produced by the product. This was done by first creating a broth. The procedure includes preparing the plate counting agar. 23.5 g of agar was suspended in 1000 cc of distilled water. Which was put to boil and it was completely diluted. It was sterilised in an autoclave at 121 degrees Fahrenheit for 15 minutes. It was then kept chilled until it reached 45–50 degrees. This sterilised agar broth was then placed in the petri dish, and the sample was ground and streaked on the petri dish using the streaking procedure.

After 48 hours in the incubator, the colony growth was counted under the microscope.

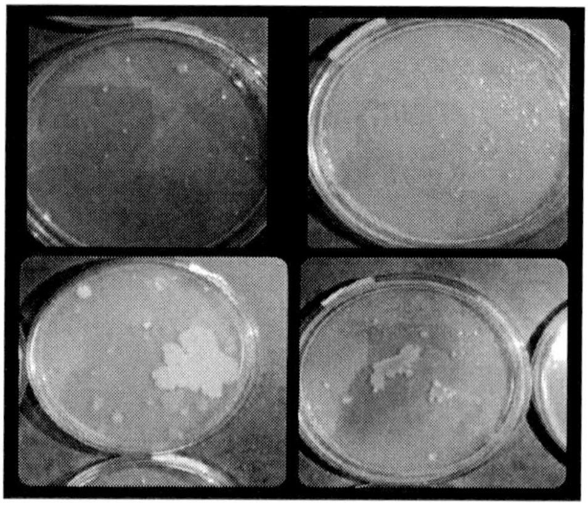

RESULTS AND DISCUSSION

The results of sensory evaluation for the millet cookie.

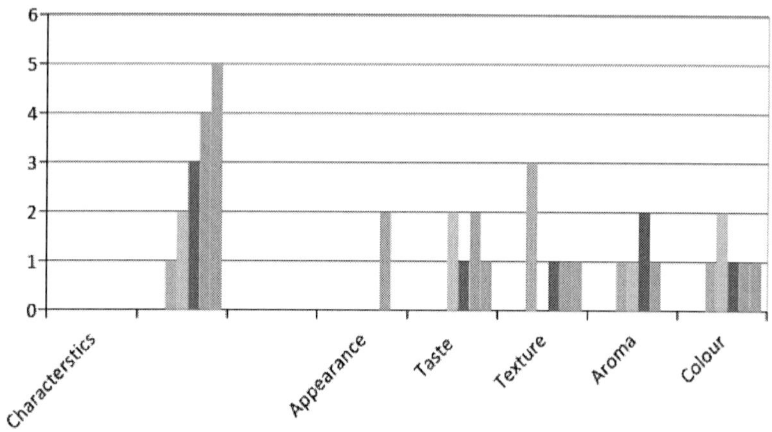

Figure 1: Sensory Evaluation of Millet-cookie

The result of sensory evaluation was depicted in Figure 1. The scale used for sensory evaluation was Likert scale on various attributes which were considered under study. Qualitative test showed an acceptance of modified product by naïve panellist.

NUTRITIVE VALUE OF MILLET-COOKIE

The above Table 1 indicates the nutritive value of millet based cookie. Cookie is high in energy that is 397 kcal. The protein value is 9 g whereas carbohydrate is 61.5 g. The fat content of the product is 12.7 g.

Table 1: Nutritive Value of Millet-Cookie

Ingredients	*Amount(g)*	*Energy (kcal)*	*Carbohydrates (g)*	*Protein (g)*	*Fat (g)*
Ragi flour	30	100	20	3	0.6
Wheat flour	15	50	10	1.5	0.3
Oats	15	50	10	1.5	0.3
Jaggery	20	80	20	–	–
Almond	8	50	1.5	3	4
Ghee	7.5	67.5	–	–	7.5
Total		**397 kcal**	**61.5 g**	**9 g**	**12.7 g**

Microbial Test

The results of the microbial test were obtained from four samples of petri dishes. A gradual decline in the colony count was noted in the petri plate. This indicated a least count in the plate.

Counts of such organisms, cultivated as colonies on or in nutrient agar, are a good way of determining a water's overall bacterial concentration. Following incubation at 20–22°C, the colony count, or plate count, indicates the diversity of bacteria present at normal environmental conditions.

CONCLUSION

Sensory evaluators approved of this product. It's a handy cookie that's easy to transport and eat. Millets, which are high in nutrients, are a better option than cereals. These goods are gluten-free, making them a better choice. This cookie is free of preservatives and additives, making it safe to ingest. It is high in energy and protein and can be consumed within two weeks of production. Because of its high nutritional value, it is easy to digest and provides an instant boost to the athlete's energy levels.

REFERENCES

[1] Asritha, V., A Review on Role of Millets in Weight Loss. *Indian J Nutri,* 2021; 8(3): 239.

[2] Anitha, S., Kane-Potaka, J., Botha, R., Givens, D.I., Sulaiman, N.L.B., Upadhyay, S., ... and Bhandari, R.K. (2021). Millets can have a major impact on improving iron status, hemoglobin level, and in reducing iron deficiency anemia—A systematic review and meta-analysis. *Frontiers in Nutrition,* 712.

[3] Aljobair, M.O. (2022). Physicochemical properties and sensory attributes of cookies prepared from sorghum and millet composite flour. *Food Science & Nutrition.*

[4] Alaunyte, I., Stojceska, V. and Plunkett, A. (2015). Iron and the female athlete: a review of dietary treatment methods for improving iron status and exercise performance. *Journal of the international society of sports nutrition, 12*(1), 38.

[5] Beals, K.A. and Manore, M.M. (2002). Disorders of the female athlete triad among collegiate athletes. *International journal of sport nutrition and exercise metabolism, 12*(3), 281–293.

[6] Cover, K., Hanna, M, and Barnes, M.R. A Review and Proposed Treatment Approach for the Young Athlete at High Risk for the Female Athlete Triad. *ICAN: Infant, Child and Adolescent Nutrition*, 2012.

[7] Gull, A., Jan, R., Nayik, G.A., Prasad, K. and Kumar, P. (2014). Significance of finger millet in nutrition, health and value added products: a review. *Magnesium (mg), 130*(32), 120.

[8] Kamatar, M.Y. (2013). Noble millet food products for quality life of all walks of life and age groups (in) International symposium on RTE Foods: Innovations in Ready-toeat products: Drivers, trends and emerging technologies held during 24–25 Sept. 2013 at Mumbai. *Maharastra, India,* pp. 24–26.

[9] M.K. and Sundgot-Borgen, J. (2005). The female athlete triad: are elite athletes at increased risk? *Medicine & Science in Sports & Exercise, 37*(2), 184–193.

[10] Hobart, J.A. and Smucker, D.R. (2000). The female athlete triad. *American family physician, 61*(11), 3357–3364.

[11] Rizwana, M., Singh, P., Ahalya, N. and Mohanasundaram, T. (2022). Assessing the awareness of nutritional benefits of millets amongst women in Bangalore. *British Food Journal*, (ahead-of-print).

[12] Sundgot-Borgen, J., Torstveit, M.K. and Trainingcenter, O. (2003). The female athlete triad-the role of nutrition. *Schweizerische Zeitschrift fur Sportmedizin und Sporttraumatologie, 51*(1), 47–52.

[13] De Souza, M.J., Koltun, K.J. and Williams, N.I. (2019). The role of energy availability in reproductive function in the female athlete triad and extension of its effects to men: an initial working model of a similar syndrome in male athletes. *Sports Medicine, 49*(2), 125–137.Torstveit

[14] Statuta, Siobhan M. MD. The Female Athlete Triad, Relative Energy Deficiency in Sport, and the Male Athlete Triad: The Exploration of Low-Energy Syndromes in Athletes. *Current Sports Medicine Reports*, 19(2), pp. 43–44, February 2020.

[15] Tumwine, G., Atukwase, A., Tumuhimbise, G.A., Tucungwirwe, F. and Linnemann, A. (2019). Production of nutrient enhanced millet based composite flour using skimmed milk powder and vegetables. *Food Science & Nutrition, 7*(1), 22–34.

[16] Yenilmez, M.I. and Palamutoglu, M.I. (2020). Critical Thoughts and Insights on the Female Athlete Triad: Precedents, Existing Challenges and Prospects. *Journal of Management Information and Decision Sciences, 23*(4), 324–331.

Development of Metal Oxide Based Bio-Nanocomposites and their Application in Increasing Shelf Life of Millets

S.R. Wadekar*, M.S. Hate and R.S. Chaughule

Ramnarain Ruia Autonomous College, Mumbai, India

ABSTRACT: Spoilage of millet plants takes place due to variety of microorganism like bacterial and fungal species or due to environmental circumstances. This spoilage cannot be completely stopped but can be minimized by increased in its shelf life through nanotechnology. Nanotechnology has revolutionized agricultural sector to food packaging industry through development of new techniques like climate-smart agriculture, increasing plant nutrients, diseases control and management etc. Till today Conventional plastic films have been used for agricultural mulching process. Mulching improves crop yield, saves water for irrigation, generate use of pesticides and promote proper growth of plants. These conventional plastics generate polyethylene residues which can degrade agricultural land leading to landfills and various environmental hazards. Biodegradable plastic (bioplastic) serves as an alternative to conventional plastic for mulching process of millets in agricultural sector. Zinc oxide nanoparticles (ZnO NPs) is considered as GRAS (generally recognized as safe) by US food and drug administration and show antimicrobial activity, therefore could be used for crop protection against various microorganism. This study reviews biosynthesis of zinc oxide nanoparticles (ZnO NPs) using bahava plant extract (*Cassia Fistula*) and combining it with the natural biopolymer forming bio-nanocomposite films. These films would show antimicrobial activity against various pathogen of millets and increasing their shelf-life extension. Also, the percentage quantity of ZnO NPs required to show antimicrobial activity would be studied.

Keywords: *Mulching, ZnO NPs, GRAS, Biopolymer, Bionanocompoite Films.*

INTRODUCTION

Agriculture is the backbone of any country, but unfortunately now, the agriculture sector is facing various global challenges like climate changes, urbanization, sustainable use of resources, and environmental issues such as runoff, accumulation of pesticides and fertilizers. Millets are the group of small-seeded annual grasses that are grown as grain crop primarily on marginal land in dry areas and belong to the "Poaceae" family. Nanotechnology has gained intense attention in the recent years due to its wide application in the recent years in several areas like medicine, medical drugs, food packaging and agricultural sectors etc. Some of the challenges like agricultural production and risk of plant related disease can be treated using nanotechnology. Many organizations, researchers and industries are coming up with novel techniques, methods and products that show a direct application of nanotechnology in agricultural sector.

Millets is the only crop that will address important issues in the future like food, feed, malnutrition, health and climate change. Millets are adapted to the wide range of ecological conditions demanding less water and fit well even in fertile soil. India is the highest producer of millets in the globe and 5th largest exporter of millets. Millets are highly nutritious and can be a solution for India's chronic disease burden. Millets are rich in dietary fiber, both soluble and insoluble. They are rich in potassium, calcium, iron, zinc, phosphorous, magnesium etc. that support kidney and heart function. Due to benefits of millets, the Government of India has approved 2018 as the national year of millets to boost production rich millets. Millets are also termed as the "miracle grain" or "crop of the future" as they can grow in any soil and in any harsh circumstances.

Mulching is the process or practice of covering the soil/ground to make more suitable conditions for plant growth, soil moisture, suppress weed growth development and efficient crop production. Mulches got attention in the late 1930, as these materials can alter the surrounding conditions of agricultural lands, forest areas, and horticultural lands. In 1941, deep mulches were also used for trees and shrub plantation because these deep-type mulches give protection to drought stress such as frost injury and freezing damage of crop plants in harsh environmental conditions. More water conservation was achieved when the same quantity of mulch material was used as mulch compared with that incorporated into the soil. However, the agricultural and forest domains are also used as mulches in late 1900s. Although mulching has many advantages it does show some negative effects. The major negative impact in agriculture is handling the plastic waste and its associated environmental impacts. The plastic films significantly reduce soil organic matter and greatly increases greenhouse gas emission. Plastic waste management is the major problem in plastic mulching process. Plastic mulching creates difficulties in dumping and emits harmful gases during burning. They promote soil degradation, causing soil water repellency and occur as potent pollutant for soil. It also becomes problematic in top dressing of fertilizers when plastic mulches are used. Sometimes natural mulches like straw, hay and seeds may develop weeds. Organic mulching material such as wood chips and bark may occur as soil acidifier. Too much organic mulch can lead to excessive moisture, creating new problems such as pests, anaerobic conditions and rotting of the roots that can damage plants. Straw mulch often contaminate the soil and deplete the seedbed nitrogen due to high carbon to nitrogen ratio. Moreover, when carbon rich materials such as straw and stalks are used for mulching, more and more nitrogen are used for its decomposing those material. Thus, nitrogen is temporarily unavailable for plant growth. In paper mulch, paper degrades more rapidly and also becomes difficult to cover large area using newspaper. Conventional plastic usage leads to environmental pollution and ends up with landfills and pollution, therefore researchers are now focusing on using an alternative for conventional plastic and came up with an idea of a biodegradable plastic (bioplastic).

(a) (b)

Figure 1(a)&(b): Mulching Process

Among various metal oxide nanoparticles, Zinc oxide nanoparticles (ZnO NPs) is considered as one of the most important metal oxide nanoparticles and are employed in various fields due to their peculiar physical and chemical properties. As the main component of various enzymes, zinc takes part in various body's metabolism and play crucial role in proteins and nucleic acid synthesis. Also, ZnO NPs is considered as "GRAS" (Generally Recognized as Safe) substance by US food and drug administration (FDA) and also shows excellent bio-compatibility, economic and low toxicity, and hence used in agriculture. These innovative bio-polymer nanocomposite films can be potentially applied in agricultural sector with improved properties and has paved ways for researchers to establish a promising route to develop metal-oxide based bio nanocomposite films.

This study aims at biosynthesis of zinc oxide nanoparticles and combining it with a biopolymer forming bio-nanocomposite films (bioplastic). Biopolymers are polymers which consist of monomeric unit and are bonded covalently forming long chain like molecules. These are Bio-degradable materials and are further broken down to organic waste like carbon-dioxide (CO_2) and water (H_2O) by naturally occurring microorganisms and bacteria. This bioplastic can be used for mulching process and to protect various millet crop from different fungal infections.

MATERIALS AND METHODS

Materials: Bahava leaves, zinc nitrate hexahydrate ($Zn(NO_3)_2 \cdot 6H_2O$), 2M NaOH, agar powder, chitosan powder from 99% shrimps, starch (potato and corn), gelatin, glycerol (all chemicals are of AR grade and 99% pure) and silicon tray.

Method

Step 1: Synthesis of zinc oxide nanoparticles.

Bahava leaves were dried at 60 degree Celsius in an oven and grinded into a fine powder. 10 g of powdered leaves is boiled with 150 ml of distilled water and stirred at 100 degree Celsius for 1 hr. After 1 hr filter the extract was filtered using whatmann filter paper no. 41 and stored at 4 degree Celsius for further use.

25 ml of plant extract was stirred at 60–70 degree Celsius and then 2.5 g of zinc nitrate hexahydrate is added and continuously stirred for 30 min. After 30 min 2M NaOH is added dropwise till pH 12 is attained (checked with pH meter). After cooling the solution was filtered using whatmann filter paper no. 41 and dried in an oven at 60 degree Celsius followed by in furnace at 500 degree Celsius for 2 hrs. The resultant product is ZnO NPs.

Step 2: Development of bio-nanocomposite films.

4 ml of Glycerol is diluted in 1000 ml of distilled water out of which 360 ml is taken in a beaker and 1 g of agar powder is added and stir well at 45 degree Celsius.

1% solution of chitosan in acetic acid is prepared and stirred for 24 hrs. 10 ml solution of chitosan is mixed with the agar solution and again stirred at 90 degree Celsius. The hot solution was poured on a silicon tray and keep it for drying in sunlight for 48 hrs.

After 48 hrs the resultant film is a biofilm.

RESULTS AND DISCUSSION

- Zinc-oxide based bio-nanocomposite film (bioplastic) could be a best alternative for protection of millet crop plant against fungal infection. The exact size of nanoparticles can be confirmed through various analytical techniques.
- These bioplastics are biodegradable and are synthesized in an eco-friendly manner.
- Biodegradation of bioplastic takes around 7 months to 1 year as compared to conventional plastic which requires 500–1000 years.
- The Zinc oxide nanoparticles and chitosan present in the films also helps in plant growth and prevention from fungal infection.

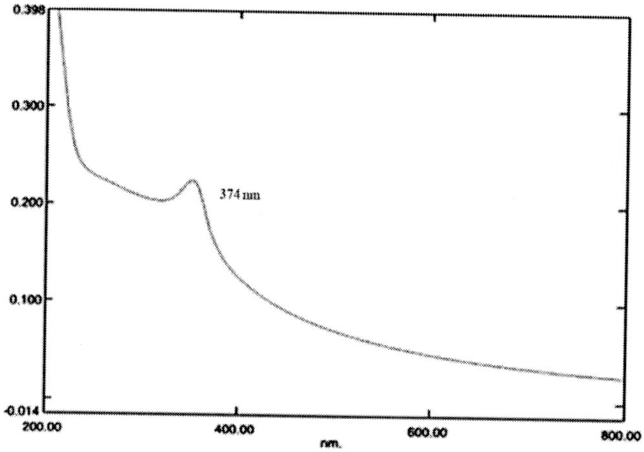

Figure 2: UV-VIS Spectroscopy of ZnO NPs

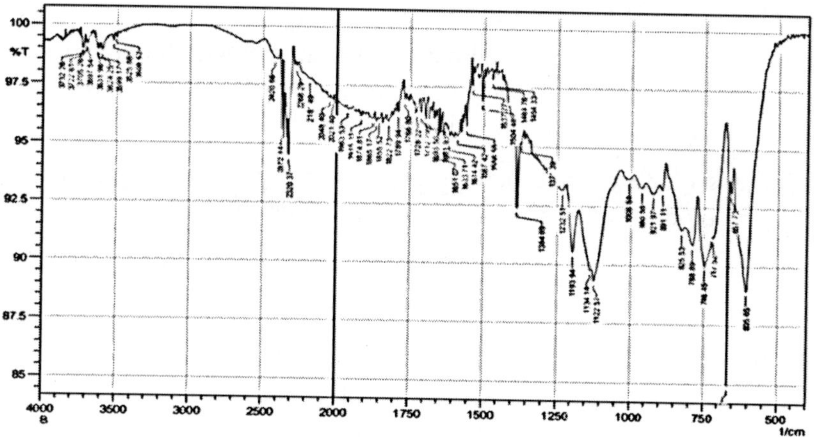

Figure 3: FT-IR Analysis of ZnO NPs

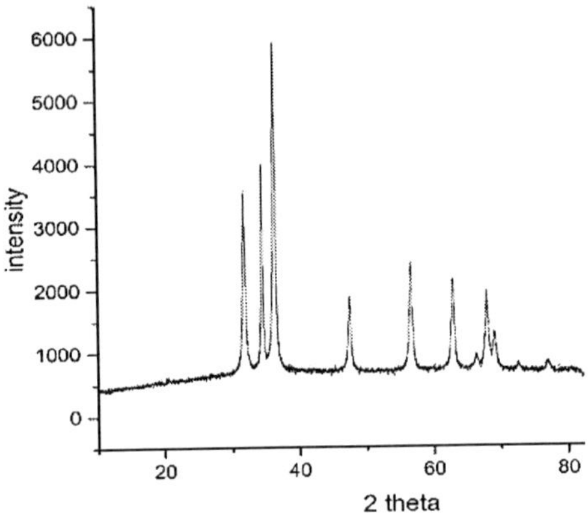

Figure 4: PXRD Analysis of ZnO NPs with Crystalline Size of 14.91 nm

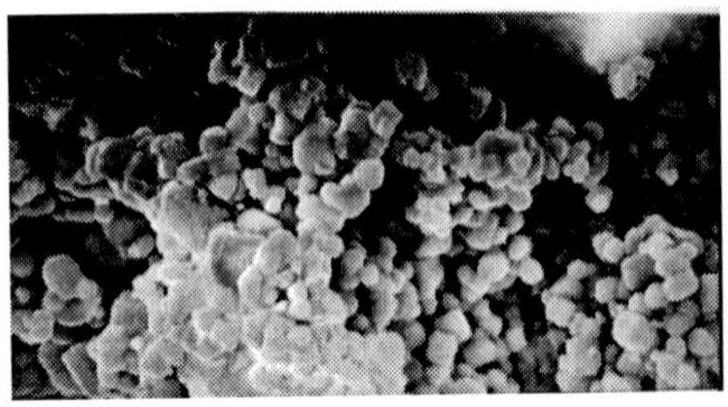

Figure 5: SEM Analysis of ZnO NPs

CONCLUSION

Green methods would be considered as safer and environmentally friendly than physical and chemical methods.

The green source acts as a stabilizing and reducing agent for the synthesis of nanoparticles of controlled size and shape.

Addition of metal-oxide nanoparticles into bio-polymer matrix would allow production of bio-nanocomposite films with increased mechanical strength, water and oxygen barrier properties and can confer other benefits including antimicrobial activity.

Zinc oxide nanoparticles are safe suggesting that exposure to nanomaterials and its associated health risk is low.

In general, combination of metal-oxide nanoparticles with biopolymers leads to bio-nanocomposite films with improved antimicrobial properties and are useful for increasing shelf life of millet plants.

Biosynthesis of Zinc oxide nanoparticles from plants is considered as eco-friendly method as compared to chemical and physical method.

Bioplastic developed are biodegradable and therefore helps in protection of plants against fungal infection.

Along with plant protection the crop yield and quality can also be enhanced.

Plasticulture is crucial to Indian agriculture in view of the changing technological scenario for boosting crop yields and productivity.

ACKNOWLEDGEMENT

Authors wish to gratefully thank ICT (Institute of chemical technology) Matunga, and Ramnathan Laboratory (Ruia college) for characterizations. Author also wish to thank IIT-Bombay (Indian institute of technology) for carrying out certain tests and analysis.

REFERENCES

[1] Gunvant H. Sonawane, Sandip P. Patil, Shirish H. Sonawane, 2018, *Applications of nanomaterials: Nanocomposites and its applications*, pp. 1–22.

[2] Santosh Kumar, Abinab Mudai, Barnali Roy, Indra Bhusan Basumatary, Avik Mukharjee and Joydeep Dutta (2020). Biodegradable hybrid nanocomposites of chitosan-gelatin and green synthesized zinc oxide nanoparticles for food packaging, 9(9), p. 1143.

[3] Chu, Zhuangzhuang; Zhao, Tianruli; Fan, Jian and Qin, Yuyue (2017). Characterization of antimicrobial poly (lactic acid)/nano-composites films with silver and Zinc oxide nanoparticles, 10(6), p. 659.

[4] Chiarakorn, Siriluk; Perpoonwiwat, Chompoonuh K. and Nanthachtchavankul, Papondhanai, Cost benefit analysis of bioplastic production in Thailand, pp. 56–85.

[5] Thakur, Sourbh; Chaudhary, Jyoti; Sharma, Bhawna; Verma, Ankit; Tamulevicius, Sigitas; Thakur, Vijay Kumar (2018). Sustainability of bio-plastic: Opportunities and Challenges. *Current opinion in green and sustainability chemistry,* 2018.

[6] Abbas, Mohsin; Buntinx, Mieke; Deferme, Wim and Peeters, Roos (2019). (Bio) polymer /ZnO Nanocomposites for packaging applications: A review of gas barrier and mechanical properties. *Nanomaterials,* 2019, 9(10), 1494.

[7] Kritiga. Narayanaswamy; Rajalakshmi, Athimoolam and Jayachitra, Aayyavoo (2015). Green synthesis of silver nanoparticles using leaf extract of Clitoria ternatea and Solanum nigrum, and study of its antibacterial effects against common nosocomial pathogens. *Journal of Nanoscience,* Article ID.928204.

[8] Steinmetz, Zacharias; Wollmann, Claudia; Schaefer, Miriam; Buchmann, Christain; David, Jan; Troger, Josephine; Munoz, Katherine; Fror, Oliver and Schaumann, Gabriele Ellen (2016). Plastic mulching in agriculture and trading short term agronomic benefits for long term soil degradation.

[9] Merino, Danila; Simonutti, Roberto; Peretto, Giovanni and Athanassiou, Athanassia (2021). Direct transformation of industrial vegetable waste into bioplastic composite intended for agricultural mulch films. *Journal of green chemistry,* 2021.

[10] Ciriminna, Rosaria and Pagliaro, Mario (2019). Biodegradable and compostable plastic: A critical prespective on the dawn of their global adoption, Vol. 9, pp. 8–13.

Utilisation of Sorghum-Stover Agro Waste as Culture Medium for *Aspergillus niger* and Extraction of Microbial Oil

Saloni Malekar, Gauri Kulkarni, Sayali Patil and Sneha Salgaonkar

Department of Biochemistry, Ramnarain Ruia Autonomous College, Matunga, Mumbai

ABSTRACT: Sorghum stover refers to residual leaves and stalks that remain after harvesting sorghum grains. This agro-waste, if disposed of unsystematically or burnt off, causes environmental hazards and promotes global warming by generating greenhouse gases. Present study is aimed at sustainable management of this stover by developing a microbial growth medium for oleaginous fungus, particularly *Aspergillus niger*. The carbohydrate-rich stover was suitable and used to cause increased microbial oil production by *A. niger*. The microbial oil was extracted from the biomass of the fungus using the Soxhlet extraction method. This microbial oil was further assessed for its biochemical parameters like acid value, and peroxide value, as well as for its antimicrobial and antioxidant potential. The extracted microbial oil can have various applications such as the production of essential oils, cosmetics, nutraceuticals, soaps and detergents. Thus, the utilisation of sorghum stover to develop a microbial-culture medium will have dual benefits namely, management of the agro waste and development of a value-added product.

Keywords: *Sorghum stover, Culture Medium, Aspergillus niger, Microbial Oil, Agro Waste.*

INTRODUCTION

India ranks sixth in total Sorghum production with a contribution of 4.4 million metric tons, according to the USDA (2022) report. Most of the sorghum produced in India comes from Maharashtra (47%), followed by Karnataka (22%), Madhya Pradesh (8%), and Rajasthan (4%). As the sorghum grain is the only component of the sorghum plant that is edible, other parts of the plants such as residual leaves and stalks–collectively called stover–add to the agricultural waste. This agro waste, if disposed of unsystematically or burnt off, has harmful effects on the environment such as the release of greenhouse gases and carbon monoxide (Maji *et al.*, 2020). One of the ways for sustainable management of the sorghum stover agro waste can be to utilize it in preparation of a value-added product.

Aspergillus niger is an oleaginous fungus, which means it can accumulate 20–80% lipid in storage form and thus can be used to extract microbial oil (Alrubayae *et al.*, 2021). Microbial oils have varied applications in biodiesel, nutraceuticals, cosmetics, essential fatty acids, soap and detergent industries and thus can be used as an alternative to plant and animal-based oils and fats (Dyal *et al.*, 2005). Oleaginous fungi, in presence of an excess carbohydrate source but a limited nitrogen source, accumulate citric acid which is further metabolized to acetyl Co-A and oxaloacetate. A continuous supply of acetyl Co-A and NADPH triggers reverse beta-oxidation to synthesise lipids. These lipids are finally stored in the form of triacylglycerols in the microbial cell (Ochsenreither *et al.*, 2016).

Sorghum stover is rich in carbohydrates, particularly cellulose and hemicellulose, but has limited nitrogen content (Sekhon *et al.*, 2016). This composition ofsorghum stover can serve as a nutritional source in the fungal growth medium thereby enhancing lipid production in the microbe. Therefore, the present study aimed to develop a culture medium using sorghum stover for *Aspergillus niger.*

MATERIALS AND METHODS

Isolation of *Aspergillus niger*

A. niger was isolated by making a suspension of the black mold found on onion bulbs (Ko *et al.*, 2002). The suspension was made in sterile distilled water, inoculated in 100 mL of Potato dextrose broth and kept at 37°C for a 4-day incubation period (Passamani *et al.*, 2014).

Identification of *A. niger*

After the four-day incubation period, a loopful of the culture was stained with Sudan Black (Burdon, 1946) and Lactophenol cotton blue stains (Benazir *et al.*, 2011); the stained mounts were observed under a light microscope. Apart from morphology, growth and colony characteristics helped in identifying *A. niger* (Moslem *et al.*, 2010). The same culture of *A. niger* was inoculated in the sustainable medium prepared from sorghum stover.

Preparation of Sorghum Stover Broth (SSB)

Sorghum-stover broth comprised (g/L): 150 g dried and powdered sorghum stover plus 20 g dextrose. 150 g of the stover was boiled in 1000 ml of distilled water for 40 minutes and the infusion was strained through cheesecloth. To the infusion, 20 g of dextrose was added and the volume was made up to 1000 ml with distilled water. The broth was autoclaved at 120°C at 15 psi for 20 minutes.

Culturing *Aspergillus niger* in SSB

A suspension of *A. niger* was inoculated in SSB and was kept at 37°C. After a 4-day incubation period, the *A. niger* colony formed on the top of the broth was separated, dried at 105°C for 1 hour (Xie *et al.*, 2007), powdered and stored in an airtight bag at room temperature until further use.

Extraction of Microbial Oil from *A. niger* Grown on SSB

Microbial oil from *A. niger* was extracted by the Soxhlet extraction method using n-hexane as the solvent (Li *et al.*, 2014). The extracted oil was stored in an airtight bottle at 4°C until further use.

Biochemical Analyses of *A. niger* Oil

The acid value and peroxide value of *A. niger* oil were determined by using the methods described by Sadasivam (1996). Further, the same analyses were performed for refined soybean oil and the values obtained from both oils were compared.

Antimicrobial and Antioxidant Properties of *A. niger* Oil

The microbial oil was checked for its antimicrobial activity against *Staphylococcus aureus* by the disc-diffusion method; 0.2 g% ampicillin was used as a positive control whereas saline was used as a negative control. The antioxidant property of the microbial oil was qualitatively determined by Thin-layer chromatography (TLC) using 0.2% DPPH in methanol as the derivatizing agent (Sethiya *et al.*, 2013) and 5% Ascorbic acid was used as the positive control; petroleum ether: diethyl ether: acetone (90:10:1) was used as the solvent system (Upadhyay *et al.*, 1993).

RESULTS AND DISCUSSION

The *A. niger* specimens stained with Sudan black and Lactophenol Cotton Blue (LCB) are given in Figure 1. Sudan black stained the hyphae as well as the vesicles and the conidiophores, whereas the hyphae remained unstained with LCB.

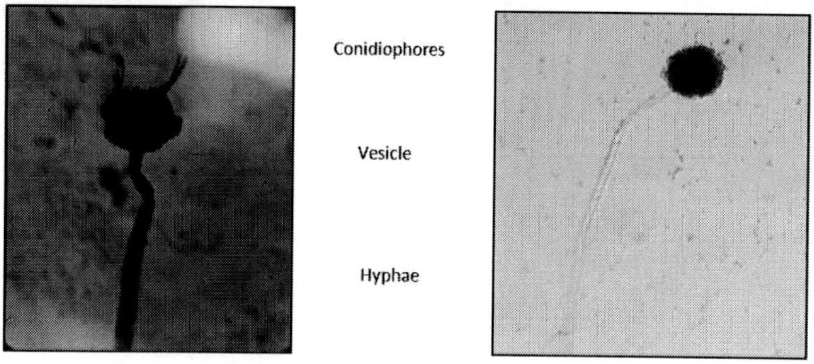

(a) (b)

Figure 1: a) *A. niger* Stained with Sudan Black. Vesicles, Conidiophores and Hyphae are Stained Black (characteristic of *A. niger*), Observed under 1000x. b) *A. niger* Vesicle Stained Blue with Lactophenol Cotton Blue, Observed Under 450x.

Unlike the characteristics of the microbial colony cultured in PDB, the one cultured in SSB showed a thinner layer with the reverse side being smooth and off-white. These unique features will aid in the screening of *A. niger* cultured in SSB.

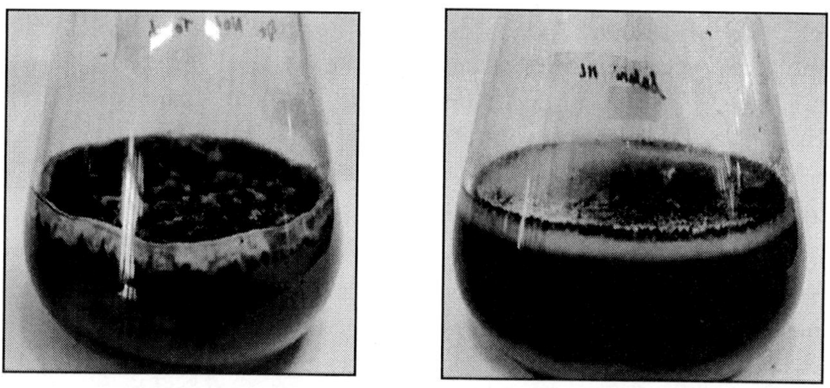

Figure 2: a) *A. niger* Culture in Potato Dextrose Broth; b) *A. niger* Cultured in Sorghum Stover Broth (after 4 days of incubation)

The average dry cell yield of *A. niger* cultured in PDB and SSB is provided in the following table. Also, the cost-benefit of SSB over PDB was analysed and is tabulated as follows.

Table 1: Cost-benefit Analysis of Sorghum Stover Broth over Potato Dextrose Broth

Culture Medium	*Requirements (Per litre)*	*Source*	*Cost (Indian National Rupees)*	*Total cost (Indian National Rupees)*
Potato Dextrose Broth (PDB)	Potatoes: 200 g	Local market	5	25.8
	Dextrose: 20 g	Commercial	20.8	
	Distilled water: 1 L	Inhouse	0	
Sorghum Stover Broth(SSB)	Sorghum driver: 150 g	Field waste	0	20.8
	Dextrose:20 g	Commercial	20.8	
	Distilled water:1 L	Inhouse	0	
Cost-benefit from using SSB as an alternative to PDB = 24.04%				

Table 2: Average Dry Cell Weight of *A. niger* Cultured in SSB and PDB (after 4-day incubation)

Culture Medium	*Average Dry-Cell Weight (g/100 mL medium)*
SSB	0.302 ± 0.05
PDB	0.734 ± 0.06

Using Sorghum Stover Broth instead of Potato Dextrose Broth will be economically profitable thereby helping in the management of agro waste. As the biomass yield of *A. niger* cultured in PDB was more than that cultured in SSB, using SSB for screening purposes would prove more beneficial.

Extracted Microbial Oil

The microbial oil extracted showed a total yield of 11.4%. At first, the appearance of the oil was pale yellow, which turned darker as days passed. The results of the biochemical analysis of the *A. niger* oil and soybean oil are given in the below table.

Table 3: Acid Value and Peroxide Value of *A. niger* Oil and Soybean Oil

	Acid Value (mg KOH/ g oil)	*Peroxide Value (mEq peroxide/kg oil)*
A. niger oil	62.05	51.24
Soybean oil	0.47	4.75

The high acid value and peroxide values may be due to the autoxidation of the microbial oil. Oils with a higher content of unsaturated fatty acids are more susceptible to autoxidation as these fatty acids are more unstable than saturated fatty acids (Buist *et al.,* 2010, Negash *et al.*, 2019). Therefore, the extracted microbial oil is rich in unsaturated fatty acids. However, due to higher peroxide value, which indicates high levels of free radicals, the microbial oil needs to be added with antioxidants to make it more stable and less susceptible to autoxidation.

A. niger oil showed antimicrobial activity against *Staphylococcus aureus* with a minimum zone of inhibition, measuring 8 mm, as compared to the positive control. The presence of a zone of inhibition indicates that *S.aureus* is sensitive to the *A. niger* oil and that the oil inhibits the growth of *S.aureus* in a certain way. The components of the oil which exhibit antimicrobial potential need to be further analysed to understand their spectrum of activity and toxicity levels.

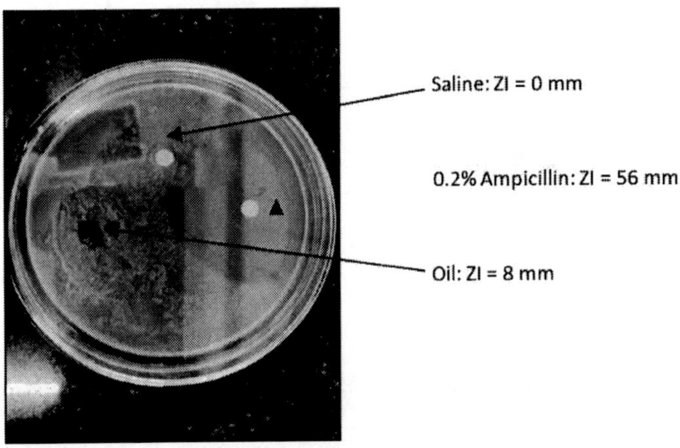

Figure 3: Antimicrobial Activity by Disc-Diffusion Method
(ZI stands for Zone of inhibition)

Thin Layer Chromatography was used to detect the presence of antioxidants in the microbial oil. Two components were separated and were visible as coloured spots before derivatization. After the addition of 0.2% methanol, the spot close to the sample application point turned yellow showing the presence of antioxidant activity.

Owing to the solvent system used, the least non-polar components such as phospholipids and fatty acids get separated faster whereas the most non-polar components such as triacylglycerides travel the farthest on the stationary phase (Bilyk *et al.*, 1991). According to research conducted by El-Nekeety *et al.* (2016), *A. niger* extract showed antioxidant activity and was positively correlated to the number of phenolic substances. This extract also consisted of eicosapentaenoic acid (EPA) which is yet another antioxidant. Considering the direction of the reducing spot on the TLC plate, the antioxidant component could be a phenolic lipid or EPA.

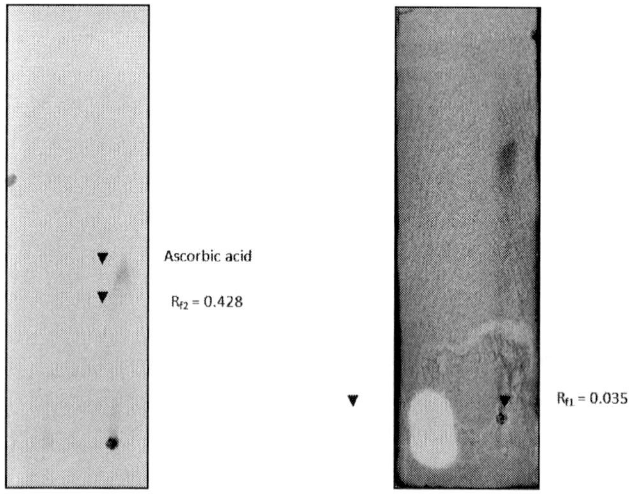

Figure 4: a) TLC Slide before Derivatization Showing Two Separated Components; b) Same TLC Slide after Derivatization showing a Discolouration of DPPH Near the Sample Application Spot

CONCLUSION

With an increase in demand for millets such as sorghum, the subsequent agricultural production to meet the demand will also rise leading to an equivalent increase in the generation of agro waste. Utilization of sorghum stover in preparation of Sorghum Stover Broth will help in the management of agro waste and will also prove economically beneficial when compared to the use of potatoes in preparation of Potato Dextrose Broth. Our study demonstrated that Sorghum Stover Broth proves to be suitable for the growth of *Aspergillus niger* as well as for the synthesis and storage of lipids in the same. These microbial lipids when extracted were shown to have antioxidant and antimicrobial components present in them, although the high acid and peroxide values (which might indicate high levels of unsaturated fatty acids) make it advisable to make certain modifications to the oil that will prevent the oxidation of fatty acids and enhance the shelf life of the oil. Further research, using techniques like GC-MS or HPTLC, will provide more data on the composition of the microbial oil. Using *A. niger* oil as an alternative to plant and animal-based oils and fats, moreover, culturing that *A. niger* in a sustainable medium like SSB, will lower the upstream processing costs, improve the energy crisis, and play a part in sustainable management of agro waste.

REFERENCES

[1] Alrubayae, I.M. and Kadhim, K.F. (2020). Mycodiesel Production from some Isolates of Oleaginous Fungi Isolated from Oil-Rich Soils in Basrah. *International Journal of Pharmaceutical Research, 12*(4).

[2] Buist, P.H. (2010). Unsaturated fatty acids. Comprehensive Natural Products II. Elsevier, Oxford, 5–33.

[3] Burdon, K.L. (1946). Fatty material in bacteria and fungi revealed by staining dried, fixed slide preparations. *Journal of bacteriology*, 52(6), 665–678.

[4] Benazir, J.F., Suganthi, R., Anjana, H., Kumar, V.R., Aswathi, M.P., Niraimathi, G., ... and Santhi, R. (2011). Bio utilization of agroindustrial waste in solid state fermentation by Aspergillus niger for the production of protease. *Asiatic Journal of Biotechnology Resources,* 2(4), 422–435.

[5] Bilyk, A., Piazza, G.J., Bistline Jr, R.G. and Haas, M.J. (1991). Separation of cholesterol, and fatty acylglycerols, acids and amides by thin-layer chromatography. *Lipids*, 26(5), 405–406.

[6] Dyal, S.D. and Narine, S.S. (2005). Implications for the use of Mortierella fungi in the industrial production of essential fatty acids. *Food research international,* 38(4), 445–467.

[7] El-Neekety, A.A., Abdel-Aziz, M.S., Hathout, A.S., Hamed, A.A., Sabry, B.A., Ghareeb, M.A., ... and Abdel-Wahhab, M.A. (2016). Molecular identification of newly isolated non-toxigenic fungal strains having antiaflatoxigenic, antimicrobial and antioxidant activities. *Der Pharm Chem*, 8, 121–134.

[8] Ko, S.S., Huang, J.W., Wang, J.F., Shanmugasundaram, S. and Chang, W.N. (2002). Evaluation of onion cultivars for resistance to Aspergillus niger, the causal agent of black mold. *Journal of the American Society for Horticultural Science*, 127(4), 697–702.

[9] Li, Y., Ghasemi Naghdi, F., Garg, S., Adarme-Vega, T.C., Thurecht, K.J., Ghafor, W.A., ... and Schenk, P.M. (2014). A comparative study: The impact of different lipid extraction methods on current microalgal lipid research. Microbial cell factories, 13(1), 1–9.

[10] Maji, S., Dwivedi, D.H., Singh, N., Kishor, S. and Gond, M. (2020). Agricultural waste: Its impact on environment and management approaches. *In Emerging eco-friendly green technologies for wastewater treatment* (pp. 329–351). Springer, Singapore.

[11] Moslem, M.A., Mashraqi, A., Abd-Elsalam, K.A., Bahkali, A.H. and Elnagaer, M.A. (2010). Molecular detection of ochratoxigenic Aspergillus species isolated from coffee beans in Saudi Arabia. *Genetics and Molecular Research,* 9(4), 2292–2299.

[12] Negash, Y.A., Amare, D.E., Bitew, B.D. and Dagne, H. (2019). Assessment of quality of edible vegetable oils accessed in Gondar City, *Northwest Ethiopia. BMC research notes,* 12(1), 1–5.

[13] Ochsenreither, K., Glück, C., Stressler, T., Fischer, L. and Syldatk, C. (2016). Production strategies and applications of microbial single-cell oils. *Frontiers in microbiology*, 7, 1539.

[14] Passamani, F.R.F., Hernandes, T., Lopes, N.A., Bastos, S.C., Santiago, W.D., Cardoso, M.D.G. and Batista, L.R. (2014). Effect of temperature, water activity, and pH on growth and production of ochratoxin A by Aspergillus niger and Aspergillus carbonarius from Brazilian grapes. *Journal of food protection*, 77(11), 1947–1952.

[15] Sadasivam, S. (1996). *Biochemical methods*. New age international.

[16] Sekhon, R.S., Breitzman, M.W., Silva, R.R., Santoro, N., Rooney, W.L., De Leon, N. and Kaeppler, S.M. (2016). Stover composition in maize and sorghum reveals remarkable genetic variation and plasticity for carbohydrate accumulation. *Frontiers in plant science*, 7, 822.

[17] Sethiya, N.K., Raja, M.M.M. and Mishra, S.H. (2013). Antioxidant markers based TLC-DPPH differentiation on four commercialized botanical sources of Shankhpushpi (A Medhya Rasayana): A preliminary assessment. *Journal of advanced pharmaceutical technology & research,* 4(1), 25.

[18] Upadhyay, A., Upadhyay, K. and Nath, N. (1993). Biophysical chemistry principles and techniques.

[19] USDA (2022). Foreign Agricultural Service. Sorghum Explorer. (https://ipad.fas.usda.gov/cropexplorer/cropview/commodityView.aspx?cropid=0459200&sel_year=2022&rankby=Production). (Accessed 18th December 2022)

[20] Xie, G. and West, T.P. (2007). Citric acid production by Aspergillus niger on condensed corn distillers solubles. *Research Journal of Microbiology,* 2(5), 481–485.

Development of Gluten-Free & Sugar-Free Cal-Mill Bhakarwadi for Young Female Adults in Central Mumbai

Maisheri Dhanvi, Mhatre Sakshi and Sengupta Rupali

Smt. Manjulaben Gunvantrai Shah Department of Post Graduate Studies, Clinical Nutrition and Dietetics, Dr. BMN College of Home Science (Autonomous), SNDT Women's University, Matunga, Mumbai

ABSTRACT: Millets are small seeds which provide a major source of energy and proteins. This study focused on developing a millet-based food product rich in multiple nutrients which will benefit undernourished female adults. The aim of the study was to develop a millet-based food product, to understand the benefits of millets on health, to understand marketing of millet-based food products. Bhakarwadi is a wheat based snack and addition of millets in bhakarwadi can enhance its nutritional quality. The process of making the bhakarwadi involves germination, preparing the dough, filling and frying. The study consisted of various phases including standardization of the product, the sensory evaluation and microbial analysis, lastly, the cost and nutritional parameters were evaluated. Amaranth incorporated in this recipe is rich in fiber and micronutrients such as Vitamins B3, B2, C, B12, Calcium and Magnesium. Jowar is rich in Fe and Ca and ingredients like Pumpkin seeds & Foxtail Millet have health benefits such as being antioxidant, anti-inflammatory, anti-diabetic, anti-hyper glycemic and antilipidemic. The recipe is gluten-free and sugar-free. This product is rich in multiple macro and micronutrients (especially calcium) which makes it beneficial for undernourished adults. Millets have additional health benefits like having gluten-free proteins, high fiber content, low glycemic index and rich in bioactive compounds. Millets are also rich in calcium, Fe, B2, B3, folic acid, beta carotene. The anti-hyperglycemic and anti-lipidemic activities of millets have also been established.

Keywords: *Millets, Food Product, Indian Snack, Anti-Hyperglycemic, Anti-Lipidemic.*

INTRODUCTION

The United Nations General Assembly (UNGA) has declared 2023 as the International Year of Millets on 5th March, 2021. Millets are a collective group of small seeded annual grasses that are grown as grain crops, primarily on marginal land in dry areas of temperate, subtropical and tropical regions. (FAO, 2009) Millets are staple food in many African and Asiatic countries. Small millets are nutritionally superior to large-grained cereals like wheat or barley, in terms of proteins, minerals, and vitamins.

Millets appear to have a wide range of nutritive and physiological advantages, according to numerous research. They are high in protein, dietary fibre, polyphenols, calcium, iron, and dietary fibre (Devi *et al.*, 2011; Kaur *et al.,* 2014). As a result, it can help feed the vegetarian population the nutrients they need. Additionally, millets are a non-gas-forming, non-gluten, low glycemic index, and readily digestible diet (Chandrasekara *et al.,* 2012) Millets are not only a great source of nutrition, but they are also good for the environment because they can withstand extreme weather. (Taylor & Awika., 2017) Millet-based food products can offer a nutritious source of food because of all these adventages. Malnutrition and bad food habits have a well-established connection.

One of the causes contributing to the rise in malnutrition around the world is the consumption of unhealthy foods and drinks that are high in calories but low in nutrients. Deficit or excess nutrient intake, nutritional imbalances, or impaired nutrient use are all signs of malnutrition. Approximately one in three people experience some sort of malnutrition (WHO, 2016). Even micronutrients, in addition to macronutrients, are crucial for illness prevention.

The goal of this endeavour is to modify "bhakarwadi" a popular traditional Indian snack. It features a fried pinwheel shape and a hot or sour flavour. Being one of the most consumed snacks in western India its filling is typically comprised of coconut, sesame, and poppy seeds, along with a sour ingredient like lemon, amchur powder, or tamarind pulp. They are typically made of gram flour, or refined flour.

Indian Statistics and Millets' Advantages

With a market share of approximately 40.62% and an anticipated production of 10.91 million tonnes in 2018–2019, India dominates the world's millets market. Despite being the world's top producer of the nutrient-dense millet and coming in second for rice and pulses, India unfortunately comes in second for child malnutrition rates. More than one-third of the malnourished children in the world live in India. In contrast, the nation has also developed into a centre for the overweight and diabetic population, which places a double burden of malnutrition on the nation. (Gurikar *et al.,* 2022)

Millets are an excellent food for the diabetic population since they have a low carbohydrate content and a high dietary fibre content that helps regulate hyperglycemia. Millets are therefore a significant component of the modern diet and a potential supply of key nutrients, particularly in developing and poor nations. Before eating, millets are typically treated to remove the unpalatable parts, lengthen the shelf life, and enhance the nutritional and sensory qualities. To make millets fit for food, primary processing methods such as dehulling, soaking, germination, roasting, drying, polishing, and milling (size reduction) are used. At the same time, millet-based value-added processed food products are created using contemporary or secondary processing techniques such as fermentation, parboiling, frying, puffing, popping, malting, baking, flaking, extrusion, etc. Although the goal of these processing methods is to increase nutrient bioavailability and digestibility, a sizable portion of nutrients are lost during subsequent processing (Gurikar *et al.,* 2022).

Benefits of Developed Food Product

Gluten-free: The raw materials used in gluten-free goods all include corn, rice, soy, cassava, and potatoes. In conventional products, these substances take the place of cereals like wheat, rye, and barley that contain gluten. Though compositional trends may differ by food type, generally speaking, gluten-free items are higher in fat, sugar, and sodium than conventional products. Sales of gluten-free products have increased as a result of media and celebrity endorsements of the gluten-free diet for weight loss. (Joye *et al.,* 2018)

Sugar-free: Limiting sugar consumption is a crucial aspect of living a healthy lifestyle and may help prevent diseases that are linked to food. Consuming too much sugar can cause type 2 diabetes and cardiovascular illnesses both directly and indirectly. The health of a young woman's offspring, commencing in the earliest stages of life, is especially dependent on her diet's reduction of sugar intake. (French *et al.,* 2018)

Nature of Millets

Compared to other cereals and millets, foxtail millet is regarded as a sustainable alternative protein source (Sachdev *et al.,* 2021). Millets are a good source of iron and zinc, which boosts immunity (Kumar *et al.*, 2018). Overall, tiny millets are promising cereals with the potential to provide long-term food security. Due to the recent Covid-19 epidemic, there has been a significant shift in emphasis over the past two years toward the frequent consumption of foods that increase immunity. Compared to other cereal crops, millets have a stronger necessary amino acid profile and easily digested proteins. Due to these qualities, millets are more nutrient-dense than other major grains and are preferred for providing food security during pandemics. (Muthamilian, 2022)

MATERIALS AND METHODS

Raw Materials and Ingredients

The ingredients that were used to make Cal-Mill Bhakarwadi were purchased from the local market of Mumbai, Maharashtra, India. It includes: (1) 7 gm Rajgira flour, (2) 7 gm Foxtail millet flour, (3) 5 gm Bajra Flour, (4) 5 gms gram flour, (5) 15 gm Dessicated coconut, (6) 1 tbsp sesame seeds, (7) 1 tbsp poppy seeds (8) 1 tbsp pumpkin seeds (9) 1 tbsp fennel seeds (10) 30 ml Oil (11) ¼ tsp turmeric powder (12) 2 pinch asafoetida (hing) (13) 1 tsp red chili powder (14) 1 tsp cumin powder (15) ½ tsp garam masala (16) ½ tsp amchur powder (17) 1 tsp salt.

Preparation of Bhakarwadi Dough

Rajgira flour, Foxtail Millet flour, Bajra flour, Bengal gram flour, turmeric powder, asafoetida, salt and oil were mixed in a bowl. The flours were kneaded into a dough with water.

Preparation of Bhakarwadi Stuffing

White sesame seeds, dessicated coconut, poppy seeds, fennel seeds, pumpkin seeds were heated in a pan on a low to medium flame. Spices like asafoetida, kashmiri red chilli powder, coriander powder, cumin powder, garam masala, amchur powder and salt were added after turning off the heat and mixed well. After cooling, the mixture was grinded into semi fine powder. This semi fine powder was mixed with tamarind pulp for a tangy taste.

Assembling and Preparation of Bhakarwadi Pinwheels

The dough was divided into two parts and rolled into a round shape. The stuffing was spread on the dough and the dough was tightly rolled into a log. This log was divided into equal parts to get the shape of a pinwheel.

Frying the Bhakarwadi

Oil was heated in kadhai and bhakarwadi pinwheels were fried in heated oil. It was then stored in an airtight container once cooled at room temperature.

For Microbial Analysis

For the microbial analysis, the pour plate technique was used. In 5.6 nutrient agar. From the 5.6 nutrient agar, half of the amount- 2.8 agar was dissolved in 100 ml distilled water in a conical flask. In another flask, 100 ml water with 1 g of sodium chloride and 0.1 g of sodium arsenic was taken. It was then autoclaved with cotton plugs for 30 mins. After 30 mins the saline solution was taken out and mixed with 1 g of the product (the appam premix) Five test tubes were taken and another 100 ml of fresh saline solution (not the one mixed with the product) was poured in each one of the test tubes and labelled 1–5, t 10 ml of the product saline was pipetted out and poured into the 1st test tube, then another 10 ml of the saline was pipetted out- now directly from the first test tube and poured into the test tube number 2. This process was repeated for another three test tubes by pipetting out 10 ml of saline from the preceding number of test tube, for example for test tube number 2 the saline is taken from test tube number 1, for test tube number 3 the saline is taken from test tube 2, and so on. After this 8 petri dishes were taken and the saline solution was poured from the 4 test tubes into the 4 petri dishes and label the dishes according to their test tubes respectively, these 4 petri dishes were covered with another 4 petri dishes. The nutrient agar solution was then mixed with the saline solution in the petri dishes with minimal possible air contact. All these dishes were incubated for 48 hours and the colony counting was done via a colony counter.

Physicochemical Analysis

Sensory Evaluation

The freshly prepared Bhakarwadi was presented to 10 members to evaluate on the basis of sensory attributes—color, flavor, aroma, texture and overall acceptability. For each sample panelists scored their liking of these characteristics using a 9-point hedonic scale (1-dislike extremely, 2-dislike

Table 1: Sensory Evaluation

Attributes → *Panel No. ↓*	*Color*	*Flavour*	*Aroma*	*Texture*	*Overall Acceptability*
1.	6	8	6	7	7.5
2.	8	8.5	7	7	9
3.	5	8	6.5	8	7.5
4.	7	8	8	6	8
5.	8	9	8	5.5	8.5
6.	8	7.5	9	9	9
7.	5	8.5	8	9	9
8.	9	9	8	8.5	8.5
9.	7	7	7.5	8	9
10.	9	7	8.5	9	9

very much, 3-dislike moderately, 4-dislike slightly, 5-neither like nor dislike, 6-like slightly, 7-like moderately, 8-like very much, 9-like extremely).

Nutrient Values

Table 2: Nutrient Value

Ingredients	*Amount (gm)*	*Energy (kcal)*	*CHO (gm)*	*Protein (gm)*	*Fats (gm)*	*Calcium (mg)*
Amaranth flour	10	35	6.15	1.33	0.5	16.2
Foxtail millet	10	33	6.09	1.23	0.43	3.1
Bajra flour	5	17.4	3.09	0.5	0.3	1.37
Gram flour	5	19.35	2.89	1.12	0.33	2.25
Dry coconut powder	15	93.61	1.2	1.09	9.49	4.8
Sesame seeds	5	25.98	0.5	1.09	2.15	64.15
Poppy seeds	5	21.13	0.6	1.02	1.52	68.6
Pumpkin seeds	5	28.7	0.74	1.49	2.45	2.6
Oil	20	180			20	
Total		**454**	**21.26**	**8.87**	**36.9**	**206.34**

Total amount of raw ingredients was 150 gms and evaluated for Energy, Carbohydrate, Protein, Fat and Calcium using IFCT Table.

RESULTS AND DISCUSSION

Nutrient Value and Product Content

The weight of raw ingredients was 150 gm and when the product was fully prepared after different cooking methods, the weight of the product was 100 gm. Energy provided by bhakarwadi is 454 kcal of energy, 21.26 gm of carbohydrate, 8.87 gm of protein, 36.9 gm of total and 206.34 mg of calcium. Table 2 depicts the analysis of major ingredients.

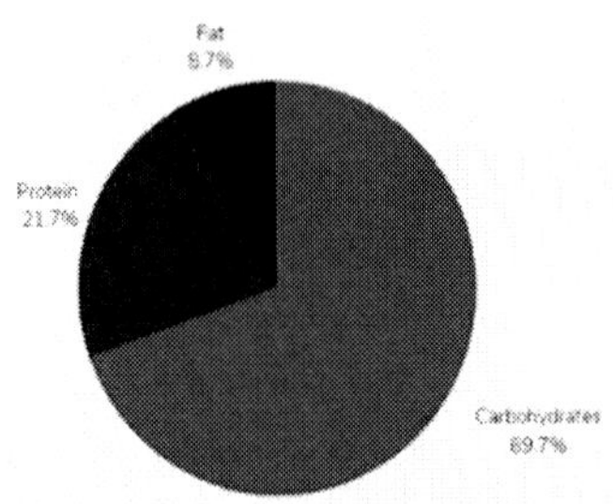

Figure 1: Nutrient Value of Cal-Mill Bhakarwadi

Sensory Evaluation

Table 1 presents the ratings given by the panelists that tasted the product.

Table 3: Analysis of Sensory Evaluation

Attributes →	*Color*	*Flavour*	*Aroma*	*Texture*	*Overall Acceptability*
Mean →	7.2 ± 1.4	8.05 ± 0.72	7.65 ± 0.91	7.7 ± 0.91	8.5 ± 0.62

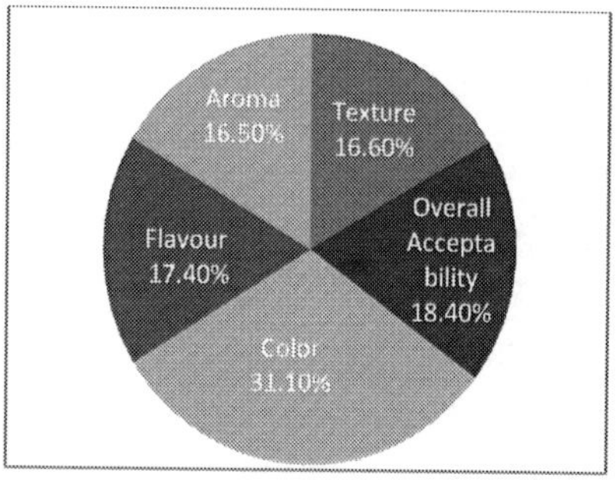

Figure 2: Sensory Evaluation

Figure 3: Optimized Product

Microbial Analysis

Table 4 present the results of microbial analysis using pour plate method.

Table 4: Microbial Analysis

Number of Plates	*Microbial Content*
1	10–12 CFU/ml
2	20–30 CFU/ml
3	40–50 CFU/ml

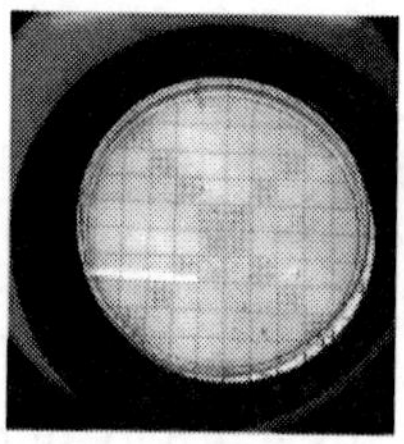

Figure 4: Pour Plate Method

Shelf Life

The shelf life of the bhakarwadi in an airtight glass jar was for 30 days. After a month the food product became soggy.

Packaging

A total number of 10 Bhakarwadi were packed in one box. It was a black colour good quality polystyrene plastic airtight box and was rectangular in shape. It prevented the product from getting in contact with microorganisms, dust and dirt. Polystyrene is a versatile plastic that can be rigid or foamed.

General purpose polystyrene is clear, hard, and brittle. It has a relatively low melting point. Typical applications include protective packaging, food service packaging, bottles, and food containers. This box was selected to pack the product as it is one of the most reliable types of boxes to pack solid food items. The lid was airtight that can provide a good way to ensure that the product was free from damage and microorganisms. The product was kept in this box for 4 weeks during the sensory evaluation and it was observed that the product inside was completely free from any environmental or physical damages. Each box of Cal-mill bhakarwadis weighed approximately 150 gms.

Figure 5: Packaging Material

CONCLUSION

Millets have a similar amount of energy as other cereals. Due to their rich fibre, mineral, vitamin, macro- and micro-nutrient, and phytochemical content, they also offer more substantial health advantages and can be used to treat chronic diseases. Millets can be a cheap, satisfying, and healthful addition to a normal diet. The nutritional fibre, mineral, and vitamin content of the majority of millets was shown to increase after germination and fermentation. By enhancing the protein digestibility and mineral bioavailability, straightforward processing methods like soaking, germination/ malting, and fermentation can assist address the issue of protein-energy deficiency. The total proteins, total dietary fibre, and micronutrients were shown to be decreased as a result of decortication, dehulling, milling, and extrusion. Therefore, care should be used when decorticating millets since excessive dehulling might reduce fibre content and cause micronutrient losses because the nutrient-dense bran and germ component are lost.

ACKNOWLEDGMENT

The authors thankfully support acknowledgments of Dr. Rupali Sengupta for supporting us. The authors acknowledges Dr. BMN College of Home Science, Matunga, Mumbai for granting for the Food product development.

REFERENCES

[1] Abah, C.R., Ishiwu, C.N., Obiegbuna, J.E. and Oladejo, A.A. (2020). Nutritional composition, functional properties and food applications of millet grains. *Asian Food Science Journal, 14*(2), 9–19.

[2] Aggarwal, P.R., Pramitha, L., Choudhary, P., Singh, R.K., Shukla, P., Prasad, M. and Muthamilarasan, M. (2022). Multi-omics intervention in Setaria to dissect climate- resilient traits: Progress and prospects.

[3] Amadou, I., Gounga, M.E. and Le, G.W. (2013). Millets: Nutritional composition, some health benefits and processing-A review. *Emirates Journal of Food and Agriculture*, 501–508.

[4] Banerjee, D., Chowdhury, R. and Bhattacharya, P. (2017). Optimization of extraction process of inulin from Indian millets (jowar, bajra and ragi)—characterization and cost analysis. *Journal of food science and technology*, 54(13), 4302–4314.

[5] Baraniak, J. and Kania-Dobrowolska, M. (2022). The Dual Nature of Amaranth—Functional Food and Potential Medicine. *Foods*, 11(4), 618.

[6] Batool, M., Ranjha, M.M.A.N., Roobab, U., Manzoor, M.F., Farooq, U., Nadeem, H.R., ... and Ibrahim, S.A. (2022). Nutritional Value, Phytochemical Potential, and Therapeutic Benefits of Pumpkin (Cucurbita sp.). *Plants, 11*(11), 1394.

[7] Chandrasekara, A., Naczk, M. and Shahidi, F. Effect of processing on the antioxidant activity of millet grains. *Food Chemistry*, 2012; 133: 1–9.

[8] Czlapka-Matyasik, M., Lonnie, M., Wadolowska, L. and Frelich, A. (2018). "Cutting down on sugar" by Non-dieting young women: An impact on diet quality on weekdays and the weekend. *Nutrients, 10*(10), 1463.

[9] Devi, P.B., Vijayabharathi, R., Sathyabama, S., Malleshi, N.G. and Priyadarisini, V.B. (2014). Health benefits of finger millet (Eleusine coracana L.) polyphenols and dietary fiber: A review. *Journal of Food Science and Technology, 51*(6), 1021–1040.

[10] El Khoury, D., Balfour-Ducharme, S. and Joye, I.J. (2018). A review on the gluten-free diet: Technological and nutritional challenges. *Nutrients, 10*(10), 1410.

[11] Gowda, N.N., Siliveru, K., Prasad, P.V., Bhatt, Y., Netravati, B.P. and Gurikar, C. (2022). Modern processing of Indian millets: A perspective on changes in nutritional properties. *Foods*, 11(4), 499.

[12] Martin, L., Messager, E., Bedianashvili, G., Rusishvili, N., Lebedeva, E., Longford, C., ... & Herrscher, E. (2021). The place of millet in food globalization during Late Prehistory as evidenced by new bioarchaeological data from the Caucasus. *Scientific Reports*, 11(1), 1–11.

[13] Sachdev, N., Goomer, S. and Singh, L.R. (2021). Foxtail millet: a potential crop to meet future demand scenario for alternative sustainable protein. *Journal of the Science of Food and Agriculture, 101*(3), 831–842.

[14] Taylor, J. and Awika, J. (Eds.). (2017). Gluten-free ancient grains: cereals, pseudocereals, and legumes: sustainable, Nutritious, and health-promoting foods for the 21st century. *Woodhead publishing.*

Development of Nutri-Rich Powder for Women

R.B. Modi and K.N. Patel

Assistant Professor, College of Food Processing Technology and Bio Energy, Anand Agricultural University, Anand, Gujarat, India

ABSTRACT: Pregnancy and lactation are the two states, which need increased nutritional demand like protein, calcium, iron etc. in women. The study is part of the effort to provide nutritional food for women. To cater the elevated demand of nutrition, moth bean and finger millet were selected as major ingredients, being richer sources of iron and calcium respectively. The processing steps like soaking and germination helped to lower anti-nutritional factors. The best germination time was optimized for both, moth bean and finger millet based upon reduction in the phytic acid and trypsin inhibitor activity and increase in amylase activity with optimum yield. The best germination period selected for moth bean and finger millet were 36 h at 30°C and 24 h at 30°C respectively with 80% RH. The formulation was optimized using Response Surface Methodology at five levels of three variables viz. moth bean flour, finger millet flour and barley malt extract. The optimized product was prepared by mixing germinated and roasted moth bean (21.98%) and finger millet (19.37%) flours with skim milk and barley malt extract (20%), followed by cooking (100°C, 5 min.), vacuum drying (70°C, 700 mm Hg), grinding and sieving. The developed nutri-rich food contained 16.74 g protein, 5.70 mg iron, 284.49 mg calcium and 1.8 mg zinc. The product cost was around 122 INR per kg and it was shelf stable up to 180 days of ambient storage.

Keywords: Finger Millet, Moth Bean, Women, Nutritive Food.

INTRODUCTION

Pregnancy and Lactation are the stages with increased nutritional demands in women for maternal foetal tissue growth. Out of the total foetal growth 75 per cent is related to maternal nutritional status (Worthington R., and Williams J., 1993). Vitamin and mineral deficiencies during Pregnancy and Lactation are associated with adverse health effects. Pregnant and Lactating women require 3200 and 3400 kcal energy per day and 75 and 78 g protein per day respectively. It has been suggested that pregnant women need micronutrients such as iron, calcium, zinc, magnesium at the levels of 35, 1200, 12 and 310 mg per day respectively (ICMR, 2009). Anemia is the most common indirect cause of maternal mortality in India. Zinc insufficiency is associated with pregnancy complications, growth retardation, congenital abnormalities, retarded neurobehavioral and improper immunological development in the fetus (Black R., 2001). Also, insufficient calcium supply during Pregnancy and Lactation can result in maternal bone loss, reduced breast-milk calcium secretion or impaired infant bone development (Prentice A., 1994).

Finger millet is considered as a rich source of calcium containing around 300–350 mg per 100 g (Singh P., and Singh R., 2012) while, moth bean is a rich source of protein (24–25%) and iron (about 9.6 mg/100 g) (Salve R. and Mehrajfatema Z., 2011), which makes them promising food ingredients for pregnant and lactating women.

Despite of ample nutrients in the cereals, they contain certain anti-nutritional factors which restrict the bio availability of nutrients. Khokhar S., and Chauhan B. (1986a) observed significant varietal differences in the phytic acid content of four varieties of moth bean. Phytate content in finger millet as observed by various authors in range of 0.679 to 0.693 g/100 mg (Antony U., and Chandra T., 1999). Kadam S., and Salunkhe D. (1986) had studied TI from moth bean for thermal stability and changes during germination and cooking and found the destruction in TI during germination followed by cooking. The germination had no effect on total protein content, but a progressive increase in free amino acid content in moth bean and in vitro protein digestibility of moth bean was increased during germination (Satwadhar *et al.*, 1981). Mbithimwikya *et al.*, (2000) observed that sprouting of finger millet resulted in lowered levels of the anti-nutrients, namely tannins, phytates and TI activity. Rao P. (1994) reported phytate content to be 149 to 150 mg/100 g in finger millet grains, malting decreased the phytin phosphorus content by 58 to 65% in brown and white finger millet varieties.

Looking in to the detailed nutritional profile of different cereals, finger millet, moth bean and barley malt extract can be utilized to develop a product, which can be promising for Pregnant and Lactating women to encounter the maternal malnutrition.

MATERIALS AND METHODS

Materials

Superior quality moth bean and finger millet were procured from the local market. Barley malt extract was procured from J.K. Malt Product Pvt. Ltd., Nadiad, Gujarat, India. Fresh mixed skim milk was collected from the Vidya Dairy, Anand, Gujarat, India.

Methods

Soaking and Germination of the Cereals

Moth bean and finger millet grains were soaked in water (1:3) (Archan *et al.*, 1998), for 12 h at room temperature (Khokhar S., and Chauhan B., 1986b). The grains were allowed to germinate in the seed germinator for 24, 36, 48 h for moth bean and 12, 24, 36 h for finger millet at 30°C temperature and 80% relative humidity (RH). The germination parameters were optimized by using dependent variables like amylase activity (AA), phytic acid content (PAC), trypsin inhibitor activity (TIA) and yield (Y). After germination, the moth bean and finger millet grains were dried at 60°C and sand roasted at 160°C for 60 s and 150°C for 60 s, respectively (Pratape V., and Kurien P., 1986) followed by milling (100 mesh).

Experimental Design

The product was formulated by Response Surface Methodology (RSM) using a Central Composite Rotatable Design (CCRD) (Table 4). The factors were moth bean (20–30%), finger millet (15–25%) and barley malt extract (10–20%) with additional skim milk to make 100%. The responses were organoleptic attributes (flavour, colour & appearance, consistency and overall acceptability) and Protein content.

Method of Preparation

Dry ingredients (as per experimental design matrix) were blended together and slowly added to the heated skim milk in order to prepare the gruel, followed by cooking at 100ºC for 5 min. the cooked gruel was vacuum dried at 70°C temperature and 700 mm of Hg vacuum till less than 8% moisture in the final product.

The product was reconstituted in standardized milk at 15% with addition of 4% sugar for sensory evaluation using 9-point hedonic scale method.

Analytical Methods

The raw materials and final products were analysed for proximate composition by adopting the standard analytical methods. The AA of moth bean and finger millet grains were measured by the procedure described by Campbell (1980). The physical/reconstitution properties of the dried product were determined by the standard methods. The microbiological analysis was also carried out by adopting the standard test methods.

The steeped, germinated and dried samples of moth bean and finger millet were estimated for PAC using the Phytic acid estimation kit (Megazyme International Ireland Limited).

Trypsin Inhibitor Activity (TIA)

TIA in moth bean and finger millet was determined according to the method given by Kakade *et al.* (1974).

Sample Preparation: 1 g of grain (ground to 100 mesh) sample was used for extraction with 50 mL of 0.01 N NaOH for 3 hours (pH: 9.6–9.8). The suspension (1 mL) was diluted with glass distilled water (make up volume 50 mL).

Tris Buffer: Tris buffer (0.05 M, pH 8.2) containing 0.02 M $CaCl_2$ was prepared by dissolving 6.05 g of tris (Hydroxymethylamino methane) and 2.94 g of $CaCl_2{\circ}2H_2O$ in 900 mL of glass distilled water. The pH was adjusted to 8.2 and the volume made up to 1 L with glass distilled water.

Substrate Solution: Benzoyl–DL-arginine-p-nitroanalide (BAPNA) hydrochloride was used as the substrate. 40 mg of the substrate was dissolved in 1 mL of dimethyl sulfoxide, diluted to 100 mL with tris buffer at 37°C.

Trypsin Solution: 4 mg of crystallized, salt free trypsin was dissolved in 200 mL 0.001 M HCl.

TIA Assay: Portions (0 and 1 mL) of sample extract were pipetted into 2 sets of test tubes and 2 mL volume adjusted with glass distilled water, further, added 2 mL of trypsin solution to each tube and placed in a water bath at 37°C. 5 mL of pre-warmed BAPNA solution was added to each tube. Exactly after 10 minutes of incubation, reaction was terminated by adding 1 mL of 30 percent acetic acid. After mixing, absorbance was read at 410 nm against a reagent blank (prepared by adding 1 mL of 30 per cent acetic acid to a test tube containing 2 mL each of trypsin solution and glass distilled water followed by 5 mL BAPNA solution).

Expression of Activity: 1 trypsin unit (TU) is defined as an increase in 0.01 absorbance unit at 410 nm per 10 mL of the reaction mixture under the conditions of assay. TIA was calculated as trypsin units inhibited (TUI) per mg of sample using the following formula:

$$\text{Trypsin inhibitor activity (TUI/mg)} = \frac{\text{Standard O.D} - \text{Blank O.D}}{0.01} \quad \ldots (1)$$

Mineral Analysis

The Fe and Zn contents were estimated with the help Atomic Absorption Spectrophotometer (AAS) using acetylene as a fuel and air as an oxidant. Specific hollow cathode lamps were used for the determination of these elements (AOAC, 1995) while calcium was estimated by dry ashing method (Ranganna S., 1986).

RESULTS AND DISCUSSION

Chemical composition of finger millet and moth bean is shown in Table 1. The compositions of moth bean and finger millet were found to be in accordance with those reported by Salve R. and Mehrajfatema Z. (2011) and Singh P., and Singh R. (2012) respectively.

Table 1: Proximate Composition of Raw Materials

Constituents/Materials	*Moth Bean*	*Finger Millet*
Moisture (%)	7.83 ± 0.21	11.00 ± 0.20
Protein (% db)	26.02 ± 0.03	8.22 ± 0.08
Fat (% db)	3.68 ± 0.01	1.46 ± 0.09
Fiber (% db)	3.82 ± 0.03	3.97 ± 0.19
Ash (% db)	3.25 ± 0.01	2.55±0.17
Carbohydrates (by diff.)	55.40 ± 0.24	72.80 ± 0.32

All values are mean ± standard deviation (n=3), % db = % dry matter basis.

Optimization of Germination Period for Moth Bean

From the Table 2, the AA was maximum, while there was maximum reduction in PAC and TIA after 48 h of germination. While, the yields obtained after 24 and 36 h of germination were significantly ($P < 0.05$) at par with each other, it was significantly lower after 48 h germination period, thus to obtain optimum nutritional profile without compromising the yield, the germination time for 36 h was optimized. At optimized germination period, there was 85.59% increase in AA, 50.04% and 39.98% reduction in PAC and TIA respectively. This increase/decreased was calculated by the formula

% increase/decrease = (B-A)/A × 100, where A is the value of raw moth bean and B is the value after 36 h of germination.

Table 2: Changes in Anti-Nutritional Factors and Yield of Moth Bean during Processing Treatments

Treatments/ Parameters	*AA (unit/g)*	*PAC (mg/100 g)*	*TIA (TUI/mg)*	*Yield (%)*
Raw moth bean	$8.05^{e} \pm 0.13$	$811.76^{a} \pm 5.74$	$13.03^{a} \pm 0.17$	$100.00^{a} \pm 0.00$
12 h soaking	$8.17^{d} \pm 0.06$	$550.09^{b} \pm 14.72$	$11.53^{b} \pm 0.38$	$81.98^{b} \pm 0.68$
24 h germination	$10.94^{c} \pm 0.25$	$441.56^{c} \pm 6.01$	$9.75^{c} \pm 0.53$	$69.43^{c} \pm 0.45$
36 h germination	$14.95^{b} \pm 0.19$	$405.51^{d} \pm 9.58$	$7.83^{d} \pm 0.19$	$68.92^{c} \pm 0.27$
48 h germination	$19.40^{a} \pm 0.35$	$310.25^{e} \pm 2.25$	$7.03^{e} \pm 0.15$	$55.01^{d} \pm 0.20$
SEM	0.05	76.40	0.10	0.20
C.D	0.28	11.20	0.41	0.59
C. V (%)	1.80	1.73	3.24	0.52

All values are mean ± standard deviation (n=4); All values are in dry matter basis, Different characters means significantly different at $p < 0.05$ (Tukey's test), TUI = Trypsin units inhibited.

Optimization of Germination Period for Finger Millet

The effect of processing steps on dependent parameters is shown in Table 3, where it is visible that, the rates of reduction in anti-nutrients were decreased,

Table 3: Changes in Anti-Nutritional Factors and Yield of Finger Millet during Processing Treatments

Treatments/ Parameters	*AA (unit/g)*	*PAC (mg/100 g)*	*TIA (TUI/mg)*	*Yield (%)*
Raw finger millet	$3.15^{e} \pm 0.03$	$770.77^{a} \pm 3.26$	$5.18^{a} \pm 0.31$	$100^{a} \pm 0.00$
12 h soaking	$3.27^{d} \pm 0.08$	$442.01^{b} \pm 5.74$	$4.08^{b} \pm 0.05$	$79.99^{b} \pm 0.88$
12 h germination	$8.04^{c} \pm 0.42$	$389.79^{c} \pm 2.49$	$3.83^{b} \pm 0.10$	$61.97^{c} \pm 0.62$
24 h germination	$10.41^{b} \pm 0.46$	$306.04^{d} \pm 5.35$	$2.68^{c} \pm 0.10$	$59.77^{d} \pm 0.63$
36 h germination	$16.92^{a} \pm 0.30$	$244.63^{e} \pm 2.59$	$1.93^{d} \pm 0.24$	$47.58^{e} \pm 0.59$
SEM	0.16	2.06	0.09	0.31
C.D	0.47	6.21	0.28	0.93
C. V (%)	3.714	0.956	5.254	0.883

All values are mean ± standard deviation (n = 4); All values are in dry matter basis, Different characters means significantly different at $p < 0.05$ (Tukey's test), TUI = Trypsin units inhibited.

while the yield started decreasing drastically after 24 h of germination, thus the optimized germination period of 24 h was selected. At optimized germination period, there was 230.48% increase in AA, 60.02% decrease in PA and 49.96% decrease in TIA with 53.96% yield achieved.

Formulation of Nutri-Rich Food using Response Surface Methodology

The experimental design matrix and the responses noted are shown in Table 4, while the ANOVA obtained is tabulated in Table 5.

Table 4: Experimental Design Matrix with Responses for Nutri-Rich Food

Trial	*A Mothbean (%)*	*B Finger Millet (%)*	*C Barley Malt Extract (%)*	*Sensory Score*				*Protein (% db)*
				Colour	*Flavour*	*Consistency*	*Overall Acceptability*	
1	30	15	20	6.9	6.6	7	7.21	17.39
2	16.59	20	15	7.15	7.05	7.07	6.86	13.08
3	25	28.41	15	7.1	6.45	7	7.3	18
4	30	25	20	7	6.55	7.3	6.8	16.2
5	30	15	10	6.95	6.7	6.8	7.3	17.5
6	20	25	20	7.7	7.05	7.5	7.09	18
7	25	11.59	15	7.35	6.65	7.2	7.1	16.9
8	25	20	15	7.5	7.1	7.2	7.28	16.7
9	33.41	20	15	6.9	6.9	7.1	7.09	16.7
10	30	25	10	6.9	7	6.63	7.4	16.8
11	25	20	15	7.3	7.2	7.3	7.34	17.2
12	20	15	20	7.05	7.35	7.32	6.8	15.22
13	25	20	6.59	6.75	6.7	6.9	7.27	16.05
14	25	20	15	7.45	7.05	7.3	7.21	17
15	20	15	10	6.9	6.9	7.4	6.81	15.5
16	20	25	10	7.1	6.85	6.85	7.3	16.23
17	25	20	15	7.15	7.2	7.1	7.21	17
18	25	20	15	7.3	7.05	7.3	7.4	16.13
19	25	20	15	7.35	7.2	7.2	7.4	16.8
20	25	20	23.41	6.9	6.7	7.15	6.91	18.11

Table 5: ANOVA for the Effect of Variables on Sensory Attributes and Protein Content of Nutri-Rich Food

Partial Coefficient	*Colour and Appearance*	*Flavour*	*Consistency*	*Overall Acceptability*	*Protein Content*
Intercept	7.34*	7.13**	7.23*	7.28*	16.8
A- Moth bean	0.10*	–0.11**	–0.094*	0.0074NS	0.66*
B-Finger millet	0.035NS	–0.032NS	0.042NS	–0.03NS	0.25NS
C-Barley malt extract	0.077NS	0.0073NS	0.14**	0.17**	0.31NS
AB	–0.100NS	0.075NS	0.063NS	0.098*	–0.67*
AC	–0.087NS	–0.156**	0.038NS	0.05NS	0.27NS
BC	0.075NS	–0.075NS	0.15*	0.015NS	0.20NS
A^2	–0.100*	–0.023NS	–0.044NS	–0.0396NS	–0.65**
B^2	–0.029NS	–0.17**	–0.039NS	–0.099**	0.26NS
C^2	–0.17**	–0.12**	–0.065NS	–0.055NS	0.13NS
Model F-value	4.54	7.16	4.19	5.32	5.49
R^2	0.80	0.87	0.79	0.83	0.83
APV	7.34	8.30	6.66	7.33	8.98
Lack of fit	NS	NS	NS	NS	NS

** = significant at 1% level (P<0.01), * = significant at 5% level (P < 0.05), NS = non-significant

Effect of Compositional Variables on Sensory Attributes of Nutri-Rich Food

The ANOVA of quadratic models indicated that model F-values were more than tabulated F-value. Furthermore, the coefficient of determination (R^2) that reflects the proportionate variability in data explained by the model. The R^2 values for colour and appearance, flavor, consistency and overall acceptability were showing good fit of models to the data (Table 5). The lack of fit test was found to be non-significant which indicated that the developed models were adequate for predicting the responses.

From Table 5, it was observed that mothbean had shown highly significant ($P < 0.01$) negative effect on flavour, while significant ($P < 0.05$) negative effect on colour and appearance and consistency of the nutri-rich food. The barley malt showed highly significant ($P < 0.01$) positive effect on consistency and overall acceptability of the nutri-rich food.

As shown in the Figure 1, when barley malt extract was kept at center point (15%), with an increase in finger millet, the colour and appearance score increased, whereas with an increase in moth bean, the colour and appearance score initially increased up to certain point and then started decreasing.

When finger millet was kept at center point (20%), with an increase in barley malt extract the flavor score of the product was increased, whereas with an increase in moth bean, the flavor scores were slightly increased (Figure 2). When moth bean was kept at center value (25%), with an increase in barley malt extract the consistency scores of the product were initially increased and decreased after certain point, whereas with an increase in finger millet the consistency scores were decreased (Figure 3). As shown in the Figure 4, when barley malt extract was kept at center point (15%), with an increase in finger millet the overall acceptability of the product was initially increased and decreased after certain point, whereas with an increase in moth bean, the overall acceptability was slightly increased.

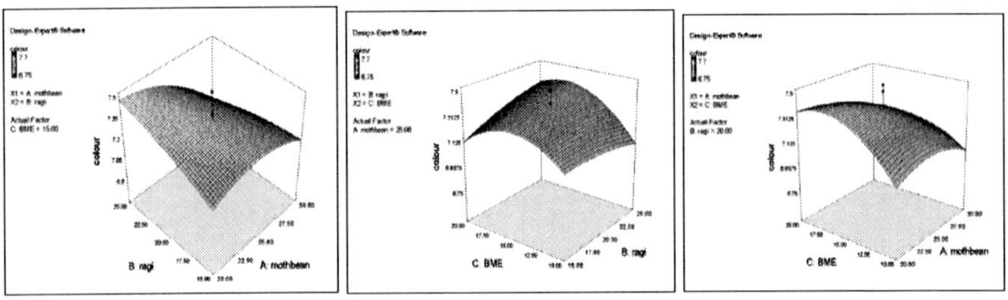

Figure 1: Response Surface Plots for Colour and Appearance Score of Nutri-Rich Food

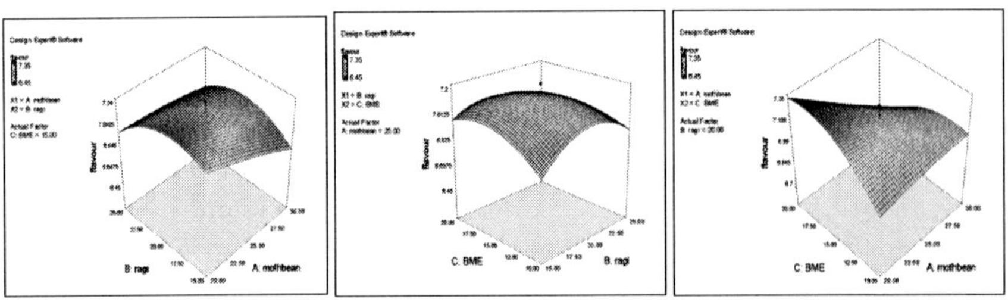

Figure 2: Response Surface Plots for Flavour Score of Nutri-Rich Food

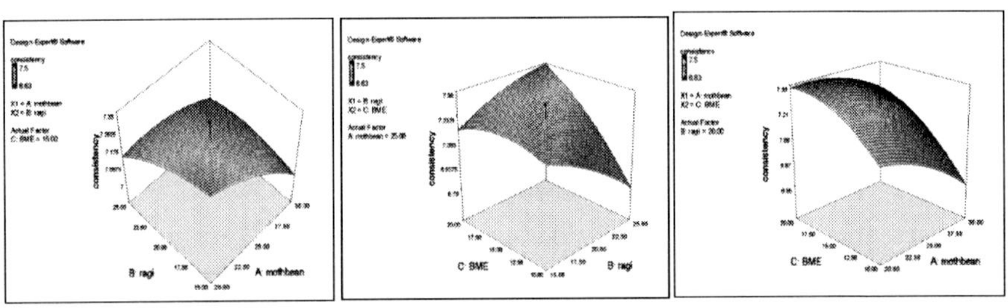

Figure 3: Response Surface Plots for Consistency Score of Nutri-Rich Food

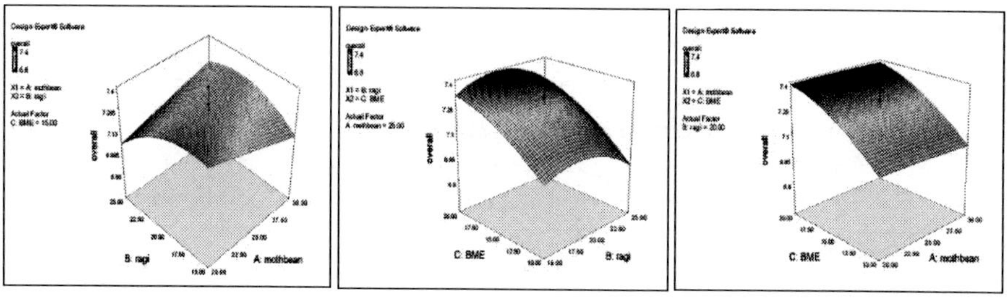

Figure 4: Response Surface Plots for Overall Acceptability Score of Nutri-Rich Food

Effect of Compositional Variables on Protein Content of Nutri-Rich Food

Protein contents ranged from 13.08% to 18.11% on dry matter basis (Table 4). The ANOVA of quadratic model (Table 5) indicated that model F-value of 5.49 was more than tabulated F-value. Furthermore, the coefficient of determination (R^2) value for protein was 0.83 and adequate precision value (APV) of 8.98 which was higher than the required (4.00) for high prediction ability showing good fit of models to the data and the lack of fit test was found to be non-significant.

Further, it can be observed in Table 5 that moth bean showed significant ($P < 0.05$) positive effect on protein content of Nutri-Rich Food. As shown in the Figure 5, when barley malt extract was kept at center point (15%), with an increase in finger millet the protein content of the product was gradually increased, whereas with an increase in moth bean, the protein content was drastically increased.

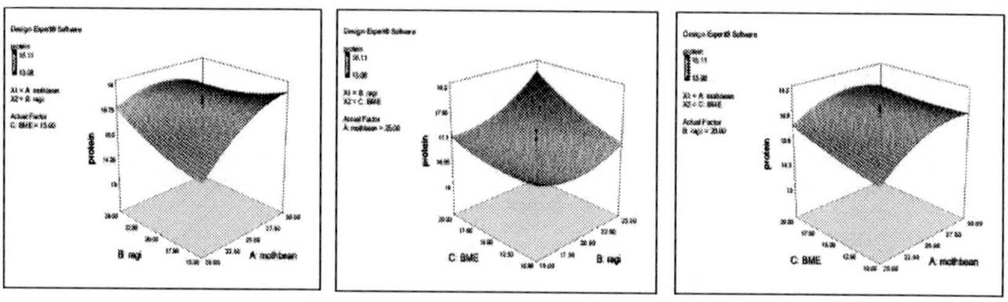

Figure 5: Response Surface Plots for Protein Content of Nutri-Rich Food

The model was optimized using the software design expert and the optimized solution had 21.98% moth bean, 19.37% finger millet and 20% barley malt extract and remaining part as skim milk. The desirability value of the model was 0.84. The predicted values of colour and appearance 7.30, Flavour 7.19, consistency 7.31, overall acceptability of 7.4 and protein content 16.66%. The predicted values were further compared using t-test with the actual values obtained by conducting final trials to validate the solution.

Profiling of Nutri-Rich Food

The physico-chemical properties, mineral profiling and microbiological analysis of the nutri-rich food are as followed:

Table 6: Physico-Chemical Properties of Nutri-Rich Food

Parameters	*Value*	*Parameters*	*Value*
Chemical properties			
Moisture (%)	8.0 ± 0.10	Protein (%db)	16.74 ± 0.17
Fat (%db)	3.01 ± 0.12	Ash (%db)	3.30 ± 0.16
Carbohydrate (%db)	75.79 ± 0.20	Crude fibre (%db)	1.16 ± 0.13
Reducing sugar (%)	1.73 ± 0.01	Total sugar (%)	8.01 ± 0.03
Physical properties			
Solubility (%)	60.62 ± 1.50	Tapped bulk density (gm/cm^3)	2.91 ± 0.012
Flowability (minute)	2.36 ± 0.006		

Table 7: Mineral Analysis of Nutri-Rich Food

Parameters	*Value*
Iron (mg/100 g)	5.70 ± 0.04
Calcium (mg/100 g)	284.49 ± 2.67
Zinc (mg/100 g)	1.79 ± 0.02

Table 8: Microbial Analysis of Nutri-Rich Food

Parameters	*Value*
SPC (log cfu/g)	3.77 ± 0.03
Coliform (log cfu/g)	NIL
Yeast and mold (log cfu/g)	1.03 ± 0.124

The product possessed good amount of protein with low levels of anti-nutrients. The suggested consumption of the product is as follows: The product shall be reconstituted in whole milk at 15% level, suggested serving size is 125 ml of reconstituted product. If four servings/day are given to pregnant/lactating women, she can acquire 34–35% of protein, 14–24% Iron, 70–71% calcium and 18–19% zinc of her daily requirements prescribed by ICMR (2009).

CONCLUSION

The optimum germination period for moth bean was 36 h while that for finger millet (ragi) was 24 h with significant reduction in the anti-nutrients. Response Surface Methodology could be effectively applied for formulating the final product with better sensory scores and protein content. The final product consisted of 21.98% moth bean flour, 19.37% finger millet and 20% barley malt extract, which scored 7.40, 7.80, 7.70 and 7.60 for flavour, colour and appearance, consistency and overall acceptability respectively. The protein content in the formulated product was 16.74 percent.

ACKNOWLEDGEMENT

The authors are grateful to the College of Food Processing Technology and Bio-Energy, Anand Agricultural University, Anand, Gujarat, India for supporting the research work in all manner.

REFERENCES

[1] Antony, U. and Chandra, T.S. (1999); Enzymatic treatment and use of starters for the nutrient enhancement in fermented flour of red and white varieties of finger millet (*Eleusine coracana); J. Agric. Food Chem.,* Vol. 47(5), pp. 2016–2019.

[2] AOAC (1995). Official methods of analysis; Association of Official Analytical Chemists. Washington D.C., USA.

[3] Archan, Salil Sehgal and Asha, Kawatra (1998). Reduction of polyphenol and phytic acid content of pearl millet grains by malting and blanching; Department of food and nutrition, CCS harayana agricultural university, Hisar, Harayana, India.

[4] Black, R.E. (2001). Micronutrients in pregnancy; British Journal of Nutrition, Vol. 85(2), pp. 193–197.

[5] ICMR (2009). Nutrient requirements and recommended dietary allowances for Indians; Indian Council of Medical Research report, Hyderabad.

[6] Kadam, S.S. and Salunkhe, D.K. (1986); Trypsin inhibitor activity in moth bean: thermal stability and changes during germination and cooking; *Plant Foods for Human Nutrition,* Vol. 36(1), pp. 43–46.

[7] Kakade, M.L., Rackis, J.J, Mcghee, J.E. and Puski, G. (1974). Determination of trypsin inhibitor activity of soya products; *Cereal Chem.*, Vol. 51, pp. 376.

[8] Khokhar, S. and Chauhan, B.M. (1986a). Antinutritional factors in moth bean (*Vigna aconitifolia*) varietal differences and effects of methods of domestic processing and cooking; *Journal of Food Science,* Vol. 51, pp. 591–594.

[9] Khokhar, S. and Chauhan, B.M. (1986b); Effect of domestic processing and cooking on in vitro protein digestibility of moth bean; *Journal of food science,* Vol. 51(4), pp. 1083.

[10] Mbithimwikya, S., Ooghe, W., Vancamp, J., Nagundi, D. and Huyghebaert, A. (2000). Amino acid profile after sprouting, Autoclaving and lactic acid fermentation of finger millet (*Elusine coracana*) and kidney beans (*Phaseolus vulgaris* L.); *J. Agric. Food Chem.*, Vol. 48(8), pp. 3081–3085.

[12] Pratape, V.M. and Kurien, P.P. (1986); Studies on puffing of Bengal gram; *Journal of Food Science and Technology*, Vol. 23(3), pp. 127–130.

[13] Prentice, A. (1994). Maternal calcium requirements during pregnancy and lactation; *American Journal of Clinical Nutrition,* Vol, 59(2), pp. 477–482.

[14] Ranganna, S. (1986). Handbook of analysis and quality control of fruits and vegetable products; New Delhi: Tata McGraw-Hill publications.

[15] Rao, P.U. (1994). Evaluation of protein quality of brown and white ragi before and after malting; *Food Chem.,* Vol. 51, pp. 433–436.

[16] Salve, R.V. and Mehrajfatema, Z.M. (2011). Effect of different pretreatment on trypsin inhibitor activity and nutritional composition of moth bean and its utilization in fortified cake; *Journal of Dairy & Food Sciences,* Vol. 6(2), pp. 212–218.

[17] Satwadhar, P., Salunkhe, D.K. and Kadam, S.S. (1981). Effects of germination and cooking on polyphenols and in vitro protein digestibility of horse gram and moth bean; *Plant Foods for Human Nutrition,* Vol. 31(1), pp. 71–76.

[18] Singh, P. and Singh, R. (2012). Finger millet for food and nutritional security, African *Journal of Food Science,* Vol. 6(4), pp. 77–84.

[19] Worthington, R.B. and Williams, J.K. (1993). Nutrition in Pregnancy and Lactation; *The journal of nutrition,* Vol. 124(85), pp. 14795–14905.

Development of Functional Food Product Ladoo using Response Surface Methodology

Minal Chauhan[1], S. Rema[2] and V.H. Patel[2]

[1]Assistant Professor, S.M. Patel College of Home Science, Sardar Patel University V.V. Nagar, Gujarat, India

[2]Professors, P.G. Department of Home Science, Sardar Patel University, V.V. Nagar, Gujarat, India

ABSTRACT: Ladoo is a traditional dessert that is enjoyed by the community. However, traditional (wheat) laddoo consisted primarily of sugar and saturated fatty acids, which are unhealthy.

The present investigation aimed to develop value-added ladoo incorporating different proportions of functional ingredients using RSM.

Methodology: In this study, malted ragi, popped amaranth seeds and sesame seeds were used, considering their Nutraceuticals and Functional properties.

For the development and optimization of ladoo formulation, response surface methodology based on a five-level-three-factor central composite rotatable design (CCRD) was used. Effects of amounts of ingredients such as malted ragi flour (15–25%), popped amaranth (15–30%), and sesame seeds (10–20%) as independent (or process) variables on, nutrient contents such as calcium, phosphorous, Ca/P ratio, and fat as the dependent(or response) variables were investigated. The effects of various percentages of malted ragi popped amaranth and sesame seeds on each response variable were examined using significant regression models.

Results: The Coefficient of determination, R^2, of all the response variables were higher than 0.80. In the present study, the adequate precision value (APV) for calcium, fat, phosphorous, and Ca/P ratio were all greater than 4. The software generated optimal solution with highest desirability (0.873) was selected as optimised recipe. Based on the given criteria for optimization, the basic formulation for the development of ladoo with desired sensory quality was obtained incorporating 20% malted ragi flour, 15% popped amaranth and 15% sesame seeds with 329.912 mg% calcium, 285.069 mg% phosphorous, and 1.15 Ca/P ratio and 16.50 g% fat as the predicted nutritional value.

Conclusion: RSM is used successfully to optimize the level of process variables for the development of value added ladoo as a functional food product. The optimum recipe was high in calcium, high in Ca/P, high in protein and highly acceptable.

Keywords: *Malted Ragi, Amaranth Seeds, RSM, Ladoo.*

INTRODUCTION

Increased consumer interest in promoting health and lowering the risk of certain diseases has driven the desire for foods that provide additional health benefits beyond their basic nutritional value. This has increased awareness of the possibilities of functional foods. The processed foods with nutrient-rich components promoting healthy bodily processes are known as functional foods. (Indu, P and Awasthi, P., 2018). Ladoo is a traditional, energy-dense sweet that people of all ages enjoy. Ladoos are comprised of various indigenous ingredients and are simple to consume. The primary ingredient of ladoo is wheat flour or bengal gram flour, which can be easily compensated by supplementing with various functional ingredients.

Increasing public awareness of nutrition and health care studies demonstrates phytochemicals' efficacy in preventing chronic diseases (Thapliyal, V., & Singh, K., 2015. Aging decreases energy needs but increases protein and micronutrient needs, demanding a more nutrient-dense diet (van Staveren, W.A., *et al.,* 2000).

In developing nations, the commercial processing of locally cultivated millets into food products with added value is a significant economic drive. (Taylor, J., 2016).

Millets are small grains that make up the main food for a large part of the population in India and Africa. However, because these grains are not readily available in ready-to-eat forms, millet is still primarily used as food by traditional consumers and people in lower socioeconomic strata (Ushakumari, S., *et al.,* 2004). Millets have the same amount of protein, energy, vitamins, and minerals as the most common cereals, and they even have more. In addition, they have a lot of fiber, phytochemicals and are also gluten-free foods. (Chakraborty S., *et al.,* 2009).

Hima B. and Sumathi S. (2003), using Foxtail millet, developed traditional Indian foods such as murukku, dosa, chapati, laddu, and payasam. All of the items were considered acceptable. In one study, Barnyard and Kodo millet-based chapati, dosa, noodles, and rusk were developed. According to reports, the acceptable levels of millet flour incorporation are 20% for noodles and 30% for rusk, chapati, and dosa (Poongodi T., *et al.* 2003).

Ragi, also known as Eleusine coracana (finger millet), has numerous health advantages. Health benefits result from a powerful antioxidant, calcium, other minerals, and dietary fiber (soluble and insoluble) (Devi, P. B. *et al.,* 2014).

Sesame seeds are an excellent source of phytochemicals like sesamin and minerals such as calcium and zinc. (Mansour, F. *et al.,* 2015).

Amaranth grains contain phenolic compounds (such as flavonoids) and a high antioxidant capability. Dietary fiber, especially lignin, has been conclusively related to preventing various diseases. (Coelho, L.M., *et al.,* 2018).

The development of functional foods is a unique potential to enhance the quality of food and the health and well-being of consumers (Wadikar *et al.* 2008).

Many food processing businesses employ statistical design techniques like response surface methodology (RSM), which are quite efficient and best suited in optimising product ingredient levels and the process parameters to formulate new food products (Wadikar *et al.* 2008). RSM is widely used in the development, formulation, and design of new products as well as in the improvement of the designs of already existing ones. (Herceg, Z., *et al.,* 2012).

The consumer has always preferred ready-to-eat products containing unique flavour and textures (Almeida Domiguez *et al.,* 1990) suggested snack food serves as a vehicle for essential nutrients while being readily accepted by the population.

The nutritional and functional quality of traditional ladoo is enhanced by incorporating functional ingredients such as ragi, amaranth seeds and sesame seeds, making it rich in calcium, protein, optimal Ca:P ratio, and moderate fat content.

Considering the above facts, the present study was undertaken to develop the functional food product ladoo using response surface methodology (RSM).

MATERIALS AND METHOD

Ingredients such as finger millet, amaranth seeds, sesame seeds, and other ingredients were purchased from the local market in Nadiad based on their high calcium content, optimum Ca:P ratio, moderate fat level, and bioactive components. To get free from impurities, all ingredients were thoroughly washed (manually) after being purchased.

Preprocessing of the Raw Ingredients

For malted ragi, the grains were thoroughly washed 5–7 times with water to eliminate dust and foreign matter. The grains were soaked overnight at room temperature in water, after which the excess water was drained—wrapped in muslin cloth at 281°C for 36 hours for germination. The malt was roasted at low temperatures till aroma release (10 to 12 minutes). Remove rootlets and grind them to make fine flour.

Sand roasting was used to pop amaranth seeds. Amaranth seeds were exposed to hot sand at a temperature of about 250°C. Due to the sudden thermal difference, the moisture within the grain vaporizes and attempts to escape through the micropores, thereby expanding the starchy endosperm.

Using a hot pan and applying dry heat, sesame seeds were roasted. After being roasted, they were powdered. The beans were roasted for 10 minutes at 60°C to reduce the moisture content, enhance the flavour, and extend their shelf life.

Nutrient Analysis of Raw Ingredients

Standard techniques were used to determine the calcium (Clark and Collip, 1925) and phosphorus (Fiske and Subbarow, 1925) contents of raw components including malted ragi, popped amaranth, and sesame seeds.

Organoleptic Evaluation

A semi-trained panel of ten judges evaluated the organoleptic qualities of ladoo based on their appearance, aroma, texture, mouthfeel, and overall acceptability. The panellists were asked to fill out a hedonic scorecard with a maximum of nine points and a minimum of one for each of the five variables, including Taste & Flavor, Texture, Appearance & Color, Mouthfeel, and Overall Acceptability.

APPLICATION OF RESPONSE SURFACE METHODOLOGY (RSM)

Optimization Process Through RSM for Laddu

Response surface methodology was applied to experimental data using a commercial statistical kit (Design expert, Trial version 12.0.1. Stat Ease Inc., Minneapolis, MN, USA) to generate a response surface plot and optimize process variables.

Response Surface Methodology (RSM) was used to optimize the proportions of malted ragi flour, popped amaranth seeds, and sesame seeds to develop value added product, laddu. The quantities of the remaining ingredients, including jaggery, khajur, and cow ghee, were kept constant.

On the basis of preliminary trials, the variables or main ingredients were selected. In the present study, nutrient contents (such as calcium, phosphorus, the Ca/P ratio, and fat) were dependent variables, while ingredients (such as ragi, amaranth seeds, and sesame seeds) were independent variables.

To enhance the quality of laddu, RSM with central Composite Rotatable Design (CCRD) was used with four variables at five levels and six replicates at the center point. The optimized level of laddu was suggested by the software based on the results of 20 trials.

Experimental Design for Laddu

To create an experimental design, malted ragi, popped amaranth seeds, and sesame seeds were the three independent variables (ingredient factors) with the corresponding codes of A, B, and C. To optimize the quality of the laddu, a factorial design with eight factorials, a central point with six replicates, and six axial points was used, yielding 20 experimental sets with four variables at five levels.

Different levels (low and high) of malted ragi flour (15–25%), popped amaranth (15–30%), and sesame seeds (10–20%) were the independent variables. The level of each variable was determined based on preliminary sensory evaluations. Each investigated independent variable has five levels: – alpha, –1, 0, +1 and + alpha.

The experimental runs and coded values for all factors are given in Table 1.

Table 1: Central Composite Design for Laddu with Coded Values and Independent Experimental Variables

Run	*Coded Levels*			*Factors*		
	A (M. ragi)	*B (P. amaranth seeds)*	*C (Sesame seeds)*	*A Malted Ragi (g)*	*B P. amaranth (g)*	*C Sesame (g)*
1	–1	–1	–1	15.0	15.0	10.0
2	+1	–1	–1	25.0	15.0	10.0
3	–1	+1	–1	15.0	30.0	10.0
4	+1	+1	–1	25.0	30.0	10.0
5	–1	–1	+1	15.0	15.0	20.0

Run	Coded Levels			Factors		
	A (M. ragi)	B (P. amaranth seeds)	C (Sesame seeds)	A Malted Ragi (g)	B P. amaranth (g)	C Sesame (g)
6	+1	–1	+1	25.0	15.0	20.0
7	–1	+1	+1	15.0	30.0	20.0
8	+1	+1	+1	25.0	30.0	20.0
9	0	0	0	20.0	22.5	15.0
10	0	0	0	20.0	22.5	15.0
11	0	0	0	20.0	22.5	15.0
12	0	0	0	20.0	22.5	15.0
13	-α	0	0	11.6	22.5	15.0
14	+α	0	0	28.4	22.5	15.0
15	0	-α	0	20.0	9.9	15.0
16	0	+α	0	20.0	35.1	15.0
17	0	0	-α	20.0	22.5	6.6
18	0	0	+α	20.0	22.5	23.4
19	0	0	0	20.0	22.5	15.0
20	0	0	0	20.0	22.5	15.0

Preparation and Nutrient Analysis of 20 Combinations of Laddu

The sensory evaluation data were collected to determine the acceptability of the products, but they were not used in RSM.

The responses such as calcium, phosphorus, Ca/P ratio (calculated), and fat were estimated and entered into the programme datasheet for each combination. Before analysis, the samples were thoroughly ground/mixed to ensure uniformity, and each sample was analyzed in triplicate.

ANOVA and Regression Analysis

The regression equation's ANOVA allowed for determining the significance of the linear, quadratic, and interaction effects and their goodness of fit. The effectiveness of the resulting models was evaluated concerning their determination coefficients (R^2) and a lack of fit analysis, which revealed that the model analysis, R^2, and fit analysis determined the models' adequacy (Bunkar *et al.,* 2014). Significant and insignificant lack-of-fit are prerequisites for a good model. The result's proportion of variability is described by R^2. The model's efficacy is significant when the R^2 value is above 0.75 (Faridah, A. & Widjanarko, S.B., 2013).

Optimal Solution Generated by the Software for Laddu

Several optimal solutions were generated by the software. As an optimized recipe, an optimum solution with the highest desirability was selected as a final product.

RESULTS AND DISCUSSION

Sensory data for laddu indicated that Runs 12, 19, and 20 had the highest mean taste and flavour score, with a value of 8 ± 0.94 indicating that the panellists liked it very much. Additionally, these three varieties of laddu scored highest in terms of overall acceptability (8.2 ± 0.63). The variation in Run no. 8 received a taste and flavour score of 7 ± 1.05 and a score of 6.8 for overall acceptability, which was the lowest score.

All sensory attributes revealed a statistically significant difference (p. 0.01).

Table 2: Organoleptic Evaluation of Laddu Prepared using Varying Combinations of Ingredients, Generated from Software

Run Nos.	*Taste and Flavour*	*Appearance and Colour*	*Texture*	*Mouth Feel*	*Overall Acceptability*
1	7.2 ± 0.63	7.3 ± 0.95	7.6 ± 0.97	7.2 ± 0.92	7.3 ± 0.82
2	7.5 ± 0.97	7.8 ± 0.63	7.7 ± 0.95	7.6 ± 0.7	7.6 ± 0.7
3	7.1 ± 0.74	7.3 ± 0.67	7.3 ± 0.67	7.7 ± 0.48	7.5 ± 0.53
4	7.4 ± 0.84	7.4 ± 0.84	7.4 ± 0.84	7.6 ± 0.7	7.6 ± 0.7
5	7.8 ± 0.79	8 ± 0.82	7.8 ± 0.79	8 ± 0.94	7.9 ± 0.88
6	7.8 ± 0.92	7.9 ± 1.1	7.8 ± 1.14	7.7 ± 0.95	7.8 ± 1.03
7	7.4 ± 0.7	7.6 ± 0.84	7.6 ± 0.97	7.5 ± 0.71	7.4 ± 0.7
8	7 ± 1.05	7 ± 1.15	6.7 ± 1.34	6.9 ± 1.1	6.8 ± 1.14
9	7.7 ± 0.82	7.9 ± 0.74	7.7 ± 0.48	7.6 ± 0.7	8 ± 0.67
10	7.7 ± 0.82	7.9 ± 0.74	7.7 ± 0.48	7.6 ± 0.7	8 ± 0.67
11	7.8 ± 0.79	7.9 ± 0.74	7.8 ± 0.42	7.7 ± 0.67	8.1 ± 0.57
12	8 ± 0.94	7.9 ± 0.74	7.8 ± 0.63	7.7 ± 0.67	8.2 ± 0.63
13	7.4 ± 0.84	7.5 ± 0.85	7.4 ± 0.84	7.6 ± 0.7	7.6 ± 0.7
14	7.6 ± 0.97	7.7 ± 0.67	7.6 ± 0.52	7.5 ± 0.71	7.7 ± 0.67
15	7.5 ± 0.97	7.8 ± 0.79	8 ± 0.82	7.7 ± 0.82	7.9 ± 0.74
16	7.8 ± 0.63	7.6 ± 0.52	7.5 ± 0.53	7.6 ± 0.52	7.7 ± 0.48
17	7.7 ± 0.48	7.8 ± 0.63	7.9 ± 0.74	7.9 ± 0.74	7.9 ± 0.74
18	7.9 ± 0.57	7.7 ± 0.82	7.8 ± 0.92	7.8 ± 0.63	7.8 ± 0.63
19	8 ± 0.94	7.9 ± 0.74	7.8 ± 0.63	7.7 ± 0.67	8.2 ± 0.63
20	8 ± 0.94	7.9 ± 0.74	7.8 ± 0.63	7.7 ± 0.67	8.2 ± 0.63
Mean	7.75 ± 0.87	7.79 ± 0.81	7.73 ± 0.82	7.72 ± 0.76	7.88 ± 0.77
F value	2.794**	2.340**	2.768**	2.695**	4.222**

Table 3 depicts the results for response variables such as calcium, fat, phosphorus, and Ca/P ratio with various combinations of independent factors for laddu preparation.

Table 3: Experimental Designs Generated and Observed Responses for Laddu

Run	*Factors*			*Responses*			
	A (*M. Ragi*)	B (*P. Amaranth*)	C (*Sesame*)	*Ca* (*mg%*)	*P* (*mg%*)	*Ca/P*	*Fat* (*g%*)
1	15.0	15.0	10.0	329.4	288.1	1.14	16
2	25.0	15.0	10.0	287.5	258.9	1.11	13.1
3	15.0	30.0	10.0	266.1	279.1	0.95	13.2
4	25.0	30.0	10.0	248.1	260.2	0.95	13.4
5	15.0	15.0	20.0	335.4	280.8	1.19	16.6
6	25.0	15.0	20.0	371.2	293.5	1.26	16.3
7	15.0	30.0	20.0	340.3	305.2	1.12	13.9
8	25.0	30.0	20.0	317.7	289.4	1.10	17
9	20.0	22.5	15.0	326.6	281.9	1.08	16.2
10	20.0	22.5	15.0	304.9	274.9	1.11	15.9
11	20.0	22.5	15.0	298.6	274.3	1.09	15.7
12	20.0	22.5	15.0	296.1	287.1	1.03	16.3
13	11.6	22.5	15.0	319.5	306.3	1.04	17
14	28.4	22.5	15.0	316.8	286.4	1.11	16.2
15	20.0	9.9	15.0	356.3	293.5	1.21	17.4
16	20.0	35.1	15.0	283.7	300.9	0.94	14.1
17	20.0	22.5	6.6	228.2	264	0.86	12.2
18	20.0	22.5	23.4	377.7	318.4	1.19	20.6
19	20.0	22.5	15.0	305.2	292	1.05	16.8
20	20.0	22.5	15.0	298.3	286.9	1.04	15

Optimization of Parameters for Regression Coefficients (R^2)

The parameters obtained by fitting calcium, fat, phosphorous, and Ca/P data are shown in Table 4, along with the model's linear regression coefficients, intercepts, and cross-product terms. The coefficient of determination indicated the model's fitness and adequacy.

R^2 is the proportion of variation in the response that can be attributed to the model as opposed to random error, and it was recommended that for a model to be a good fit, R^2 should be at least 80% (Gan *et al.*, 2007). Calcium, fat, phosphorus, and Ca/P all have high R^2 values, indicating that the model suits the data well.

A signal-to-noise ratio, which must be larger than 4.0, is a measure of adequate precision.

In the present research, the adequate precision value (APV) for calcium, fat, phosphorous, and the Ca/P ratio were all greater than 4, indicating the model's adequacy for design navigation.

And that there is no significant lack of fit in all the responses. In addition, statistical analysis suggests that the model fitted the data well.

Table 4: Regression Coefficients of Predicted Quadratic Models of Laddu

Coefficient	*Calcium*	*Phosphorous*	*Ca/P*	*Fat*
Intercept	305.05	284.93	1.06	16.14
Linear				
A: Malted ragi	–3.75*	–6.20#	0.0092*	–0.0912*
B: P.amaranth	–20.02#	1.83*	–0.0768#	–0.7359#
C: Sesame seeds	35.51#	12.75#	0.0771#	1.63#
Quadratic				
A^2	4.18*	1.67*	0.0134*	–0.0766*
B^2	4.84*	1.97*	0.0148*	–0.3722*
C^2	–1.19*	–0.1476*	–0.0040*	–0.1473*
Cross product				
AB	–4.31*	–2.27*	–0.0068*	0.8125^{ns}
AC	9.14*	5.63#	0.0107*	0.6875*
BC	6.76*	3.50*	0.0125*	0.0625*
R^2	0.9122	0.8818	0.9474	0.8229
APV	11.4546	11.2633	15.5113	7.3681
F-value	10.39	7.46	18.00	4.65
CV %	5.23	2.44	3.01	7.33
Suggested Model	Quadratic	Quadratic	Quadratic	Quadratic

*p ≥ 0.1; #p < 0.05; APV = Adequate Precision Value; R^2 = Coefficient of determination. NS = Non-significant.

Effect of Process Conditions on the Calcium Content of Laddu

Figure 1 depicts the response surface plot linear with sesame seeds for the dependent variable calcium, indicating that the calcium content increased as the quantity of sesame seeds increased.

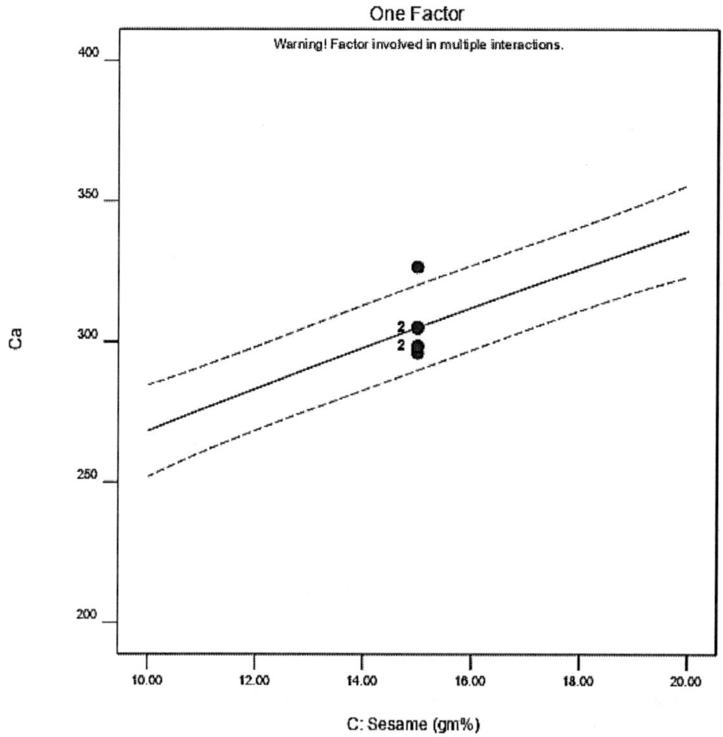

Figure 1: Response Surface of the Calcium Content of Laddu as Influenced by the Level of Sesame Seeds

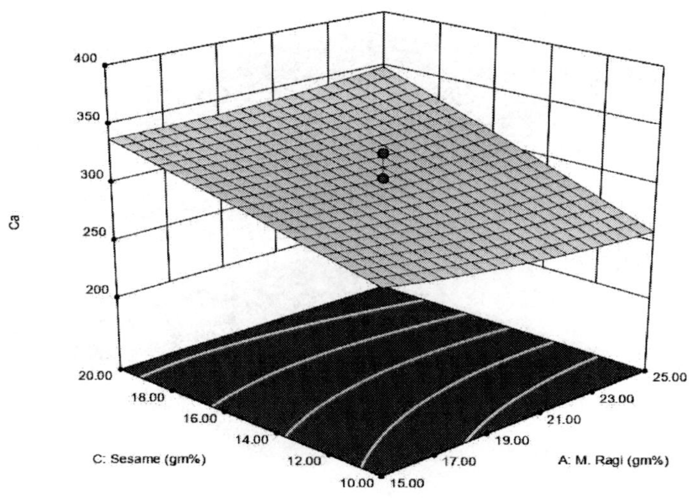

Figure 2: Response Surface Plot for the Ca Content of Laddu as Influenced by the Level of Sesame Seeds and Malted Ragi Flour

Figure 2 depicts the interaction between malted ragi and sesame seeds (AC), which increased the calcium content. The following multiple regression equation was created to estimate the calcium content in terms of coded factors affected by different factors.

$$\text{Calcium} = 305.05 - 3.75 \times A - 20.02 \times B\ 35.51 \times C - 4.31 \times AB\ 9.14 \times AC\ 6.76 \times BC\ 4.18 \times A^2\ 4.84 \times B^2 - 1.19 \times C^2$$

Jideani (2009) examined the impact of sprouted soybean flour on wheat bread. According to the quadratic polynomial regression model, adding 5% sprouted soybean to white bread could significantly increase its protein, fiber, and ash content.

Effect of Process Conditions on the Phosphorous Content of Laddu

Phosphorous content had a non-significant beneficial effect at the linear level with popped amaranth as the dependent variable ($p \geq 0.1$). The phosphorus content increased significantly ($p < 0.05$) as the quantity of sesame seeds increased. At a quadratic level, malted ragi and amaranth had positive but non-significant ($p \geq 0.01$) effects on the phosphorous content. The interactive effect of the amount of sesame seeds and malted ragi (AC) on the response variable calcium (Figure 3) revealed that the highest phosphorus level was obtained with a high sesame content and decreased with increasing ragi content.

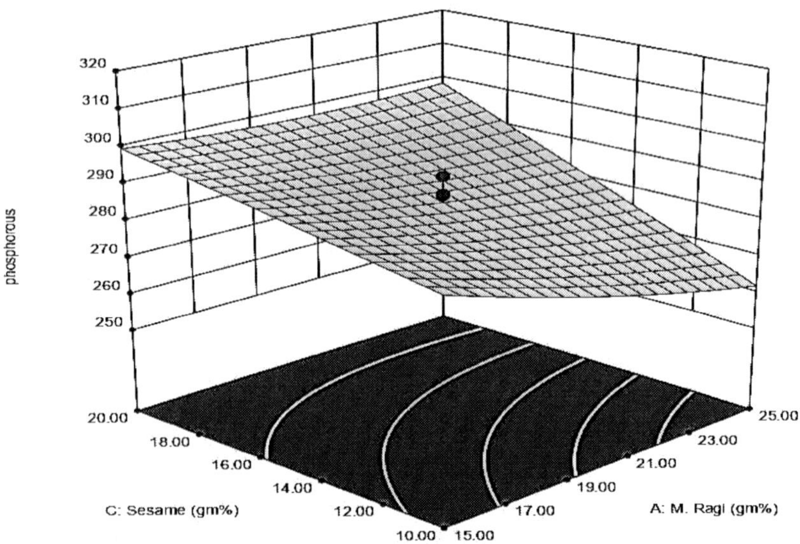

Figure 3: Response Surface Plot of the Phosphorous Content of Laddu as Influenced by the Level of Sesame Seeds and Malted Ragi Flour

Multiple regression equation generated to predict the phosphorous as affected by different factors in terms of coded factors is as follows:

$$\text{Phosphorous} = 284.93 - 6.2A\ 1.83B\ 12.75C - 2.27AB\ 5.63AC\ 3.5BC\ 1.67A^2\ 1.97B^2 - 0.1476C^2$$

Effect of Process Conditions on Ca/P ratio

Malted ragi had a non-significant positive influence on Ca/P on a linear scale ($p \geq 0.01$), whereas popped amaranth had a significant negative effect on a linear level. The Ca/P ratio exhibited a statistically positive effect ($p \geq 0.1$) on sesame seeds. Figure 4 displays the linear association between sesame seeds and the Ca/P ratio.

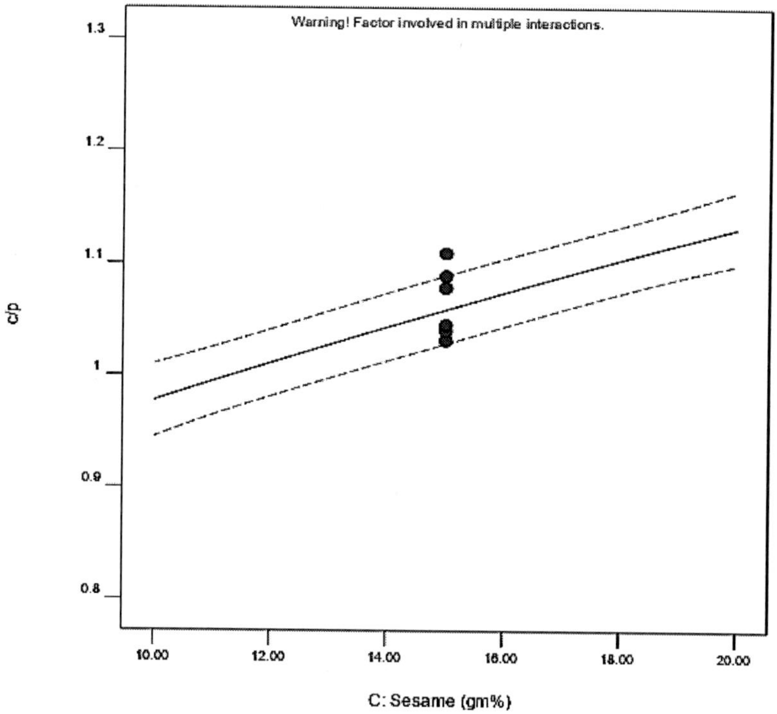

Figure 4: Response Surface of the Ca/P Ratio of Laddu as Influenced by the Level of Sesame Seeds

Multiple regression equation generated to predict the Ca/P as affected by different factors in terms of coded factors is as follows:

$$Ca/P = 1.06\ 0.0092 \times A - 0.0768 \times B - 0.0768 \times B\ 0.0771 \times C - 0.0068 \times AB\ 0.0107 \times AC\ 0.0125 \times BC\ 0.0134 \times A^2\ 0.0148 \times B^2 - 0.004 \times C^2$$

At the quadratic level, malted ragi and popped amaranth also showed a non-significant and positive relationship. At the interaction level, malted ragi with sesame seeds and popped amaranth with sesame seeds found nonsignificant positive ($p \geq 0.1$) relationship. At the quadratic level, both malted ragi and popped amaranth exhibited a non-significant positive correlation with respect to the Ca/p ratio.

Effect of Process Conditions on the Fat Content of Laddu

According to the data presented in Table 4, malted ragi (non-significant) and popped amaranth (significant) had a negative effect on fat in linear terms.

At the linear level, varying sesame seeds had a significant positive effect on fat ($p < 0.05$), shown in Figure 5 (A) and (B).

A non-significant positive effect of varying ragi and sesame (AC) and popped amaranth and sesame (AC) on the fat content (p 0.1) was obtained. For each of the three process variables, quadratic terms had a negative effect.

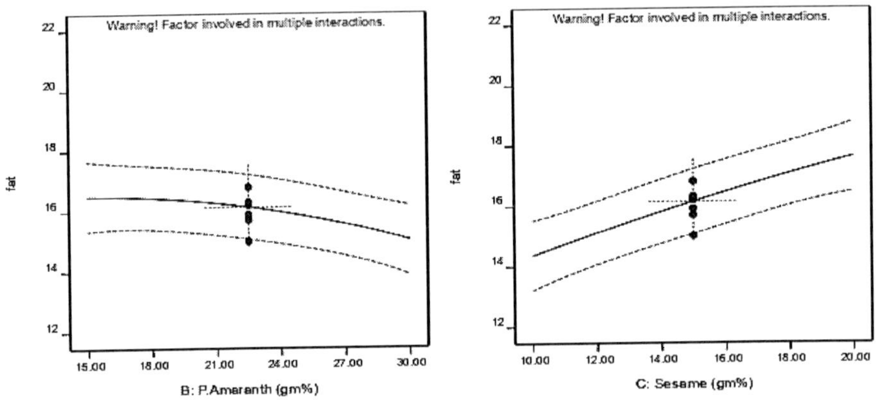

(A) (B)

Figure 5: Response Surface of the Fat Content of the Laddu as Influenced by the Level of Amaranth Seeds (A) and Sesame Seeds (B)

The following equation was derived using multiple regression to predict the fat content of laddu as influenced by several coded factors:

$$\text{Fat} = 16.14 - 0.0912A - 0.7359B\ 1.63C\ 0.8125AB\ 0.6875AC\ 0.0625BC - 0.0766A^2 - 0.3772B^2 - 0.1473C^2$$

The quality of the laddu is not dependent on a single primary component, according to the findings cited above. All independent variables played a significant role in defining the characteristics of laddu. In order to produce the desired final product qualities, a suitable combination of independent/process factors has to be found. Therefore, in order to produce the best product, numerical optimisation was done for the process parameters. For each factor and response, desired goals were chosen.

The software generated five solutions for independent variables with predicted response values. Once the best solution among these optimal solutions was identified, the optimized laddu was formulated and selected.

Optimal solution desirability ranged between 0.83 and 0.87 (see Appendix).

The optimal recipe consisted of 20 g malted ragi flour, 15 g amaranth, and 15 g sesame seeds, with estimated nutritional values of 329.9 mg% calcium, 16.5 g% fat, 285.7 mg% phosphorous, 1.15 Ca/P ratio, and 16.5 g% fat.

CONCLUSION

The process factors for laddu were malted ragi, popped amaranth, and sesame seeds, while the response variables were calcium, phosphorus, the Ca/P ratio, and fat. Initially, the RSM software generated twenty different laddu combinations. The addition of sesame seeds increased the Ca, P, Ca/P ratio, and fat content of Laddu significantly ($p < 0.05$) at a linear level whereas the addition of malted ragi increased the Ca/P ratio (NS). Phosphorous content had a non-significant positive effect at the linear level with popped amaranth as the dependent variable ($p \geq 0.1$). The phosphorous content increased substantially ($p < 0.05$) as the quantity of sesame seeds increased. At a quadratic level, malted ragi and amaranth had positive but non-significant ($p \geq 0.01$) effects on the phosphorous content. Both malted ragi and popped amaranth had a positive effect on the Ca, P, and Ca/P ratio at the quadratic level (NS).

A combination of malted ragi and sesame seeds increased the Ca, P, Ca/P ratio and fat content, as indicated by the interactive effect of ingredients (NS). The addition of amaranth and sesame seeds increased all four response variables (NS). The R^2 values ranged from 82 to 95%, showing that the variations in the model could be attributed to the process ingredients and that the model suited the data well.

After entering the minimum and maximum limits of the process variables and the response variables, the RSM software was later used to optimize the process parameters. The software generated five solutions with a desirability range of 0.838 to 0.873. The solution first with the highest desirability generated by the software was selected for ladoo.

The RSM-optimized ladoo included 329.91 mg of calcium, 285.07 mg of phosphorus, a Ca/P ratio of 1.15, and 16.5 g of fat per 100 g. Laddu had 20 g of malted ragi flour, 15 g of amaranth, and 15 g of sesame seeds. RSM was able to successfully optimise a functional and highly palatable food product.

REFERENCES

[1] Almeida-Dominguez, N.G., Valencia, M.E. and Higuera-Ciapara, I. (1990). Formulation of corn-based snacks with high nutritive value: biological and sensory evaluation. *Journal of food science*, *55*(1), 228–231.

[2] Bunkar, D.S., Jha, A. and Mahajan, A. (2014). Optimization of the formulation and technology of pearl millet based 'ready-to-reconstitute'kheer mix powder. *Journal of food science and technology*, *51*(10), 2404–2414.

[3] Chakraborty, S.K., Singh, D.S. and Chakraborty, S. (2009). Extrusion: a novel technology for manufacture of nutritious snack foods. *J. Beverage Food World*, *42*, 23–26.

[4] Coelho, L.M., Silva, P.M., Martins, J.T., Pinheiro, A.C. and Vicente, A.A. (2018). Emerging opportunities in exploring the nutritional/functional value of amaranth. *Food & function*, *9*(11), 5499–5512.

[5] Das, R., Biswas, S. and Banerjee, E.R. (2016). Nutraceutical-prophylactic and therapeutic role of functional food in health. *J Nutr Food Sci*, *6*(527), 2.

[6] De Groot, L.M. and van Staveren, W.A. (2000). SENECA's accomplishments and challenges. *Nutrition (Burbank, Los Angeles County, Calif.)*, *16*(7–8), 541–543.

[7] Devi, P.B., Vijayabharathi, R., Sathyabama, S., Malleshi, N.G. and Priyadarisini, V.B. (2014). Health benefits of finger millet (Eleusine coracana L.) polyphenols and dietary fiber: a review. *Journal of food science and technology*, *51*(6), 1021–1040.

[8] Faridah, A. and Widjanarko, S B. (2013). Optimization of multilevel ethanol leaching process of porang flour (Amorphophallus muelleri) using response surface methodology. *International Journal on Advanced Science Engineering Information Technology*, *3*(2), 74–80.

[9] Gan, H.E., Karim, R., Muhammad, S.K.S., Bakar, J.A., Hashim, D.M. and Rahman, R.A. (2007). Optimization of the basic formulation of a traditional baked cassava cake using response surface methodology. *LWT-Food Science and Technology*, *40*(4), 611–618.

[10] Herceg, Z., Juraga, E., Sobota-Šalamon, B. and Režek-Jambrak, A. (2012). Inactivation of mesophilic bacteria in milk by means of high intensity ultrasound using response surface methodology. *Czech Journal of Food Sciences*, *30*(2), 108–117.

[11] Hima Bindu, N. and Sumathi, S. Organoleptic evaluation and nutrient composition of Foxtail millet products. In Souvenir, 5th International Food Convention, Innovative Food Technologies and quality systems, strategies for global competitiveness 2003 (Vol. 121).

[12] Indu, P.C. and Awasthi, P. (2018). Development and evaluation of cereal-legume based ladoo supplemented with Ashwagandha (Withania somnifera). *The Pharma Innovation Journal, 7*(7), 358–362.

[13] Jideani, V.A. and Onwubali, F.C. (2009). Optimisation of wheat-sprouted soybean flour bread using response surface methodology. *African Journal of Biotechnology, 8*(22).

[14] Mansour, F., Elhabeby, M., Nooh, H., El-Mehi, A.R.A. and Refaey, A.A.M. (2015). The possible protective role of sesame on experimentally induced osteoporosis. *Journal of American Science, 11*(3s), 14–20.

[15] Poongodi, T., Vijayakumar, J.B., Nazni, P. and Rajeshwari, M. Value addition for minor millets and its glycemic load among normal and type 2 diabetic subjects. *In 39th national Nutritional Conference of NSI*, Hyderabad 2003.

[16] Taylor, J.R.N. (2016). Grain production and consumption: Africa. *Encyclopedia of Food Grains, 4*, 359–366.

[17] Thapliyal, V. and Singh, K. (2015). Finger millet: potential millet for food security and power house of nutrients. *International or Research in Agriculture and Forestry, 2*(2).

[18] Ushakumari, S.R., Latha, S. and Malleshi, N.G. (2004). The functional properties of popped, flaked, extruded and roller-dried foxtail millet (Setaria italica). *International journal of food science & technology, 39*(9), 907–915.

[19] Wadikar, D.D., Majumdar, T.K., Nanjappa, C., Premavalli, K.S. and Bawa, A.S. (2008). Development of shelf stable pepper based appetizers by response surface methodology (RSM). *LWT-Food Science and Technology, 41*(8), 1400–1411.

APPENDIX

Solutions for Laddu

Number	*M. Ragi*	*P. Amaranth*	*Sesame*	*Ca*	*Phosphorous*	*c/p*	*Fat*	*Desirability*	
1.	20.000	15.000	15.000	329.912	285.069	1.150	16.503	0.873	Selected
2.	20.000	15.000	15.065	330.287	285.190	1.151	16.523	0.872	
3.	20.000	15.000	15.160	330.829	285.365	1.152	16.553	0.871	
4.	20.021	15.074	15.000	329.623	285.032	1.149	16.499	0.871	
5.	20.000	15.000	15.529	332.937	286.045	1.157	16.667	0.864	
6.	20.000	15.427	15.000	328.238	284.955	1.144	16.503	0.862	
7.	20.279	15.000	15.000	329.955	284.855	1.151	16.452	0.838	

A Study on Development of Millet based Spread as an Healthy Alternative to Market Based Spreads and Energy-Rich Powders

Bendre Pratiksha, Mavani Dhara, Sengupta Rupali and Manjulaben Gunvantrai Shah

Department of Post Graduate Studies, Dr. BMN College of Home Science (Autonomous), SNDT Women's University, Matunga, Mumbai, India

ABSTRACT: The present study was carried out in order to develop millet based spreads a healthy alternative to combat malnutrition in school going adolescents aged upto 12 years. The objective was to develop a substitute for chocolate spreads for school going adolescents to inculcate healthy eating choices in comparison with market-based spreads. The spread is made from dry roasted ragi flour, dates, groundnuts and gingelly seeds. The dates used balance out the sweet taste in the product while groundnuts make it slightly oily and creamy. In addition to it unsweetened cocoa added makes it chocolaty. Healthy fats like use of garden cress seeds enhances the texture and composition of the product. The millet spresd can be stored up to 15 days in an airtight container hence reducing risk of spoilage. This product is enriched with essential macro and micronutrients which aid in fulfilling the demand of optimum health in school going adolescents. The recipe can be used in maintaining normal body composition and providing optimum nutrition required to maintain health. This product can be therapeutically used in combating under nutrition and malnutrition status.

Keywords: *Finger Millet, Undernourished, Protein Source, Calcium-Rich, Millet-based Spreads.*

INTRODUCTION

Hidden hunger or micronutrient deficiency is another malnutrition faced globally by many countries. This increases the prevalence of many infections and diseases. Nutrition is the cheap and easy way to enhance the disordered condition. Millets are nutritionally enriched grains which are easy to grow and come with a composition of healthy macro and micro nutrients.

Millets are nutri-rich cereal crops of poor people living in the developing countries of Asia and Africa. Amongst all millets, finger millet i.e. *Eleusine coracana (L)* is the most extensively consumed species in India. In India, Karnataka is the leading producer of finger millet accounting to 58% of its global production, yet only a few Indians are aware about its health benefits and nutritional value. The production area of finger millet in India stands sixth after wheat, rice, maize, sorghum and bajra (O'Kennedy *et al.*, 2006).

Non-consumers of millets have benefited from the concept of processing and product development of regional and traditional millets with their specific nutrients, especially the phytochemical and micronutrients, and the processed products, ready to eat products; health foods are limitedly produced at industrial level. Earlier it was believed that polyphenols, phytates, tannins and dietary fiber contents of finger millet act as anti-nutrients because of their metal chelating and enzyme inhibition activities but now it has been confirmed that these constituents can contribute to antioxidant activity, which is an important factor in resisting aging and metabolic diseases.

Also, in this product the use of white gingelly seeds, brown dates and peanut powder add to its nutritional value. Sesame seeds are rich in fat, protein, minerals, vitamins, and dietary fiber. In vitro protein digestibility of sesame protein isolate using pepsin pancreatic enzyme systems shows that the sesame protein isolate has digestibility of 89.57%. Dates bring the sweetness and peanut powder adds to the protein content of the value added product.

Hence, the aim of the present study is to develop a healthy product in the form of premix by incorporating finger millet as a primary ingredient which can be modified to different healthy snack alternatives and study the food development process through sensory evaluation, microbial testing and shelf life estimation of the premix.

MATERIALS AND METHODS

Standardization of the Product

Ingredients Used to Prepare the Recipe

Ragi flour (30 gram), seedless dates, brown (20 gram), Peanut powder (15 gm) and white gingelly seeds (5 gm) are the ingredients used to make the recipe of millet spread. The dates are used to balance out the sweet taste in the product while the groundnuts make it slightly oily and creamy. Healthy fats like white gingelly seeds are used to enhance the texture and composition of the product.

Method of Recipe Making

In a kadhai (pan), dry heat the ragi flour until an aroma appears and a slight change in the color is observed. In a pan, add the peanuts and roast them. Once they are properly roasted, let them cool and rub the nuts between your hands which will knock the skins off and then grind it into powder and weigh it as per needed measure. In the same pan, add the white gingelly seeds and roast them. Once all the ingredients are cooled down, add to a mixer grinder. Add the peanut powder and dates into the same and grind it into a fine powder.

Sensory Evaluation

In order to evaluate a food product, a five-point likert scale was used to understand the perspective and view-point of consumers towards the product. The scale or ranking by the consumer was based on five sensory aspects. The scale was centered on the number of 3 for the value of 'average'. The sensory aspects considered were taste, colour, texture, innovation, mouthfeel and overall acceptability of product mentioned in the scorecard. The appearance or the visual aspect was the first pathbreaker in the eyes of the consumer. In spite of being a healthy product, visual appearance of the product, appealing to the consumer, was considered important. The texture of the product was determined to decide if the consumer considers the product as desirable and appealing. Other aspects like colour and flavor (taste) were considered dependent on each other and could decide if the product has the potential of what it claims to be (Pambo *et al.*, 2018).

Microbial Pour Plate Method

The Pour Plate Method technique was established in the laboratory of Robert Koch is simple, less resource-consuming, easy, and economical; however, it requires the sample to be in liquid or suspension. The pour plate method is a microbiological laboratory technique for isolating and counting the viable microorganisms present in a liquid sample, which is added along with or before molten agar medium prior to its solidification. This technique is used to count viable microorganisms in the given sample by enumerating the total number of colony-forming units (CFUs) within and/or on the surface of the solid medium. It is mostly used for enumerating bacteria; however, Actinobacteria, molds, and yeasts can also be isolated and enumerated (Clark, 1965). Prior to performing the pour plate technique, 1 ml of the sample is supposed to be serially diluted to make the microbial load in the sample between 20–300 CFU/mL (suitable colony counting range is 20–200). Certain points to be noted while preferring this technique are: If the sample is liquid, then it can be serially diluted with sterile distilled water or sterile broth. If the sample is solid or semisolid, it must be first emulsified and then serially diluted to reduce microbial load up to the permitted range (Clark, 1965).

The sample was added to the Petri plate and then the 15 mL of specific molten agar medium of about 40–45°C (less than 50°C) was poured in the Petri plate. After pouring in the Petri plate, the plate was swirled quickly to properly mix the sample with the medium. The medium is allowed to solidify and is incubated, usually at 37°C for 24–48 hours to grow the microorganisms present in the sample. Following the incubation, the viable microorganisms in the sample will grow into visible colonies on the surface of and within the medium. The visible colonies can be counted and CFU/mL can be calculated using the following formula:

CFU/ml = Total number of colonies obtained × dilution factor/
Volume of specimen used (aliquot)

Certain points noted are, if the colonies are uncountable or fused or more than 300 CFU/mL or less than 20 CFU/mL, it is recommended to repeat the process for getting the optimum count (Clark, 1965).

The general procedure for performing the pour plate method can be summarized as follows:

1. ***Sample Preparation:*** The sample was in solid form, therefore it was dissolved in sterile distilled water (1 gm mixed 9 ml of solvent to get concentration of 10–1 gm/ml) to make the microbial load to the range of 20–300 CFU/mL (Clark, 1965).
2. ***Media Preparation:*** Suitable media (general-purpose media like Nutrient Agar and Plate Count Agar for bacteria, and Potato Dextrose Agar or Sabouraud Dextrose Agar for fungi) were prepared and autoclaved. The media was allowed to cool to about 40–45°C (maximum up to 55°C). The sterile Petri plates were arranged and labeled at the edge of the bottom of the plate with required information (Clark, 1965).
3. ***Inoculation:*** 1 ml of diluted sample was dispensed in the center of the Petri plate using a sterile micropipette or calibrated pipette. The lid of the bottle was opened and the mouth was flamed. 15 mL of sterilized molten media was poured at the appropriate temperature above the sample. The lid of the plate was closed then the sample in the media was mixed properly by gently swirling the plate. The plate was swirled in an "S" or "8" shape (Clark, 1965).

Shelf Life of Product

Shelf life was determined by assessing the physical, sensory aspects and microbial content estimations. A small amount of the spread was kept in an airtight glass jar. It was kept in the refrigerator for evaluating the shelf life of the product to ensure the best period for consumption without any side-effects (Corradini, 2018).

RESULTS

Execution of the Idea

The idea of developing a millet premix was executed by modifying the commercially available spreads and powders with the use of easily available ingredients. The basic aim of this product is to develop a healthy substitute for children to inculcate healthy eating choices in their diet. Millets which are consumed in Indian households on a common basis are incorporated in this recipe to increase the nutritional benefits. The ingredients were selected inorder to execute a sustainable product. Dates and peanuts were selected as the ingredients to ensure perfect binding and texture of the recipe. Firstly, the required amount of dry ingredients was weighed using a weighing scale. Furthermore, according to one serving, ingredients were cooked together to formulate a fine textured, sweet in taste, palatable premix. It also serves as an excellent option for those having a sweet tooth.

Modification of the Product

Nutritional Content

According to Indian Food Composition Table 2017, the nutritional composition of the product is mentioned in the Table 1.

Table 1: Nutritive Value of Product

Ingredients	*Amount (g/ml)*	*Energy (kcal)*		*Carbohydrates (g)*		*Proteins (g)*		*Fats (g)*		*Calcium (mg)*	
		per 30 g	*Per 100 g*	*Per 30 g*	*per 100 g*	*per 30 g*	*per 100 g*	*per 30 g*	*per 100 g*	*per 30 g*	*per 100 g*
Ragi flour	30	96	321	20	66.8	2.15	7.1	0.58	1.9	109.2	364
Peanuts roasted	15	78	520	2.5	17.2	3.55	23.6	5.94	39.6	8.1	54
Dates, brown	20	64	320	14.9	74.9	0.49	2.4	0.07	0.3	14.2	71.2
White gingelly seeds	5	26	519	0.54	10.8	1.08	21.7	2.15	43	64.1	1283
Total		**264 kcal**	**1680 kcal**	**37.94 g**	**169.7 g**	**7.27 g**	**54.8 g**	**8.74 g**	**84.8 g**	**195.6 mg**	**1772.2 mg**

Also in association to main macronutrients, micronutrients like calcium are found to be comparatively high which depicts the product being an excellent source of protein, energy and calcium. Further, ragi is packed with essential amino acids and antioxidants. Also white gingelly seeds and peanuts add to the healthy combination of fats to the product.

Sensory Evaluation

Sensory evaluation was done by 5 naive panel members. The recipe was highly acclaimed by the testers assessing food quality, sustainability and practicality to express the therapeutic effect of the product. The objective of the food evaluation included characteristics such as 'taste', 'color', 'texture', 'innovation', mouth feel, and 'overall acceptability. Likert scale was used for evaluation of sensory aspects of the product. Scores ranging from 1 to 5, where 1 being the lowest and 5 being the highest was assigned to such characteristics. The innovation of the product received the highest average score. The product was approved for further product specification process.

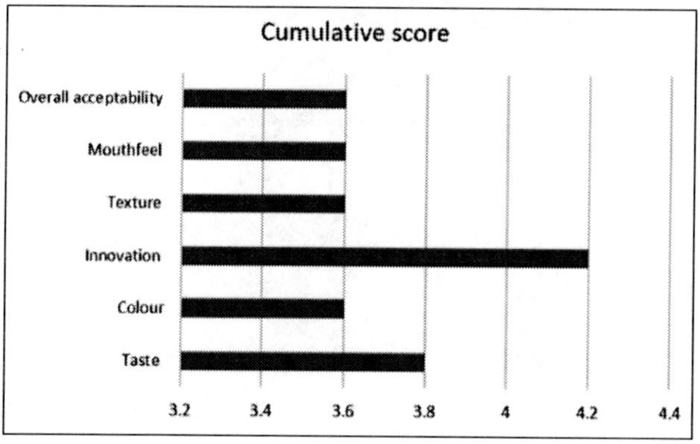

Figure 1: Sensory Evaluation of Product

Product Specification

(a) ***Microbial Analysis:*** In order to check microbiological safety, analysis was performed by the pour plate method. Minute sample of the product was placed in a tight container preventing microorganism exposure. Afterwards the sample was placed in a selective culture medium and incubated at 25 degree celsius. The microbial tests revealed that the product is sustainable for future use. The pH found was about 3.45 and bacteria count was about 25–250 CFU/gm. Therefore, microbial analysis revealed that further speculation of assessing shelf life and sustainability can be predicted.

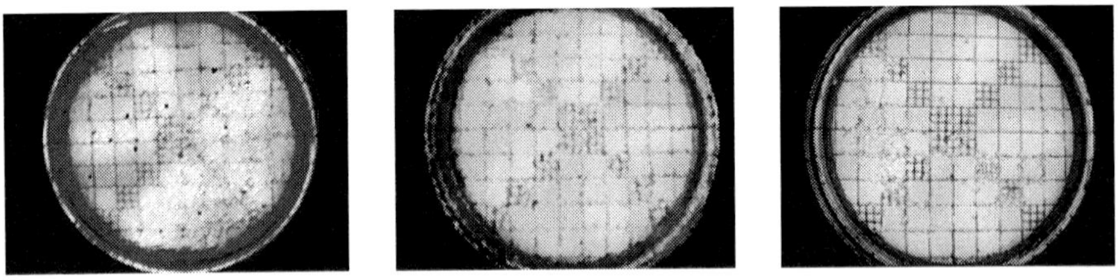

Figure 2: Microbial Analysis of Product

(b) ***Shelf Life:*** The formulated product was stored in a tight closed glass jar at the refrigerator to avoid any contamination and test the shelf life. It was found that the product lasted for approximately 20 days after which there was a pungent smell.

Figure 3: Pictorial Presentation of Testing Shelf Life of Product

(c) ***Label Information:*** The label information should include specification of preserving the product by refrigerating and details regarding storage like shelf life of the product which was found to be 20 days.

DISCUSSION

While developing a product, preserving its nutritional value is important. So when a study was conducted on nutritional security of finger millet it showed that the process of milling reduced the nutritional value of the millet whereas processes like roasting and grinding contributed in gaining grain digestibility without any loss of nutritional components. The entire process of development of food product starting with execution of the idea by considering practicality and benefits, making of the product by trial and error of the recipe many a times, modification for better adaptability, sensory evaluation to judge the view of consumers, microbial analysis to interpret the bacterial content to take the process forward and finally estimating the shelf life of the product gives detailed overview of generation of this food product. Due to widespread awareness of the benefits of millets, it has now shifted the perception of people from "poor man's food" to a "healthy alternative". As the prevalence of malnutrition is increasing in school-going adolescents hence the product was formulated by keeping their nutritional requirements of protein and calcium as a healthy substitute. The market spreads and energy-rich powders are fortified but also contain sugars and flavor-added substances which enhances sweetness. Our product provides the sweetness and more protein per serving in comparison to market based energy-rich powders.

CONCLUSION

To the best of knowledge, this is the first product as a healthy alternative for market based spreads which primarily comes in a premix form. The value added premix consumption can be done in many forms. Healthy energy balls and cookies can be formed. Instead of market based powders, a tablespoon of this can be added in milk and can be given as a milkshake to the children. Keeping in mind the increasing prevalence of protein energy malnutrition majorly in school going adolescents, our attempt of making this premix aims at fulfilling the nutritional requirements required to combat malnutrition.

Limitations and Future Scope

Our study was limited to just curating a energy rich powder and intervention to school-going adolescents was not done due to time constraint and unavailability of consent from parents of those children. Due to time constraint the shelf life of the product was not studied for a longer period of time hence the accurate shelf life was not mentioned of this product.

ACKNOWLEDGEMENT

We students of SYMSc CND feel highly obliged to the opportunity provided by Prof. Mala Pandurang, principal of Dr. BMN College of Home Science. We would also like to express our gratitude to our teacher, Dr. Rupali Sengupta, who guided us throughout this research. We wish to acknowledge the help provided by the staff in the Microbiology and Food labs department of Dr. BMN College of Home Science. Lastly, we would like to thank our friends and family who supported us and offered deep insight into the study.

REFERENCES

[1] Ashik Somarajan, K. and Morya, S. (2022). A review on finger millet properties, processing, health benefits, and applications.

[2] Clark, F.E. (1965). Agar-plate method for total microbial count. *Methods of Soil Analysis: Part 2 Chemical and Microbiological Properties, 9*, 1460–1466.

[3] Corradini, M.G. (2018). Shelf life of food products: From open labeling to real-time measurements. *Annu. Rev. Food Sci. Technol, 9*, 251–269.

[4] Maharajan, T., Ceasar, S.A. and Ajeesh Krishna, T.P. (2022). Finger Millet (Eleusine coracana (L.) Gaertn): Nutritional Importance and Nutrient Transporters. *Critical Reviews in Plant Sciences, 41*(1), 1–31.

[5] Mohammed, S. and Pattan, N. An Overview on Nutritional Composition and Therapeutic Benefits of Sesame Seeds (Sesamum indicum).

[6] Pambo, K.O., Okello, J.J., Mbeche, R.M., Kinyuru, J.N. and Alemu, M.H. (2018). The role of product information on consumer sensory evaluation, expectations, experiences and emotions of cricket-flour-containing buns. *Food Research International, 106*, 532–541.

[7] Pragya, S. and Rita, S.R. (2012). Finger millet for food and nutritional security. *African Journal of Food Science, 6*(4), 77–84.

[8] Srivastava, S. and Arya, C. (2021). Millets: Malnutrition and Nutrition Security. In *Millets and Millet Technology* (pp. 81–100). Springer, Singapore.

[9] Wadmare, V.B., Pawar, V.S. and Shinde, I.N. (2022). A review of finger millet (Eleusine coracana (L.) its varieties, nutrient composition, cooking qualities and health benefits. *J Pharm Innov, 11*, 3223–3227.

[10] Wei, P., Zhao, F., Wang, Z., Wang, Q., Chai, X., Hou, G. and Meng, Q. (2022). Sesame (sesamum indicum l.): A comprehensive review of nutritional value, phytochemical composition, health benefits, development of food, and industrial applications. *Nutrients, 14*(19), 4079.

Modification of Commercial Soan Papdi to Mi-Calci Bytes a Healthy Substitute and Diabetic Friendly Product

Manjulaben Gunvantrai Shah[1], Mettu Rishika, Sawant Lakshana and Sengupta Rupali*

[1]Department of Post Graduate Studies MSc. in Clinical Nutrition and Dietetics, Matunga, Mumbai, India

Dr. BMN College of Home Science (Autonomous)
SNDT Women's University, Matunga, Mumbai, India

ABSTRACT: Millets are a variety of small edible grasses belonging to the grass family. Finger millet (ragi) and barnyard are important millets grown majorly in many parts of India. Ragi has various macro and micronutrients such as calcium, protein, carbohydrates, and fiber minerals as compared to other staple cereals like rice, wheat, etc. It is also rich in calcium and iron. The Barnyard millet or Sanwa rice consists of nutrients like protein, calcium, iron, minerals, and vitamin B complex in high amounts. Ragi and Barnyard both can be ground into flour and this can be used to prepare biscuits, cookies, chapatis, cheela, etc. From the evidences, both the millets have anti-oxidant and anti-diabetic properties. However, being a rich source of micronutrients with additional benefits its consumption is often overlooked. The authors aimed to develop a healthy substitute for soan papdi by utilizing ragi and barnyard as a core component in a form of bytes (Mi-Calci bytes) making it a diabetic friendly snack. Ragi flour and barnyard flour were roasted in ghee, once cooked roasted sesame seeds and dates were added into the mixture. The mixture was allowed to set. Once the product was developed it was assessed through sensory evaluation. Results and Discussion—The texture of the product was soft, crisp and flaky. The product has the potential to fulfil the sweet craving among diabetic population based on literature review. Conclusion-The product is a healthy substitute for Soan papdi and is cost-effective. Mi-Calci bytes are nutrient dense and could be consumed by all individuals as a healthy snack option. Along with being gluten free, it is also a low glycaemic index product and hence is diabetic friendly.

Keywords: *Finger Millet, Barnyard Millet, Nutrient Composition, Ragi and Barnyard Products, Diabetes.*

INTRODUCTION

Majority of the people depends upon cereal like wheat, rice & maize as their staple food. Millets, also referred to as coarse cereals, belong to the Paniceae grass family. They are usually small seeded, annual cereal grasses which grow in tropical and arid climates and are characterized by their ability to survive in less fertile soil. There are many varieties of millets like finger millet, barnyard millet, pearl millet, etc. (Gupta *et al.*, 2017) Millets are great sources of many nutrients like carbohydrates, protein, fatty acids, minerals, etc. In India different types of millets such as Ragi and Barnyard are majorly grown and consumed. (Thapliyal *et al.*, 2015) and (Shobana *et al.*, 2013)

Ragi also called finger millet is extensively grown in various regions of India. Around 60% of finger millet is produced in India which makes it the major producer. It is grown among the states of Karnataka, Andhra Pradesh, Tamil Nadu, and North India. It is the staple food of the low- income group people living in East Africa and South-East-Asia. Nutrient composition of ragi per 100 g includes carbohydrate (72.6 g), protein (7.7 g), fat (1.5 g), crude fiber (3.6 g), calcium (344 mg), phosphorous (250 mg), iron (6.3 g), Manganese (3.5 mg) and magnesium (130 mg). (Gull *et al.*, 2014)

As it is nutritionally rich power packed cereal it is beneficial for certain health concerns as anti- diabetic by lowering the blood glucose levels and improving insulin response because it has low glycemic index, antioxidant due to its role in scavenging free radical and decrease the lipid peroxides, antibacterial, etc. (Thomas *et al.*, 2019) Being an ancient millet it has been utilized to prepare healthy foods for geriatric, infants in different forms. Ragi can be processed by using different methods like soaking, cooking, germination, malting, milling, fermentation, etc. Various products such as breads, rotis, porridge, bytes etc. can be made by converting the millet into flour. Malted form is considered as nourishing food for infants and diabetic patients. (Gull *et al.*, 2014)

Barnyard millet is a classical small millet containing two species *Echinochloa esculenta* (Japanese Barnyard millet) and *Echinochloa frumentacea* (Indian Barnyard millet). Indian barnyard millet is also known by various names such as Sanwa, Jhangora, Shamul, Sama, etc. It is mainly cultivated in India, among Orissa, Maharashtra, Punjab, Gujarat, etc. (Kaur *et al.*, 2020) This cereal does contain macronutrients in proper quantity but above all it also contains micronutrients in higher amounts majorly iron and zinc comparatively to other similar cereals. The nutrient composition per 100 g includes carbohydrate (55–65.5 g), protein (6–13 g), fat (2–4 g), crude fiber (9.5–14 g), calcium (11–27.1 mg) and iron (15–19.5 mg). (Singh *et al.*, 2022)

A nutritionally enriched grain with various health benefits like it helps in preventing constipation, bloating and cramping, reduces hyperglycemia, helps treat celiac disease, act as antioxidant, anticarcinogenic, anti-inflammatory, antimicrobial, etc. It is grown for human and as fodder for animals (Sood *et al.,* 2014). Intake of the grains by cooking it in the form of rice, porridge, cakes, biscuits, etc. and can be considered as functional food.

Despite of millets being nutritionally rich and having various health benefits, they are underutilized because of lack of awareness and knowledge (Renganathan *et al.*, 2020). Therefore, this study aims to use ragi and barnyard flour to develop a healthy replacement for sweetened snacks like soan papdi in a form of bytes (Mi-Calci Bytes).

MATERIALS AND METHODS

An innovative food product was thought that could be a healthy substitute for regular sweet snacks like soan papdi by using ragi and barnyard flour as main ingredients. It was prepared in the form of bytes (MI-CALCI BYTES) making it a healthy snack option. It has been made in such a way that it could be consumed by all age groups especially by type 2 diabetic individuals.

Table 1: Raw Materials and Ingredients

Ingredients	*Amount*
Ragi flour	15 g
Barnyard flour	15 g
Sesame seeds	5 g
Dates	15 g
Ghee	2 tsp

Procedure

Following are the steps and cooking methods used respectively:

1. ***Preparation of Dates Paste: Grinding***

Take about 3–4 dates into a grinder and grind it onto a form of paste.

2. *Preparation of the Base Mixture: Roasting of Ingredients*

In a cooking pot, add 2 tsp of ghee and let it melt. Then add the ragi and barnyard flour and let it cook well. Once the flour is cooked add 1 tsp of roasted sesame seeds and 1 pinch of cardamom powder and stir it well.

3. *Cooling the Prepared Mixture: Adding the Paste*

Once the mixture is cooked well, turn off the flame, add the dates paste and mix it well. Place the mixture on a plate having greater density and let it cool down for about 30 mins. Add 1 tsp of roasted sesame seeds from top over the mixture to set well.

4. *Preparation of Equal Sized Mi-Calci Bytes: Slicing and Cutting*

Once the mixture has cooled down, slice it by using a knife into equal square shapes. Let the bytes be exposed to air for some time to hold its shape completely.

5. *Final Execution: Plating*

After the bytes are evenly cut and cooled down properly it's ready to be served.

Figure 1: Mi-Calci Bytes

EVALUATION AND ANALYSIS METHODS

Sensory Evaluation

It was executed by using attributes such as appearance/colour, taste, aroma, texture, overall acceptability, and innovation was evaluated by a trio of panelists. The average of these attributes was taken to formulate a result. In the trials, conducted prior to developing the product, only ragi flour was used as a base ingredient and jaggery was added for sweetness. After gaining certain suggestions the product Mi-Calci Bytes was finalized due to its overall benefits.

Microbial Analysis

It was conducted wherein a pour plate method was used. This technique was used to count the no. of microorganisms in a mixed sample, to which a molten agar medium was added prior to its solidification. Molten agar was cooled at 44 degrees Celsius before plating as it may lead to death of desired organisms if not done correctly. The process resulted in colonies uniformly throughout the solid medium when the appropriate sample dilution was plated. Three repetitive microscopic findings were carried out. The materials used were sample, sterilized petri-plate, sterilized nutrient medium, flame and glass marker.

RESULTS AND DISCUSSION

The data collected was tabulated and analyzed statistically. The results from the analysis are presented and discussed in the following sequence.

The Mi-Calci Bytes developed by incorporating ragi and barnyard millet flour as a core ingredient in equal amounts were subjected to sensory evaluation by 3 panelists using Scoring Scale.

Table 2: Sensory Evaluation of Mi-Calci Bytes Sensory Evaluation [out of 5]

Panelist	*Appearance/Color*	*Taste*	*Aroma*	*Texture*	*Overall Acceptability*	*Innovation*
1	3	3	3	3	3	3
2	3	3	3.5	3	3	3
3	4	4	4	4	4	4
Average	3.33	3.33	3.50	3.33	3.33	3.33

The following Table 2 shows that Aroma received an average score of 3.50 while the Appearance, Taste, Texture, Innovation and overall acceptability received an average score as 3.33.

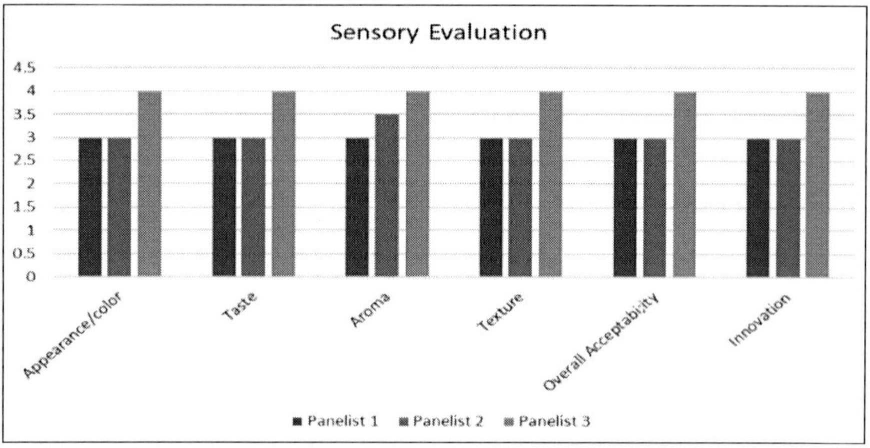

Figure 2: Graphical Representation of Sensory Evaluation of Mi-Calci Bytes

Table 3: Nutritive Value

Ingredients	*Amount (ml/tsp/g)*	*Energy (kcal)*	*Carbohydrate (g)*	*Protein (g)*	*Fat (g)*	*Calcium (mg)*	*Fiber (g)*
Ragi flour	15	48	10	1.07	0.29	54.6	1.68
Barnyard flour	15	46	9.83	0.93	0.33	3	1.47
Sesame seeds	5	26	0.54	1.08	2.15	64.15	0.85
Dates	15	47	10.9	0.36	0.05	9.92	1.37
Ghee	10	90	–	–	10	–	–
Total	60 g	257 kcal	31.27 g	3.44 g	12.82 g	131.67 mg	5.37 g

Total cooked weight-50 gm total no. of bytes prepared 6.

1 serving = 2 Mi-Calci Bytes

From the Table 3, it is observed that out of the total 257 kcal obtained by the bytes, 31.27 g (12.16%), 3.44 g (1.3%) and 12.82 g (4.98%) are contributed by carbohydrates, protein and fats respectively. Additionally, calcium and fiber contributed to 131.67 mg (51.2%) and 5.37 g (2.08%) respectively. The millets ragi and barnyard provide both soluble and insoluble fiber making it a high fiber food with a shelf-life of one month.

POUR PLATE METHOD ANALYSIS

For the microbial count, this technique was used to find the microbial colonies. After 3 repetitive microscopic findings, it was found.

Table 4: Microscopic Findings

Microscopic Findings	*Microbial Count of Colonies*
1.	20 cfu/ml
2.	10 cfu/ml
3.	30 cfu/ml

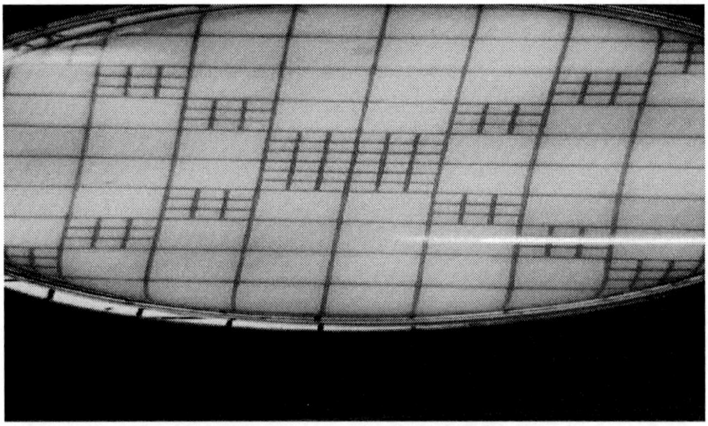

Figure 3: Pour Plate Method Analysis

CONCLUSION

Millets are grains that are historically grown and consumed by Indians for over many years. They contain high amount of nutrients & are rich in proteins, vitamins, minerals, and fibers. Basically, millets are a powerhouse of nutrients. Millets are available in a variety of types like ragi, bajra, foxtail, barnyard, buckwheat, and many more having various health benefits. Weight loss, boosting immunity, maintaining blood sugar levels, reducing cardiovascular risk are few benefits associated with these millets. Ragi and barnyard also known as finger millet and sanwa are one of the millets cultivated in India, are nutritionally dense and has an evident role in maintaining blood sugar levels. However, consumption of these millets is at lower rate despite of the its associated health benefits. Therefore, a product called Mi-Calci Bytes is developed ensuring usage of ragi and barnyard millets in equal ratios as bytes as a substitute for sweetened snacks like soan papdi. The product has been found to fulfil the aspects of taste, texture and color by the panelists. It can beconsumed by all age groups and being a good source of fiber, it can be recommended to type 2 diabetic individuals as a healthy snack option in controlled amounts.

REFERENCES

[1] Gull, A., Jan, R., Nayik, G.A., Prasad, K. and Kumar, P. (2014). Significance of finger millet in nutrition, health and value-added products: A review. *Magnesium (mg),* 130(32), 120.

[2] Gupta, S.M., Arora, S., Mirza, N., Pande, A., Lata, C., Puranik, S., ... and Kumar, A. (2017). Finger millet: a "certain" crop for an "uncertain" future and a solution to food insecurity and hidden hunger under stressful environments. *Frontiers in plant science,* 8, 643.

[3] Kaur, H. and Sharma, S. (2020). An overview of Barnyard millet (Echinochloa frumentacea). *Journal of Pharmacognosy and Phytochemistry,* 9(4), 819–822.

[4] Renganathan, V.G., Vanniarajan, C., Karthikeyan, A. and Ramalingam, J. (2020). Barnyard millet for food and nutritional security: Current status and future research direction. *Frontiers in genetics,* 11, 500.

[5] Shobana, S., Krishnaswamy, K., Sudha, V., Malleshi, N.G., Anjana, R.M., Palaniappan, L. and Mohan, V. (2013). Finger millet (Ragi, Eleusine coracana L.): A review of its nutritional properties, processing, and plausible health benefits. *Advances in food and nutrition research,* 69, 1–39.

[6] Singh, A., Bharath, M., Kotiyal, A., Rana, L. and Rajpal, D. (2022). Barnyard millet: the underutilized nutraceutical minor millet crop. *J. Pharm Innov,* 11(6), 115–128.

[7] Sood, S., Khulbe, R.K., Gupta, A.K., Agrawal, P.K., Upadhyaya, H.D. and Bhatt, J.C. (2015). Barnyard millet—a potential food and feed crop of the future. *Plant Breeding,* 134(2), 135–147.

[8] Thapliyal, V. and Singh, K. (2015). Finger millet: potential millet for food security and power house of nutrients. *International or Research in Agriculture and Forestry,* 2(2).

[9] Thomas, N. and Karuna, M. Finger millet (Ragi, Eleusine coracana L.). A review of health benefits, processing, and product developments.

Millets and Markets-Need for Networking and Integration—A Step Towards Better Supply Chain Management

Jyotsana Suraj Agarwala[1] and Vishal R. Tomar[2]

[1]Vedanta College, Vitthalwadi

[2]Smt. M.M.K. College of Commerce and Economics, Mumbai, India

ABSTRACT

Purpose: The main purpose of this paper is to understand the key factors behind the declining trend of millet production, and big challenges in the processing of millets and to find out solutions through better supply chain management.

Design/Methodology/Approach: The study is based on primary data and secondary data, and it is conducted around Maharashtra state.

Finding: Research shows that there is a need for awareness programs at the micro level towards the uses of millet as a substitute for wheat and Rice. Local-level and decentralized procurement processing is needed to include diversified millets in the fair-price outlet.

Research Limitation: As the research is based on secondary data there may be a lack of accuracy depending on the sources and the reporting stages. There may be a lack of relevance to the problem and the objective of the research. The study is restricted by time and financial constraints further study needs to be collected to get a better insight into the research problem. Therefore, more investigation is required before making any broad judgments.

Keywords: *Millet, Supply Chain Management, Networking and Integration.*

INTRODUCTION

Millet means small-grained cereals. Millets are considered to be "future crops". As it's the need of an hour, the world must think of an alternate food that's how Millet was acknowledged as a Future food. The world is looking forward to the awareness of Millet. The Indian government proposed to United Nations for declaring the year 2023 as the International Year of

Millets (IYOM). 72 countries and the United Nations General Assembly (UNGA) decided to declare 2023 as the International year of Millets which is on 5th March 2021. For mass awareness, India has involved Millets in the following manner. The previous initiative by the government of India such as Mid-Day Meals in schools and anganwadis at least one day a week Millets should be provided and Buffets at hotels to have at least one millet dish during 2023. The Chefs should give online training modules- to build up one platform for sharing dishes with everyone. The appeal should be made to consume the millets at least once a week, the food marketers and influencers must share their best experiences with millets and to Promote "Vrat" (fasting). On an online platform, the agents must deliver promotions to popularize millet to the masses. The events should be organized on various platforms (using NSS, NCC SHGs/FPOs) in schools, colleges, universities, industry, and civil societies. FSSAI needs to develop standards parameters. Monthly 'Millet Challenge' to bring awareness among masses. The Government Ministries to serve only millet snacks in 2023. The list of Millets are Jowar, Bajra, Ragi/Mandua, Minor Millets i.e. Cheena, Kodo, Kutki and two Pseudo Millets Kuttu and Sawa/Sanwa/Jhangora, Chaulai, Kangani/Kakun. The top 5 States of Millet production are Rajasthan Bajra/Sorghum, Maharashtra Ragi/Jowar, Uttar Pradesh Bajra, and Karnataka Jowar/Ragi. Millets do not require chemical fertilizers.

Millets grow better in the absence of chemical fertilizers under dryland conditions. Most of the farmers grow millet using farmyard manure under purely eco-friendly conditions. In recent years farmers have started using bio-fertilizers like vermicompost produced in their backyard and growth promoters such as panchagavya, Amrit pani etc. These practices make millet production not only eco-friendly but stay under the control of farmers. Millets grow under non-irrigated conditions in low rainfall regimes. Thus, they can also face water stress and grow. Millets are the best resource for dozens of nutrients in large quantities. Millets include major micronutrients needed by the human body. Hence, they can help people withstand malnutrition.

To bring millet from the farmer's field, it is essential to create supply chain management. Supply chain. A supply chain is a network between a company and its suppliers to produce a specific product and distribute it to the buyers. The Supply chain management includes producers, vendors, warehouses, transportation companies, distribution centers, and retailers. The essential function of the supply chain is product development, marketing, operations, distribution, finance, and customer services. The model generates results regarding low costs and fast production. The supply chain involves 5 players which include the producers, FPOs, processors, retailers, and consumers.

The Objective of the Study:

1. To understand the supply chain management of Millet.
2. To understand the impact of different factors at the micro level on the uses of millet as a substitute for wheat and Rice.

Problem Statement

As per the statistics from 1950 to 2020 the area of Millet production decreased from 35,000 Hector to 14,000 Hector and the production of Millets is not stable in Nature as in 1950 the production was between 14,000 to 15,000 tonnes but during 2015–2020 the production still at the same level. Whereas the yield in kg per hectare increased from 1950 to 2020. This data express that there is a drastic reduction in the area of production and there is no noticeable change in the quantity of production in the last 70 years. As the population is growing but the productivity of Rice and wheat is not increasing at the same ratio. To meet the consumption of grains, millet is the better substitute for India. But According to the above data, there is less popularity, awareness, and use of Millets which resulted in less opportunity in the production, processing, selling, and distribution of Millets. Due to this Government of India declared the year 2023 as the International Year of Millets.

Quinquennial Mean Area, Production & Yield of Millets in India

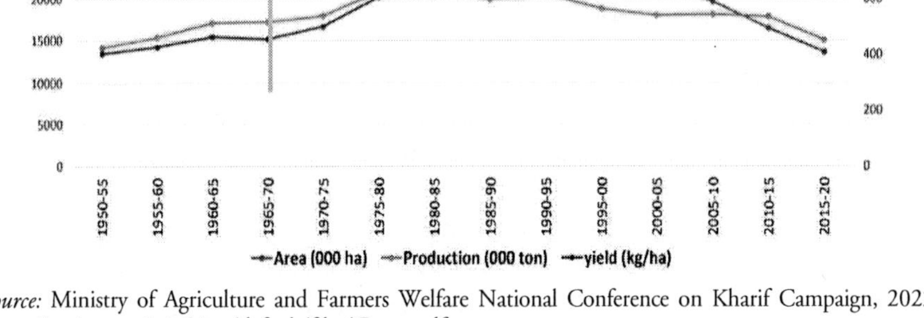

Source: Ministry of Agriculture and Farmers Welfare National Conference on Kharif Campaign, 2022. https://agricoop.nic.in/sites/default/files/Crops.pdf

LITERATURE REVIEW

Objective: To understand the supply chain management of Millet.

- N.A. Nanje Gowda, Kaliramesh Siliveru (N.A Nanje 2022) In this paper it is focused on the impact of the processing of millets, there is still a need to focus on optimizing the processing techniques of millets to make them more acceptable without compromising the health benefits. Food insecurity and malnutrition, awareness needs to be created at both commercial and household levels regarding the impact of processing methods on the nutritional properties of millets and the health benefits of millets.
- U. Uma Gowri and K.M. Shivakumar 2020 (Gowri and K.M. 2020). In this paper analysis done on the decadal trend of the area, production, and productivity of millets in India, also it is examined the cost comparativeness of millets with other crops and across states in India. Suggestions are given on suitable policy measures to improve millet crop cultivation in India. Considering the economic significance of the crops, developmental efforts should be made through demonstrations and training programs, thus popularizing the cultivation and processing of these crops. The gradual decline in net availability of cereals and the trend is more towards consumption of rice and wheat i.e. the consumer dietary cereals are being slowly replaced by major cereals like rice and wheat. This is the perspective from which millet cultivation and its promotion must be regarded.
- B. Dayakar Rao, and K. Nirmal Reddy (B. Rao and Reddy 2013) In this paper, the synergic efforts resulted in to increase in the yield and income of sorghum farmers and their marketable surplus, various nutritionally rich sorghum product technologies were developed and successfully commercialized on a pilot scale at Hyderabad. Apart from licensing private enterprises for the production and marketing of these technologies, collaborative works are also taken up for product development. Effective promotional strategies and policy sensitization attracted entrepreneurs and policymakers to consider sorghum as a priority. Sunit Das (Sunit, 2020). In this paper, reviews previous studies on Millet Farmer Producer Organization. There are many other domains like machinery Access, Traders/Middlemen, and training and demonstration where the framer producer organizations/financing farmer producer companies in Odisha face a nominal percentage of the problem, but the domains like transportation and logistics and financial support in FPO in Odisha can be improved, if Odisha Millet mission could take the initiative in reaching

out to the start-up companies and can put forward their ideas about the establishment of new FPO in the state.

Ahmed S.M. Saleh, Qing Zhang, Jing Chen, Qun Shen (Ahmed, Zhang 2013) The study reviews the recent advances in the evaluation of the nutritional quality and potential health benefits of millet grains. Processing technologies used for improving the edible and nutritional characteristics of millet as well as challenges, limitations, and future perspectives to promote millet utilization as food for a large and growing population are also discussed.

RESEARCH METHODOLOGY

The research is based on primary data and secondary data. The Primary data is collected through a structured questionnaire. The questionnaire had two parts- demographic questions such as occupation, age, and income. The second part of the questionnaire measures the model variables. A random sampling method is used. The sample size is 96 respondents.

Secondary data is gathered from various sources like journals, theses, books, websites, seminar papers, conference papers, manuals, working papers, reports, etc. which are relevant to the study.

Hypothesis

- H(0): There is no significant impact of factors at the micro level on the uses of millet as a substitute for wheat and Rice.
- H(1): There is a significant impact of factors at the micro level on the uses of millet as a substitute for wheat and Rice.

Scope of the Study

The study is focused on the awareness of millet as a substitute for Wheat, Rice, and supply chain management. The findings will help to know the consumer mindset towards the Millet and will motivate them to purchase Millets. The Doctor's suggestions and recommendations will motivate the consumers to Purchase Millet.

Limitation of the Study

- Geographic limitation
- The research is a pilot study. The sample size is selected less and its generalization or representation may not be done further study needs to collect to get a better insight into a research problem.
- Time constraints.

Data Collection

The data is collected based on primary data through structured questionnaires and secondary data is collected through the reference of Journals, periodicals, magazines, and books.

Research Test Tools

The analysis is done through charts and diagrams and the results are expressed.

Data Analysis and its Interpretation

Data analysis is described as the process of bringing order, structure, and meaning to the collected data. Data interpretation is the process of assigning meaning to the processed and analysed data. It enables us to make informed and meaningful conclusions, and implications, infer the significance between the relationships of variables, and explain the patterns in the data.

The below table shows that out of 96 respondents how many factors impact at the micro level the uses of millet as a substitute for wheat and rice.

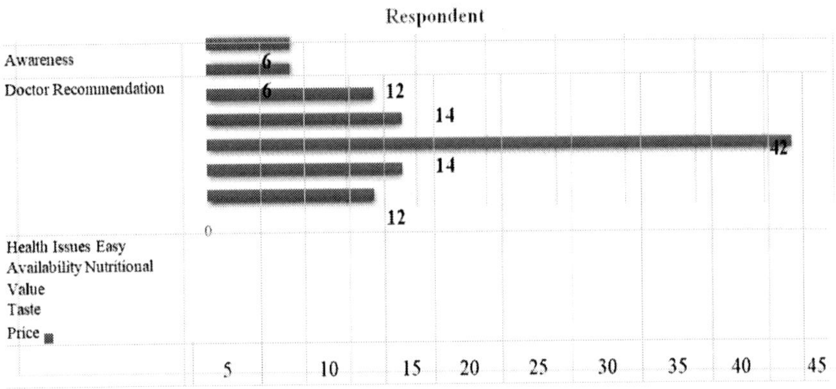

Table 1: Factors that Impact the Uses of Millet

	Price	*Taste*	*Nutritional Value*	*Easy Availibility*	*Health Issues*	*Doctor Recommendation*	*Awareness*
Respondent	12	14	42	14	12	6	6

Source: Primary Data.

Out of the 7 factors, the most impactful factor on the use of millet is the Nutritional value. Health issues, easy availability, taste, and price also impact the use of millet as a substitute for wheat and rice.

Whereas the most impactful factor that impacts not using millet is the low awareness level.

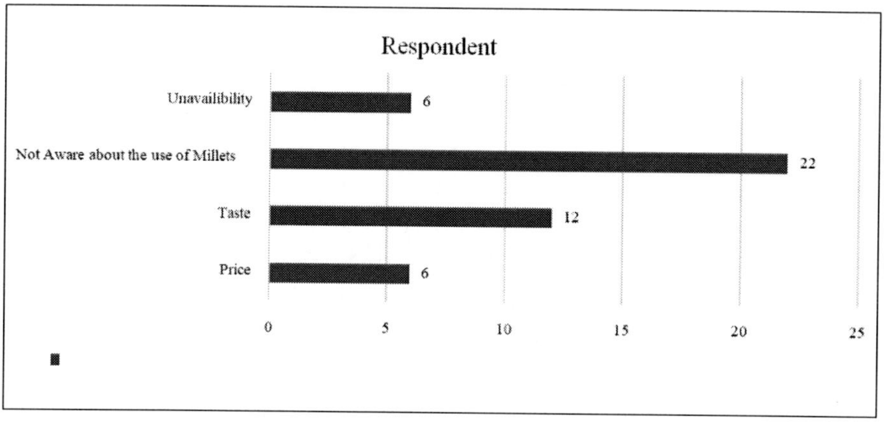

Table 2: Factors Impact Not Consuming Millets

	Price	*Taste*	*Not Aware about the Use of Millets*	*Unavailibility*
Respondent	6	12	22	6

Source: Primary Data.

FINDINGS

There should be a capacity-building program for farmers the increase the productivity of the millet. Need to build the eco-system for the farmer so that the supply of millet is available for maximum consumers. Infrastructure plays a vital role; it needs to develop. The processing, post-harvest, and storage process need to be supported financially.

Research shows that there is a need for awareness programs at the micro level towards the uses of millet as a substitute for wheat and Rice. Procurement and local-level processing should be decentralized, and it should include diversified millets in fair-price shops

CONCLUSION

The main aspects behind reducing the trend of millet production include high labor intensity, low crop productivity, difficult post-harvest operations, and lack of attractive farm gate prices. The easy availability of rice and wheat has contributed to an increase in food patterns in millet-producing regions. Small millets and finger millets are one of the factors discouraging local producers. Other demotivating factors include inadequate investment in product development commercialization, and the persisting perception of low social status associated and consumption. Lack of knowledge always to use small millets in the daily diet is also one of the reasons. The poor availability of millet foodstuffs in local markets, coupled with high prices for their products are also limiting popularity. Different interventions can be made to facilitate the access by supply chain actors to processing plants on the end and by consumers to processed millet products on the other. The lack of processing units close to millet fields forces local producers to take their produce to distant places which increases the cost. Less awareness also needs to be addressed by conducting different awareness campaigns. So, 2023 is announced as the 'International Year of Millets'.

SUGGESTION/RECOMMENDATION

- A policy measure to allow the sale of millet processing equipment across the country, accompanied by a possible exemption tax on their purchase, would encourage the development of the millet value chain. Removing restrictions for the transportation of processing equipment can also help to spread the technology across states.
- Decentralized procurement and local-level processing are needed to include diversified millets in fair-price shops.
- More specifically, there is a critical need to optimize technology for the dehulling of different small millet species, which have different seed sizes.
- More research is needed for improving the separation mechanism in hullers to reduce the removal of grits and other usable materials along with the husk. Improving the sieving efficiency of graders is also needed large-scale equipment is available for this operation but equipment tailored community level and small and medium enterprise level is needed and would be most relevant for supporting.

REFERENCES

[1] International Year of Millets (IYoM) – 2023, National conference on Kharif Campaign, 2022, 19th April 2022, Ministry of Agriculture and farmers welfare. WWW. agricoop. nic.in Gowda NAN, Siliveru K, Prasad PVV, Bhatt Y, Netravati BP, Gurikar C. Modern Processing of Indian Millets; A Perspective on changes in Nutritional Properties. Foods. 2022 Feb. 9; 11(4):499. doi: 10.3390/foods11040499. PMID: 35205975; PMCID: PMC8871339.

[2] Gowri, M. Uma and Shivakumar, K.M. Millet Scenario in India. Economic Affairs, Vol. 65, No. 3, pp. 363–370, September 2020 DOI: 10.46852/0424-2513.3.2020.7

[3] Rao, Benhur and Reddy, K. (2013). Development of Supply Chain Management in Sorghum/Millets. 10.13140/RG.2.1.2539.9847.

[4] Das, Sunit, Supply chain of Millets: An FPO perspective (with a Special reference to Odisha) *Journal Global Values*, Vol. XI, No. II Article No. 29, pp. 234–257.

[5] Ahmed, S.M. Saleh, Qing Zhang, Jing Chen, Qun Shen. Millets Grains: Nutritional Quality processing and Potential Health Benefits. *Comprehensive Reviews in foods science and Food Safety,* 08 April 2013.

A Study on Development of Gluten Free Millet Based Jackor Canapes for Diabetes Mellitus Patients

Khan Zobiya,* Pandey Shreya* and Sengupta Rupali

Dr. BMN College of Home Science (Autonomous),
SNDT Women's University, Matunga, Mumbai, India

ABSTRACT: The study aimed to create canapés using jackfruit flour and jowar flour, both from the poaceae and moraceae families respectively, with the goal of providing a nutritious and affordable option for diabetes patients. Jackfruit flour was chosen due to its high nutritional value, including macronutrients, micronutrients such as vitamin B and C, iron, dietary fiber, and wound-healing properties. Jowar flour was also included to further boost the nutritional content. The end product was evaluated for sensory properties, microbial content, and proximate principles such as energy, carbs, protein, ash, fat, and fiber. The resulting canapés had a pleasing texture and color and were found to be a promising option for addressing hyperglycemic conditions.

Keywords: Canaoes, Diabetes, Nutrition, Glycemic, Fiber.

INTRODUCTION

Canapes are a type of appetizer that originated in France and typically consist of a small piece of bread or pastry topped with meat, cheese, or vegetables. Although a variety of bases can be used, wheat flour is commonly used. However, the healthiest variant can be made from a combination of jackfruit flour and jowar flour, which can be topped with a small amount of chopped salad for added fiber. To make canapes, a dough is created by combining water, jackfruit flour, and jowar flour, with sorghum being a key ingredient. Sorghum is a type of grass that belongs to the Andropogoneae tribe in the Poaceae family and is the fifth most significant cereal crop in the world. It can be easily substituted for wheat flour and is rich in minerals like phosphorus, zinc, iron, calcium, magnesium, potassium, and salt, as well as vitamins A, B and E. Sorghum also contains bioactive chemicals that provide health benefits such as hypocholesterolemia, antifungal, anti-obesity, and antidiabetic properties.

Another key ingredient in canapes is jackfruit flour, which is derived from the jackfruit, a fruit that is native to several tropical Southeast Asian countries and belongs to the Moraceae family. While it is more nutritious than banana, it is less so than mango in terms of protein, calcium, thiamine, riboflavin and carotene. Jackfruit is a good source of dietary fiber, which can act as a bulk laxative and protect the mucous membrane of the colon by reducing exposure time and binding to cancer-causing substances. It also contains vitamin A and flavonoid pigments that work together to support antioxidant and visual functions. Additionally, jackfruit is rich in vitamin C and vitamin B-complex, including niacin, riboflavin, folic acid, and vitamin B-6 (pyridoxine). Vitamin C aids in the development of immunity to infectious agents and scavenging of free radicals, while the B-complex vitamins have numerous health benefits.

Jackfruit is a good source of the antioxidant vitamin C, which constitutes approximately 13.7 mg or 23% of the recommended daily intake (RDA). Jackfruit contains high concentrations of essential B-complex vitamins, including niacin, riboflavin, folic acid, and pyridoxine (vitamin B-6). The metabolic disorder called diabetes mellitus (DM) results from a problem with insulin secretion, action, or both, causing chronic hyperglycemia and metabolic problems with protein, lipid, and carbohydrate.

DM is a chronic noncommunicable disease (CNCD) that has become a major global health issue, and a risk factor for kidney failure, vascular brain disorders, blindness, and limb amputations. The International Diabetes Federation's (IDF) Atlas guideline study indicates that 352 million people with impaired glucose tolerance are at high risk of developing diabetes.

In 2017, an estimated 425 million individuals aged 20–79 had DM, and this number is projected to rise to 629 million by 2029.

IDF Atlas reports Ethiopia's predicted national diabetes prevalence for 2017 among individuals aged 20–79 was 5.2%.

The pancreatic islets of Langerhans contain two types of endocrine cells: beta cells producing insulin and alpha cells producing glucagon, and the secretion of hormones by these cells is altered based on the glucose environment to maintain normal glucose levels. In diabetes mellitus (DM), hyperglycemia occurs due to insulin deficiency or resistance. In type 1 diabetes (T1DM), autoimmune destruction of beta cells in the pancreas leads to an absence of insulin.

T2DM develops more gradually with an imbalance between insulin levels and sensitivity, often caused by factors such as aging and excess fat.

Genetic factors are a significant risk factor for both types of DM. Many genetic loci have been identified that increase the risk of developing DM, with major histocompatibility complex (MHC) and human leukocyte antigen (HLA) polymorphisms specifically associated with T1DM.

The hereditary component of T2DM is more prominent than that of T1DM, with evidence suggesting that lifestyle choices and genetics both play a role in the development of the disease. In many cases, individuals with T2DM have at least one parent who also has the condition.

METHODOLOGY

Ingredients

Table 1: Provides the Complete Ingredients List along with their Amounts

Ingredients	*Amount*
Jackfruit flour	20 g
Jowar	20 g
Rice	10 g
Roasted Bengal gram	20 g
Roasted peanuts	20 g
Dry coconut	10 g
Sesame seeds	10 g
Curry leaves	Few
Salt	To taste
Red chilli powder	To taste
Jeera	5 g
Baking soda	Pinch
Arrowroot flour	20 g
Oil	2 ml

Formulation of Products

To prepare the canapes, start by grinding the following ingredients into a fine powder: curry leaves, dry coconut, jeera, roasted peanuts, and Bengal gramme. Then, combine jackfruit flour, jowar flour, rice flour, arrowroot powder, the prepared masala, sesame seeds, salt, and red chilli powder to make the dough."

Oil is heated before being gradually added to the mixture. Bind it appropriately adding water gradually as needed and combining the mixture to form a dough. With a rolling pin, flatten the dough and form it into canapes. Now bake it in a 100° oven for 20 minutes. The canapes can now be served. Keep dry and in an airtight container.

Sensory Evaluation

The "Jackor Canapes" were prepared using various formulations of jackfruit flour and jowar flour, and were then subjected to a taste test. A panel of judges evaluated the canapes using a nine-point hedonic rating scale to assess their color, texture, flavor, mouthfeel, and overall acceptability. The results of the sensory analysis were presented in a table, which showed the scores for each attribute of the canapes. The score that received the highest rating for color, texture, taste, and mouthfeel was considered to be the most acceptable.

Table 2: Sensory Analysis of Jackor Canapes

Attributes	*9 Like Extremely*	*8 Like Very Much*	*7 Like Moderately*	*6 Like Slightly*	*5 I Neither Like Nor Dislike*	*4 Dislike Slightly*	*3 Dislike Moderately*	*2 Dislike Very Much*	*1 Dislike Extremely*
Appea rance		✓							
Flavor/taste	✓								
Aroma		✓							
Texture	✓								
Mouth feel	✓								
Overall acceptability		✓							

Nutritional Value

Table 3: Nutritional Value of Total Ingredients

Ingredients	*Amount*	*Energy (kcal)*	*CHO (g)*	*Protein (g)*	*Fat (g)*	*Fibre (g)*
Jackfruit flour	20 g	74	15	1.5	0.2	3.5
Rice flour	10 g	33	6	1	0.2	0.5
Jowar flour	20 g	67	13	2	0.4	1
Bengal gram	20 g	66	6.5	2.5	0.2	2.2

Ingredients	*Amount*	*Energy (kcal)*	*CHO (g)*	*Protein (g)*	*Fat (g)*	*Fibre (g)*
Roasted peanuts	20 g	100	3.4	4.8	7	2
Dry coconut	10 g	66	0.8	0.6	6	2
Sesame seed	10 g	20	–	1.5	1	1.2
Oil	2.5 ml	22.5	–	–	–	–
Arrowroot flour	20 g	72	18	–	–	1
Total		**520 kcal**	**63 g**	**14 g**	**15 g**	**14 g**

RESULTS AND DISCUSSION

A delightful and wholesome option for a snack, "Jackor Canapes" have a flaky, crunchy, and smooth texture and are a light brown hue. Its physicochemical characteristics, as shown in Tables 2 and 3, show that it is a healthy choice because it is high in fibre, protein, and carbohydrates and contains gluten-free proteins. It also has a low glycemic index and a lot of bioactive compounds. While jackfruit flour is a nutrient-rich source of carbohydrates, proteins, vitamins B and C, as well as dietary fibre, thiamine, riboflavin, Vitamin A, and phytochemicals that have anti-carcinogenic, anti-diabetic, antimicrobial, antifungal, anti-inflammatory, wound healing, and hypoglycemic effects, jowar flour, which is high in iron, can improve heart health and regulate blood sugar levels.

The snack's high fibre content aids in digestion, prevents constipation, and lowers levels of low-density lipoprotein (bad) cholesterol, which in turn lowers overall blood cholesterol levels.

Microbial Analysis

Microbiology is the study of all the minuscule living things that are essential to every process on Earth. These are the microorganisms, which also comprise algae, viruses, fungus, bacteria, archaea, and protozoa. Biological, biochemical, molecular, and chemical approaches are used in the identification and analysis of microorganisms in food products. To identify the presence of microorganisms in this instance, the product underwent microbiological investigation.

Spread Plate Method

Using a sanitised spreader with a flat metal or glass surface, a solution containing a small amount of germs is applied to a plate using the spread plate technique. For the agar to absorb the bacteria, the plate must be dry and at room temperature; this produces a discernible and countable number of isolated bacterial colonies that are evenly scattered around the plate, signifying a successful technique. This method has the benefit of not affecting heat-sensitive bacteria.

Pour Plate Method

Pour plate culture is a popular technique for growing bacteria and fungi and can produce isolated colonies. The initial sample is diluted several times to lower the microbial population in order to produce separate colonies. Most bacteria and fungi can endure brief contact with warm agar and won't be destroyed, enabling the development of clean cultures. To avoid contamination from colonies developing on or below the surface, it is crucial to utilise fresh media.

Result Obtained

Petri plate A displayed the TLTC (Too low to count) Petri dish B displayed no or few colonies.

Shelf Life

To prevent contamination and test the shelf life, the created product was kept in an airtight container. It was discovered that the product lasted for 2 months before the texture and colour started to fade.

CONCLUSION

Ingredients from traditional Indian cuisine were used. The product is a terrific snack option to have it sprout stuffing make it a chaat or as a whole alone as a healthy alternative to other canapés on the market.

Low oil content contributes to the low fat content. It is inexpensive since all of the ingredients are less expensive. There are no extra ingredients. Jowar flour, jackfruit flour, roasted Bengal gramme flour, and peanuts are used to make the product. It is fortified with a lot of protein and fibre. Acceptable as a snack. Exceptionally long shelf life.

ACKNOWLEDGEMENT

We, the authors would like to express their gratitude to Dr. Rupali Sengupta for her unwavering assistance. The development of food products was supported by Dr. BMN College of Home Science in Matunga, Bombay, which is acknowledged by the authors.

REFERENCES

[1] Aynaiem, S. and Zeleke, A. (2018). Prevalence of diabetes mellitus and it's risk factors, A cross sectional study, Volume 2018.

[2] Goswami *et al.* (2011). Assessment of physicochemical properties of jackfruit; Volume 15(3); pp. 26–31.

[3] Mondal *et al.* (2018). Product development from jackfruit and analysis of nutritional quality of processed products, Volume 4, pp. 76–84.

[4] Olawole *et al.,* (2018). Preadministeration of fermented sorghum diet provides protection against hyperglycemia induced oxidative stress and suppressed glucose utilisation in alloxan induced diabetic rats, Volume 2018.

[5] Olokoba *et al.* (2012). Type 2 Diabetes Mellitus; A review of current trends, Volume 27(4), pp 269–273.

[6] Park *et al.* (2012). Sorghum extract exerts an antidiabetic effect by improving insulin sensitivity via PPAR-r in mice fed diet a high fat diet; Volume 6(4), pp. 322–327.

[7] Poquette *et al.* (2014). Grain sorghum muffin reduce glucose and insulin responses in men; Volume 5, pp. 894–899.

[8] Raidu, V. (2013). Effect of processing and cooking on glycemic index of jowar varieties, Volume 2, pp. July–September (2013).

[9] Shehin, P. and Gupta, Prerna (2019). Product development; Snacks prepared from jackfruit seeds flour and corn flour; Volume 22, pp December 2019.

[10] Sowmaiya, R. and Swarnalatha, A. (2022). Development and characterization of functional food products from jackfruit seeds, Volume 10.

[11] Swami *et al.* (2012). Jackfruit and its many functional components as related to human health, Volume 11, pp. 565–576.

Development of an Energy Dense Millet-Based Indian Snack: Chakrimill Cones

Sawliwala Arwa*, Vichare Simran, Sengupta Rupali[1] and Manjulaben Gunvantrai Shah[2]

*Dr. BMN College of Home Science (Autonomous), SNDT Women's University, Matunga, Mumbai, India

[1]Head of Department, MSc Clinical Nutrition and Dietetics, Dr. BMN College of Home Science (Autonomous), SNDT Women's University, 338, Matunga, Mumbai, India

[2]Department of Post Graduate Studies, MSc Clinical Nutrition and Dietetics, Dr. BMN College of Home Science (Autonomous), SNDT Women's University, Matunga, Mumbai, India

ABSTRACT: The main aim of the study was to develop a millet-based, energy dense Indian snack enriched with pulses, cereals and oilseeds. Objectives included standardizing the product, conducting sensory evaluation, microbial analysis, evaluating its nutritional value and assessing it on undernourished adults. The study consisted of various phases including formulation of the product, preparation and standardization of the product, followed by the sensory evaluation and microbial analysis, analysis of nutritional parameters and testing it on 20 undernourished adults by providing the cones daily for 2 weeks as snacks. A flaky, crispy product with a soft inner core was developed. The sample developed by us was tested and sensory evaluation was conducted on a naive panel. This product received the highest overall acceptable sensory score based on five parameters- taste, texture, smell, appearance and innovation. The microbial analysis was done and approved. Nutritional analysis of the product showed that this product was rich in protein (20%), fiber, amongst other micronutrients, especially iron. The ingredients used in this product like millets, pulses, sesame seeds, etc. are known to help in improving gut function, have antioxidant and immune modulating effects and have an overall impact on the protein status of the people consuming it. The product has a long shelf life when kept in a hermetically controlled condition. The Chakrimilll Cones so developed were energy dense and free of preservatives and other additives without compromising on the taste and texture in comparison to commercially available products. The nutrient packed Chakrimill Cones might help in improving protein quality in undernourished adults (z score <3) and will be best suited as a snack for children, adolescents, and adults.

Keywords: *Millet, Traditional Indian, Standardization, Food Product, Chakrimill.*

INTRODUCTION

Cereals and millets are important for the human diet all over the world. After rice and wheat, millets like ragi, are the main sources of carbohydrate in the Indian diet. Both rice and ragi do not have gluten and therefore become ideal replacements for gluten and wheat products especially for people with gluten intolerance. In Ancient India, ragi or finger millet was well cultivated and was traditionally called nachni in Maharashtra, umi in Bihar, etc. (Patekar *et al.*, 2017).

Indian cuisine is influenced by traditional culture. Indian cuisine is known for its diversity and we have widespread options for appetizers, hors d'oeuvres, and snacks (Agrawal & Sengupta, n.d.). Snacking influences other food choices that individuals consume throughout the day. (Julie Hess, 2017). Indian snacks are popular and are made from a variety of foods like pulses, cereals, or combination of two. (Patekar *et al.*, 2017) The preference for snacks has increased in recent years as people prefer to carry light weight, easy to consume foods in place of an elaborate one or two course meal. Snack items can be carried anywhere and acts as a replacement for main meals. (Kolhe *et al.*, 2021) Chakli is an Indian snack made up of either a single flour or combination of flours and is preferred over other snacks for its crispy and friable texture. (Patekar *et al.*, 2017) A kachori is an Indian snack option popular in Uttar Pradesh, arid regions of Gujarat and Rajasthan and Maharashtra, it is a stuffed ball of flour. The stuffing in a kachori depends on the region it originates from. (Agrawal & Sengupta, n.d.) In a study conducted by Steyn *et al.*, 2011 showed an increase in fast food consumption and is threatening the role of a public health advisor. Most Indian snacks are deep fried and pose a serious threat to the country's health status.

Traditionally the main ingredients in chakli and kachori are refined flour, moong dal, besan, urad dal, carom seeds, fennel, methi and other spices and the main method of cooking is deep frying. We have modified it and replaced the main ingredient—refined wheat flour with ragi flour, rice flour, besan and soya flour. Also, we have switched to baking and sun drying as a form of cooking in place of deep frying.

The main objective of our study was to develop a millet based, nutritionally and energy dense product which is much healthier and can be a suitable snack option for all age groups. In the newly developed product, we have used rice flour, ragi flour, besan and white soya bean flour as the base for the chakli and the main ingredients in the filling are moong dal, urad dal and besan. Complete removal of wheat and other gluten flours makes our product a gluten free, high protein snack. Soya bean is a worthy alternative to meat as it is a rich source of essential amino acids and thus helps in completing the protein content of any product.

Hence, keeping all these points in mind, we have developed a new product out of two already existing Indian snacks which might be more beneficial in terms of palatability, nutritional content, shelf life and stability.

MATERIAL AND METHOD

Ingredients

Good quality Finger millet flour, Soy flour, Besan, Moong dal, urad dal flour, Spices, and Rice bran oil was procured from the local market of Matunga, Mumbai.

Equipment

Various equipment was used in the procedure which includes large plates, a chakli maker, steel cones, kadhai, a perforated spoon, measuring spoons, a ladle, and a weighing scale.

Chemicals

The final product was free from any additives and preservatives. For microbial analysis, the chemicals were used from the Microbiology Laboratory of Dr. BMN College of Home Science.

Methodology

The Chakrimill Cones was innovated and inspired by traditional millet chakli and traditional kachori filling. Initially, for preparing the filling.

The ingredient mentioned in Table 1 was weighed and dry-roasted well in a pan on a low flame until the aroma is released from the spices. Besan is then added to the pan and roasted until the mixture changes color. Grounded Moong dal paste is added after a few minutes and mixed well.

For preparing the dough

On a plate add all the dry flour, ginger garlic paste, spices, and salt as mentioned in Table 2.

Add warm water (40 degrees) to form a dough. Add ghee to the dough and knead it well. Cover the dough with a clean cloth and keep it aside for 10 min for a smooth consistency.

For preparing the cones

Greese the chakli maker with ⅓ tsp of oil and transfer the dough to a chakli maker and roll out a circular cone-like shape around the steel cones. In a Kadhai add 250 ml of oil and warm it at 184-degree celsius for 10 minutes. Once the oil is warm enough, put the glass flame at medium flame and add steel cones. Fry for a few minutes. Do not let it cook completely. Remove the cones halfway and let them cool at room temperature.

For preparing the final product

Remove the cone-shaped chakli from the steel cones. Fill those cones with the prepared filling and stuff well. Fry the stuffed cones again until they turn crisp for the next 2–3 minutes. Remove the cones from the oil and place them on the paper towel and let them cool.

Sensory and Organoleptic Evaluation of Chakrimill Cones

Sensory and organoleptic evaluation of the chakrimill cones was done by the naive judge panel using the 5-point Likert scale and recommendations were given to improve the protein content of the cones. The quality of the product was evaluated using the characteristics such as appearance, taste, texture, and flavor. According to the sensory scorecard and suggestions from the panel, required modifications were made to the product.

Characteristics of chakrimill cones:

1. *Taste:* spicy, fresh, and minty
2. *Smell:* aromatic
3. *Texture:* crunchy, flaky, ridged
4. *Color:* reddish brown
5. *Sound:* crispy, cracky when broken

Figure 1: Chakrimill Cones

Modification

To improve the recipe's nutritive value the following ingredients were changed, Jowar flour was replaced by urad dal flour and soya flour was added to improve the protein content of the recipe.

Sugar was replaced by kharik powder and additional spices such as garam masala were added to the recipe. The procedure for preparing the chakrimill cones remained the same. In the final stage instead of frying the cones, it can be placed in the oven and later fried after filling for less than one minute.

Comparison between Unmodified Ingredients (S0) and Modified Ingredients (S1)

Table 1

Ingredients (S0)	*Amounts*
For Dough	
Ragi Flour	20 g
Rice flour	25 g
Jowar flour	20 g
Besan	20 g
Ginger-garlic paste	1 tsp
Sesame seeds	1 tsp
Ajwain	1 tsp
Red chili powder	1 tsp
Salt	As per taste
For Filling	
Moong dal (cooked and ground)	30 g
Besan	15 g
Fennel seeds	15 g
Turmeric powder	¼ tsp
Coriander powder	½ tsp
Amchur powder	½ tsp
Cumin seeds	2 tsp
Chili powder	1 tsp
Sugar	2 tsp
Oil	40 ml

Table 2

Ingredients (S1)	*Amounts*
For Dough	
Ragi flour	20 g
Rice flour	25 g
Soya flour	20 g
Besan	20 g
Ginger-garlic paste	1 tsp
Sesame seeds	1 tsp
Ajwain	1 tsp
Red chili powder	1 tsp
Salt	As per taste
For Filling	
Moong dal (cooked and ground)	30 g
Besan	15 g
Urad dal flour	15 g
Fennel seed	15 g
Turmeric powder	¼ tsp
Coriander powder	½ tsp
Amchur powder	½ tsp
Cumin seed	2 tsp
Chili powder	1 tsp
Garam masala	1 tsp
Khareek	2 tsp
Oil	15 ml

Shelf Life and Microbial Analysis

The shelf life of the chakrimill cones was analyzed by keeping the content in an airtight container for 1 month. After one month of period, the cones did not retain moisture and the texture remained the same. Microbial Analysis was conducted using the pour plate method for 24 hours in the microbiology laboratory of Dr. BMN College of Home Science.

RESULTS

The main aim of modifying the chakrimill cone was to make it an energy dense, millet based Indian snack which can be enjoyed by all age groups. This aim of the experiment was successfully achieved. The following table gives the Nutritive value of the recipe Table 3 and Nutritive value per serving Table 4.

Table 3: Nutritive Value of the Final Recipe

Ingredients	*Amount (g/ml)*	*Energy (kcals)*	*CHO (g)*	*Protein (g)*	*Fat (g)*	*Calcium (mg)*	*Iron (mg)*	*Total Fiber (g)*
Ragi flour	20	64	13.3	1.43	0.38	72.8	0.92	2.24
Soya flour	30	113.22	3.05	11.34	5.83	58.5	2.47	6.47
Rice flour	10	35.64	7.82	0.79	0.05	0.75	0.07	0.28
Besan	30	135	20.3	7.7	2.45	13.5	1.46	3.24
Moong dal	30	97	15.7	7.16	0.41	12.94	1.18	2.81
Urad dal flour	15	48.61	7.65	3.46	0.25	8.35	0.7	1.79
Sesame seeds	5	33.3	1	2	2.6	64.15	0.7	0.85
Khareek powder	5	16.01	3.75	0.12	0.02	3.56	0.16	0.45
Oil	15	135	–	–	15	–	–	–
Total	**160**	677.78	72.57	34	26.99	234.55	7.66	18.13
% Energy			43%	21%	36%			

Table 4: Nutritive Value per Serving

Macronutrient	*Amount (per 160 g)*	*Amount (per serving = 3 cones = 50 g)*
Energy (kcals)	677.78	211.8
Carbohydrate (g)	72.57	22.6
Protein (g)	34	10.62
Fat (g)	26.99	8.43
Calcium (mg)	234.55	73.29
Iron (mg)	7.66	2.4
Total FIber (g)	18.13	5.66

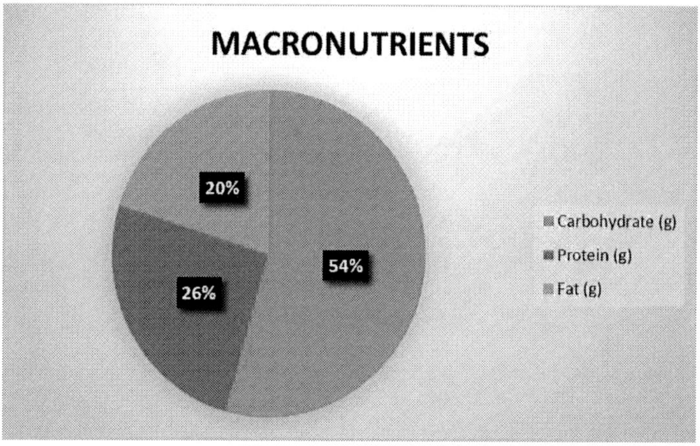

Table 5: Comparison of Nutritive Values of Chakrimill Cones and Chakli and Kachoris of Generic Brands in the Market

Nutritive Value	*Chakrimill Cones (50 g)*	*Generic Chakli (50 g)*	*Generic Kachori (50 g)*
Energy (kcals)	211.8	125	222
Carbohydrates (g)	22.6	27.78	24.1
Protein (g)	10.62	3	3.7
Fat (g)	8.43	16.6	13

Carbohydrates and protein play a very important role in maintaining the energy balance and maintaining the healthy body weight. The chakrimill cones provide 42% of carbohydrate from total energy and 20% protein from total energy making the recipe energy dense and high protein snacks.

Chakrimill cones are also high in dietary fibre (5.66 g/serving) as shown in Table 4, which is indicated by the presence of millet flours as part of the composition of the product that helps in providing a good digestive system and preventing metabolic diseases. These cones are a good source of Calcium (73.29 mg/serving) and Iron (2.4 mg/ serving).

Sensory Analysis

On the sensory and organoleptic analysis done on a panel consisting of 10 members, it was found that the chakrimill cones scored highest points for appearance and innovation as shown in the Table 6.

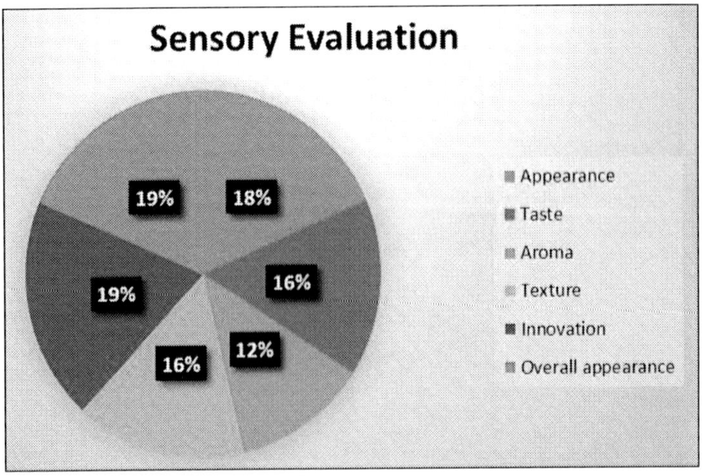

Table 6: Sensory Scorecard

Sr. No.	*Appearance*	*Taste*	*Aroma*	*Texture*	*Innovation*	*Overall Appearance*
1.	4	3	2.5	4	4	4
2.	3.5	3.5	3	3.5	4	3.5
3.	4	3	2	2	4	4
Avg	3.8	3.2	2.5	3.2	4	3.8

Microbial Analysis

Microbial analysis was conducted using the pour plate method for 24 hours. An experiment plate containing the product and a control containing nutritional agar were set simultaneously, it was observed that in the experimental plate, the TLTC level was too low to count.

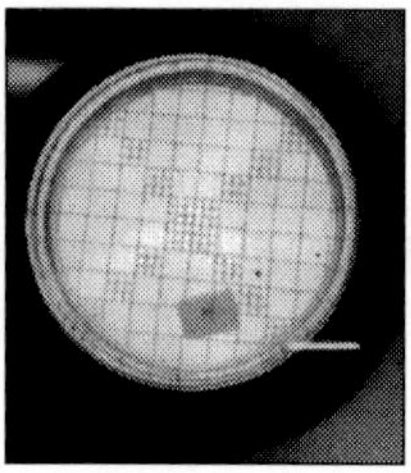

Figure 2: TLTC Count

Figure 3: Pour Plate Method

DISCUSSION

Modification and fusion of the traditional products was seen successful in the given experiment. This helps us to understand that the traditional recipes of millet chakrimill and kachori can be fusioned to make the product more appealing, tasty and nutrient dense by replacing some of the millets with pulses and soya flour.

Soya flour has proved to reduce various metabolic chronic illness. (Kumar Anil 2020) The process of frying can be replaced by the process of dehydrating the product in the sun. The product so formed can be used in rasam and can be consumed as a dumpling. Dehydrating foods helps to increase the shelf life of the product by reducing the moisture content in the food (Digvir *et al.*, 2016). This would help to reduce the fat content of the food product.

Addition of soya flour and pulse flour helps to improve the protein content in the recipe.

The traditional chakrimill is made from rice flour and kachori is made from processed flour however the modified product is made from finger millets, besan and soya flour with the filling of traditional kachori. Finger millets have polyphenols which helps to reduce the incidence of metabolic diseases (Palanisamy Bruntha Devi *et al.,* 2011).

CONCLUSION

The project was undertaken to develop and standardize the production of chakrimill cones. An investigation was undertaken to standardize the processing for chakrimill cones which are multigrain, energy dense and protein rich. The multigrain flour prepared with ragi, jowar, rice flour was found to be acceptable as compared with other level of proportion. The sensory score of chakrimill cones prepared from multigrain flour was found to be rated high as compared to chaklis which are available in market, this was in terms of innovation, taste, flavour, texture and overall acceptability of the product. The microbial analysis performed on the product showed results that the microbes were 'too low to count' which makes the product sustainable hence increasing its shelf life. Hence, it is concluded that the chakrimill cones prepared from using multigrain flour and pulse mixture as a filling was found to be evidently more satisfactorily acceptable.

REFERENCES

[1] Amerine, M.A., Pangborn, R.M. and Rossler, E.B. (1965). Laboratory Studies: Quantity-quality evaluation. Principles of sensory evaluation offood. Academic in Press, New York, 349–397

[2] Chen, Y., Zhang, R., Xu, J. and Ren, Q. (2022). Alteration of intestinal microflora by the intake of millet porridge improves gastrointestinal motility. *Frontiers in nutrition*, *9*, 965687. https://doi.org/10.3389/fnut.2022.965687

[3] Digvir S. Jayas, Food Dehydration, Reference Module in Food Science, 2016 *Erin R. Sanders, J. Vis Exp*, 11 May 2012; (63): 3064.

[4] Jagdale Yash *et al.*, Development of Innovative flour based Indian Traditional Product: Multigrain Chakli, IRJET, 2020, Vol. 7, Iss. 5.

[5] Julie M. Hess, What is a Snack, Why do We Snack, and How can We Choose Better Snacks? A Review of the definitions of Snacking, Motivations to Snack, Contributions to Dietary Intake, and Recommendations for Improvement 1,2, Advances in Nutrition journal, 2017 March 10; 8(2): 398.

[6] Kolhe, S., Mishra, S. and Giri, S. Development of a Nutridense Low Fat Finger Millet Chakli. *IJCRT* (2021); (Vol. 9), ISSN: 2320–2882.

[7] Kumar, Anil *et al.*, Soybean and Its Products: Nutritional and Health Benefits, 2020, 1. 22–29.

[8] Lakshmi Kumari, P. and Sumathi, S. Effect of Consumption of Finger Millet on Hyperglycemia in Non-Insulin Dependent Diabetes Mellitus. *NIDDM.* 2002 Fall; 57(3–4):205–13.

[9] Patekar, S.D., More, D.R. and Satwadhar, P.N. (2017). Studies on Preparation and Nutritional Quality of Sorghum—Fingermillet Chakli. *IJCMAS*, *6*(7), 1381–1389.

[10] Pradhan, Chandan, Gupta, Deepika and Kaushik, Rahul (2011). Health benefits of Indian aromatic plant ajwain. *International Journal of Pharmacy and Technology*, 3, 1356–136.

[11] Vasundhara, A. and Sengupta, R. *Baked Kachori, IJFANS*, 2014; Vol. 3, Iss. 3.

Comparative Analysis of Effect of Baking and Frying on Millet Based Biscuits (Shankarpali), an Indian Tea-Time Snack

Shivani Niphadkar, Mitali Kamble, Manasi Kadam, Gaurav Thombare, Anushka Joshi, Sandhya Menon and Sachin Palekar

Department of Bioanalytical Sciences, Ramnarain Ruia Autonomous College, Mumbai, India

ABSTRACT: Finger millet (*Eleusine coracana*) locally known as Ragi, Nachni, or Mandua is commonly consumed in Indian households, traditionally as a part of a regular diet. It is known to be a highly nutritious millet, also rich in mineral content. Indian millets in general are reported to be nutritionally superior, with a low glycemic index. Finger millet also shows the presence of anti-nutritional factors that contribute to antioxidant properties. To promote its utilization in food, it can be used as a replacement for maida and wheat in some products. As per the current trend of our society, there is a preference for processing techniques such as baking over the traditionally fried methods. This research attempts to compare fried and baked forms of finger millet-based products (Shankarpali/mini biscuits), with respect to their nutritional content and other properties. The observed values stated that the carbohydrates, and lipids content in the fried Shankarpali was found to be greater than that in baked form. The protein content was found to be almost comparable in baked and fried form.

Keywords: *Finger Millet, Nutritional Content, Baking, Frying, Comparative Analysis.*

INTRODUCTION

Indian millets comprise a group of highly nutritious, drought-tolerant crops. They are mostly grown in the arid and semi-arid regions of India. Belonging to the family Poaceae, these small-seeded grasses are an important source of food and fodder. Millets are commonly known as "cereals of the poor" (Indian millets, APEDA 2023). Indian Millets are three to five times more nutritious than most grains and do not contain gluten which makes them ideal for people with celiac disease or diabetes. Ragi is high in carbohydrates and is ranked at the top of grain products by most nutritionists.

Ragi cannot be polished like other grains because it is too tiny, making it possible for us to consume it in its purest form (Ragi: health benefits and tasty recipes for losing weight, Pharm Easy blog, 2023). A snack is a proportion of food smaller than a regular meal (snack, ecdingredients). Snack foods are generally eaten between meals; thus, called convenience foods. They are designed to be quick, portable and satisfying, usually including foods such as cookies, chips, biscuits etc. Indian cooking culture focuses a lot of time into the cooking procedure for any food. Frying and baking are two of the major cooking methods. Baking involves the process of cooking foods in the dry heat of an oven. During baking, moisture within the food is converted to steam, which combines with the dry heat of the oven to cook the food (food safety article, 2010). Baking has been reported to decrease the vitamin content on some foods, but not significantly impact other nutrients. Frying is a process of immersing food in hot oil with a contact between oil, air, and food at a high temperature (150°C–190°C). The simultaneous heat and mass transfer of oil, food, and air during deep fat frying produces the desirable and unique quality of fried foods (Sharma R., Mann B., food chemistry, module 13). Frying oil acts as a heat transfer medium and leads to even cooking, better texture, and taste; but increases calories and fat content. The present work attempts to identify the effect of different methods of cooking on millets and the suitability of such methods on finger millet.

Objectives

- Determining the suitability of baking and frying for millet-based foods.
- Evaluating the effects of baking and frying techniques.
- Making a healthy and tasty millet product.

Methodology

Recipe

Ingredients	Quantity
Ragi flour	1 cup
Ghee (in baked)	½ cup
Ghee (in fried)	¼ cup
Jaggery	½ cup
Dried ginger powder(Suntha)	Small pinch
Water	2-3 tbsp
Cooking oil (for frying)	As required

Figure 1: Ingredients Used

Ghee and jaggery were melted on a low flame, with 1–2 tbsp water. The mixture was allowed to cool slightly. Suntha powder and Ragi flour was added to the mixture and kneaded into a smooth dough; covered and rested for 45 mins. The dough was slightly kneaded again with 1–2 tbsp water; rolled out and cut into small square pieces. For fried form, these pieces were directly fried in hot oil until cooked. For baking, the pieces were baked for 6–8 mins in a preheated microwave at 200°C in a microwave dish greased with ghee.

Figure 2: Fried Shankarpali

Figure 3: Baked Shankarpali

EXPERIMENTAL METHODS

Anthrone Method for Estimation of Carbohydrates

Carbohydrates are one of the three macronutrients in the human diet, along with protein and fat. These molecules contain carbon, hydrogen, and oxygen atoms. Carbohydrates play an important role in the human body. They act as an energy source, help control blood glucose and insulin metabolism, participate in cholesterol and triglyceride metabolism, and help with fermentation. The digestive tract begins to break down carbohydrates into glucose, which is used for energy, upon consumption. In estimation with the anthrone reagent, carbohydrates are hydrolysed into simple sugars using hydrochloric acid. Glucose gets dehydrated to form hydroxymethyl furfural in acidic medium. After reaction with anthrone, the resultant compound forms a green coloured product with an absorption maximum at 630 nm. Anthrone reagent was prepared by dissolving 200 mg anthrone in 100 mL of ice-cold 95% H_2SO_4; prepared fresh before use. For extraction of carbohydrates from the samples, 100 mg of each sample (powdered) was weighed into a boiling tube, and hydrolysed with 5 mL of 2.5 N HCl, by keeping in a boiling water bath for 3 hours. The extract was cooled to room temperature and neutralized by addition of anhy. Sodium carbonate till the effervescence ceased. Neutralized extract was diluted up to 100 mL and centrifuged. 0.1 mL of the collected supernatant was diluted up to 1 mL and used for analysis. Standard series of 20, 40, 60, 80, 100 ppm was prepared using standard glucose. 4 mL of anthrone reagent was added to 1 mL test sample on ice, kept for 8 mins in a boiling water bath and cooled. Graph of Absorbance at 630 nm against concentration (ppm) using a colorimeter gave $R^2 = 0.9819$.

Lowry Method for Estimation of Proteins

Proteins are highly complex substances present in all living organisms. Proteins are of great nutritional value, directly involved in the chemical processes essential for life. A protein molecule is large, consisting of many amino acids joined together to form long chains. They are required for the structure, function, and regulation of the body's tissues and organs. They perform a vast array of functions within the organisms including catalysing metabolic reactions. The Lowry reaction for protein estimation is an extension of the biuret method, about 10 times more sensitive. The principle behind the Lowry method depends upon the presence of peptide bonds. The compounds with two or more peptide bonds form a purple coloured complex with Copper [II] ions under alkaline condition. Intensity of the purple-violet color developed depends upon the concentration of soluble proteins. The sodium potassium tartrate (Na-K-tartrate) prevents the precipitation of $CuSO_4$ as Copper (II) hydroxide under alkaline conditions. Protein when added, displaces the tartrate to form Cuprotein complex. Thus, the color intensity developed is proportional to protein concentration.

Reagents used: Solution A-2% Na_2CO_3 in 0.1 N NaOH; Solution B-0.5% $CuSO_4.5\ H_2O$ in 1%. Na-K Tartrate; Solution C-Mix 50 ml of Solution A and 1 ml of Solution B; Folin Ciocalteu reagent (FCR)-1 part Folin-Phenol (2 N): 1 part water.

Extraction of protein from the sample was done by grinding 0.5 g of the sample with 10 ml of distilled water in a pestle & mortar. Supernatant was obtained after centrifugation at 3000 rpm for 10 minutes. 0.4 mL of the supernatant was diluted to 1 mL and used for protein estimation. Bovine Serum Albumin (BSA) was used as standard. BSA (50 mg) was accurately weighed into a 50 mL volumetric flask, dissolved in 50 mL distilled water. From which 10 mL was pipetted out in 50 mL volumetric flask with distilled water to obtain 200 μg/mL i.e. 200 ppm working standard. Standard series was made by addition of 0.2, 0.4, 0.6, 0.8, 1.0 ml of the working standard solution series into the test tubes and dilution up to 1 ml to give 40, 80, 120, 160, 200 μg/mL of BSA.1 ml of test sample was added and mixed with 5 ml of solution C. The tubes were allowed to stand for 10 minutes at room temperature. 0.5 ml of FCR was introduced into each tube, mixed well and incubated in dark for 30 minutes. The absorbance was read at 660 nm using

Colorimeter. Graph of Absorbance at 660 nm against concentration (ppm) using a colorimeter gave $R^2 = 0.9909$.

Soxhlet Extraction for Estimation of Lipids

Lipids are important fats that serve different roles in the human body. The three main types of lipids are triacylglycerols, phospholipids, and sterols. The body stores energy as adipose tissue (fatty tissue/fat). When food is broken down in the gut, fat released gets combined with fat-soluble nutrients. The combination of the fat and the nutrients allows the nutrients to be digested more easily and absorbed into the body. Dietary fats can also increase the bioavailability of compounds known as phytochemicals—non-essential plant compounds considered beneficial to human health. Dietary lipids are primarily oils (liquid) and fats (solid). Commonly consumed oils are corn, olive, peanut, safflower, soy, and sunflower oil. The method of choice for extraction of lipids from any food or plant-based product is Soxhlet Extraction. A nonpolar solvent such as petroleum ether, chloroform or diethyl ether are used as extracting solvent for lipid extraction.

20 g of test sample (baked and fried separately) was accurately weighed and placed inside a thimble made from thick filter paper and loaded into the main chamber of Soxhlet extractor. 100 mL of petroleum ether was added to the flask with round bottom to which Soxhlet extractor were placed. In the main chamber, 100 mL of petroleum ether was added, allowing it to pass through the thimble. The completion of the extraction process took place between 23 hrs at 25–30°C. The solvent extracted was collected in a pre-weighed beaker ('a' g) and evaporated using a water bath (60–70°C). The residue (extracted lipid) was dried and cooled and weighed again ('b' g). Calculations for lipid content were done as follows: Weight of lipids

(g) = (b–a)

% Total lipids = Weight of lipids (g)/Weight of sample (g) × 100.

RESULTS AND DISCUSSION

Following results were obtained for the estimation of carbohydrate content. Using a dilution factor 10 for tested samples, 1000 ppm of the original non-extracted fried sample was found to contain 826.59 ppm carbohydrates. Amount in baked form was found to be 518.08 ppm.

For the estimation of proteins, 255 ppm protein in Fried sample and 338.33 ppm in Baked samples was obtained for 50000 ppm of original undiluted sample. Dilution factor 2.5 was taken into consideration.

Results of lipid estimation determine the lipid content to be 4.4 g in fried and 2.92 g in baked, per 20 gm product.

These are the results obtained from preliminary testing, although analysis needs to be repeated to establish statistical significance.

Following figure compares the nutrition content in baked and fried forms, in ug per mg of sample.

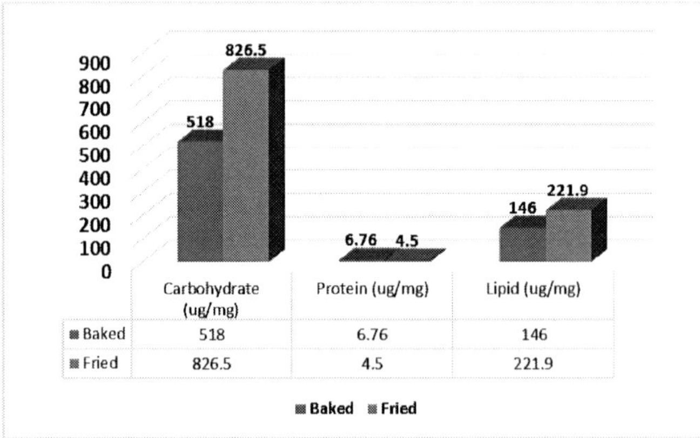

Figure 4: Comparison of Nutrient Content in Baked and Fried

The process of frying is known to impart more calorie content to the food. Presence of Carbohydrates and lipids in higher quantities in the fried form can be attributed to this. The difference in the protein content in baked and fried form is very minor. Therefore protein content in both baked and fried form are almost comparable. Though preliminary analysis results suggest a difference in the biochemical profile of the product under evaluation when subjected to baking and frying, further analysis for evaluation of specific biomolecules needs to be carried out to suggest which cooking technique is better.

REFERENCES

[1] Choe, E. and Min, D.B. (2007), Chemistry of Deep-Fat Frying Oils. *Journal of Food Science,* 72: R77-R86.

[2] https://doi.org/10.1111/j.1750–3841.2007.00352.x

[3] Gull, Amir; Jan, Romee; Nayik, Gulzar; Prasad, Kamlesh and Kumar, Pradyuman (2014). Significance of Finger Millet in Nutrition, Health and Value added Products: A Review. *Journal of Environmental Science, Computer Science and Engineering & Technology,* 03. 1601–1608. Retrieved from: https://www.researchgate.net/publicati on/262732070

[4] Hassan, Z.M., Sebola, N.A. and Mabelebele, M. (2021). The nutritional use of millet grain for food and feed: a review. *Agriculture & food security,* 10(1), 16. https://doi.org/10.1186/s40066-020-00282–6

[5] Omah, Esther Chinelo (2012). Production and evaluation of baked and extruded snacks from blends of millet, pigeon pea and Cassave cortex flour.

[6] Retrieved from: https://www.unn.edu.ng/publications/files/images/OMAH, %20ESTHER%20CHINELO.pdf"

[7] https://apeda.gov.in/apedawebsite/SubHead_Products/Indian_Millets.htm

[8] https://open.maricopa.edu/nutritionessentials/chapter/lipids/#:~:text=Food%20 Sources%20of%20Lipids,%2C%20seeds%2C%20and%20some%20fish.

[9] https://www.hsph.harvard.edu/nutritionsource/what-should-you-eat/protein/

[10] https://www.newagepublishers.com/samplechapter/000091.pdf

Comparative Nutritional Analysis and Development of Nutribar of Millets

Akanksha Radhasyam Panigrahy, Piyusha Milind More, Shreyas Prashant, Sooraj Somasundaran Nair and Kanchan Chitnis

Department of Life Science, Ramnarain Ruia Autonomous College, University of Mumbai, India

ABSTRACT: Millets are also called "superfoods" due to their high nutritive value. Promotion of millets as a healthy alternative can help meet the sustainable development goal of food security (SDG Goals 2 and 3). The objective was to collect millets from different regions across India and compare their nutritive value, perform microscopic analysis, and perform biochemical tests. Furthermore, based on the biochemical findings, to develop a healthy and cost-effective millet nutribar, targeting different stakeholders like binge eaters, diabetic patients, the working class, students, and athletes. Questionnaires were circulated via Google Forms, and responses were recorded. On analysing the responses, an inclination towards the need for a healthy, cost-effective nutribar on a regular basis was observed. From the analyses conducted, it was observed that the starch granules in Bajra (Maharashtra) were large pentagonal shaped, whereas the starch granules in Bajra (Tamil Nadu) were small, irregularly shaped, and uniformly distributed. Bajra (Tamil Nadu) and Ragi (Tamil Nadu) had high protein content (50.5 mg/gm). Jawari (Tamil Nadu) had the highest Blue Value (0.0052 mg/dl), and Ragi (Maharashtra) had the lowest (0.0012 mg/dl). The Blue Value indicates amylose-to-amylopectin ratio, which in turn indicates starch digestibility. Thus, Ragi has more amylopectin, less digestibility, slow release of glucose, and thus low glycaemic index. Based on the biochemical analysis, millet nutribars catering to different stakeholders and their requirements have been developed and tested organoleptically.

Keywords: *Millets, Nutrition, Nutribar, Diabetes, Starch, SDG, Ragi.*

INTRODUCTION

Millets are small, round-shaped coarse grains that have an impressive nutrient profile. In recent times, millet-based foods have been increasingly recommended for a healthy diet and to solve many health issues

(Anitha S. *et al.*, 2022). Millet consumption could also help meet SDG 3 (good health and well-being). Modern food systems may not offer essential nutrients. A change in food habits is crucial for increasing millet consumption among consumers. It has already been reported that transformation in the food system, including the promotion of local foods like millets, is essential to achieving many SDGs (Pradhan *et al.*, 2021). Because many other SDGs are interlinked, promoting millet cultivation and consumption could help achieve many of the UN's SDGs.

Arun A. and Faraday M.K. (2019), describe recipes using pearl millet and discuss their sensory evaluation in their paper. Gopalan C *et al.* (2011) enlist all the nutritional values of different millets in their book. Nutritional aspects of millets have been reported in research papers, but what is lacking is comparative nutrition analysis across regions of India. Also, the starch in the millets has not been characterised for its granules and amylose/amylopectin ratio, which is very important in determining its digestibility.

As a result, the primary goal of this study was to select and collect millets from various regions of India and perform biochemical analyses on them, including microscopic analysis of starch granules in millets, extraction and estimation of total proteins, and estimation of the amylose: amylopectin ratio via Blue Value determination. Also, the aim of this study was to conduct a survey among the stakeholders about their requirements for healthy millet-based nutrition. Based on biochemical analysis and survey results, the objective was to create a millet-based nutritional bar suitable for stakeholders such as diabetics, athletes, students, and working-class people, followed by the organoleptic and sensory evaluation of the millet nutribar.

MATERIALS AND METHODS

Selection and Collection of Millets from Different Regions in India

Millets from local markets in Mumbai (Maharashtra) and Chennai (Tamil Nadu) were collected. Samples used were Ragi Maharashtra/Tamil Nadu (*Eleusine coracana*), Jowar Maharashtra/Tamil Nadu (*Sorghum bicolor*), pseudo millet Rajgira (*Amaranthus*), Varai Maharashtra/Tamil Nadu (*Echinochloa frumentaceae*), Bajra Maharashtra/Tamil Nadu (*Pennisetum glaucum*), Foxtail millet Maharashtra/Tamil Nadu (*Setaria italica*), Proso millet (*Panicum miliaceum*).

Biochemical Analyses

Microscopic Examination of Millet Starch Granules

The samples were soaked overnight in water and then sectioned, stained by dilute iodine, and observed under the microscope (45X) for analysis of starch granule shape and abundance.

Extraction and Estimation of Total Proteins

The millets were powdered. 1 g of each sample was weighed, 4 mL of hexane was added, and the mixture was kept for 4 hours. This was the defatting step. Then they were centrifuged, the supernatant was removed, and in the pellet, 8 ml distilled water was added and kept for 4 hours. Again, the samples were centrifuged, and the supernatants were estimated for their protein content by Biuret method (Plummer DT. 2013).

Estimation of the Amylose: Amylopectin Ratio by Determination of the Blue Value (Nwokocha et al. 2014)

Millets were soaked overnight and then ground in mortar and pestle using distilled water, then filtered through muslin cloth. The starch settled in the filtrate, the filtrate was decanted, and the starch was then dried.

0.1 g dry starch sample was weighed in a tube, 1 ml ethanol (95%) was added followed by 9 ml of 1 M NaOH solution and heated in a boiling water bath for 40 min to solubilize the starch. The starch solution was cooled and transferred into a 100 ml standard volumetric flask, and the volume was made up to the 100 ml mark with distilled water. 2.5 ml of starch solution was taken into a 50 ml standard flask; 0.5 ml of 1 M acetic acid was added, followed by 1 ml of stock iodine (0.2 g I2 and 2.0 g KI per 100 ml), and the solution was made up to the 50 ml mark with distilled water. The color was allowed to develop for 20 minutes, and then the absorbance reading was measured at 620 nm using a UV/visible spectrophotometer. In the reference cell, an iodine solution of the same concentration as above but without starch was used. The blue value was calculated according to the method of Gilbert and Spragg (1964) using the formula:

Blue value = Absorbance at 620 nm × 4/concentration (mg/dl).

Online Survey

To better understand the needs of various stakeholders (students, working class, binge eaters, diabetic patients), questionnaires were distributed to them.

Vital data was collected with respect to the varied individualistic tendencies of different stakeholders.

Recipes Formulation

Based on the biochemical analysis and the conclusions derived from the surveys conducted, two recipes for a millet bar were formulated.

Sensory and Organoleptic Evaluation

An organoleptic survey using a questionnaire was conducted among the stake-holders.

RESULTS AND DISCUSSION

Biochemical Analyses

Microscopic Analysis of Starch Granules

Table 1 shows that there was a geographical difference in the shape, size, and abundance of starch granules in the samples collected from Maharashtra and Tamil Nadu.

Table 1: Comparative Microscopic Analysis of Starch Granules of Millets from Maharashtra and Tamil Nadu (45X)

Sample	*Maharashtra*	*Tamil Nadu*
Ragi	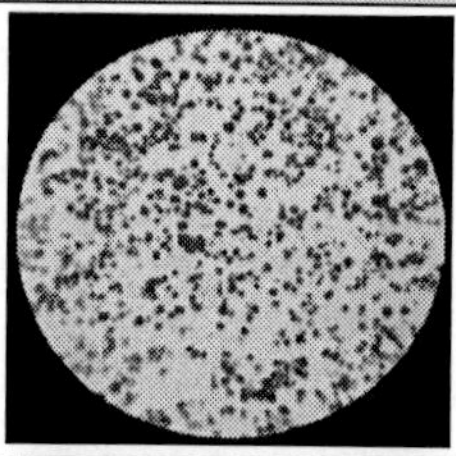	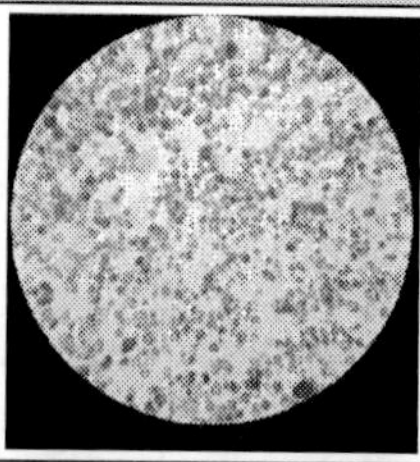
Jowari	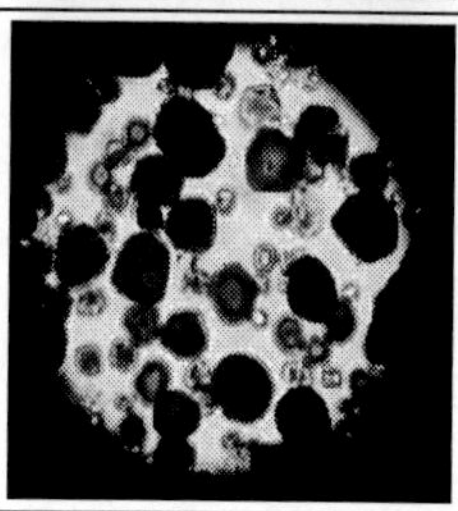	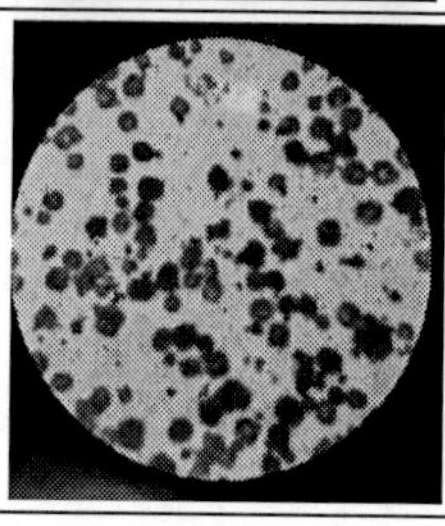

Sample	*Maharashtra*	*Tamil Nadu*
Varai	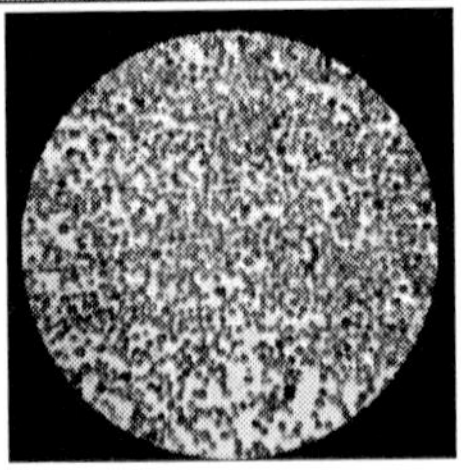	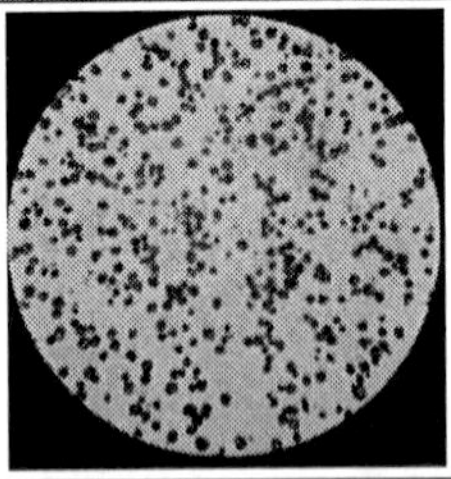
Bajra	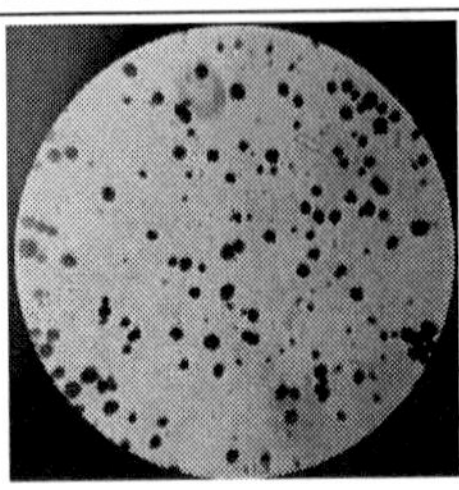	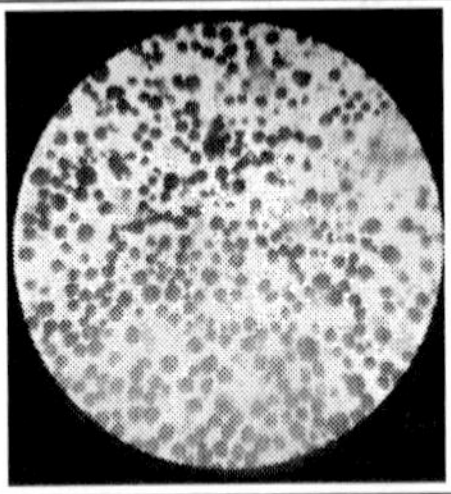
Foxtail	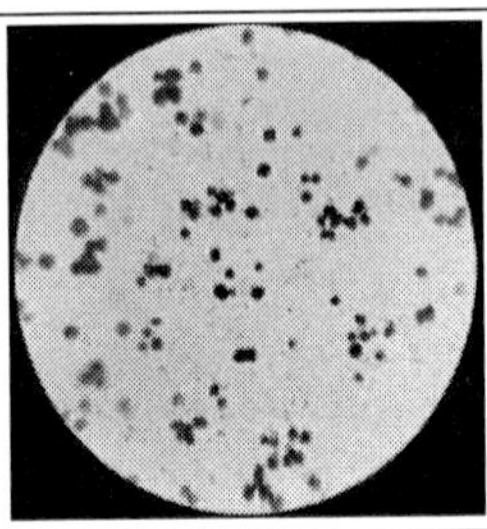	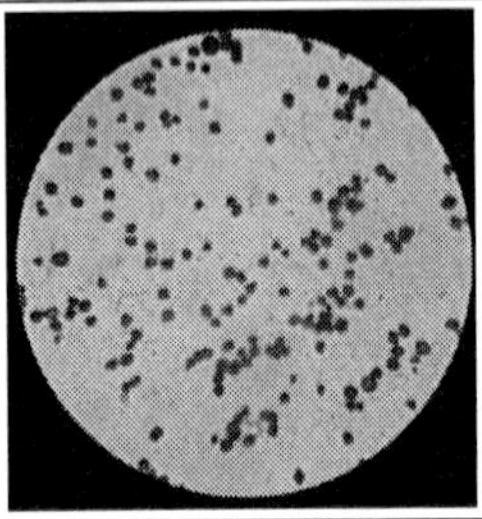

Extraction and Estimation of Total Proteins

Std graph of protein was obtained using Bovine serum albumin 10 mg/ml as in Figure 1.

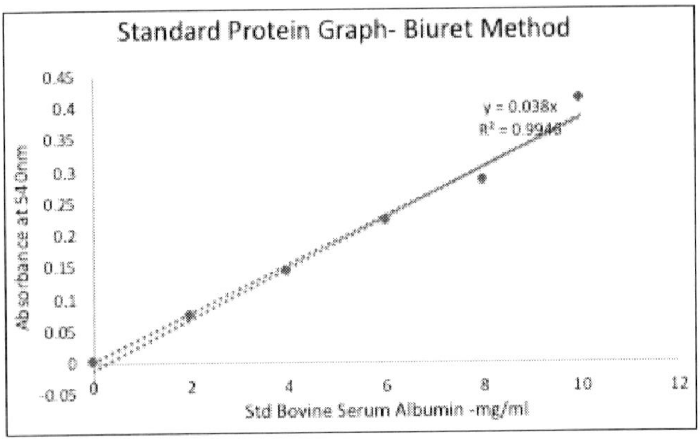

Figure 1: Std Protein Graph

Table 2: Total Protein Estimated in Millets

Sample	*Vol. of Supernatant*	*Biuret Reagent*	*Abs at 530 nm*	*Mg of Protein Calculated by Using Std Equation y = 0.038x*	*Mg/gm Protein*
Ragi M	1 ml	3 ml	0.18	4.74	37.90
Ragi T	1 ml	3 ml	0.24	6.32	50.53
Jowar M	1 ml	3 ml	0.22	5.79	46.32
Jowar T	1 ml	3 ml	0.16	4.21	33.68
Varai M	1 ml	3 ml	0.03	0.79	06.32
Varai T	1 ml	3 ml	0.04	1.05	08.42
Bajra M	1 ml	3 ml	0.21	5.53	44.21
Bajra T	1 ml	3 ml	0.24	6.32	50.52
Foxtail T	1 ml	3 ml	0.08	2.11	16.84
Foxtail M	1 ml	3 ml	0.02	0.53	04.21
Proso	1 ml	3 ml	0.03	0.79	06.32

Table 2 depicts that the highest protein content was found in Bajra and Ragi, followed by Jowar. Varai had a lower protein content. There were distinct geographical variations in protein content in especially Ragi, Jowar and Foxtail.

Estimation of Amylose: Amylopectin Ratio by Determination of Blue Value

Table 3: Blue Value of Millets

Sample	*Absorbance at 620 nm*	*Calculated Blue Value (Abs x 4)/100 = Abs x 0.04*
Ragi M	0.03	0.0012
Ragi T	0.12	0.0048
Jowar M	0.04	0.0016
Jowar T	0.13	0.0052
Varai M	0.07	0.0028
Varai T	0.09	0.0036
Bajra M	0.05	0.0020
Bajra T	0.11	0.0044
Foxtail T	0.06	0.0024
Foxtail M	0.05	0.0020
Proso	0.10	0.0040

As per Table 3, Blue Value for Jowar (Tamil Nadu) was (0.0052 mg/dl) and for Ragi (Maharashtra) it was (0.0012 mg/dl). Blue Value indicates the amylose to amylopectin ratio, which in turn tells about starch digestibility. Ragi has more amylopectin, less digestibility, slow glucose release, and hence Ragi has low glycaemic index. Thus, Ragi is most suitable to be consumed by diabetics. Distinct regional variation in blue values in observed in Ragi, Jowar and Bajra.

Online Survey

(a) On being given an option between the following macronutrients: carbohydrates, proteins, fats, multivitamins, and proteins, 63.3% respondents chose "proteins." (Figure 2).

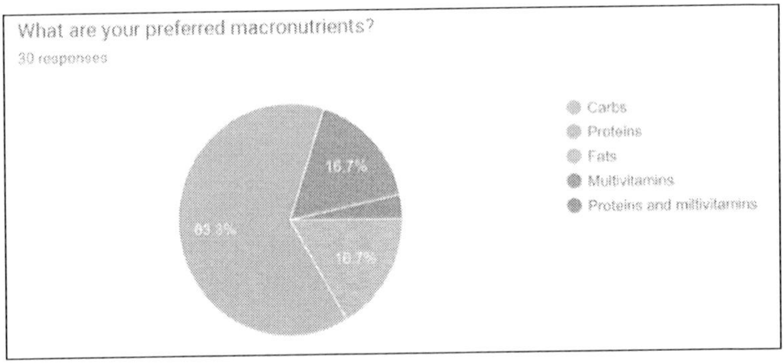

Figure 2: Preferred Macronutrients

(b) A large majority of college going respondents believe that they need a healthy alternative to the junk food they consume multiple times a week (Figure 3).

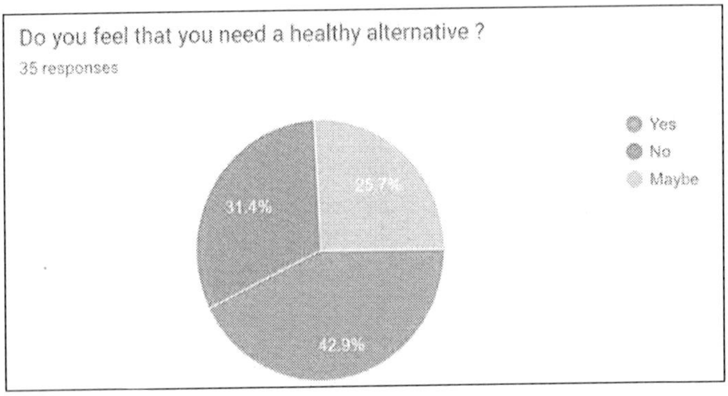

Figure 3: Need for Healthy Alternative

(c) A huge majority of diabetic respondents claim that they consume food to gain energy (Figure 4).

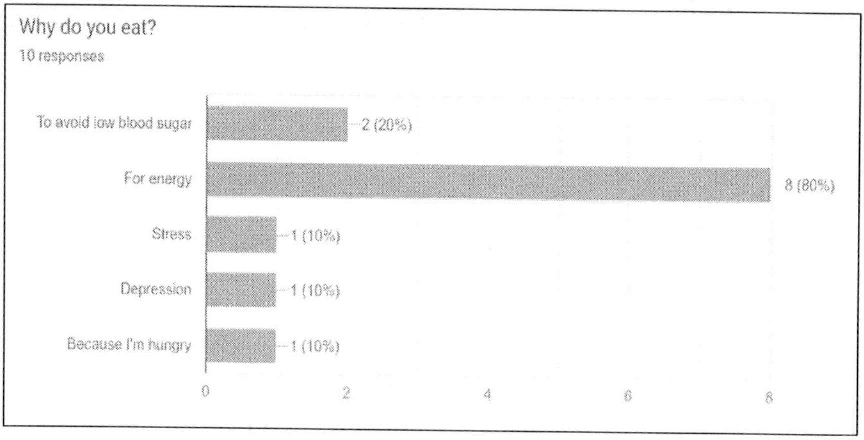

Figure 4: Why Do You Eat

(d) The following chart depicts the behaviour of individuals during episodes of overeating (Figure 5).

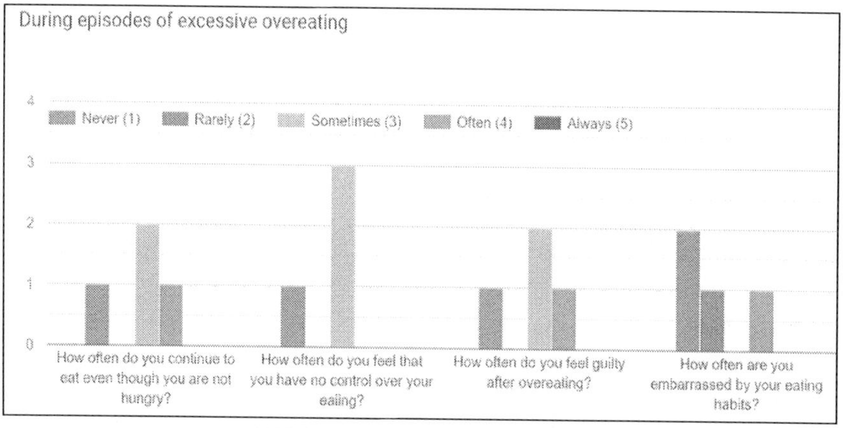

Figure 5: Episodes of Overeating

(e) The expectations and requirements of the stakeholders were also understood. The college students wanted affordability, proteins, and superior taste. The working class expected a quick source of energy while also being a time saver. The diabetics wanted a source of energy with lower level of sugar intake. The main aim for binge eaters was to create a healthier, satiating alternative to what they usually consume.

(f) By analyzing the responses to the question, “On the scale of 1 to 5, how do you rate the need for a healthy and affordable nutrient bar to compensate for one of your meals on a regular basis?” It was observed that 45.1% of respondents felt that they needed a healthy and affordable nutrient bar to compensate for a meal. 31.9% of respondents were unsure about the need for a healthy and affordable nutribar, while the rest, 23.1%, did not feel the need for a healthy and affordable nutribar in their diet (Figure 6).

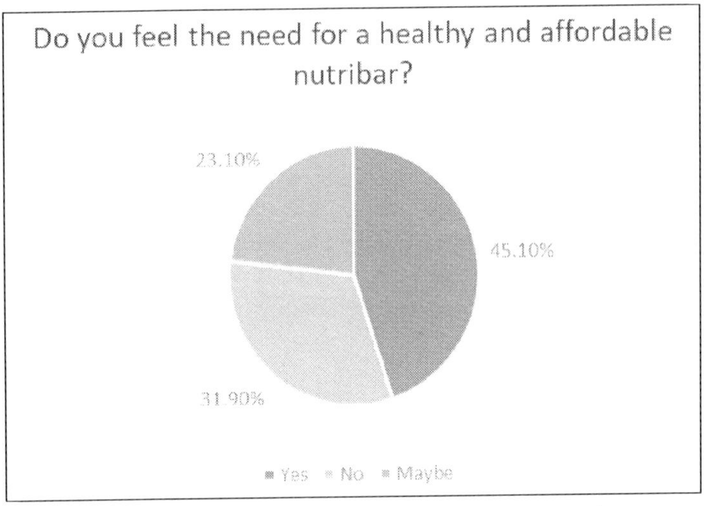

Figure 6: Need for Healthy and Affordable Nutribar

Nutribar Development

Based on biochemical analyses and the conclusions derived from the survey, recipes for the nutribar were formulated.

(a) **Ragi-dates Nutribar:** Deseeded dates (250 g) were coarsely ground. Ragi (250 g) was ground into a coarse powder and dry roasted. Dry fruits, almonds (15–20) and cashew nuts (15–20) were chopped, roughly and dry roasted along with few pumpkin seeds. Ground dates were cooked until they softened. Ragi powder was added in it in small batches and mixed well. Chopped dry fruits and the pumpkin seeds were then added along with puffed rajgira and cooked until the whole mixture turned into a ball. The mixture was spread evenly in the tray after greased with a drop of ghee, allowed to cool, and then cut into pieces.

(b) **Coffee-cocoa Nutribar:** 2 tablespoons of coffee was brewed with hot water. Deseeded dates (250 g) were ground coarsely. Coarse powder of

Ragi was made and roasted until the colour changed to dark brown. Pistachios (15–20) were chopped and roasted along with dry coconut. Dates were cooked until they softened and ragi powder was added into them in portions. To this were added roasted pistachios, dry coconut along with puffed rajgira. The mixture was cooked until it formed a ball. The brewed coffee and 2 tablespoons of cocoa powder were added to it. This mixture was spread evenly on a greased tray and cut into pieces once cooled.

Cost Effectiveness of Nutribar

The current market analysis of on-sale nutribars in general showed an average shelf life of 4 to 8 months with a price tag ranging from Rs. 40 to 60 for a 10 g bar.

Whereas the nutribars that were formulated in this study costed Rs. 1.32/10 g of portion. The cost of packaging being an additional Rs.10, the overall 10 g of nutribar costed Rs. 11.32, which is far more cost effective.

Sensory and Organoleptic Evaluation of the Nutribars

After the creation of the millet bar, it was sent out to individuals along with an organoleptic form. This was done to ascertain multiple aspects of the nutribar: aroma, texture, and satiety, as well as whether stakeholders would purchase or recommend the nutribar to other individuals.

The summarized responses were as follows:

(a) On being asked to rate our nutrition bar on a scale of 1 to 5, with regards to taste, most individuals chose 5 (Figure 7).

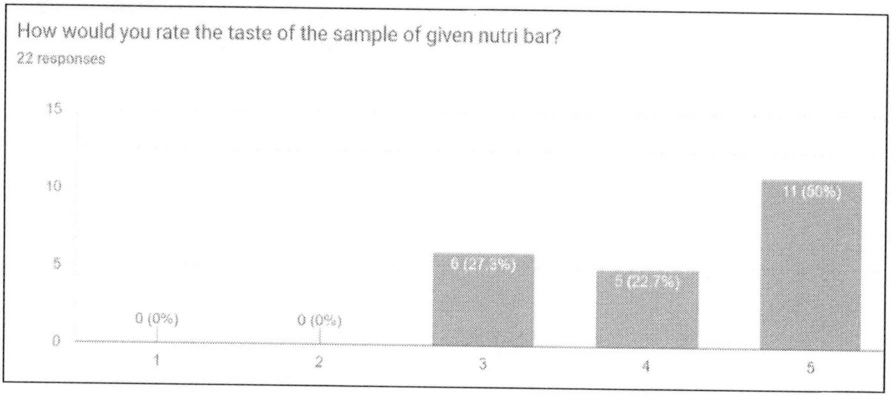

Figure 7: Taste Evaluation

(b) On being asked to rate our nutrition bar on a scale of 1 to 5, in regards to aroma, most individuals chose 4 (Figure 8).

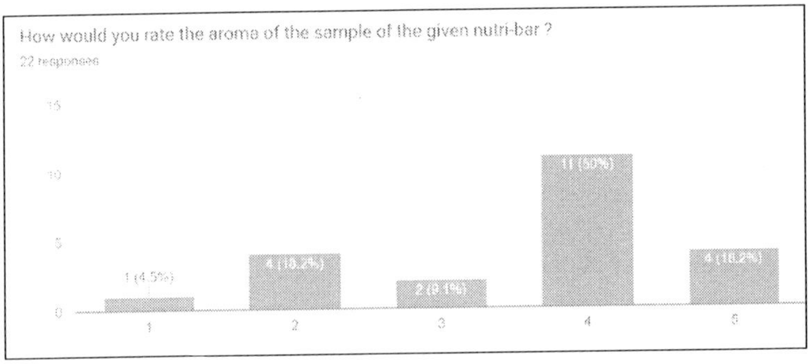

Figure 8: Aroma Evaluation

(c) On being asked to rate our nutrition bar on a scale of 1 to 5, in regards to texture, most individuals chose 4 (Figure 9).

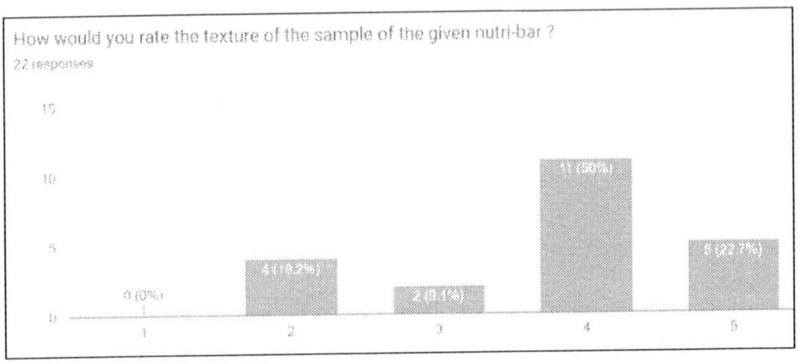

Figure 9: Texture Evaluation

(d) A large proportion of respondents responded positively to being asked if they would buy this nutribar if it were for sale (Figure 10).

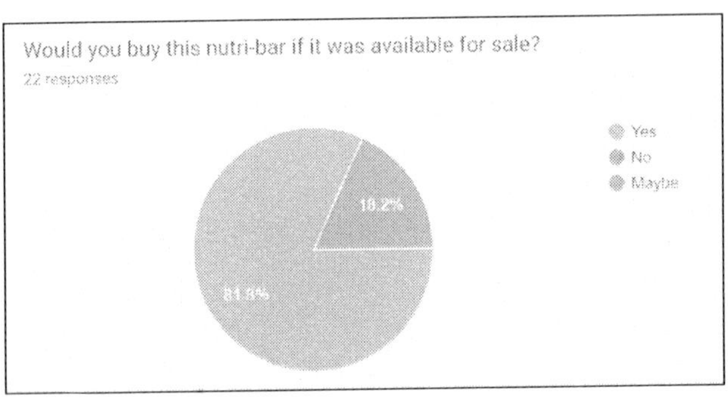

Figure 10: Would You Buy the Nutribar

From the above charts, we can infer that the respondents were mostly satisfied with the nutribar in terms of taste, aroma, and texture. A large proportion of individuals were ready to purchase the nutribar if it were available for sale.

CONCLUSION

There was a geographical difference in the shape, size, and abundance of starch granules in the samples collected from Maharashtra and Tamil Nadu. The highest protein content was found in Bajra and Ragi, followed by Jowar. Varai had a lower protein content. There were distinct geographical variations in protein content in especially Ragi, Jowar and Foxtail. Jowar (Tamil Nadu) showed the highest Blue Value (0.0052 mg/dl) whereas Ragi (Maharashtra) showed the lowest (0.0012 mg/dl). Hence, we concluded that Ragi had more amylopectin, hence low digestibility with slow glucose release, and hence low glycaemic index. Thus, it was inferred that Ragi was most suitable to be consumed by diabetics. Distinct regional variation in Blue Values was observed in Ragi, Jowar and Bajra.

From the survey it was observed that 45.10% of respondents felt there was a need for a healthy and affordable nutritional bar that could be added to their regular diet. The preferred macronutrient of majority of respondents was protein (88.9%). 80% of diabetic patients ate to gain energy.

On studying the overall preferences and choices made by the stakeholders, it was observed that college students wanted affordability, proteins, and superior taste. The working class expected a quick source of energy while also being a time saver. The diabetics wanted a source of energy with lower level of sugar intake. The main aim for binge eaters was to create a healthier, satiating alternative to what they usually consume.

Based on the biochemical and microscopic analysis performed and the Google forms circulated, two different recipes were formulated. Ragi-dates nutribar made from ragi flour, dates, almonds, cashew nuts, pumpkin seeds, puffed rajgira. Coffee-cocoa nutribar composed of brewed coffee, cocoa powder, ragi flour, dates, pistachios, dry coconuts and puffed rajgira.

Both the nutribars were found to be cost effective and were accepted well for taste, aroma and texture.

REFERENCES

[1] Anitha, S., Tsusaka, T.W., Botha, R., Potaka, J.K., Givens, D.I., Ranjendran, A. and Bhandari, R.K.; Are Millets More Effective in Managing Hyperlipidaemia and Obesity than Major Cereal Staples? A Systematic Review and Meta-Analysis; *Sustainability* 2022, *14*(11), 6659; https://doi.org/10.3390/su1411 6659

[2] Arun, A. and Faraday, M.K.; A perspective study to endorse the Nutritional aspects of Pearl Millet (*Pennisetum glaucum L*) and formulated Recipes; *Research J. Pharm. and Tech 2020; 13(2):.911–914.* DOI: 10.5958/0974-360X.2020.00 172.9

[3] Gilbert, G.A. and Spragg, S.P. (1964). Iodometric Determination of Amylose. In: Methods in Carbohydrate Chemistry. R.L. Whistler (ed.), *Academic Press, Orlando*, Florida, 168–169.

[4] Gopalan, C., Rama Shastri, B.V. and Balasubramanian, S.C. (2011). Nutritive Value of Indian Foods, *National Institute of Nutrition (ICMR),* Hyderabad.

[5] Nwokocha, L.M., Gabriel, B. and Ogunmola, G.B. (2014). Colour of starch-iodine complex as index of retrogradability of starch pastes. *African Journal of Pure and Applied Chemistry,* 8: 89–93.

[6] Plummer, D.T. (2013). An introduction to Practical Biochemistry, 3rd Ed. Mc Graw Hill Education (India) Private Limited.

[7] Pradhan, P., Sapkota, T.B. and Kropp, J.P. (2021). Why food systems transformation is Crucial for achieving the SDGs. Ruralia, 21(55), 10–12.

Development and Nutritional Evaluation of Malted Finger Millet Milk

Saee Bobhate, Jessy Pius and Bhavna R. Narula

Department of Botany, Ramnarain Ruia Autonomous College, L.N. Road, Matunga, Mumbai, India

ABSTRACT: Tackling hunger and feeding the world population are two of the biggest challenges of the modern world. From the beginning of human civilization, Milk has been an essential component of the daily diet. Scientific consensus shows that animal husbandry is the second largest contributor to human-made greenhouse gas emissions after fossil fuels, being responsible for at least 14.5% of global greenhouse gas emissions and causing significant environmental degradation. Hence the demand for milk alternatives is rising. These substitutes are called plant-based milk, which are made from legumes, cereals, pseudocereals, or nuts. Over the years, the global market for these products has become a multi-billion-dollar business. Unlike the widely used alternatives such as legume and nut milk, milk made with finger millet-ragi (*Eleusine coracana*) can be excellent, as it is rich in calcium, iron and antioxidants. It is also gluten-free and lactose-free. This study aims to develop and determine the nutritional value of malted ragi milk. The techniques used were soaking, sprouting, drying and extraction of the milk. When compared to cow milk and soy milk, it showed thrice the higher amount of Calcium. Thus, it may be a potential sustainable source for calcium in India.

Keywords: *Finger Millet, Ragi Milk, Malting, Plant-Based Milk Substitute, Calcium.*

INTRODUCTION

There are many challenges associated with feeding the world such as hunger, food insecurity and malnutrition. The Sustainable Development Goals of the United Nations have recently established precise aims to end hunger by 2030, yet there has been little advancement in reaching these goals. To effectively tackle the issue of hunger and nutrition deficiency, there must be a concerted effort from governments, international organizations and communities.

Milk is the secreted fluid of the mammary glands of female mammals. It contains nearly all the nutrients necessary to sustain life of their young offspring. From the beginning of human civilization, milk has been an essential component of the daily diet. It is a nutritionally dense food. The sustainability and environmental effect of the dairy business have come under increased scrutiny in recent years. According to the IPCC (2013), greenhouse gas (GHG) emissions that warm the atmosphere are the main culprits behind global climate change (Stocker, 2014). According to Gerber *et al.* (2013), the livestock industry is responsible for 14.5% of the world's greenhouse gas emissions, which eventually leads to land degradation, pollution of air and water, and reduction of biodiversity. However, despite improved productivity and intensification efforts, the overall amount of cultivated land worldwide has remained stable since 1991 (O'Mara, 2012), which will put more stress on the land resources, making it unstainable over course of time. Additionally, milk consumption leads to a number of health risks. Despite being nutritionally rich, it is not a food easily digested and absorbed by all. Leading to metabolic diseases such as lactose intolerance (LI) and cow's milk protein allergy (CMPA). About 75% of the world's population has symptoms of LI (Gasparin *et al.*, 2010).

Plant-based milk substitutes are fluids produced by breaking down (size reduction) plant material (cereals, pseudo-cereals, legumes, oilseeds, and nuts) extracted in water. These fluids are then homogenized, mimicking the consistency and appearance of cow's milk (Sethi *et al.*, 2016). However, it differs from it in terms of sensory qualities, stability, and nutritional makeup (Sethi *et al.*, 2016). Its preparation involves breaking down the raw plant material after it has been extracted in water, followed by homogenization to produce emulsions of the raw material's constituent parts in the water phase (Cruz *et al.*, 2007, Sethi *et al.*, 2016). The particle size and stability of the final product depend on the nature of the raw material, the extraction method and the storage conditions (Cruz *et al.*, 2007).

The *Poaceae* family of grasses includes millets as a minor cereal. According to Verma *et al.*, (2020), millets provide a number of desirable qualities, including drought and insect resistance, vitamin and mineral richness that eliminates malnutrition, possible prebiotic and probiotics that have numerous health benefits, etc. Ragi or finger millet (*Eleusine coracana L.*) is one of the common millets in several regions of India. For thousands of years, it has been a part of the staple diet in India, areas of eastern and central Africa. According to Malathi *et al.* (2007), finger millet was traditionally processed in India by means such as grinding, malting, and fermentation for foods such as drinks, porridges, idli (an Indian fermented steamed cake), dosa (an Indian fermented pan cake), and roti (an unleavened flat bread). The extensive use of Ragi in diet is also because of its rich nutritional profile.

Among cereals, it is noteworthy that finger millet has been shown to be consistently high in calcium regardless of the variety (364 ± 58 mg/100 g). It requires few resources to grow. It is resistant to stress and is ideal for sustainable agriculture. The calcium concentration is ten times higher than that of brown rice, wheat, or maize, and three times that of cow milk (Puranik *et al.*, 2017 and Longvah *et al.*, 2017). Besides calcium, finger millet is also very rich source of iron, amino acids like methionine, slowly digestible starch and phytochemicals like polyphenols. It is a gluten-free, low-fat cereal which is non-allergic and easily digestible (Kumar *et al.*, 2016). However, it also has the highest phytate content (306 mg/100 g) among cereals (Puranik *et al.*, 2017) which restricts the calcium bioavailability. According to Mbithi *et al.* (2000), after four days of germination, the phytic acids content reduces to undetectable levels and enhances calcium extractability. Malting is the process in which the grains are germinated followed by drying to halt the progress when the sprouting begins. This will enhance the bioavailability of calcium and reduction in phytic acid content of the grains, making it a suitable candidate as calcium rich diet. Also finger millet has the best malting qualities of any tropical cereal. Attempts to research the malting qualities of finger millet were performed as early as 1939, when Sastri, (1939) reported the parameters for producing good grade malt. The tiny size of finger millet grain was beneficial for consistent germination as well as kilning. Malt flour is a maltodextrin alternative that can be mixed with milk and spray dried to make newborn food (Malleshi, 2007). In India, ragi malt is traditionally given to babies while introducing them to solid foods.

In the view of above-mentioned facts, this study investigates the potential of malted finger millet milk later mentioned as ragi milk as a sustainable additional source of nutrition alongside milk to meet calcium requirement. Its nutritional composition was then further compared with the most commonly consumed bovine milk and plant-based milk i.e. cow milk and soy milk respectively.

MATERIALS AND METHOD

The finger millet was procured from local markets of Mumbai. Malted finger millet milk was extracted using conventional methods as follows: The seeds were cleansed and three times rinsed with water. The seeds were soaked in water for 12–18 hours before germination at 30°C–35°C for 4 days. Sprouted finger millet seeds were sundried for 4 hours and then roasted (kilning) at 180°C for 5 minutes. The seeds were ground in a food processor with water in a 1:5 millet-to-water ratio, then filtered through muslin cloth. The nutritional composition of the samples (carbohydrates, protein, fat, sugar, calcium, and energy) was determined using AOAC procedures (Horwitz *et al.*, 2010). The USDA database was used to acquire nutritional information for cow milk and soy milk.

RESULTS AND DISCUSSION

Table 1: Physical Characteristics of Ragi Milk

Moisture	87.38%
Total Ash	0.24%
Mineral Content (acid insoluble ash)	0.03%
pH	6.4

Table 2: Nutritional Comparison between Ragi, Cow and Soy Milk

Nutritional Parameter	*Ragi Milk*	*Cow Milk*	*Soy Milk*
Total carbohydrate	10.07%	5.51%	4.92%
Total sugar	9.55%	5.08%	3.65%
Protein	1.12%	3.2%	2.6%
Total fat	2.11%	3.3%	1.47%
Calcium	3.44%	1.23%	0.25%
Energy	63.75 kCal	61 kCal	43 kCal

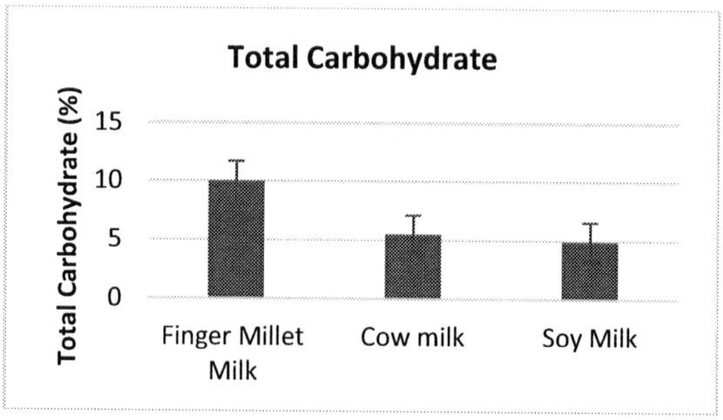

Figure 1: Comparison of Total Carbohydrate Content

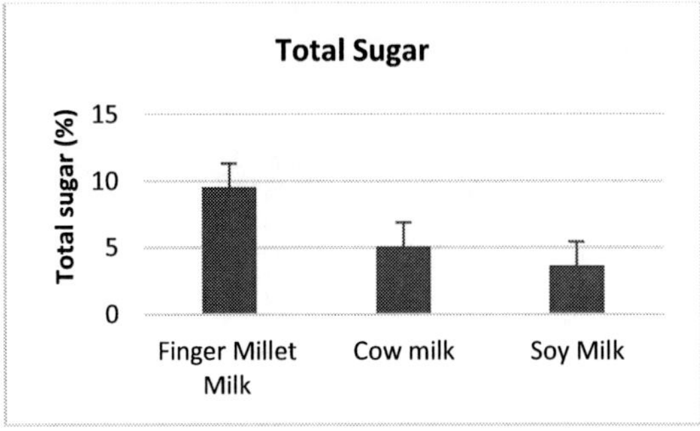

Figure 2: Comparison of Total Sugar Content

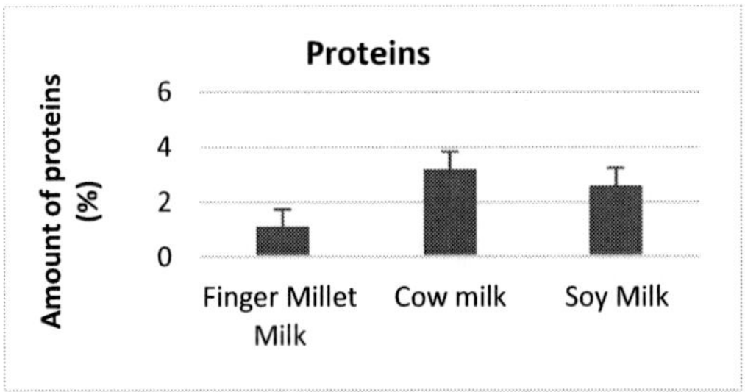

Figure 3: Comparison of Protein Content

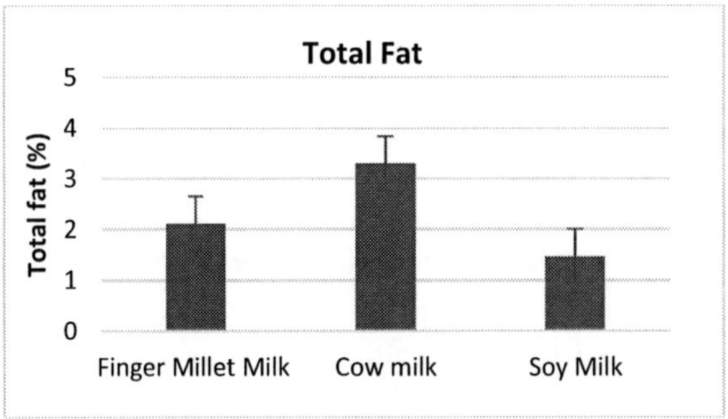

Figure 4: Comparison of Total Fat Content

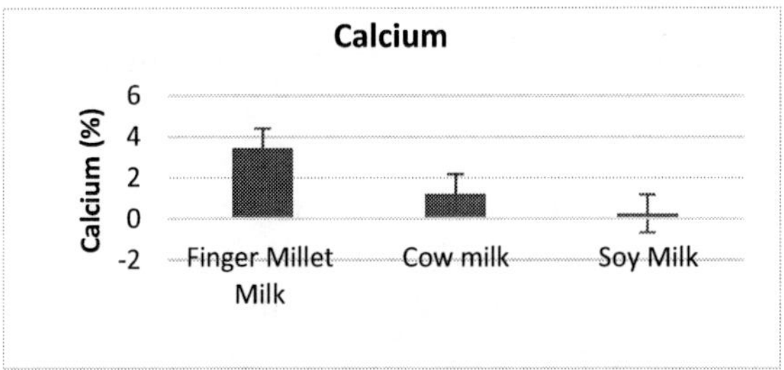

Figure 5: Comparison of Calcium Content

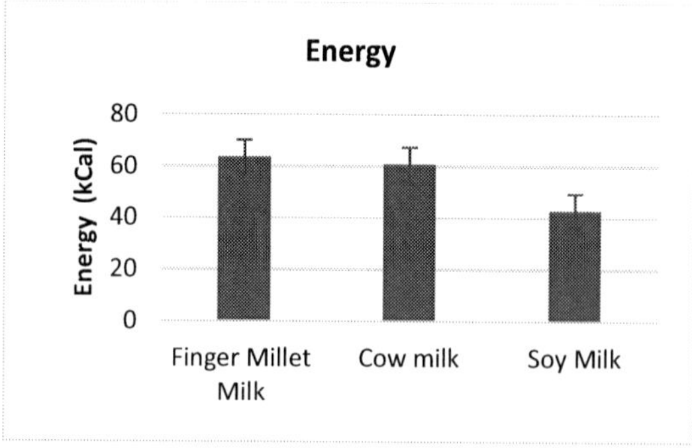

Figure 6: Comparison of Energy Content

The comparison of nutritional analysis has been tabulated in Table 1. Ragi milk (10.07%) contains nearly twice as much total carbohydrate as cow milk (5.51%) and soy milk (4.92%). Ragi milk has a higher total sugar level (9.55%) than the other two. However, the protein level of ragi milk is substantially lower (1.12%) than that of cow milk (3.2%) and soy milk (2.6%). Ragi milk contains more total fat (2.11%) than soy milk (1.47%) but less than cow milk (3.3%). The amount of calcium in ragi milk is three times (3.44%) than that of cow milk (1.23%) and nearly 14 times that of soy milk (0.25%). The energy content of ragi milk is slightly higher (63.75 kCal) than that of the cow milk (61 kCal).

Calcium and vitamin D are essential for good bone health. Dietary adjustments that include calcium-rich foods, such as dairy products, appear to be the most effective method to prevent calcium deficiency. However, convincing people to eat a diverse diet is difficult. For example, 65% of the world's population is lactose intolerant, therefore they cannot get their Ca from dairy products, majority of them live in Asian and African countries. (Curry, 2013) They are largely dependent on agriculture-based economies. Plant-based calcium such as that found in traditional staple grains is important for diets in many countries. As a result, many cultures in these locations, especially vegetarians and vegans, require alternate sources of calcium.

According to Harinarayan *et al.* (2018), dietary calcium intake in rural, tribal, and urban India has decreased over the last half-century. Despite being the world's largest producer of milk, non-dairy items are the primary source of calcium in India. Rural and tribal subjects consumed the most cereals and the least milk and milk products, whereas urban and metropolitan subjects consumed the same amount of cereals, milk, and milk products. The bioavailability of calcium derived from dairy sources was one of the explanations for lower calcium consumption.

CONCLUSION

Germinated finger millet has the potential to be an effective and sustainable source of calcium as compared to the mostly consumed cow milk and plant-based milks. Hence it can be used as a supplement to bovine milk to some extent to meet the daily calcium requirements of the human body.

Further research has to be done to understand the phytic acid content of the ragi milk as well as shelf life, palatability and effect of different processing techniques such as pasteurization, heating, freezing etc. on its nutritional profile.

ACKNOWLEDGMENT

The authors thank the Principal and Teaching & non-teaching staff of Department of Botany of Ramnarain Ruia Autonomous College, Mumbai, Maharashtra for providing the infrastructure to carry out this work.

REFERENCES

[1] Cruz, N., Capellas, M., Hernández, M., Trujillo, A.J., Guamis, B. and Ferragut, V. (2007). Ultra high-pressure homogenization of soymilk: Microbiological, physicochemical and microstructural characteristics. *Food research international,* Vol. 40(6), pp. 725–732.

[2] Curry, A. (2013). Archaeology: the milk revolution. When a single genetic mutation first let ancient Europeans drink milk, it set the stage for a continental upheaval. *Nature,* Vol. 500, pp. 20–22.

[3] Gasparin, F.S.R., Carvalho, J.M.T. and de Araujo, S.C. (2010). Alergia à proteína do leite de vaca versus intolerância à lactose: as diferenças e semelhanças. *Saúde e Pesquisa,* 3(1).

[4] Gerber, P.J., Steinfeld, H., Henderson, B., Mottet, A., Opio, C.. Dijkman, J., Falcucci, A.; Tempio and G. (2013). Tackling climate change through livestock—A global assessment of emissions and mitigation opportunities; *Food and Agriculture Organization of the United Nations (FAO), Rome.*

[5] Harinarayan, C.V., Ramalakshmi, T., Prasad, U.V. and Sudhakar, D. (2008). Vitamin D status in Andhra Pradesh: A population based study; *Indian Journal of Medical Research,* Vol. 127(3), pp. 211–218.

[6] Horwitz, W. (2010). Official methods of analysis of AOAC International organisation of agricultural chemicals, contaminants, drugs, Vol. 1.

[7] Kumar, A., Metwal, M., Kaur, S., Gupta, A.K., Puranik, S., Singh, S. and Yadav, R. (2016). Nutraceutical value of finger millet [*Eleusine coracana (L.) Gaertn.*], and their improvement using omics approaches; *Frontiers in plant science,* Vol. 7, pp. 934.

[8] Longvah, T., Ananthan, R., Bhaskarachary, K. and Venkaiah, K. (2017). Indian Food Composition Table; *National Institute of Nutrition: Hyderabad, India.*

[9] Malathi, D. and Nirmalakumari, A. (2007). Cooking of small millets in Tamil Nadu. Food uses of small millets and avenues for further processing and value addition, pp. 57–63.

[10] Malleshi, N.G. (2007). Nutritional and technological features of ragi (finger millet) and processing for value addition; Food uses of small millets and avenues for further processing and value addition, pp. 9–19.

[11] Mbithi-Mwikya, S., Camp, J.V., Yiru, Y. and Huyghebaert, A. (2000). Nutrient and antinutrient changes in finger millet (*Eleusine coracana*) during sprouting; *Lebensm Wiss Technol*, Vol. 33(1), pp. 9–14.

[12] O'Mara, F.P. (2012). The role of grasslands in food security and climate change, *Annals of Botany*, Vol. 110(6), pp. 1263–1270.

[13] Puranik, S., Kam, J., Sahu, P. P., Yadav, R., Srivastava, R.K., Ojulong, H. and Yadav, R. (2017). Harnessing finger millet to combat calcium deficiency in humans: challenges and prospects; *Frontiers in Plant Science*, Vol. 8, pp. 1311.

[14] Sastri, B.N. (1939). Ragi, *Eleucine coracana Gaertn*—A new raw material for the malting industry; *Current Science*, Vol. 1, pp. 34–35.

[15] Sethi, S., Tyagi, SK. and Anurag, R.K. (2016). Plant-based milk alternatives an emerging segment of functional beverages: A review. *Journal of Food Science and Technology*, Vol. 53(9), pp. 3408–3423.

[16] Stocker, T. (2014). Climate change 2013: The physical science basis: Working Group I contribution to the Fifth assessment report of the Intergovernmental Panel on Climate Change. *Cambridge university press.*

[17] USDA National Nutrient Database for Standard Reference, Release 28 (2015). US Department of Agriculture, Agricultural Research Service, Nutrient Data Laboratory.

[18] Verma, V.C., Acharya, S. and Verma, B.C. (2020). Millets for Sustainable Agriculture and Nutritional Security.

THEME-4

Bioprospecting and Innovative Sustainable Cultivation Techniques for Millets

Comparative Analysis of Age-Dependent Phytochemical Variation of Sorghum and Sweet Sorghum Using GC-MS

M. Nandini, P. Saloni, S. Rajvi, G. Nandini and K. Sayali

Department of Bioanalytical Sciences,
Ramnarain Ruia Autonomous College, Matunga (East)

ABSTRACT: Sweet sorghum (*Sorghum bicolor L. Moench*) is commonly known as 'hurda', 'poonkh', etc. It is the tender form of jowar (*Sorghum bicolor*). 'Hurda party' is celebrated in different Indian states, in order to celebrate the successful cultivation of jowar during the winter season. Hurda contains crude protein, crude fat, carbohydrate, and crude fibres. The secondary metabolites of hurda are yet to be explored thus this study focuses on phytochemical aspects of sweet sorghum grain and age-dependent phytoconstituent variability in hurda and jowar. The presence of Bioactive components was identified using GC-MS and other qualitative tests.

Keywords: *Sweet Sorghum, Age-Dependent Variability, Phytochemical Analysis, Bioactive Components, GC-MS.*

INTRODUCTION

Millets are packed with nutrients. In addition to being gluten-free, millets improve weight loss and promote good health. They are available all year round in various forms, each with unique health advantages. Sorghum bicolor commonly known as jowar is a member of the Poaceae genus of grass. According to the data from Chamodini *et al.,* 2021, Sorghum is a significant grain crop that is widely farmed in tropical and subtropical areas of the world. It is utilized in the production of alcohol, sugar, starch, and biscuits. It contributes to various health benefits such as the presence of vitamin B is abundant in sorghum (Samarth *et al.,* 2018) and is essential for metabolism, nerve cell formation, and maintaining healthy hair and skin. Sorghum also has a lot of flavonoids, phenolic acids, and tannins, which are powerful anti-oxidants (Jairam *et al.,* 2018). The tender form of sorghum, i.e. *Sorghum bicolor L. Moench* commonly known as Hurda, Ponkh, leli jowar, etc. It is cultivated in Gujarat, Tamil Nadu, and Maharashtra, India, and is accessible during the winter season. It is consumed in various Indian delicacies like chatt, fitters, etc. Its high sugar content in the stalk it's used in the production of ethanol and syrup. Apart from its sweetness, it is a promising target for bioethanol production. It is a C4 crop with a low input requirement and accumulates high levels of sugar in the stalks. (Mathur *et al.*, 2017)

Phytochemicals are plant-derived chemicals with bioactive components having various health benefits. Knowing the phytochemicals will give us information on medical and nutritional science since many have antioxidant, anti-carcinogenic, neuroprotective, or anti-inflammatory properties. These phytoconstituents can be estimated using Qualitative and Quantitative tests. Techniques like TLC and instruments like HPLC and GC-MS enable us to qualitatively and quantitate the components present. The objective of the research is to screen phytoconstituents present in sweet sorghum qualitatively using different tests and to study the age-dependent phytoconstituent variability in Sorghum & Sweet sorghum.

Dragendroff's Test for Alkaloids	1 ml extract + few drops of Dragendr off's reagent	White ppt
Tannins	1 ml extract + few drops of freshly prepared 0.1% $FeCl_3$	Blue-black ppt
Keller-Killani Test for Glycosides	5 ml extract + 2 ml glacial acetic acid + 1 ml conc. H_2SO_4 from the sides	Brown ring at the inter-phase
Terpenoids	1 ml extract + 2 ml $CHCl_3$ + few drops of conc. H_2SO_4	Brown ring at the base
Flavonoids	1 ml extract + 10 ml distilled water + 5 ml dilute NH_3 + conc H_2SO_4	Yellow colour

Figure 1: *Sweet Sorghum* (hurda)

Figure 2: *Sorghum* (Jowar)

METHODOLOGY

Collection of Sample

Sorghum and sweet sorghum samples were procured from the local market.

Sample Extraction

Sample extracts were prepared Methanol, sonicated and macerated for 24 hours. After 24 hours extracts were filtered using Whatman filter paper.

Qualitative Tests

Preliminary tests were performed to find the phytoconstituent composition in Sorghum and sweet sorghum.

Table 1: Qualitative Tests

Phytoconstituents	*Test*	*Results*
Mayer's Test for Alkaloids	1 ml extract + few drops of Mayer's reagent	Orange ppt

- *GC-MS:* GC-MS was performed to confirm the presence of phyto-constituents.

Observations

Qualitative Tests:

Keywords: +: Present,
-: Absent

Table 2: Results for Qualitative Tests

Tests	*Sorghum Extract*	*Sweet Sorghum Extract*
Mayer's Test for Alkaloids	–	–
Dragendorff's Test for Alkaloids	–	–
Test for Saponins	+	+
Keller-kitiani's Test for Glycosides	+	+
Test for flavonoids	–	–
Test for Tannins	–	–

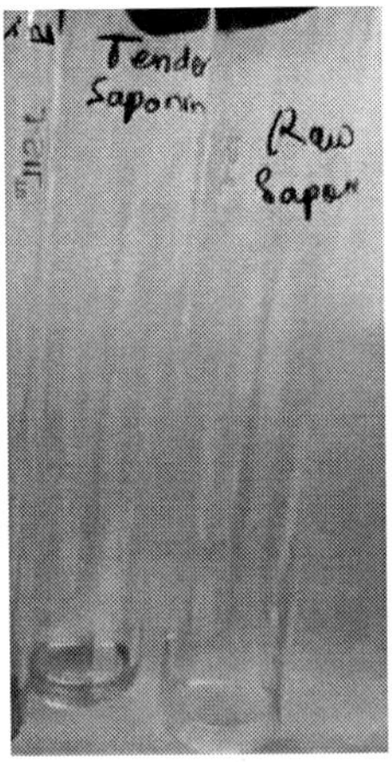

Figure 3: Qualitative test for Saponins

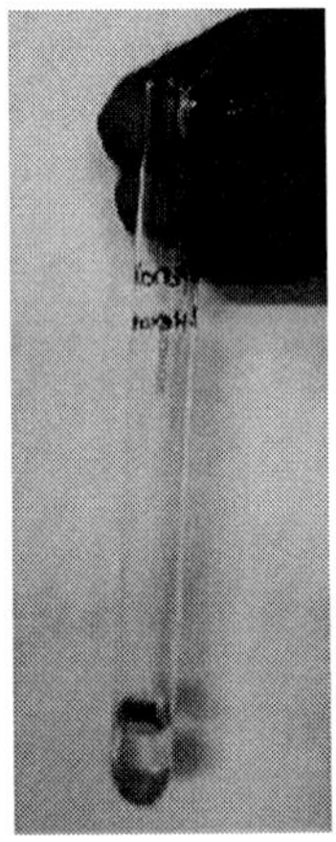

Figure 4: Qualitative Tests for Terpenoids

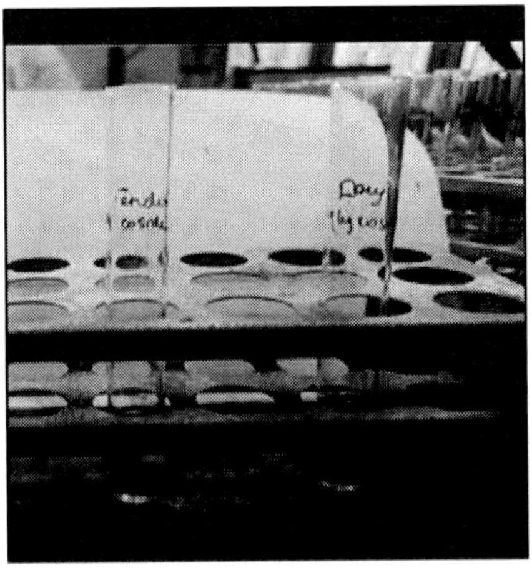

Figure 5: Qualitative Test for Glycosides

GC-MS Data

GC-MS Analysis Chromatogram for Sorghum

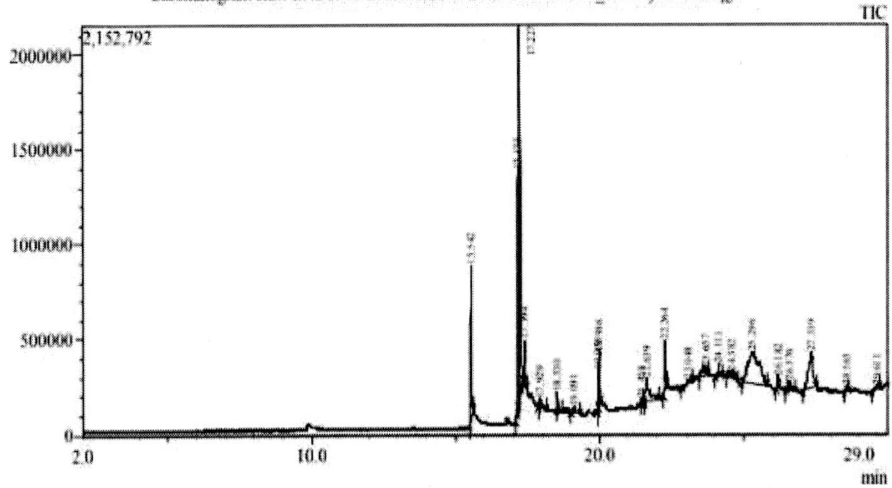

Peak#	R.Time	Area	Area%	Height	Name
1	15.542	1941842	8.12	805660	n-Hexadecanoic acid
2	17.175	4118293	17.23	1258423	9,12-Octadecadienoic acid (Z,Z)-
3	17.227	5850063	24.48	1981052	9-Octadecenoic acid, (E)-
4	17.394	432284	1.81	232767	Octadecanoic acid
5	17.929	143120	0.60	48015	9,12-Octadecadienoic acid (Z,Z)-
6	18.530	178480	0.75	94416	Hexadecanoic acid, 1-(hydroxymethyl)-1,2-ethanedi
7	19.091	112729	0.47	32377	9,12-Octadecadienoic acid (Z,Z)-
8	19.955	382045	1.60	210082	9,12-Octadecadienoyl chloride, (Z,Z)-
9	19.986	503841	2.11	283958	9-Octadecenoic acid, 1,2,3-propanetriyl ester, (E,E,I
10	21.458	56072	0.23	20481	9,12-Octadecadienoic acid (Z,Z)-
11	21.639	890633	3.73	123450	9-Octadecenoic acid, 1,2,3-propanetriyl ester, (E,E,I
12	22.264	471977	1.97	256259	13-Docosenamide, (Z)-
13	23.048	56661	0.24	25252	9-Octadecenoic acid, 1,2,3-propanetriyl ester, (E,E,I
14	23.657	567803	2.38	47201	9-Octadecenoic acid, 1,2,3-propanetriyl ester, (E,E,I
15	24.111	232584	0.97	67578	9-Octadecenoic acid, 1,2,3-propanetriyl ester, (E,E,I
16	24.582	545593	2.28	40773	9-Octadecenoic acid, 1,2,3-propanetriyl ester, (E,E,I
17	25.296	4415395	18.47	163878	Dodecanoic acid, 1,2,3-propanetriyl ester
18	26.182	399159	1.67	73130	7-Hexadecenal, (Z)-
19	26.576	227898	0.95	50278	9,12,15-Octadecatrienoic acid, 2-(acetyloxy)-1-[(ace
20	27.339	2098559	8.78	191933	Ethyl iso-allocholate
21	28.565	154864	0.65	32961	E,E,Z-1,3,12-Nonadecatriene-5,14-diol
22	29.611	120222	0.50	16379	9-Octadecenoic acid, 1,2,3-propanetriyl ester, (E,E,I
		23900117	100.00	6056303	

GC-MS Analysis Chromatogram for Sweet Sorghum

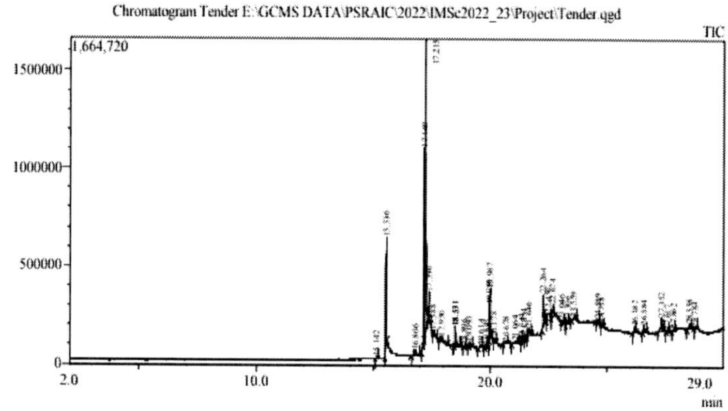

Peak#	R.Time	Area	Area%	Height	Name
1	15.142	24201	0.19	12282	Hexadecanoic acid, methyl ester
2	15.536	1257506	9.88	574389	n-Hexadecanoic acid
3	16.806	123502	0.97	28466	7-Octadecenoic acid, methyl ester
4	17.169	2803897	22.03	994179	9,12-Octadecadienoic acid (Z,Z)-
5	17.215	4320219	33.94	1515716	9-Octadecenoic acid, (E)-
6	17.390	273590	2.15	149609	Oleic Acid
7	17.588	85192	0.67	19569	9,12-Octadecadienoic acid (Z,Z)-
8	17.930	34529	0.27	25938	9,12-Octadecadienoic acid (Z,Z)-
9	18.531	122393	0.96	92613	Hexadecanoic acid, 1-(hydroxymethyl)-1,2-ethanedi
10	18.531	259174	2.04	101208	Hexadecanoic acid, 1-(hydroxymethyl)-1,2-ethanedi
11	18.896	22165	0.17	9042	9,12-Octadecadienoic acid (Z,Z)-
12	19.093	81012	0.64	21337	9,12-Octadecadienoic acid (Z,Z)-
13	19.614	39338	0.31	9254	9,12-Octadecadienoic acid (Z,Z)-
14	19.845	44780	0.35	14406	Cyclopentadecanone, 4-methyl-
15	19.955	347571	2.73	195249	9,12-Octadecadienoyl chloride, (Z,Z)-
16	19.987	474972	3.73	256967	9-Octadecenoic acid, 1,2,3-propanetriyl ester, (E,E,I
17	20.178	26864	0.21	17501	9-Octadecenoic acid, 1,2,3-propanetriyl ester, (E,E,I
18	20.678	59744	0.47	10556	9,12-Octadecadienoic acid (Z,Z)-
19	21.064	73644	0.58	12087	Z-(13,14-Epoxy)tetradec-11-en-1-ol acetate
20	21.344	29537	0.23	12659	9,12-Octadecadienoic acid (Z,Z)-
21	21.454	85577	0.67	25250	Z-(13,14-Epoxy)tetradec-11-en-1-ol acetate
22	21.646	178780	1.40	40509	9-Octadecenoic acid, 1,2,3-propanetriyl ester, (E,E,I
23	22.264	296770	2.33	155225	Bis(cis-13-docosenamido)methane
24	22.456	284083	2.23	38711	9-Octadecenoic acid, 1,2,3-propanetriyl ester, (E,E,I
25	22.674	275230	2.16	67768	9-Octadecenoic acid (Z)-, 2-hydroxy-3-[(1-oxohexa
26	23.046	33857	0.27	9149	9-Octadecenoic acid, 1,2,3-propanetriyl ester, (E,E,I
27	23.306	44702	0.35	9758	9-Octadecenoic acid, 1,2,3-propanetriyl ester, (E,E,I
28	23.559	47200	0.37	11603	9-Octadecenoic acid, 1,2,3-propanetriyl ester, (E,E,I
Peak#	R.Time	Area	Area%	Height	Name
29	24.589	85075	0.67	18017	9-Octadecenoic acid, 1,2,3-propanetriyl ester, (E,E,I
30	24.758	42902	0.34	13688	E,E,Z-1,3,12-Nonadecatriene-5,14-diol
31	26.182	136583	1.07	33416	Ethyl iso-allocholate
32	26.584	137138	1.08	35644	Ethyl iso-allocholate
33	27.352	348715	2.74	64477	Ethyl iso-allocholate
34	27.617	30204	0.24	8700	7-Hexadecenal, (Z)-
35	27.862	37183	0.29	9859	Z-(13,14-Epoxy)tetradec-11-en-1-ol acetate
36	28.558	128509	1.01	25304	Z-(13,14-Epoxy)tetradec-11-en-1-ol acetate
37	28.784	30944	0.24	7528	9-Octadecenoic acid, 1,2,3-propanetriyl ester, (E,E,I
		12727282	100.00	4647633	

RESULTS

By GCMS profiling, it was possible to identify 9-Octadecenoic acid, Hexadecenoic acid, 4-Methyl Cyclopentadecanone, 9, 12-Octadecadienoic acid (Z, Z), and Z-(13, 14 Epoxy) tetradec-11-en-1-ol acetate. The tender jowar contains traces of 4- Methylcyclopentadecanone, which gives it a faint fragrance. Hexadecenoic acid, octadecenoic acid, and Z-(13, 14 Epoxy) tetradec-11-en-1-ol acetate are renowned for their antibacterial, antioxidative, and anti-inflammatory effects. Additionally, octadecanoic acid decreases LDL cholesterol, a well-known risk factor for coronary heart disease. Qualitative tests indicate age-dependent variability though further confirmatory tests are required.

DISCUSSION

According to the GC-MS study, 9, 12-Octadecadienoic acid predominates as a fatty acid which indicates the presence of glycoside in sorghum and sweet sorghum. Sweet sorghum has a higher concentration of terpenoids than sorghum, which may be one of the reasons it is only available during the winter. Qualitative tests revealed the presence of glycosides, saponins and terpenoids which was further confirmed by GC-MS. There is a need for quantification and further analysis of the extracts for determining the variation in concentration.

REFERENCES

[1] Chamodini, R., Thilakarathna, N., Devage, G., Prabhath, M., Senevirathne, M. and Navaratne, B. (2021). Potential food applications of sorghum (*Sorghum bicolor*) and rapid screening methods of nutritional traits by spectroscopic platforms. https://ift.onlinelibrary.wiley.com

[2] Jairam, K.P. Vanamala; Aaron, R. Massey; Srinivas Rao, Pinnamaneni; Lavanya, Reddivari and Kenneth, F. Reardon (2018). Grain and sweet sorghum (*Sorghum bicolor* L. Moench) serves as a novel source of bioactive compounds for human health, Critical Reviews in Food Science and Nutrition, 58:17, 2867–2881, DOI: 10.1080/10408398.2017.1344186

[3] Mathur, S.; Umakanth, A.; Tonapi, V.; Sharma, R. and Sharma, M. (2017). Sweet sorghum as biofuel feedstock: Recent advances and available resources. https://biotechnologyforbiofuels.biomedcentral.com/articles/10.1186/s13068-017-0834-9.

[4] Samarth, A.G.; More, D.R. and Imran, Hashmi (2018). Studies on physico-chemical properties and nutritional profile of sweet sorghum *International Journal of Chemical Sciences;* 6(2): 2826–2828.

Effect of Combination of Aries Soil Application and Foliar Application Nutrients on Quality and Productivity of Rainfed Finger Millet

Mohanasundaran

Aries Agro Ltd.

ABSTRACT: A critical stage of the plant's life cycle is germination and insufficient seedling emergence. Poor soils, inadequate moisture, and poor management practices contribute to the lower productivity of finger millet. Normally soil application and foliar application improves seedling emergence, reduces stand establishment time, improves seedling germination and increasing quality and productivity of crop. A nutrient based management strategy is being considered for improving the physiological efficiency of finger millet to produce higher yields during drought condition. Foliar application minimizes fertilizer loss due to absorption, leaching or other processes that are involved with` soil application under optimum moisture availability or soil condition. A field experiment was conducted at farmer field, Salem during Kharif season to evaluate the response of effective farming practice for micronutrient application for finger millet on crop growth and productivity. Six treatment combinations involving Control (No manure) (R_1), RDF (40:20:20 N, P and K kg ha^{-1}) (R_2), RDF (40:20:20 N, P and K kg ha^{-1}) + zinc sulphate (25 kg ha^{-1}) (R_3), RDF (40:20:20 N, P and K kg ha^{-1}) + Agromin soil application (25 kg ha^{-1}) (R_4), RDF (40:20:20 N, P and K kg ha^{-1}) + Aries chelamin foliar spray (250 gm ha^{-1}) @ vegetative stage (R_5) and RDF (40:20:20 N, P and K kg ha^{-1}) + Agromin soil application (25 kg ha^{-1}) + Aries chelamin foliar spray (250 gm ha^{-1}) (R_6) treatments were replicated thrice in Randomized Block design (RBD). The results of the study revealed that, among the six treatments of finger millet RDF (40:20:20 N, P and K kg ha^{-1}) + Agromin soil application (25 kg ha^{-1}) + Aries chelamin foliar spray (250 gm ha^{-1}) @ vegetative stage (R_6) recorded significantly higher growth and yield attributing characters compared to the rest of the treatments, yet, it was found on par with (R_4).

Keywords: *Finger Millet, Soil Application, Chelated Zinc, Micronutrient and Sustainability.*

INTRODUCTION

Millets are primarily grown as a rainfed crop, and they thrive in difficult ecosystems where other crops fail to produce a satisfactory harvest. Millets are one of the essential staple food crops for millions of people in the world's arid and semi-arid regions because of the higher pest and disease tolerance and more adaptability to a broad range of environments. Their higher yielding capacity, ability to withstand high levels of salinity, short growing season, resistance to water logging, drought tolerance.

To encourage the production of nutrient-rich millets, the government has stated celebrating 2018 as the "National Year of Millets". The United Nations General Assembly declare 2023 as the "International Year of Millets".

Finger millet (*Eleusine coracana* L.) is one of the most important millet crops in India, along with sorghum and pearl millet. It belongs to the poaceae family and is extensively cultivated (Dass *et al.*, 2013). India is the world's major producer of finger millet, accounting for roughly 60 per cent of global production. Finger millet includes 9.2 per cent protein, 76.3 per cent carbohydrate, 2.2 per cent minerals, 1.3 per cent fat and 3.9 per cent ash as well as vitamins like A and B. The grains are high in phosphorus, potassium and amino acids and they are the highest source of energy calcium (410 mg/100g grain), which is especially essential for developing elderly people (Tomar *et al.*, 2011).

Finger millet contain higher protein level than rice and a well-balanced amino acid composition. It is high in essential important vitamins including thiamine, riboflavin, folic acid and niacin as well as methionine and lysine. Because of its therapeutic properties, it is called poor man's food and is also suggested as a food for diabetics.

Finger millet cultivation area in India is 1.02 million ha, with a total production of 1.38 million tonnes and an average productivity of 1.4 t/ha. In Tamil Nadu, finger millet is the most significant indigenous millet crop with 0.11 million tonnes cultivated and a yield of 1.9 t/ha providing food and nutritional security to marginal and small farmers (INDIASTAT, 2017).

Micronutrients are essentially as important as macronutrients to have better growth, yield and quality in plants. There requirement by plants is in trace amounts. Boron, iron, copper, zinc, manganese, magnesium and molybdenum constitute main micronutrients required by different crops in variable quantities. Others such as selenium (Se) and Co, which are needed in specific cases, are commonly referred to as beneficial elements. The requirement of micronutrients is partly met from the soil or through chemical fertilizer or through other sources. Various physical and metabolic functions are governed by these mineral nutrients.

Micronutrients are essential elements that are used by plants in small quantities. Yield and quality of agricultural products increased with micronutrients application, therefore human and animal health is protected with feed of enrichment plant materials.

Foliar application is associated with the advantages of rapid and effective nutrient use, reduction of losses due to leaching and fixation and aids in controlling nutrient uptake by plants (Manonmani and Srimathi, 2009). Micronutrient deficiencies in humans and crop plants are hard to detect, therefore the problem is referred to as "hidden hunger" (Stein *et al.*, 2008). While zinc (Zn) deficiency being the most prevalent nutritional deficiency next after iron and iodine. According to the world health organization, Zn deficiency is the fifth major cause of illness among juveniles and old age peoples in developing countries. Majority of the Indian soils are reported to be Zn deficient, therefore food crops cultivated in those soils contain minimum level of Zn nutrient. Crop species have shown significant genetic variability in sustaining growth and yield in Zn deficient conditions.

MATERIALS AND METHOD

A field research experiment was conducted in the farmers field, Salem during Kharif season to study the response of finger millet variety CO 15 to varied levels of RDF (Recommended dose of fertilizer) with Aries agromin soil application, chelated foliar application and zinc sulphate. The study was conducted during 2022. The experiment was laid out in Randomized Block design (RBD. All the treatments and replications were randomized to reduce the experimental error. Six treatment combinations involving Control (No manure) (R_1), Recommended dose of fertilizer (40:20:20 N, P and K kg ha^{-1}) (R_2), RDF (40:20:20 N, P and K kg ha^{-1}) + zinc sulphate (25 kg ha^{-1}) (R_3), RDF (40:20:20 N, P and K kg ha^{-1}) + Agromin soil application (25 kg ha^{-1}) (R_4), RDF (40:20:20 N, P and K kg ha^{-1}) + Aries chelamin foliar spray (250 gm ha^{-1}) @ vegetative stage (R_5) and RDF (40:20:20 N, P and K kg ha^{-1}) + Agromin soil application (25 kg ha^{-1}) + Aries chelamin foliar spray (250 gm ha^{-1}) (R_6). The fertilizers were applied to the experimental field as per the recommended manurial schedule of 40:20:20 kg N, P_2O_5 and K_2O ha^{-1}. Urea (N), Single super phosphate (P_2O_5) and Muriate of potash (K_2O) fertilizers were used to supply N, P and K nutrients, respectively. In the final puddling at the base in accordance with the treatment schedule or two days prior to seed planting depending on the treatments, zinc sulphate at 25 kg ha^{-1} and

Aries agromin soil application at 25 kg ha^{-1} were applied. Biometric observations on plant height at harvest, LAI at harvest, number of tiller m^{-2} at harvest and number of productive tillers m^{-2} at harvest and yield components *viz.,* grain yield and straw yield at harvest. The data was subjected to statistical scrutiny to arrive at meaningful for the effect of treatments on finger millet crop.

RESULTS AND DISCUSSION

Growth Attributes and Yield of Finger Millet

The investigations with treatments revealed a substantial difference in the treated as well as control plot of the observations shown in Tables 1 and 2. The efficiency of finger millet with treatment (R_6) was higher to that of the control plot (R_1) of finger millet in all growth components, including plant height at harvest, LAI at harvest, the number of tillers m^{-2} at harvest and the number of productive tillers m^{-2} at harvest.

Among the treatments ($R_{6)}$-RDF (40:20:20 N, P and K kg ha^{-1}) + Agromin soil application (25 kg ha^{-1}) + Aries chelamin foliar spray (250 gm ha^{-1}) recorded highest mean of Plant height at harvest (150.8), LAI at harvest (5.9), number of tiller m^{-2} at harvest (232.2) and number of productive tillers m^{-2} at harvest (225.6), grain yield (2880.2 kg ha^{-1}) and straw yield (6918.2 kg ha^{-1}) followed by (R_4) treatment (144.6, 5.1, 209.9, 202.5, 2777.9 and 6540.9) at plant height at harvest, LAI at harvest, number of tiller m^{-2} at harvest and number of productive tillers m^{-2} at harvest, grain yield (kg ha^{-1}) and straw yield (kg ha^{-1}) respectively.

According to Imtiaz *et al.* (2010), the appropriate amounts of micronutrients are necessary for increased agricultural product quality, productivity, growth, flowering, fruit set, and post-harvest survival, according to Imtiaz *et al.* (2010). Micronutrient deficiencies are a significant problem in soil and plants everywhere Isreal Zewide *et al.* (2021).

In plants, zinc functions as an enzyme activator and acts directly in the production of auxin, which results in the formation of more cells and dry matter that may be stored in seeds. The development of total cells and dry matter increased significantly in experiments on rice, bread wheat, and rabi sorghum. On the other hand, increasing concentrations had a detrimental effect on the development of finger millet.

Zinc has been found to improve seed germination, seedling vigour, early flowering and leaf chlorophyll content. In low concentrations, they also observed that chelated nutrients improved peanut plant growth, development and yield. But, at higher concentrations, chelated zinc had the opposite impact of the bulk nutrients (Prasad *et al.,* 2012).

CONCLUSION

The study summarized that varied levels of RDF (Recommended dose of fertilizer) with Aries agromin soil application, chelated foliar application and zinc sulphate recorded greater growth parameters, resulting in an increase in finger millet crop yield. Therefore, using a small quantity of micronutrient fertilizers may reduce the amount of fertilizer, fertilizers waste, environmental risks and improve the efficiency of nutrient utilization. Effective crop nutrition might be one of the management techniques used to increase millet yield. The utilization of organic nutrient sources along with the optimum NPK and micronutrient fertilizer as part of balanced nutrition, as compared to depending only fertilizer, boosts finger millet production.

REFERENCES

[1] Anand, R.K.R.; Kamatar, M.Y.; Mummigatti, U.V. and Basavaraj, B. (2008). "Evaluation of rabi sorghum genotypes for seed zinc content and yield in high regimes." *Karnataka Journal of Agricultural Sciences,* 21 (04):568–569.

[2] Dass, A.; Sudhishri, S. and Lenka, N. (2013). "Integrated nutrient management to improve finger millet productivity and soil conditions in hilly region of Eastern India." *Journal of crop improvement,* 27(5):528–546.

[3] Genc, Y.; McDonald, G.K. and Graham, R.D. (2006). "Contribution of different mechanisms to zinc efficiency in bread wheat during early vegetative stage." *Plant and Soil,* 281(1): 353–367.

[4] Imtiaz, M., Rashid, A., Khan, P., Memon, M.Y. and Aslam, M. (2010). The Role of Micronutrients in Crop Production and Human Health. *Pak. J. Bot.*, 42(4), 2565–2578.

[5] Isreal, Zewide and Abde, Sherefu (2021). Review Paper on Effect of Micronutrients for Crop Production. *J. Nutrition and Food Processing,* 4(7).

[6] Manonmani, V. and Srimathi, P. (2009). "Influence of mother crop nutrition on seed yield and quality of blackgram." *Madras Agricultural Journal,* 96 (1/6):125–128.

[7] Ministry of Agriculture and Farmers Welfare Govt. of India (2017). Selected State-wise area, production and productivity of finger millet, Govt. of India. In: https://www.indiastat.com.

[8] Ozkutlu, F., Torun, B. and Cakmak, I. (2006). "Effect of zinc humate on growth of soybean and wheat in zinc-deficient calcareous soil." *Communications in Soil Science and Plant Analysis,* 37 (15–20):2769–2778.

[9] Prasad, T., Sudhakar, P., Sreenivasulu, Y., Latha, P., Munaswamy, V., Reddy, K.R., Sreeprasad, T., Sajanlal, P. and Pradeep, T. (2012). "Effect of nanoscale zinc oxide particles on the germination, growth and yield of peanut." *Journal of Plant Nutrition,* 35(6):905–927.

[10] Rehman, A., Yasin, M., Akram, M. and Awan, Z. (2002). "Response of Zn applied and N sources in calcareous soils." *Science,* 8:100–104.

[11] Slaton, N.A.; Wilson, C.E., Ntamatungiro, S., Norman, R.J. and Boothe, D.L (2001). "Evaluation of zinc seed treatments for rice." *Agronomy Journal,* 93(1):152–157.

[12] Stein, A.J., Meenakshi, J., Qaim, M., Nestel, P., Sachdev, H. and Bhutta, Z.A. (2008). "Potential impacts of iron biofortification in India." *Social Science & Medicine,* 66(8):1797–1808.

[13] Tomar, G., Taunk, S. and Choudhary, J. (2011). Science of Crop Production: Part I. Kharif Crops: Kushal Publications and distributors.

Gas Chromatography-Mass Spectrometry (GC-MS) Analysis and Phytochemical Screening of Finger Millet and Foxtail Millet in Ultrasonic Bath Sonicator (UBS) Extract

Parth B. Trivedi, Nitin Gupta, Pravith P. Warrier and Madhavi R. Badole

Department of Chemistry, Ramnarain Ruia Autonomous College, Matunga, Mumbai

ABSTRACT: *Eleusine coracana* (FGM) and *Setaria italica* (FXM), more generally referred to as finger millet and foxtail millet, are regarded as nutrient-dense grains. These grains are a valuable source of vitamins, minerals, vital amino acids, protein, and dietary fibre. Also, studies have demonstrated that they may have health advantages, including the management of weight and the protection of cardiovascular health and diabetes prevention. The use of ultrasonic bath sonication (UBS) as a method for extracting phytochemicals from these grains is an innovative approach. Furthermore, the extraction of FGM and FXM was done successively. The gas chromatograph mass spectrometer (GC-MS) analysis of the extracts revealed the presence of 9,12-Octadecadienoic acid (Z, Z)- and Pentadecanoic acid, two important phyto-components that are known to have health benefits. The phytochemical screening data also showed the presence of flavonoids, carbohydrates, and phenols in the extracts, which are known to have antioxidant and anti-inflammatory properties. In conclusion, the use of UBS to extract phytochemicals from *Eleusine coracana* (FGM) and *Setaria italica* (FXM) is a promising approach for further study. The identification of phyto-components and the presence of other beneficial phytochemicals in the extracts highlights the potential health benefits of these grains. Further research is needed to fully establish the value of these grains in promoting health and preventing disease.

Keywords: *Finger Millet, Foxtail Millet, Ultrasonic Bath Sonicator, GC-MS.*

INTRODUCTION

Eleusine coracana (FGM) and *Setaria italica* (FXM), both Poaceae family millets, are monocotyledons. They are primarily grown in India's

Sahelian region. FGM, also known as ragi and mandua in India, is a minor grain native to Ethiopia, but widely grown in India and Africa. India is the leading producer of millets in the world. (Chandra *et al.,* 2016; Pandey *et al.,* 2013)

Millets are nutrient-dense super grains that were the first type of grain available to humans. People have long ignored these crops because they are so focused on eating rice and wheat. Many people still rely on millets because they grow in hot, dry, hilly areas with little rainfall. (Huchchannanavar *et al.*, 2019) Millets are nutritionally superior to wheat and rice because they contain more calcium, dietary fibre, and protein. Millets have a dietary fibre content ranging from 15% to 20%, a protein content ranging from 5% to 10%, and a carbohydrate content ranging from 65% to 75%. Millets are known to be superior in terms of nutrient content and antioxidant activity to other cereals such as rice and wheat. (Kharat *et al.,* 2019) Because of their physicochemical and nourishing properties, major millets can be processed into a wide range of value-added food products, such as baby foods, dietetic foods, and ready-to-eat snacks, making them suitable for widespread use and feasting around the world. It's especially interesting to see how they're used in flour and grain products (Subramanian and Viswanathan, 2003). Millets are superior to other important muesli crops such as rice and wheat because they contain higher concentrations of calcium, magnesium, fat, vitamins, dietary fibre, iron, protein, and other minerals. (Krishna *et al.,* 2018; Vetriventhan, 2016)

Karnataka is India's largest producer of FGM, accounting for 58% of the country's annual revenue, despite the fact that few Indians are aware of its nutritional value and health benefits. After wheat, rice, maize, sorghum, and bajra, FGM is India's sixth most important crop. (Huchchannanavar *et al.* 2019) Carbohydrates in FGM are composed of starch (75–80%), free sugars (1–2%), and non-starchy polysaccharides such as cellulose and hemicellulose. It contains a lot of polyphenols, minerals, and dietary fibre. One reason FGM has such a long shelf life could be due to its low-fat content. (Mathanghi and Sudha, 2012) FGM is a very important crop for agriculture in addition to being very nutrient-dense when compared to other cereal crops like rice, sorghum, and FXM. In addition to essential and sulfur-containing amino acids, it is a good source of calcium, iron, dietary fibre, polyphenols, and calcium. (Avashthi, 2020; Subramanian and Viswanathan 2003) They contain biologically active substances such as saponins, terpenoids, alkaloids, flavonoids, phenols, tannins, and phenolics, which are linked to antioxidant activity and are a good source of polyphenols, minerals, and dietary fibre. (Mounika *et al.*, 2022) According to the research report, tannins and terpenoids were present in every solvent extract of whole grains and bran-enriched fractions, whereas alkaloids, reducing sugars, and flavonoids were only found in methanol and aqueous extracts. (Jelku *et al.,* 2019)

The study claims that while both FXM and FGM have higher calcium contents than rice, FXM has nearly twice as much protein (11.2%) and fat (4%) as rice. (Saleh *et al.,* 2013) Plant development for millets, particularly FXM, is slow at first. (Reddy *et al.,* 2014) FXM contains a lot of phytochemicals, micronutrients, and antioxidants like glycated flavonoids and phenolic acids. Because of their high levels of dietary fibre, low glycaemic index, and water-soluble gum content viz. β-glucan, FXM have been shown to improve glucose metabolism. Millet grains have antioxidative properties due to higher levels of phytochemicals such as phenolics, tannins, phytates, and microminerals, among others. According to the study, FXM is a grain that is used as a nutraceutical due to its high nutritional content and useful compounds. (Huchchannanavar *et al.*, 2019)

Millets are employed as nutraceuticals due to their abundance in antioxidants and are believed to be useful in the treatment of heart attacks, migraines, blood pressure, diabetes, asthma, and atherosclerosis. (Mounika *et al.*, 2022)

The present study aims to extract phytoconstituents from FGM and FXM using methanol and n-hexane solvents, and then identify these compounds using preliminary and instrumental analyses (GC-MS). This could potentially provide valuable information about the chemical composition and nutritional properties of FGM and FXM, which could have important implications for their use as food sources or nutraceuticals. It will be interesting to see the results of this study and how they contribute to the existing body of knowledge on millets and their potential health benefits.

EXPERIMENTAL

Procurement of Raw Material

FGM (*Eleusine coracana*) and FXM (*Setaria italica*) were purchased from the local shop located in Dadar, Mumbai-400 019 (India). All the solvents and reagents were of analytical grade and GC-MS grade. Solvents and reagents were used by SRL company, having a purity of 99.0% and 99.5%, including reagent Dragendroff's reagent and Benedict's reagent.

Preparation of Extract

The FGM (*Eleusine coracana*) and FXM (*Setaria italica*) were cleaned with clean cloth and ground separately using mortar and pestle. 5 g of FGM (*Eleusine coracana*) and FXM (*Setaria italica*) each were weighed separately and 50 ml of methanol was added to the flask. The mouth of the flask was covered and the flask was wobbled for a few minutes, using an Ultrasonic bath sonicator (UBS) to extract the powder FGM and FXM in methanol. It was sonicated for the first 5 minutes, held for 1-minute and then sonicated again for 5 minutes. This was repeated two more times and then the extract was filtered with Whatman filter paper 41, the liquid extract was stored in the closed-packed cleaned bottles for further tests. The remaining residue was scraped with a spatula and transferred to another conical flask for both millets separately. 50 ml of n-hexane was added to FGM and FXM residue obtained after extraction with methanol and the same procedure was followed using Ultrasonic bath sonication (UBS) by sequential extraction technique. (Laware, 2015)

Gas Chromatography-Mass Spectrometry

The GC-MS analysis of the sample was performed on a Shimadzu GCMS-QP2020 NX gas chromatograph-mass spectrometer, fitted with an Rtx-5 fused silica capillary column (30 × 0.25 mm, with 1 μm film thickness). The oven temperature was set to hold at 60°C for 5 minutes, 140°C for 3 minutes, 210°C for 2 minutes, and 300°C for 2 minutes, respectively. Helium was used as a carrier gas at a flow rate of 1 mL/min. The injection volume was 2 mL, and the injection temperature was 250°C. Ion source temperature was 200°C, interface temperature was 300°C, split ratio was 1/80, and pressure was 57.4 kPa. The complete analysis time was 44 min for each sample, at a scan speed of 1666 amu/s, and a scan range of 35 to 500 m/z.

Phytochemical Screening Tests

Qualitative phytochemical tests were conducted on the methanol and n-hexane extracts of FGM and FXM to identify the presence of different classes of phytochemicals. (Laware 2015; Raaman 2006; De Silva *et al.,* 2017; Singh and Kumar 2017)

RESULT AND DISCUSSION

By using the UBS method for phytochemical screening, it was revealed that flavonoids, saponins, carbohydrates, phenols, tannins, and proteins were present in FGM in the methanol extract. However, the presence of flavonoids, carbohydrates, phenols, and proteins was entirely positive. Alkaloids, flavonoids, saponins, carbohydrates, and proteins are all present in the FXM methanol extract, with the presence of carbohydrates and proteins being the most prevalent. The results of FGM and FXM in methanol extract are shown in Table 1. In the extract of FGM in n-hexane, flavonoids, carbohydrates, and phenol are present, whereas in the extract of FXM, just flavonoids and carbohydrates are present. The results of FGM and FXM in n-hexane extract are shown in Table 2.

Table 1: Methanol Extract of FGM and FXM by UBS

Sr. No.	*Test*	*FGM (Eleusine coracana)*	*FXM (Setaria italica)*	*Procedure*	*Reference*
1.	Alkaloids	–	+	Add 1 mL of filtrate along with few drops of Dragendroff's reagent	(De Silva *et al.*, 2017)
2.	Flavonoids	+	+	Add 1 mL of extract along with few drops of 10% lead acetate solution	(De Silva *et al.*, 2017)
3.	Saponins	+	+	Few mLof extract and add 20 mL of distilled water shake for 15 minutes	(Venu *et al.*, 2022)
4.	Carbohydrate	+	+	Add 1 mL of filtrate with few drops of alcoholic α-naphthol in test tube and add 1 mL of concentrated sulphuric acid	(Raaman 2006)
5.	Reducing Sugars	–	–	Add 1 mL of filtrate along with few drops of Beneddict's reagent and boil for few minutes	(Raaman 2006)
6.	Phenol	+	–	Add 1 mL of extract along with few drops of 5% ferric chloride solution	(Singh and Kumar, 2017)
7.	Tannins	+	–	Add 1 mL of extract along with few drops of lead acetate solution	(Singh and Kumar, 2017)
8.	Proteins	+	+	Add 2 mL of filtrate and then 10% of sodium hydroxide solution in test tube with 0.1% of copper sulphate solution	(De Silva *et al.*, 2017)

+ = Present; – = Absent.

Table 2: n-Hexane Extract of FGM and FXM by UBS

Sr. No.	*Test*	*FGM (Eleusine coracana)*	*FXM (Setaria italica)*	*Procedure*	*Reference*
1.	Alkaloids	–	–	Add 1 mL of filtrate along with few drops of Dragendroff's reagent	(De Silva *et al.*, 2017)
2.	Flavonoids	+	+	Add 1 mL of extract along with few drops of 10% lead acetate solution	(De Silva *et al.*, 2017)
3.	Saponins	–	–	Few mLof extract and add 20 mL of distilled water shake for 15 minutes	(Venu *et al.*, 2022)
4.	Carbohydrate	+	+	Add 1 mL of filtrate with few drops of alcoholic α-naphthol in test tube and add 1 mL of concentrated sulphuric acid	(Raaman 2006)
5.	Reducing Sugar	–	–	Add 1 mL of filtrate along with few drops of Beneddict's reagent and boil for few minutes	(Raaman 2006)
6.	Phenol	+	–	Add 1 mL of extract along with few drops of 5% ferric chloride solution	(Singh and Kumar, 2017)
7.	Tannins	–	–	Add 1 mL of extract along with few drops of lead acetate solution	(Singh and Kumar, 2017)
8.	Proteins	–	–	Add 2 mL of filtrate and then 10% of sodium hydroxide solution in test tube with 0.1% of copper sulphate solution	(De Silva *et al.*, 2017)

+ = Present; – = Absent.

The presence of bioactive chemical compounds in the solution was determined by further analysis of all four extracts using a GC-MS equipment. Following the mass spectral analysis, the chromatogram showed the existence of peaks corresponding to chemicals found in the extracts. GC-MS of Chromatogram of the methanolic extract of FGM showed 6 peaks (Figure 1) and have been identified after comparing to the mass spectra with the NIST library, indicating the presence of 5 phytocomponents. According to the results, the main constituents of the extract were 1,2-Cyclopentanedione, Azulene, Pentadecanoic acid, 9,12-Octadecadienoic acid (Z,Z)- (also known as linoleic acid), and 9-Octadenoic acid. Table 3 lists the phytochemicals that contribute to the plant's therapeutic properties. 1,2-Cyclopentanedione is stated to be an antioxidant and antimicrobial activity, (Alghamdi *et al.,* 2018; Kavitha 2021) Azulene having antiviral, antimicrobial, anti-inflammatory, antiulcer and antidiabetic activity, (Bakun *et al.,* 2021) Pentadecanoic acid having antibacterial, antifungal and anticancer activity, (Chowdhury *et al.,* 2016) Linoleic acid having anti-adipogenic and antimicrobial activity, (Krishnaveni, Dhanalakshmi, and Nandhini 2014; Martins *et al.,* 2015) 9-Octadecenoic acid having antimicrobial and antibacterial activity. (Ghavam *et al.*, The FXM methanolic extract displayed 4 peaks (Figure 2), and subsequent phyto-component identification revealed 3 phytocomponents. According to the results, the main constituents of the extract were Pentadecanoic acid, 9,12-Octadecadienoic acid (Z,Z)-methyl ester and 9,12-Octadecadienoic acid (Z,Z)- (also known as linoleic acid). Table 4 list the phytochemical that contribute to the plant's therapeutic properties. Pentadecanoic acid having antibacterial, antifungal and anticancer activity, (Chowdhury *et al.,* 2016) 9,12-Octadecadienoic acid (Z,Z)-methyl ester having uricosuric activity, anti-inflammatory and antioxidant activity, (Hadi *et al.*, 2016) Linoleic acid having anti-adipogenic and antimicrobial activity. FGM in n-hexane extract displays 4 peaks (Figure 3), which represents 1 peak compound with a phytocomponent is 9,12-Octadecadienoic acid (Z,Z)- (also known as linoleic acid) having anti-adipogenic and antimicrobial activity (Krishnaveni *et al.*, 2014; Martins *et al.,* 2015) shown in Table 5. FXM in n-hexane extract displays 5 peaks out of which 1 peak (Figure 4) has phytocomponent with therapeutic activity is anti-adipogenic and antimicrobial activity (Krishnaveni *et al.*, 2014; Martins *et al.,* 2015) shown in Table 6.

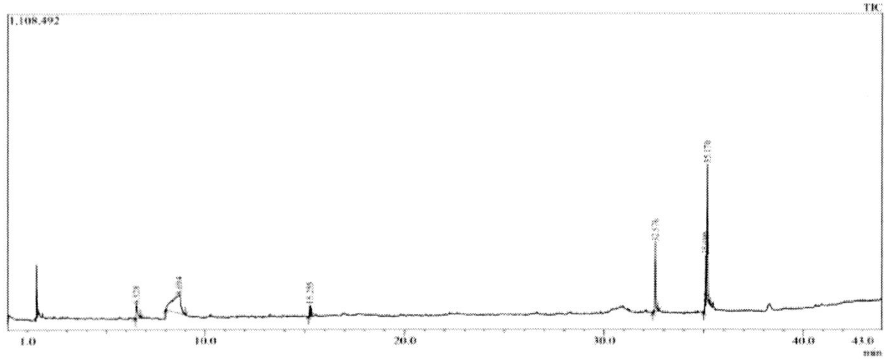

Figure 1: GC-MS Chromatogram of FGM in Methanol Extract

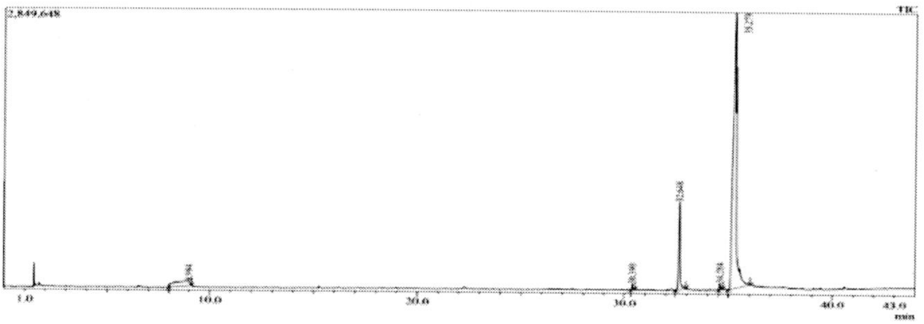

Figure 2: GC-MS Chromatogram of FXM in Methanol Extract

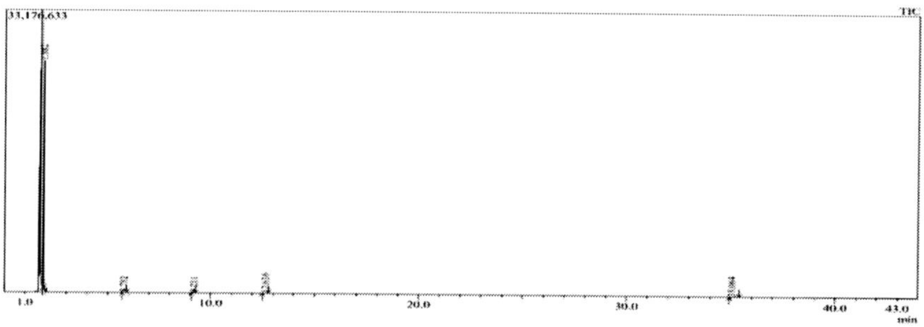

Figure 3: GC-MS Chromatogram of FGM in n-Hexane Extract

Table 3: FGM in Methanol Extract Phytocomponent Therapeutic Activity

Sr No.	*Compound Name/Synonyms*	*MW*	*Area %*	*Activity*	*Reference*
1.	1,2-Cyclopentanedione	98	5.68	antioxidant and antimicrobial activity.	(Alghamdi *et al.*, 2018; Kavitha 2021)
2.	Azulene	128	2.24	antiviral, antimicrobial, anti-inflammatory, antiulcer and antidiabetic activity.	(Bakun *et al.*, 2021)
3.	Pentadecanoic acid	242	15.04	antibacterial, antifungal and anticancer activity.	(Chowdhury *et al.*, 2016)
4.	9,12-Octadecadienoic acid (Z,Z)-/Linoleic acid	280	6.95	anti-adipogenic and antimicrobial activity.	(Krishnaveni *et al.*, 2014; Martins *et al.*, 2015)
5.	9-Octadenoic acid	282	26.89	antimicrobial and antibacterial activity	(Ghavam *et al.*, 2021)

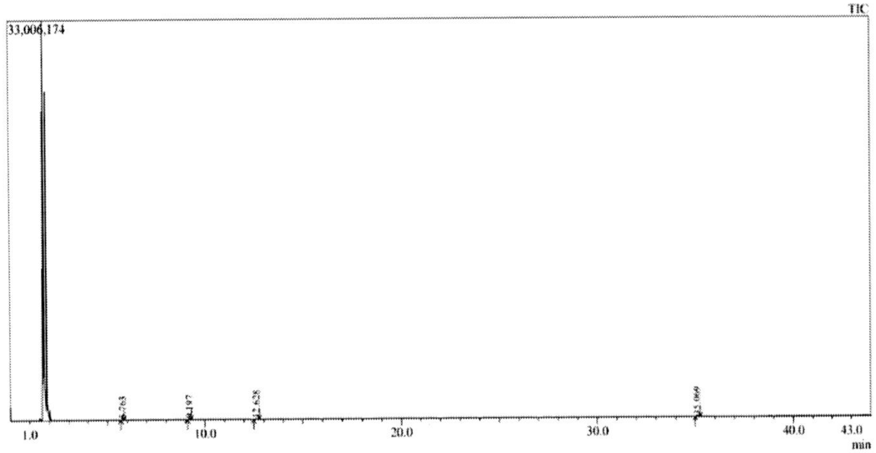

Figure 4: GC-MS Chromatogram of FXM in n-Hexane Extract

Table 4: FXM in Methanol Extract Phytocomponent Therapeutic Activity

Sr No.	*Compound name/Synonyms*	*MW*	*Area %*	*Activity*	*Reference*
1.	Pentadecanoic acid	242	9.99	antibacterial, antifungal and anticancer activity.	(Chowdhury *et al.*, 2016)
2.	9,12-Octadecadienoic acid (Z,Z)-methyl ester	294	0.48	uricosuric activity, anti-inflammatory and antioxidant activity.	(Hadi *et al.*, 2016)
3.	9,12-Octadecadienoic acid (Z,Z)-/Linoleic acid	280	82.86	anti-adipogenic and antimicrobial activity.	(Krishnaveni *et al.*, 2014; Martins *et al.*, 2015)

Table 5: FGM in n-Hexane Extract Phytocomponent Therapeutic Activity

Sr No.	*Compound name/Synonyms*	*MW*	*Area %*	*Activity*	*Reference*
1.	9,12-Octadecadienoic acid (Z,Z)-/Linoleic acid	280	48.89	anti-adipogenic and antimicrobial activity.	(Krishnaveni *et al.*, 2014; Martins *et al.*, 2015)

Table 6: FXM in n-Hexane Extract Phytocomponent Therapeutic Activity

Sr No.	*Compound name/Synonyms*	*MW*	*Area %*	*Activity*	*Reference*
1.	9,12-Octadecadienoic acid (Z,Z)-/Linoleic acid	280	0.52	anti-adipogenic and antimicrobial activity.	(Krishnaveni *et al.*, 2014; Martins *et al.*, 2015)

It is interesting to note that both millets contain a common component 9,12-Octadecadienoic acid (Z,Z)- also known as linoleic acid. The methanolic extract contained pentadecanoic acid, which was present in both millets.

CONCLUSION

The current study work elaborates on the prospective bioactivity of 9,12-Octadecadienoic acid (Z,Z)- and Pentadecanoic acid in two varieties of millets, including its potential to be anti-adipogenic, antimicrobial, antibacterial, antifungal, and anticancer activity. While the majority of peaks in the polar extract of FGM and FXM were detected by GC-MS. The bioactive molecule found in 9,12-Octadecadienoic acid (Z,Z)- and Pentadecanoic acid emphasises the use of these compounds in medicine and nutraceuticals. This crop may be utilised as a key component of the daily nutraceutical diet or even as medicine based on the phytocomponents identified and their claimed biological activities. However, detailed research is required to further assess the bioactivity and toxicity profile of these millets.

ABBREVIATION

FGM – Finger millet
FXM – Foxtail millet
GC-MS – Gas chromatography-mass spectrometry
UBS – Ultrasonic Bath Sonicator

REFERENCES

[1] Alghamdi, Salem S. (2018). "Comparative Phytochemical Profiling of Different Soybean (*Glycine Max* (L.) Merr) Genotypes Using GC–MS." *Saudi Journal of Biological Sciences,* 25(1): 15–21.

[2] Avashthi, Himanshu (2020). "Comparative Analysis of ROS-Scavenging Gene Families in Finger Millet, Rice, Sorghum, and Foxtail Millet Revealed Potential Targets for Antioxidant Activity and Drought Tolerance Improvement." *Network Modeling Analysis in Health Informatics and Bioinformatics* 9: 1–23.

[3] Bakun, Paweł; Beata, Czarczynska-Goslinska; Tomasz, Goslinski and Sebastian, Lijewski (2021). "In Vitro and in Vivo Biological Activities of Azulene Derivatives with Potential Applications in Medicine." *Medicinal Chemistry Research,* 30: 834–46.

[4] Chandra, Dinesh, Satish Chandra, Pallavi and Sharma, A.K. (2016). "Review of Finger Millet (Eleusine Coracana (L.) Gaertn): A Power House of Health Benefiting Nutrients." *Food Science and Human Wellness,* 5(3): 149–55.

[5] Chowdhury, Rajiv; Marinka, Steur; Pinal, S. Patel and Oscar, H. Franco (2016). "Individual Fatty Acids in Cardiometabolic Disease." In *Handbook of Lipids in Human Function*, AOCS Press Elsevier Inc., 207–318.

[6] Ghavam, Mansureh; Afsaneh Afzali and Maria, Letizia Manca (2021). "Chemotype of Damask Rose with Oleic Acid (9 Octadecenoic Acid) and Its Antimicrobial Effectiveness." *Scientific Reports,* 11(1): 1–7.

[7] Hadi, Mohammed Yahya; Ghaidaa, Jihadi Mohammed and Imad, Hadi Hameed (2016). "Analysis of Bioactive Chemical Compounds of Nigella Sativa Using Gas Chromatography-Mass Spectrometry." *Journal of Pharmacognosy and Phytotherapy,* 8(2): 8–24.

[8] Huchchannanavar, Shilpa; Yogesh, L.N. and Prashant. S.M. (2019). "Nutritional and Physicochemical Characteristics of Foxtail Millet Genotypes." *Int. J. Curr. Microbiol. App. Sci.,* 8(1): 1773–78.

[9] Jelku, Jelang (2019). "Phytochemical Screening for Antioxidant Properties of Germinated Foxtail Millet." *Current Journal of Applied Science and Technology,* 38(6): 1–6.

[10] De, Silva; Gusthinnadura, Oshadie; Achala, Theekshana Abeysundara; Malamige, Minoli Aponso and Weroshana (2017). "Extraction Methods, Qualitative and Quantitative Techniques for Screening of Phytochemicals from Plants." *American Journal of Essential Oils and Natural Products,* 5(2): 29–32.

[11] Kavitha, R. (2021). "Phytochemical Screening and GC-MS Analysis of Bioactive Compounds Present in Ethanolic Extract of Leaf and Fruit of Trichosanthesis Dioica Roxb." *International Journal of Pharmaceutical Sciences and Research,* 12(5): 2755–64.

[12] Kharat, Swapnil (2019). "Extrusion Processing Characteristics of Whole Grain Flours of Select Major Millets (Foxtail, Finger, and Pearl)." *Food and Bioproducts Processing,* 114: 60–71.

[13] Krishna, Ajeesh T.P. (2018). "Microsatellite Markers of Finger Millet (Eleusine Coracana (L.) Gaertn) and Foxtail Millet (Setaria Italica (L.) Beauv) Provide Resources for Cross-Genome Transferability and Genetic Diversity Analyses in Other Millets." *Biocatalysis and Agricultural Biotechnology,* 16: 493–501.

[14] Krishnaveni, Marimuthu; Ravi, Dhanalakshmi and Nagaraj, Nandhini (2014). "GC-MS Analysis of Phytochemicals, Fatty Acid Profile, Antimicrobial Activity of Gossypium Seeds." *International Journal of Pharmaceutical Sciences Review and Research,* 27(1): 273–76.

[15] Laware, Shankar L. (2015). "Sequential Extraction of Plant Metabolites." *International journal of Current Microbiology and Applied Sciences,* 4(2): 33–38.

[16] Martins, Susana V. (2015). "Adipocyte Membrane Glycerol Permeability Is Involved in the Anti-Adipogenic Effect of Conjugated Linoleic Acid." *Biochemical and Biophysical Research Communications* 458(2): 356–61.

[17] Mathanghi, S.K. and Sudha, K. (2012). "Functional and Phytochemical Properties of Finger Millet (Eleusine Coracana L.) For Health." *International journal of Pharmaceutical, Chemical and Biological sciences,* 2(4): 431–38.

[18] Mounika, D., Sangeetha, U. and Sireesha, G. (2022). "Estimation of Phytochemicals in Millets and Selected Millet Products." *Indian Journal of Applied & Pure Biology,* 37(3): 810–20.

[19] Pandey, Garima (2013). "Genome-Wide Development and Use of Microsatellite Markers for Large-Scale Genotyping Applications in Foxtail Millet [Setaria Italica (L.)]." *DNA Research,* 20(2): 197–207.

[20] Raaman, N. (2006). *Phytochemical Techniques.* 1st ed. New India Publishing Agency.

[21] Reddy, Seshadri S. (2014). "Tolerance of Foxtail, Proso and Pearl Millets to Saflufenacil." *Crop Protection* 57: 57–62.

[22] Saleh, Ahmed S.M.; Qing, Zhang; Jing, Chen and Qun, Shen (2013). "Millet Grains: Nutritional Quality, Processing, and Potential Health Benefits." *Comprehensive Reviews in Food Science and Food Safety* 12(3): 281–95.

[23] Singh, Vandana and Ramesh, Kumar (2017). "Study of Phytochemical Analysis and Antioxidant Activity of Allium Sativum of Bundelkhand Region." *International Journal of Life-Sciences Scientific Research,* 3(6): 1451–58.

[24] Subramanian, Shinoj and Viswanathan, R. (2003). "Thermal Properties of Minor Millet Grains and Flours." *Biosystems Engineering,* 84(3): 289–96.

[25] Venu, S. (2022). "A Comparative Analysis of the Phytochemical Screening, Antioxidant, and Antibacterial Properties of Millets." *International Journal of Research and Analytical Reviews,* 9(4): 45–51.

[26] Vetriventhan, Mani (2016). "Finger and Foxtail Millets." In *Genetic and Genomic Resources for Grain Cereals Improvement.*, Elsevier Inc., 291–319.

Increase Shelf Life of Pearl Millet Flour by Using Different Heat Treatments

V.D. Deshmukh and K.A. Sukase*

Mgm Krishi Vigyan Kendra, Gandheli, Aurangabad, Maharashtra, India

ABSTRACT: Biochemical composition of pearl millet indicates that it is a good source of energy, protein, vitamins and minerals but pearl millet flour when stored developed the problems of bitterness and rancidity which limited its shelf life. These problems occurred due to the activity of lipase enzymes which caused the breakdown of glycerides and subsequent increase of free fatty acid profile. The major objective of this study is to increase the shelf life of pearl millet flour at ambient conditions by inactivating the lipase activity of pearl millet grains.

Mahatma Gandhi Mission Krishi Vigyan Kendra at Gandheli Tq. Dist. Aurangabad (MH) conducted an on-farm trial on Increase shelf life of Pearl Millet flour by using three different heat treatments in the year 2022 at adopted village Puri of MGM KVK. At the end of this study, it was found that microwave treatment was very effective to increase shelf life of pearl millet flour but it is not feasible for rural women to conduct this method. Second was the blanching method. This method also increases the shelf life of pearl millet flour. Third one was a traditional method of rural women that millets were heated in sun rays followed by milling, but it was not an effective method to store flour for more days. Heat treatment found to be more effective to increase shelf life of flour and increase processing of pearl millet in India.

Keywords: *Pearl Millet Flour, Shelf Life, Heat Treatment, Rancidity.*

INTRODUCTION

Pearl millet is also known as Bajara. It is the most grown variety and gaining importance as a climate-resilient and healthpromoting nutritious crop. Pearl millet (Pennisetum glaucum), one of the world's hardiest cereal crops, can grow under harsh climatic conditions. It is grown in the drylands, mostly by smallholder farmers and forms part of the staple diet for millions of resource-poor people in the drylands. Pearl millet forms a major part in diet in some parts of India and in African region but traditionally it has been considered as a "poor man's bread" but now situation has been changed, which is made possible due to its climate supportive nature (tolerance to drought, heat and soil salinity and high-water use efficiency) and additional nutritional rich properties like high level of calcium, iron, zinc, lipids and high-quality proteins (Nambiar *et al.*, 2011) but development of off-flavour in pearl millet flour upon storage after a short time is the major hindrance for its acceptability by the consumers. Thus, use of pearl millet in the food industry and consequently its consumption by urban population is very low. Due to the high fat content and lipase activity, the pearl millet (Pennisetum typhoides) flour had limited shelf life at ambient conditions. This is caused by an active lipase enzyme which is responsible for the breakdown of glycerides and consequent increase of free fatty acids (Pruthi 1981 and Arora *et al.,* 2002). These chemical changes are usually manifested as off-flavours during storage, especially under conditions of moderately high moisture and oxygen exposure (Nantanga *et al.,* 2008). Various treatments were applied to pearl millet to prevent the development of rancidity in the stored flour, including decorticating the grains (Abdebrahman *et al.,* 1983), the use of antioxidants (Kapoor and Kapoor 1990), application of dry heat treatment to grain (Chavan and Kachare 1994) or flour (Kapoor and Kapoor 1990), acid soaking and hot water blanching (Chavan and Kachare 1994, Archana *et al.,* 1998), use of malting (Archana *et al.,* 1998), preservation of pearl millet flour by refrigeration (Mohamed *et al.,* 2011), use of different containers for storage like gunny sacks, earthen pots, tin cans and polythene bags (Chaudhary and Kapoor 1984); application of thermal treatment for producing shelf- stable and also pre-cooked flour (Nantanga *et al.,* 2008), use of pre-milling treatments, viz. pearling, heat treatment and fermentation Biochemical composition of pearl millet [Pennisetum glaucum (L) R Br] indicates that it is a good source of energy, protein, vitamins and minerals (Osman 2009). It contains a higher content of triglycerides, which are rich in unsaturated fatty acids. Effects of thermal treatments on pearl millet flour which increased storage life of pearl millet flour up to 6 days without undue deterioration in quality at ambient condition (Tiwari *et al.,* 2014). The available review clearly indicates that shelf-life of pearl millet flour was very limited and heat treatment can increase the shelf life of pearl millet. Therefore, the objective of this study was to evaluate the effects of dry and wet heat treatments of storage life to inactivate anti-nutritional content of pearl millet grains through some pre-milling treatments like blanching and hydrothermal treatment and thus evaluate the storage stability of flour at different storage conditions.

MATERIALS AND METHODS

MGM Krishi Vigyan Kendra, Gandheli, Aurangabad II (MH) conducted on farm trial on Assessment of Increase shelf life of Pearl millet by using three treatment, farmer method (T1), micro-oven drying (T2) and blanching method (T3) in adopted village Puri, Tq. Gangapur, Dist. Aurangabad. The pearl millet was carefully cleaned and freed from foreign materials. After cleaning, the treatments like roasting and blanching were applied to the grain. Dry heating can be done by heating the grains in hot air oven at 100±2°C for 120 min for 60 to 120 min, followed by rapid cooling and finally milling to obtain hole meal (Kadlag *et al.,* 1995) and blanching is done by boiling water at 98°C, followed by immersing of the grains in the boiling water (1:5 ratio of seeds to boiling water) for 90 sec (Bhati *et al.* (2016).

RESULTS AND DISCUSSION

The research paper entitled "Increase shelf life of pearl millet flour by using different heat treatments" shows Effect of pre-milling heat treatments on shelf life of pearl millet flour. Table 1 explained that Treatment 1 Farm women dry grain in sunlight for 5 to 6 hr and mill the grains. It means, direct sun drying of pearl millet grain and then milling. By using this treatment, it is stated that pearl millet flour can store only 6.14 days and has an organoleptic score of 4.71. This method cannot effectively store flour for more days. The second treatment was blanching grains at 98^0 c for 90 seconds and then milling. It shows that the pearl millet can be stored for 21.43 days and has an organoleptic score of 3.42. Second method is effective and can store flour for maximum days. At last, the third treatment i.e., heat drying in micro-oven. This method's result indicates that the pearl millet flour can store for 27.71 days and has an organoleptic score of 2.71. Third method is more effective than the other two but this method is not feasible for rural women to conduct.

Table 1: Observations of Treatment Conducted to Record Shelf Life of Pearl Millet Flour

Observations	*Treatment 1*	*Treatment 2*	*Treatment 3*
Shelf life of Bajra in days	6.14	21.43	27.71
Organoleptic score	4.71	3.42	2.71

CONCLUSION

Processing methods and their impact over the shelf life of pearl millet flour were reviewed in this study. It can be concluded that heat treatments are necessary for improving the nutritional availability and storage stability of flour. Therefore, for encouraging the commercial utilization of pearl millet grains in food formulations and to achieve the better food security feature, it is necessary to use appropriate processing methods.

REFERENCES

[1] Abdelrahman, A., Hoseney, R. and Varriano-Marston, E. (1983). Milling process to produce low fat grits from pearl millet; *Cereal Chemistry,* 60(3): 189–91.

[2] Archana, S. and Kawatra, A. (1998). Reduction of polyphenol and phytic acid content of pearl millet grains by malting and blanching; *Plant Foods and Human Nutrition,* 53: 93–8.

[3] Bhati, D., Bhatnagar, V. and Acharya, V. (2016). "Effect of pre-milling processing techniques on pearl millet grains with special reference to in-vitro iron availability"; Asian *Journal of Dairy and Food Research;* 35(1):76–80.

[4] Chavan, J.K. and Kachare, D. (1994). Effects of seed treatment on lipolytic deterioration of pearl millet flour during storage; *Journal of Food Science and Technology,* 31: 80–1.

[5] Chaudhary, P. and Kapoor, A. (1984). Changen in the nutritional value of pearl millet flour during storage; *Journal of the Science of Food and Agriculture,* 35: 1 219–24.

[6] Kadlag, R., Chavan, J. and Kachare, D. (1995). "Effects of seed treatments and storage on the changes in lipids of pearl millet meal"; Plant Foods for Human Nutrition (Dordrecht, Netherlands). 47(4):279–285.

[7] Nambiar, V., Dhaduk, J., Sareen, N., Shahu, T. and Desai, R. (2011). Potential functional implication of pearl millet (Pennisetum glaucum) in health and diseases; *J Appl Pharm Sci,* 1: 62–67.

[8] Nantanga, K., Seetharaman, K., Kock, H. and Taylor, J. (2008). Thermal treatments to partially pre-cook and improve the shelf life of whole pearl millet flour; *Journal of the Science of Food and Agriculture,* 88: 1 892–9.

[9] Osman, M. (2009). Effect of germination on the nutritional quality of pearl millet; *Journal of Agricultural Science and technology,* 3(8): 1–6.

[10] Pruthi, T. (1981). Free fatty acid changes during storage of bajra (Penmselum typhoideum) flour; *Journal of Food Science and Technology,* 18(6): 257–8.

[11] Simwemba, C., Hoseney, R., Varrino-Marston, E. and Zeleznak, K. (1984). Certain B vitamin and Phytic acid contents of pearl millet 95 (Pennisetum americanum L Leeke); *Journal of Agriculture and Food Chemistry,* 32: 31–4.

[12] Tiwari, A., Jha, S., Pal, R., Sethi, S. and Krishan, L. (2014). Effect of pre-milling treatments on storage stability of pearl millet flour; *Journal of Food Processing and Preservation,* 38: 1 215–23.

Bio-Stimulating Activity of Seaweed Liquid Fertilizer on Seed Germination and Productivity of Millets Crop

Ravinder Sah, Akshada Parab, Shreya Pabrekar, Manasi Pawar and Sunil Shankhadarwar

Department of Botany, Ramnarain Ruia Autonomous College, Matunga, Mumbai, India

ABSTRACT: Millets are annual, small-grained cereals in the grass family grown in the semi-arid tropical region. They are a rich source of dietary fibre, proteins, vitamins, and minerals. Urbanization and population explosion have led to a tremendous increase in demand for food and fertile land. Excessive use of chemical fertilizer is being carried out worldwide to increase millets yield. There is a race to find organic fertilizer as an alternative to chemical fertilizer. Recently seaweeds are getting explored as organic fertilizers due to their bio-stimulating properties. In the present study, the effects of Seaweed Liquid fertilizers (SLF) were checked on commonly grown millets for their seed germination rate, plumule and radicle growths as well as mean germination time. Seaweed liquid fertilizers of red, green and brown seaweeds were tested at various concentrations, and the optimal concentration was found to be 2%, which shows the maximum germination rate prepared with brown seaweeds. The emergence of radicles and plumule were observed after 24 and 48 hours respectively in millets treated with Seaweed liquid fertilizer. Seed grains treated with chemical fertilizer and water as control showed an average germination rate. Thus, Bio stimulating effects of these serve as a preferable option over chemical fertilizer, which helps in promoting the growth and development of millets.

Keywords: *Millets, Seaweed Liquid Fertilizer, Biostimulant Activity, Phytohormones.*

INTRODUCTION

Seaweeds are nowadays getting recognized due to their bioactive properties. Seaweeds have huge potential for sustainability and replacing the harmful source of various substances. As seaweeds are reported to possess, endogenously found bioactive compounds (Sharifuddin *et al.,* 2015), their demand in the field of agriculture, pharmaceutical, cosmetics and food is increasing day by day. In agriculture, seaweed fertilizer has already reached heights from ancient Chinese and Romans till now to each corner of the world (El Boukhari *et al.*, 2020). Seaweeds have been reported to present phytohormones, amino acids, vitamins, secondary metabolites like phenol, flavonoids, tannins etc. along with essential minerals which help in growth and development (Thirumaran *et al.*, 2009). These bioactive compounds are extracted using various techniques of preparing Seaweed liquid fertilizer (SLF). There are many reports on justifying the use of seaweed liquid fertilizer for plant growth enhancement (Khan *et al.*, 2009). Plants which are difficult to grow in normal conditions or are grown using excess chemical fertilizers can utilize seaweed liquid fertilizer. Millets are one such crop grown using excess fertilizers to increase yield. These are the world's most important crops after rice and wheat. In semi-arid tropical regions, maize, barley, and other grains are significant staple foods whereas, in drier areas of the world, where other crops are harder to grow due to inadequate soil fertility and minimal rainfall result in limited availability of key food crops, they serve as food and nutrition for poor people. (Das, 2016) Since millet contains minerals and phenolic compounds that are beneficial to health, it is ideal as food and feed (Hassan *et al.,* 2021). Currently, the globe is experiencing multiple global concerns, including the expanding population, changing climatic conditions, escalating food prices, water depletion, chemical spills, and other socioeconomic factors due to which the agricultural land is widely affected and ultimately affecting the productivity of crops (Yusuf, *et al.,* 2021). Farmers use chemical fertilisers to increase agricultural output to satisfy the growing demand, but because these fertilisers are harmful, they pose a serious environmental problem, so there is an increasing need for biofertilizers which can minimize the use of chemical fertilizers, improve soil fertility, crop production, without affecting the environment. Seaweed liquid fertilizers are already used on millets, and *Padina boergesenii* and *Gracilaria edulis* extracts were used for checking the growth effects on foxtail millet (Rathinapriya *et al.*, 2020). Similarly, the effects of *Ascophyllum* seaweed extracts which are commercially available in the market were also tested for checking the agronomic growth of three millets (Moradnezhad *et al.*, 2022). Field trials of pearl millet using seaweed as manure was conducted which showed positive results (Bokil *et al.*, 1972). However, the present study plans in using seaweed liquid fertilizer prepared from local seaweed species found on the west coast of Maharashtra. Three millets are selected of which seed germination and seedling growth parameters are recorded. Further to justify, biochemical assays are carried out of treated plants. Seaweeds contain a high number of phenols which are detected, thereby concluding the use of seaweed liquid fertilizer as a substitute for chemical fertilizer.

MATERIALS AND METHODS

Seaweed Collection

Ulva rigida C. Agardh, *Spatoglossum asperum* J. Agardh and *Centroceras clavulatam* C. Agardh collected from the west coast of Maharashtra, India. Selected seaweeds were washed properly and cleaned to remove macroscopic debris like shells, marine organisms as well as other seaweed attached to them. For identification and authentication, samples were sent to the Botanical Survey of India, south region campus, Coimbatore. Washed seaweeds were shade dried at room temperature, further dried at 40^0C in an oven to remove moisture and made into powder using a mixer grinder. Further, the powder was used for the preparation of Seaweed Liquid fertilizers (SLF).

Preparation of Seaweed Liquid Fertilizer

2 g of dried seaweed powder was taken along with 100 ml of distilled water, it was sonicated for 10 min and then autoclaved for 20 mins at 120°C at 20 psi (Bhosale *et al.*, 1975). The extract prepared was further cooled and filtered using a muslin cloth. The filtrate obtained is considered a 2% seaweed liquid fertilizer. It was further used for checking germination rate and bioactive compounds.

Seed Selection and Treatment

Millets namely *Pennisetum glaucum, Setaria italica* and *Sorghum bicolour* were collected from local farmers. Seed grains were surface sterilized using a standard protocol, by washing with distilled water, followed by 1% mercuric chloride, and again with sterile distilled water. Seed grains were further placed in a humidity chamber prepared using the petri dish germination method (Chanthini *et al.*, 2019). The germination rate was checked for 72 hrs, the study was continued till 96 hrs for observing root and shoot development. 10 seeds were selected for each seaweed liquid fertilizer set as well as for control (distilled water) and chemical fertilizer (0.3% NPK). The seed germination test was carried out in triplicates to avoid standard errors. Seedlings were grown apart from the germination test. 10 seeds were selected which were washed as mentioned above and sown in the soil. Seedlings were observed for two weeks and used for analysing biochemical parameters. Seeds were soaked in seaweed liquid fertilizer for 24hrs before testing for both studies. It was added to Petri dishes after 72hrs while it was foliar sprayed on seedlings once a week.

Growth and Biochemical Parameters

The rate of germination was calculated followed by measuring the shoot and root lengths. The characteristic observation of each test plantlet was observed and noted. Seedlings were monitored physically; chlorophyll content and protein content of seedlings were estimated using Arons and Lowrey's method respectively.

Estimation of Total Phenols Fromseaweed Liquid Fertilizer

Phenols are compounds present in seaweed liquid fertilizer even after high-pressure high-temperature extraction. They were estimated using Phenol Coicalteu reagent. Gallic acid was used as standard phenol (w/w). 1 ml of extract was used with dilution in a 1:1 ratio with distilled water. Estimation was carried out at 725 nm on a UV spectrophotometer. The concentration of phenols from seaweed liquid fertilizer was determined using a standard concentration curve.

RESULTS AND DISCUSSION

Percent Germination Rate

From to Table 1, it was observed that the seeds treated were showing a higher germination rate compared with the control. *Setaria* treated with *S. asperum* extracts had the maximum germination rate among all the seaweed liquid fertilizers and control. While in *Sorghum*, *U. rigida* showed a greater germination rate of 9.67% and in *Pennisetum*, treated plants have values close to control plants respectively. The rate of germination was studied till 48 hrs, mentioned results are carried out in triplets, by applying standard deviation.

Table 1: Germination Rate of Millets Grown Using Seaweed Liquid Fertilizer and Control after 48 hrs

Germination Rate	*Pennisetum glaucum (± STDEV)*	*Sorghum bicolour (± STDEV)*	*Setaria italica (± STDEV)*
Control	6.33 ± 1.15	7.33 ± 0.58	7.00 ± 1.00
Spatoglossum asperum	7.67 ± 1.53	9.00 ± 1.00	8.33 ± 1.15
Ulva rigida	4.33 ± 2.31	9.67 ± 0.58	6.33 ± 2.89
Centroceras clavulatam	7.67 ± 0.58	9.00 ± 1.00	5.67 ± 1.15

Lengths of Root and Shoot

As shown in Table 2, after 72 hrs of germination, seeds were treated with seaweed liquid fertilizer. The initiation of root and shoot development was already observed after 72 hrs of the treatment. Seeds were kept for 96 hrs shows proper epicotyl and hypocotyl formation. In *Pennisetum*, U. rigida showed an 8.13 cm root length compared with the control which was 6.17 cm. In *Sorghum*, all three were showing root lengths more than that of the control and chemical-treated ones. In *Setaria,* treated seeds were showing maximum root development in seeds treated with *C. clavutam*. Similarly, epicotyl formation in treated seeds was initiated first. Studies were conducted in triplicates to avoid standard errors, while statistical methods like standard deviation were used.

Table 2: Root Length, Shoot Length and Leaves Length of Germinated Seeds after 72 hrs Treated with Seaweed Liquid Fertilizer and Control

Growth Parameters	*Millets*	*Control (± STDEV)*	*Spatoglossum asperum (± STDEV)*	*Ulva rigida (± STDEV)*	*Centroceras clavulatam (± STDEV)*
Root Length	*Pennisetum glaucum*	6.17 ± 0.30	7.73 ± 0.15	8.13 ± 0.06	2.03 ± 0.15
	Sorghum bicolour	3.47 ± 0.45	6.00 ± 0.10	6.23 ± 0.25	8.03 ± 0.15
	Setaria italica	2.53 ± 0.23	2.07 ± 0.11	2.43 ± 0.15	3.17 ± 0.12
Shoot Length	*Pennisetum glaucum*	2.20 ± 0.20	1.77 ± 0.05	3.52 ± 0.12	2.37 ± 0.06
	Sorghum bicolour	5.17 ± 0.28	1.73 ± 0.15	1.87 ± 0.15	3.40 ± 0.10
	Setaria italica	4.07 ± 0.37	2.27 ± 0.20	3.60 ± 0.10	1.35 ± 0.05
Leaves Length	*Pennisetum glaucum*	1.50 ± 0.10	0.21 ± 0.03	0.30 ± 0.03	3.23 ± 0.15
	Sorghum bicolour	1.63 ± 0.15	0.59 ± 0.01	1.23 ± 0.15	1.70 ± 0.10
	Setaria italica	2.17 ± 0.30	0.91 ± 0.01	1.93 ± 0.15	2.43 ± 0.06

Biochemical Parameters

Protein and chlorophyll estimation of seedlings after two weeks of germination was carried out, treated plants along with chemical fertilizer (NPK) and D/W as control. Of which showed maximum protein content treated plants. In *Pennisetum*, the highest protein content was found in *U. rigida-treated* plants. In *Setaria,* it was maximum in *Control followed by U. rigida* while it was maximum in *Sorghum* treated with control, followed by *U. rigida* treated plants.

Table 3: Protein Content in ug/g of Millets Grown Using Seaweed Liquid Fertilizer, Control and NPK

Protein Content	*Pennisetum glaucum (ug/g ± STDEV)*	*Sorghum bicolor (ug/g ± STDEV)*	*Setaria italica (ug/g ± STDEV)*
Control	7.40 ± 0.35	17.06 ± 0.21	14.10 ± 0.28
NPK	10.61 ± 0.26	10.84 ± 0.25	13.82 ± 0.17
Spatoglossum asperum	10.16 ± 0.18	10.61 ± 0.30	12.85 ± 0.23
Ulva rigida	11.75 ± 0.17	14.59 ± 0.26	13.75 ± 0.22
Centroceras clavulatam	9.56 ± 0.21	14.27 ± 0.09	12.60 ± 0.17

Chlorophyll content in treated plants is higher than carotenoid content as shown in Table 4.

Table 4: Pigment Analysis of Treated Millet Plants Using Seaweed Liquid Fertilizer, Control and NPK

Sample Name		*Chlorophyll a (mg/g)*	*Chlorophyll b (m/g)*	*Total Chlorophyll (mg/g)*	*Carotenoids (mg/g)*
Pennisetum glaucum	*Control*	38.90	45.59	84.46	4.92
	NPK	16.84	18.56	35.39	2.82
	S. asperum	28.90	30.41	59.29	4.66
	U. rigida	25.82	30.84	56.64	3.38
	C. clavulatam	28.83	31.72	60.53	4.46
Sorghum bicolor	*Control*	24.84	28.63	53.45	3.72
	NPK	61.61	65.70	127.27	9.99
	S. asperum	20.63	19.25	39.87	4.79
	U. rigida	46.04	49.39	95.40	6.44

Sample Name		*Chlorophyll a (mg/g)*	*Chlorophyll b (m/g)*	*Total Chlorophyll (mg/g)*	*Carotenoids (mg/g)*
	C. clavulatam	47.07	55.22	102.25	6.27
Setaria italica	*Control*	27.68	110.30	137.93	4.36
	NPK	48.98	59.91	108.86	8.05
	S. asperum	33.11	37.21	70.30	6.61
	U. rigida	3.67	4.09	7.75	0.64
	C. clavulatam	3.59	4.03	7.62	0.64

Phenol Estimation

Phenolic content from seaweed was estimated to check its bio-stimulating activity. Brown seaweed showed maximum phenol content of 18.96 mg/g followed by red seaweed at 3.93 mg/g and a minimum in green seaweed of about 2.12 mg/g.

Table 5: Total Phenol Content from Seaweed Liquid Fertilizer

Seaweeds	*Phenol Content mg/g ± stdev*
Spatoglossum asperum	18.96 ± 0.54
Ulva rigida	2.12 ± 0.12
Centroceras clavulatam	3.93 ± 0.37

DISCUSSION

Millets are still an underutilized cereal crop (Saini *et al.*, 2021). It was reported in 2022, that recently the production of millet has not increased much. Farmers in India hesitate to grow millet compared to rice and wheat even though they are unable to fulfil the nutrient requirements (Gupta *et al.*, 2013). Millets are grown using chemical fertilizers which are harmful in the long term. Nitrogen requirement is essential as it affects productivity (Jagdish, 2022; Maitra *et al.*, 2020). There are many reports on using organic fertilizers which help in increasing the nutrient availability of plants and maintaining soil health (Shyamsunder and Menon, 2020). Seaweed Fertilizers are already used since ancient times, above research study thus describes the bio-stimulating effects of Seaweed liquid fertilizers on common millets.

Results from seed germination rate show Seaweed Liquid Fertilizer pre-treated seeds are germinated earlier and more in quantity compared with water as control. For *Pennisetum glaucum,* seeds treated with 2% extracts of *Spatoglossum and* 2% *Centroceras* show the highest germination rate of *7.67%* individually. While in *Sorghum bicolour 2% Ulva extracts show maximum germination rates. In Setaria italica, 2% Spatoglossum extract* has an 8.33% germination rate which is the highest. Similarly, when sown in the soil, treated seeds had grown in lesser time compared with control and chemical fertilizers respectively, 2% *Spatoglossum extracts* showed promising results in all three selected millets compared to other Seaweed Liquid Fertilizers and control. It was reported in an experiment carried out on foxtail millets using *Padina* and *Gracilaria* extracts showed similar results in which Seaweed Liquid Fertilizer-treated seeds showed maximum germination rate, also lower concentrations triggered more growth (Rathinapriya *et al.,* 2020). Seed germination rates were promoted with *Sargassum polycystum* extracts which were 1.5%, thereby lower concentrations show more growth-enhancing effects (Erulan 2009).

In this research work, shoot and root lengths of germinated seedlings were observed, 72 hrs after germination when the stored nutrients from the seeds are utilised, bioactive compounds from the seaweeds are uptaken by radicle and plumule thereby developing lateral and hairy growth in roots as well as increased in plumule length. Thus, growth regulators present in Seaweeds play an important role in seed germination (Osuna *et al.,* 2015). In all three millets, control seeds were showing the least root length compared with *Ulva*-treated *Pennisetum* had 8.13 cm, while *Sorghum* and *Setaria* had 2% *Centroceras* with root lengths of 8.03 cm and 3.17 cm respectively.

The biochemical mechanism is enhanced when the foliar application of seaweed liquid fertilizer is carried out. Several research reports show an increase in protein content and pigment concentration in treated plants. *Gracilaria-treated Vigna radiata* showed an increase in seed protein content by twice in amount (Huda *et al.*, 2022). In the current work, treated plants showed similar results with more than or equal to control and chemical-treated ones.

As, seaweed contains various micro-macro nutrients, endogenously present phyto hormones, secondary metabolites etc. (Farghali, *et al.,* 2022). There are certain bio-active compounds which play a very important role in plant growth and development. Phenols are secondary metabolites which are formed in plants to overcome biotic and abiotic stresses. Phenolic compounds are found maximum in brown seaweeds and have proven to be useful as biostimulants (Lopez *et al.,* 2021). It is reported, as phenolic compounds present in seaweed extracts help in promoting adventitious root and root lengthening in plants. It acts as a regulator of phytohormone indole-3-acetic acid (IAA) which is an auxin (Aremu *et al.*, 2015).

In the present study, the effect of phenolic compounds is seen with root lengthening and the formation of root hairs in the seedling stages of millets. Chlorophyll content in plants has also been influenced due to the effects of treatment. Thereby, seedling growth is also enhanced.

CONCLUSION

Seaweed extracts show enhancing properties. Millets which are grown traditionally using chemical fertilizer can be now grown using seaweed bio-stimulant. From the study, it is observed, initially, the germination rate is low for 24 hrs study, but once the seeds are germinated after 72 hrs they imbibe the extracts thereby increasing the growth. As reported, phenolic compounds are present in seaweeds and are extracted during high-pressure high temperature, these phenolic compound acts as one the bio-stimulating compounds on seeds. Agronomic characters are enhanced by seaweed extracts, these bioactive compounds trigger seed physiology which further helps in the growth and development of seedlings. From the present study, three seaweeds selected show promising effects on seed germination and activities, thereby concluding the use of Seaweed Liquid Fertilizer on large scale to reduce the harmful effects of chemical fertilizer.

ACKNOWLEDGEMENT

The authors would like to thank Ramnarain Ruia autonomous college, for encouraging research study and assisting whenever needed.

REFERENCES

[1] Aremu, A.O., Masondo, N.A., Rengasamy, K.R., Amoo, S.O., Gruz, J., Bíba, O. and Van Staden, J. (2015). Physiological role of phenolic biostimulants isolated from brown seaweed *Ecklonia maxima* on plant growth and development. *Planta*, 241, 1313–1324.

[2] Bhosle, N.B., Untawale, A.G. and Dhargalkar, V.K. (1975). Effect of seaweed extract on the growth of *Phaseolus vulgaris, L. Indian J Mar Sci.*

[3] Bokil, K.K., Mehta, V.C. and Datar, D.S. (1972). Seaweeds as manure III. field manurial trials on *Pennisetum typhoids* SH (pearl millet) and *Arachis hypogea* (Groundnut).

[4] Chanthini, K.M.P., Stanley-Raja, V., Thanigaivel, A., Karthi, S., Palanikani, R., Shyam Sundar, N. and Senthil-Nathan, S. (2019). Sustainable agronomic strategies for enhancing the yield and nutritional quality of wild tomato, *Solanum Lycopersicum* (l) var Cerasiforme Mill. Agronomy, 9(6), 311.

[5] Das, I.K. (2016). Biotic Stress Resistance in Millets || Millets, Their Importance, and Production Constraints. 3–19. doi:10.1016/B978-0-12-804549-7.00001-9.

[6] El Boukhari, M.E.M., Barakate, M., Bouhia, Y. and Lyamlouli, K. (2020). Trends in seaweed extract based biostimulants: Manufacturing process and beneficial effect on soil-plant systems. Plants, 9(3), 359.

[7] Erulan V, Soundrapandian P, Thirumaran G, Ananthan G. Studies on the effect of *Sargassum polycystum* (C. Agardh.) extract on the growth and biochemical composition of *Cajanus cajan* (L.) Mill sp. *Am Eur J AgricEnviron Sci,* 2009; 6(4): 392–399.

[8] Farghali, M., Mohamed, I.M., Osman, A.I. and Rooney, D.W. (2022). Seaweed for climate mitigation, wastewater treatment, bioenergy, bioplastic, biochar, food, pharmaceuticals, and cosmetics: a review. *Environmental Chemistry Letters,* 1–56.

[9] Gupta, A., Sood, S., Agrawal, P.K. and Bhatt, J.C. (2013). Under-utilized food crops of Himalayan region: Utilization and prospective. *Newer Approaches to Biotechnology,* 101–120.

[10] Hassan, Z.M., Sebola, N.A. and Mabelebele, M. (2021). The nutritional use of millet grain for food and feed: a review. *Agriculture & food security,* 10, 1–14.

[11] Huda, M.N., Mannan, M.A., Bari, M.N. and Rafiquzzaman, S.M. (2022). Red Seaweed (*Gracilaria tenuistipitata* var. liui) Liquid Fertilizer Increase Yield and Seed Protein of Mungbean (*Vigna radiata). Sustainable Agricultural Innovations for Resilient Agri-Food Systems,* 358.

[12] Jagdish (2022, March 27). Best Fertilizer for Finger Millet (Ragi): Organic, Bio-fertilizers, NPK, How and When to Apply.

[13] Jimenez-Lopez, C., Pereira, A.G., Lourenço-Lopes, C., García-Oliveira, P., Cassani, L., Fraga-Corral, M., ... and Simal-Gandara, J. (2021). Main bioactive phenolic compounds in marine algae and their mechanisms of action supporting potential health benefits. *Food Chemistry,* 341, 128262.

[14] Khan, W., Rayirath, U.P., Subramanian, S., Jithesh, M.N., Rayorath, P., Hodges, D.M. and Prithiviraj, B. (2009). Seaweed extracts as biostimulants of plant growth and development. *Journal of plant growth regulation,* 28, 386–399.

[15] Maitra, S., Pine, S., Shankar, T. and Pramanick, B. (2020). Nutrient management in foxtail millet: a review. *Indian Journal of Natural Sciences,* 10(60), 23156–23160.

[16] Moradnezhad Hesari, V., Sadrabadi Haghighi, R., Bagherzadeh Charjoie, A. and Ghasemi, M. (2022). Investigation of Plant Density and Biological Fertilizer Effects on Agronomy Characteristics, Yield Seed and Yield Components of Commercial Cultivars of three Millet Species. *Applied Field Crops Research,* 35(1), 87–72.

[17] Osuna, D., Prieto, P. and Aguilar, M. (2015). Control of seed germination and plant development by carbon and nitrogen availability. *Frontiers in Plant Science,* 6, 1023.

[18] Rathinapriya, P., Satish, L., Pandian, S., Rameshkumar, R., Balasangeetha, M., Rakkammal, K. and Ramesh, M. (2020). Effects of liquid seaweed extracts in

improving the agronomic performance of foxtail millet. *Journal of Plant Nutrition*, 43(19), 2857–2875.

[19] Saini, S., Saxena, S., Samtiya, M., Puniya, M. and Dhewa, T. (2021). Potential of underutilized millets as Nutri-cereal: an overview. *Journal of Food Science and Technology*, 1–13.

[20] Sharifuddin, Y., Chin, Y.X., Lim, P.E. and Phang, S.M. (2015). Potential bioactive compounds from seaweed for diabetes management. *Marine drugs*, 13(8), 5447–5491.

[21] Shyamsunder, B. and Menon, S. (2020) Review on Biofertilizer technology and millet production.

[22] Thirumaran G, Arumugam M, Arumugam R, Anantharaman P. Effect of seaweed liquid fertilizer on growth and pigment concentration of *Cyamopsis tetrogonolaba*(L.) Taub. *American-Eurasian Journal of Agronomy*. 2009 Jan; 2(2):50–56.

[23] Yousaf, L., Hou, D., Liaqat, H. and Shen, Q. (2021). Millet: A review of its nutritional and functional changes during processing. *Food Research International*, 142, 110197.

THEME-5

GO's, NGO's and Policies

Role of Women Farmers, FPOs and NGOs in Mainstreaming Millets: A Case Study of Odisha Millets Mission

Tulika Rani Talukdar and Rajshree Bedamatta
Research Scholar, Department of Humanities and Social Sciences, IIT Guwahati, India

ABSTRACT: Millets, an indigenous food crop, were displaced in the quest of achieving food security during the 1970s-80s, shifting the dietary diversity primarily towards wheat and paddy, owing to India's historical policy neglect. However, the National Food Security Act of 2013 has given a major push for the revival of millets, with a policy focus on enhancing production and consumption. In the context of evaluating Odisha Millets Mission initiated by the Government of Odisha in 2016, this paper found that the collaboration and coordination among women farmers, NGOs and FPOs is strengthened through the government intervention of mainstreaming a lost crop. The integration between the village level institutions such as FPOs, Women Self Help Groups supported and facilitated by the NGOs plays an important role in ensuring a sustainable model of market integration along with strengthening of the local tribal community. The program promotes women entrepreneurship through engagement of women SHGs in millet promotion through establishing millet tiffin centres, millet kiosks. The study uses the qualitative research method, wherein in-depth interviews of women millet farmers, NGOs and focus group discussions of FPOs and WHGs were undertaken. Koraput district, a tribal dominated region, in Odisha is selected for the study. A comprehensive policy is the need of the hour to ensure that emphasis on millets does not become a divergence from policy imperativeness towards diversification. Reviving an indigenous crop through the engagement of village-level institutions such as FPOs, WSHGs and NGOs is a significant learning towards creating a sustainable model, which needs to be integrated by other states in their respective agricultural policy.

Keywords: *Odisha Millets Mission, Multi-stakeholder Convergence, Institutions, Farmer Producer Organizations.*

INTRODUCTION

Millets form the staple diet and are a part of mixed crop systems in India, especially in communities belonging to the rain-fed regions of the country. In addition to, being used as a food crop for humans, it serves a variety of purposes such as feed for poultry, fodder for livestock, and substantiating bio-fuel needs (Patnaik, 2020). Millets are coarse cereal food grain crops and are cultivated in diverse soil types ranging from heavy sandy to infertile soil, moisture stress, and tropical and sub-tropical climate. This makes it a climate-resilient crop, which is hardy and contributes substantially to the food, nutrition and livelihood security of the population in rain-fed areas. There are two categories of millets crop, produced in the country. The major millet includes sorghum (jowar), finger millet (ragi) and pearl millet (bajra). The minor millet category comprises of 6 kinds of small millets-foxtail millet (kangni), little millet (kutki), barnyard millet (sawan), proso millet (cheena), kodo millet, and brown top millet (korale) (Raina, *et al.*, 2022).

In spite of the numerous benefits of millets, its cultivation started declining in quest of India's struggle to achieve food security in the 1960–70. The introduction of neo-liberal economic processes led commercial or market-oriented farming has replaced the traditional food and livelihood system causing a sharp rise in food prices, and massive displacement of rural poor (Lodhi, 2013; Kumbamu, 2018). This has resulted in growing inequality between different classes of people along with having a disempowering effect on women from their traditional roles in agriculture (Padulosi, *et al.*, 2015; Patel, 2012 and Schutter, 2013).

Indian agriculture system is characterized by input-intensive industrial agriculture. Input-intensive farming systems dominated by crops like wheat, rice and maize were given scientific, technological and policy attention with the intention to provide food and nutrition security to the starving millions. Lack of production support in terms of providing incentives to increase the yield and technological support for post-harvest processing technologies, lack of public procurement and marketing support are a few of the other factors behind the decline of millet cultivation and consumption (Revalorising Small Millets in Rainfed Regions of South Asia, 2012).

Various alternate food system models have emerged, largely focusing on agroecology, localization (Martinez *et al.*, 2010), equitable access (Chebrolu and Dutta, 2021) and building sustainable Social and Solidarity Economies (Kumbamu, 2018), Zero Budget Natural Farming (Khadse and Rosset, 2019), Food Sovereignty (Wittman *et al.*, 2010) and Sustainable Agriculture.

These alternate models have emerged in resistance to social and economic injustice and environmental degradation at the hands of neoliberal industrial agriculture. These initiatives advocate work with small and marginal farmers, indigenous peoples, farm workers, and unite them on the basis of undertaking responsibility and decision-making power to reclaim control over local food and agriculture system. These models emphasize taking in the context of local, place and network based, issues and possibilities arising from regional food and agriculture system. The central philosophical underpinning guiding these models is founded on the principles of agro-ecology, democracy, inclusiveness, social justice, changing relations, institutional innovations and socio-economic sustainability (Kumbamu, 2018; Chebrolu and Dutta, 2021).

Against the backdrop of declining yield and production and declining area under millet cultivation in the last five decades since 1960s, National Food Security Act (NFSA) of 2013, has given a major push for revival of millets, with a policy focus on enhancing production, consumption and inclusion of millets in PDS. In recent years, revival of millets cultivation has received a lot of attention from state and national government, including the state of Odisha. This paper studies the government program of revitalizing millets, 'Odisha Millets Mission', launched by the Government of Odisha in 2017 to revive millets in farms and on plates.

The program was initiated with the objective to support and revive neglected crops like millet, primarily grown by tribal communities in the rainfed region of Odisha. The program envisages a comprehensive revival of millets by focusing not just on improving the productivity of millets, which has been the focal point of other government agricultural schemes, but also simultaneously on consumption, supporting ragi growers with MSP, marketing and making millets a part of mainstream culture by popularizing it in urban and rural areas. The program, in an attempt to address the malnutrition problem, works towards the inclusion of millets in State Nutrition Programmes. The program is operational in 19 districts and 143 blocks of the state.

MATERIALS AND METHODS

The study uses the qualitative exploratory research method, wherein Stakeholder Analysis is performed. Primary methods were employed to perform Stakeholder Analysis. We have selected Koraput district, a tribal dominated region, in Odisha. Five blocks of Koraput district, Lamtaput, Nandapur, Kundra, Boipariguda, and Borigumma, where the program is being implemented were selected for the study. By doing the Stakeholder Analysis, we have understood the kinds of implications resulting from concerted efforts of different stakeholders at different levels (state and community) towards sustaining and mainstreaming millets. To understand the institutional framework emergent from the program design, Stakeholder Analysis was conducted in two ways.

First, literature review was conducted. In this, reports, news articles, publications and proceeding reports of government meetings were studied and analyzed. Online websites of the two key stakeholders out of three, that is, Odisha Millets Mission (OMM) and an academic institute, Nabakrushna Choudhury Institute of Development Studies (NCDS) were studied and analyzed. Second, stakeholders were identified, In-depth interviews were taken and Focus Group Discussions with were conducted. For this particular study, 30 in-depth interviews with individual women farmers, 8 in-depth interviews with board level representatives of Farmer Producer Organizations (FPOs) in the five blocks and 20 FGDs with Women Self-Help Groups spread across the five blocks were conducted. These stakeholders are categorized as state level actors (Government departments, Research and Program Secretariat), block level actors (Facilitating Agency and ATMA) and sub-block level actors (FPOs, individual women farmers, Women Self Help Groups, and Community Resource Persons).

RESULTS AND DISCUSSIONS

Women Farmers-Engagements, Challenges and Opportunities

Women's immense investment of labour and time has sustained millet for many decades ensuring availability of different varieties of millet to consume at household level. Women's engagement with the millet ecosystem is central to reviving millet cultivation initiative with the support of Facilitating Agencies and the Government of Odisha. Women's involvement is marked along the pre-production, production, weeding, harvesting, processing and marketing stage. Women's traditional knowledge about seed preservation, harvesting and organic manure making, through seed extracting, storing and treatment practices, contribute to preservation of agro-biodiversity and crop diversification. The value of millets can be estimated from the fact that most of the women interviewed for the study agreed that either by exchanging millets or by selling millets in the local haat (market) they acquire other products needed for households. It is primarily by selling millets that these women farmers purchase vegetables and other essential items required for consumption at home. Millet represents the multi-cropping biodiverse system of agriculture pre-dominant in rain fed agriculture areas, and is particularly, argued to favour women farmers as different kinds of crops are yielded at different times, which helps in ensuring household food security.

One of the pertinent challenge and reason behind decline of consumption of millets amongst rural households is women's drudgery involved in the production and post-harvest processing of millets. Lack of availability of good quality seeds, as most of the seeds stored at homes, either gets eaten due to lack of other food items to consume or gets mixed with other grain seeds, which reduces its potent fertility. This along with inadequate knowledge on appropriate agronomic practices for better productivity impedes in millet cultivation resulting in low productivity. This adds to the diminishing traditional knowledge associated with various local variety seeds and their cultivation practices.

Lack of local processing units at village and block level is one of the biggest challenge faced by women farmers when it comes to convert the harvested grain into ready to eat food. Traditionally, women manually perform millet cleaning tasks, which adds to their drudgery. Women thresh the grain with their feet, use a pestle and mortar to dehusk and finally use a stone quern to grind the grain into flour (Padulosi *et al.,* 2015). Ten Ragi threshers have been provided to ten Women Self Help Groups, in Borigumma, Boipariguda, Lamtaput, Nandapur and Kundra block of Koraput district, to manage and operate the functioning of ragi threshers. Women farmers from nearby villages come to thresh their ragi according to the availability and season. SHGs responsible for managing the operation of threshers generate income each month by leasing out threshers operation on an hourly basis to women farmers. Establishing processing unit consisting of de-stoner, de-huskers, and pulverizers suited to the varying shape, size and structure of minor millets, would require joint collaboration of technological knowledge of agriculture universities and research institutes, expertise of CSOs and state government at the village level. There are other difficulties with establishing processing unit at the block level. Because of lack of processing facilities, especially for minor millets like little millet and foxtail millet, highly nutritious millet in the millet family, is getting produced by farmers purely for commercial purposes.

There are few challenges in the production front. Initially at the time of launching of OMM, farmers were reluctant to adopt the improved methods of production because of various reasons. People used to think that improved methods is time consuming, more labour will be required, task of weeding will be complex and an expensive process. However, with the efforts of FA, farmers are now able to understand and adopt the improved methods. The state government is incentivizing the use of agronomic practices by farmers for increasing the productivity of millets. For providing Minimum Support Price to the ragi growers, the Government of Odisha started procuring finger millet in kharif 2018–19 through Tribal Development Cooperative Corporation of Odisha Limited (TDCCOL) in eight districts of the state.

However, women farmers have opined that it becomes difficult to survive for the rest of the year since government procurement window is open only from January to March each year, which is insufficient to sustain them throughout the year. Women farmers resort to selling millets to support household requirements throughout the year. One time opening of procurement channel for a short period, leaves no other choice for women farmers but to sell their millets in local market at low prices before and after the procurement closes. It is been reported that there is a delay in getting the payment into farmer's account through the Direct Benefit Transfer measure. But selling ragi to traders in local market would fetch farmers an instant cash. Traders and middlemen exploit farmers with low rate for the produce and a farmer is left without much choice than to sell his product at a low rate.

Women's engagement with the millet value chain, right from production to post-harvest and marketing presents an opportunity to utilize women's collective strength towards encouraging farmers to increase their millet cultivation area and promoting its consumption among rural households. Women Self-Help Groups are roped in for four levels of activities to be undertaken under the Millets Mission program. To operationalize these four levels of activities, OMM forged the convergence with Mission Shakti, Directorate of WSHGs, Department of Women and Child Development. Mission Shakti is an umbrella organization with a WSHG membership of 70 lakh women members in Odisha (Balam, 2020). Convergence is forged to leverage the strength and expertise of women collectives under Mission Shakti associated with supplying Take Home Ration (THR) to anganwadi centres under ICDS scheme. Four levels of engagement is sought with the convergence. They are:

1. Massive awareness campaigns to be undertaken by WHSGs of Mission Shakti in 14 districts, about the nutritional value of millets targeting household consumption of millets.
2. Preparation of millet recipes and inclusion of ragi ladoos in ICDS.
3. Operating primary processing, and post-harvest units.

4. Promotion of millet entrepreneurship through setting up millet tiffin centres and millet cafes in rural and urban areas.

Women self-help groups manage and sustain community institutions such as the Community Seed Bank, Custom Hiring Centre, operation of ragi threshers, etc. By involving WSHGs cultivating millets in their fields, women's role and contribution in sustaining an ancient crop is strengthened. Through these initiatives, in rural areas, demand for millets is created, raising awareness about consuming millets at household level.

Emergence of FPOs

The nodal CBO in each block gets registered in the form of a FPO. The FPO is a collective representation of farmers, formed to provide services to farmers in terms of improving production, facilitate processing and value addition, and creating market linkage for farm products. Beyond aggregation, FPOs play an important role in the value chain, and contribute to the input and output supply chain of millets and other farm related activities. Formation of community-based institutions such as Millet Processing Unit, Custom Hiring Centre and Community Managed Seed Systems, established and promoted under OMM have created a benchmark that ensures the community leadership in sustaining millet cultivation, processing, marketing and value addition.

The FPO is a registered entity of a collective of farmers, that apart from performing aggregation related work and dealing with farmer's challenges and constraints, also undertakes supply of products and services. These supply of products include timely supply of local and quality seeds, supply of bio-inputs, processed grains and value-added products. Supply of services include primary processing of millet grains, farm implements of hiring basis, storage of seeds and other value-added products (Directorate of Agriculture and Food Production, Odisha, Bhubaneswar, 2019). These FPOs anchor different community institutions such as Community Managed Seed System (CMSS), Custom Hiring Centre (CHC), Farmer Producer Organization (FPO) at the village or Gram Panchayat (GP) level, by linking with WSHGs and other collective entities. This indicates the decentralization of millet production and consumption to the block level. Apart from these, saving and sharing seeds, facilitating decentralized processing facilities, procurement of grain, marketing, extending cultivation knowledge and providing handholding support to farmers through Community Resource Persons (CRPs) and mobilizing farmers to cultivate millets are the other activities entrusted by the FPO. 76 FPOs have been formed under OMM since its inception in 2017 till June 2021 (Garg *et al.,* 2023).

Engagement of the Facilitating Agencies (CSOs) with Program, Millet and Community

The significance of the CSOs in the program formulation and implementation lies in the Program Secretariat (a CSO) working in close collaboration with the state departments and institutions at the state level, district functionaries and Facilitating Agencies at the block level. These collaborations between different actors, at various levels, in a top down approach, shows the commitment of the state government in acknowledging the role of various agencies associated with mainstreaming a rainfed crop to strengthen the local food system of the state. Association of state government, line departments, CSOs, and academia, demonstrate the acknowledgment and mutual respect of knowledge about local agronomy, economy and culture among the actors collaborating to guide the program policy and monitor the implementation design.

Facilitating Agency (FA) acts as a project facilitator at the block level. FA is a CSO, active in the block and chosen through a process of three-step evaluation taken up jointly by district level committee and Program Secretariat (WASSAN). The selection process of these NGOs to work as FAs, is based on their prior experience of working on agriculture, particularly with millet farming communities. NGOs selected as FAs are not new to the block or to the work of millets in Odisha. These FAs were working in their full capacity around mobilizing communities not to shift to cash crop cultivation entirely, and cultivate and consume millet for better nutrition. The focus of such works was mainly on ensuring food security through increasing yield, doubling income of farmers through improved agronomic practices, providing technical support, know-how and improved agricultural practices. FAs have proved to be instrumental in their efforts of working with millets and community from much before the government launched the state program of Millets Mission.

In some of the blocks in Koraput district, like Kundra, Lamtaput, Dasmantapur and Boipariguda, larger organizations, like MSSRF, PRADAN, DHAN Foundation, CYSD, respectively, have a significant impact in their respective blocks in terms of mobilizing the community to cultivate and consume millets and bringing technology and improved methods of cultivation to the farmers.

Community Based Organization can be considered as another NGO that works at the sub-block level. FPO take over the work of implementing the millets program. Any farmer group or FPO or SHG or any other community institution, based out of village, can be a CBO. The Facilitating Agency, VAW (Village Agriculture Worker), and Assistant Agriculture Officer (AAO) assist the CBO in implementing the program at the sub-block level. The FA plays a pivotal role by extending support to Community Based Organizations by conducting technical trainings on events ranging from productivity enhancement, farm mechanization, millet recipe development, easing the drudgery, encouraging and investing in small-scale processing infrastructure, market linkage, exposure visits, and FPO establishment.

CONCLUSION

With the central government, state government and public attention on revival of millets, health benefits of consuming millets to mitigate the nutritional requirements of the rural population as well as the "healthy food" conscious urban elite is in the limelight. However, the current initiative of the Government of Odisha is also argued in the framework of transition to sustainability and resilient agri-food system having a positive impact on the environment, creation of employment and nutrition. The inclusivity of state actors, non-state actors such as CSOs and a participatory and decentralized Institutional Framework indicates an innovative institutional change that challenges centrally controlled national policy-making regime (Raina *et al.*, 2022; Chebrolu and Dutta, 2021).

The Odisha Millets Mission (OMM) lays an innovative institutional framework to revive and mainstream an indigenous crop, but a comprehensive policy is the need of the hour to ensure that emphasis on millets does not become a divergence from policy imperativeness towards diversification. Millet, is often intercropped with other pulses and oilseeds to ensure crop diversity and availability of food at the household level. However, it needs to be reiterated, that value chain development of other minor millets such as little millet, kodo, barnyard, prosso and foxtail millet is extremely crucial to ensure crop diversification. Millet, being a hard grain, technological innovation regarding processing machines to suit the varying shape, size and texture of minor millets is an area, which, require concentrated efforts among various agriculture and technical universities and government. Easy access to post-harvest processing machineries and facilities, value-addition facilities and entrepreneurial opportunities supporting the livelihood of small and marginal farmers are the ways in which mainstreaming of millets can be successful.

REFERENCES

[1] Akram-Lodhi, A.H. (2013, July). Food sovereignty: A critical dialogue. In comunicação apresentada na XIV Internacional Conference–How to Build Food Sovereignty, realizada no âmbito do Programa Agrarian Studies, Yale University (pp. 14–15).

[2] Balam, D. (2020). "Millet Shatki Cafes-Quick Service Restaurant By WSHGS" To Be Launched By Mission Shakti And Odisha Millets Mission. Environment and Ecology. (https://vikalpsangam.org/article/millet-shakti-caf%C3%A9s-quick-service-restaurant-by-wshgs-to-be-launched-by-mission-shakti-and-odisha-millets-mission/).

[3] Chebrolu, S.P. and Dutta, D. (2021). Managing sustainable transitions: Institutional innovations from India. Sustainability, 13(11), 6076.

[4] DHAN Foundation and WASSAN. Supported by IDRC and CRDI (2012). Supporting Millets in India: Policy Review and Suggestions for Action. Part of Revalorising Small Millets in Rainfed Regions of South Asia (RESMISA). Supported by IDRC and CRDI.

[5] Directorate of Agriculture and Food Production, Odisha, Bhubaneswar (2019). OMM FPO strategy note approved by Director AFP. Promotion of Farmer Producer Organizations (FPOs) under Odisha Millets Mission.

[6] Garg, S., Muthukumar, M., Balam, D. and Mohanty, B. (2023); Odisha Millet Mission: A transformative food system for mainstreaming sustainable diets. In Routledge Handbook of Sustainable Diets (pp. 340–352). Routledge.

[7] Khadse, A. and Rosset, P.M. (2019). Zero Budget Natural Farming in India–from inception to institutionalization. Agroecology and Sustainable Food Systems, 43(7–8), 848–871.

[8] Kumbamu, A. (2018). Building sustainable social and solidarity economies: Place-based and network-based strategies of alternative development organizations in India. Community Development, 49(1), 18–33.

[9] Martinez, S. (2010). Local food systems; concepts, impacts, and issues. Diane Publishing.

[10] Padulosi, S., Mal, B., King, O.I. and Gotor, E. (2015). Minor millets as a central element for sustainably enhanced incomes, empowerment, and nutrition in rural India. Sustainability, 7(7), 8904–8933.

[11] Patel, R.C. (2012). Food sovereignty: power, gender, and the right to food. PLoS medicine, 9(6), e1001223.

[12] Raina, R., Mishra, S., Ravindra, A., Balam, D. and Gunturu, A. (2022). Reorienting India's Agricultural Policy: Millets and Institutional Change for Sustainability. Journal of Ecological Society.

[13] Saxena, L.P. (2020). Community self-organisation from a social-ecological perspective: 'Burlang Yatra' and revival of millets in Odisha (India). Sustainability, 12(5), 1867.

[14] Schutter, O. (2013). Food Sovereignty: A Critical Dialogue.

[15] Wittman, H., Desmarais, A. and Wiebe, N. (2010); The origins and potential of food sovereignty. Food sovereignty: Reconnecting food, nature and community, 1–14.

Entrepreneurship Development through Pearl Millet Processing for Better Livelihood

Rohini Bharad*, Vasant Deshmukh[2] and Krushna Kardile[3]

Deendayal Research Institutes, Krishi Vigyan Kendra—
Beed 1 District Beed–431517 (Maharashtra), India

ABSTRACT: Bajra or Pearl Millet one of the oldest millets used by ancestors, along with wheat flour even bajra was included in the regular diet. It is known to have a very high fibre content and excellent source of micronutrients like iron and zinc which makes it healthier. Despite of plenty of health benefits, pearl millet is majorly used in area of cultivation so the focus on bajra processing will aids to financial enhancement of self-help group and bajra millet will be known to the greater areas.

Deendayal Research Institute, Krishi Vigyan Kendra, Ambajogai at GondanpatitandaTq. Ambajogai district Beed conducted study on Pearl millet processing (Bajra Kharwadi) to elevate self-help group's financial income enhancement during the year 2020 and 2021. A self-help group which used to migrate for sugarcane cutting was selected and various training like bajra kharwadi preparation, packing, labelling of product was imparted to the group. Pearl millet processing (Bajra Kharwadi) has been found to be a successful profit-making avenue for the self-help group. This has not only created additional employment and enhanced income of the group members but also played significant role in increasing the demand of pearl millet. Continuous inclusion of Bajra Kharwadi in consumption as a snack will help in improvement in children and women malnutrition, and further increased demand may help in preventing migration of sugarcane cutters.

Keywords: *Pearl Millet, Entrepreneurship.*

INTRODUCTION

Millets are one of the oldest foods known to humans and possibly the first cereal grain to be used for domestic purposes. Pearl millet is a multifunctional cereal crop which belongs to the poaceae family. It is generally referred to in various local Indian languages as bajra, bajri, sajje, kambu, kamban, sajjalu, etc. It is widely used for food and forages (Arora P.D. *et al.*, 2003). Pearl Millet is highly valuable in semi-arid regions because of higher productivity under heat and drought conditions. Rajasthan, Maharashtra, Gujarat, Uttar Pradesh and Haryana are the largest pearl millet growing states in the country (Singh N. *et al.*, 2018). Due to higher nutrient content, pearl millet has been recognized by the Ministry of Agriculture, Government of India as one millet under "Nutri-Cereals" (GOI cereals.) It has a higher content of macronutrients and is considerably rich in resistant starch and soluble and insoluble dietary fibre in contrast to other millets (Ragaee S. *et al.*, 2006). Basically, millets is much more adaptive and sustainable yield can be achieved under changing agro climate condition than other crops. Pearl millet has a large root structure, which absorbs soil nutrients and has a higher importance for nutrition than other cereal crops, including wheat, rice, maize, and sorghum. A high degree of iron, zinc, magnesium, copper, manganese, potassium, and phosphorus is found in the mineral. This is a strong energy source with a calorific value of 361 Kcal/100 g and a high amount of fibre (1.2 g/100 g) (Singh N. *et al.*, 2018). It is higher (Taylor J.R.N. and Emmambux M.N., 2018) and is a healthy source for vitamin B, vitamin A, folic acid, calcium, and magnesium (Pattanashetti S.K., *et al.*, 2016). Pearl millet is nutritionally comparable and even superior to major cereals with respect to protein, energy, vitamins, and minerals. Dieticians and many health professionals recommended pearl millet because of its various health benefits and good effect on the body. Pearl millet is underutilized food grains in many countries despite its nutritive value. However, the processing rate of pearl millet is very low as compared to other. Therefore self-help group has an opportunity to start small scale business on processing of pearl millet.

Pearl Millet can be processed and value-added into various products like pearl millet flours, multigrain flour, biscuits, pearl millet-based snacks etc. It creates income generation opportunity to the rural women and increasing thereby the economic and social strength of women. Linking of farmers to the markets through efficient value chains would reduce the use of intermediaries in the chain and strengthen the value adding activities by better technology and inputs, upgraded infrastructure and processing and exports. This process can raise the income of farmers and will provide incentive for improving their management practices towards higher farm productivity. The income of the farmers can be enhanced by increasing production, value-addition to the product, and better marketing options. The present paper describes the efforts made by Deendayal research institute, Krishi Vigyan Kendra, Beed-I to establish village level enterprise on processing and value-addition to millet products with an objective of self-employment and income generation to the rural women and for sugarcane cutters.

MATERIAL AND METHODS

Krishi Vigyan Kendra, Beed-I conducted four skill development training programmes for rural and sugarcane cutter women, each of two days duration on processing and value addition of pearl millets at villages and at the campus during the year of 2020–201. Fifty-Five women participated in these four training programmes from different villages of Beed district. Frontline demonstration on improved package of practices for bajra cultivation was planned to popularise, create awareness, curtail the cost of production and bring ease in availability of raw material i.e. grains for processing. The women farmers of SHG were supplied with input (biofortified hybrid bajra seed: cultivar AHB1200) which is rich in iron (88 mg/kg) and zinc (43 mg/kg) recommended for cultivation by VMNKV, Parbhani (M.S.).

Trainings and demonstrations: Interactive lectures coupled with hands-on experience on preparation of pearl millet biscuits, kharwadi, papad and multigrain atta flour preparation and composition etc. Two self-help groups were selected from these trainees—one was from the village Nagpur to Parali and other from village Gondanpatti tanda which prefer to migrate for sugarcane cutting. The said groups were sensitized on nutrition education, importance of value addition in food products, handling of processing and value-addition unit, maintaining hygiene while handling food products, labelling, packing, licensing and financial management.

The Flow Chart of the Bajra Kharwadi

Add Oil to Pan & Heat

↓

Add Mustard Seeds

↓

Add water & allow to boil

↓

Add cumin seed, Garlic, chilli powder, salt & coriander pest

↓

Add pearl millet coarse flour to boiling water & allow to cook for 45 minutes (till it cooks properly)

↓

Close pan with lid & allows to cook for 45 minutes (till it cook properly)

↓

Allow to dry in small parts in sun dryer/electronic dryer by hand mould or hand picking

↓

Allow to sundry completely & store to serve.

Processing and Value Addition: The aim of processing and value addition to pearl millet was to convert the grains into convenient food, to make the product nutritionally superior, to market it easily and to have a shelf life of minimum one year. Pearl millet kharwadi was selected for preparation and marketing. Finished product was assessed for nutrient composition by computation method using 'Nutritive Value of Indian Foods' (Gopalan *et al.*, 2004).

Establishment of Processing Unit: To bring systematic and regular production, there is a need for establishment of their own processing unit for preparation of pearl millet kharwadi. Hence, a plan was developed for establishment of processing unit with minimum necessary machineries.

PACKAGING AND LABELLING

Packaging serves as a silent salesman. To promote the products and to get better market opportunity, farmer women were supported for development of attractive and suitable packaging with nutrition label.

Table 1: Nutritional Information of the Selected Bajra Kharwadi Per 100 gm

Sr. No.	*Nutrient*	*Value*
1.	Energy	378 kcal
2.	Carbohydrates	72.8 g
3.	Dietary Fibre	8.5 g
4.	Fat	4.2 g
5.	Protein	1.1 g

RESULT AND DISCUSSION

Table 2: Socio-Demographic Profile of Selected Two Group Each Group having to Women (n=20)

Sr. No.	*Variables*	*Category*	*Number*	*Percentage*
1.	Age	20–25 Years 26–35 Years 36–40 Years	8 7 5	40 35 25
2.	Education	Illiterates Primary High School/ above	6 10 4	30 50 20

Sr. No.	Variables	Category	Number	Percentage
3.	Occupation	Housewife Labour Farming	3 8 9	20 45 35
4.	Type of family	Nuclear Joint	7 13	35 65
5.	Family size	2–4 members 5–7 members	7 13	35 65
6.	Monthly Income of family	1000 to 5000 5001 to 10000 10000 to 15000	4 11 5	20 55 25

SOCIO-DEMOGRAPHIC PROFILE

The socio-demographic profile of the selected women is presented in Table 2. The age profile indicated that majority of farmer women belonged to the age group of 20–25 years (40%) followed by 26–35 years (35%) and 36–40 (25%) respectively. With respect to literacy, 50% of women were educated up to primary school level and (30%) were illiterate and only 20% percent were educated up to high school. The occupation pattern indicated that majority (45%) of them were farm labour followed by Farm labour (35%). Majority of these women were from joint family (65%) and the family size was 5–7 members. With respect to the monthly income of the family, majority of the families earn 5001 to 10000 Rs (55%) per month followed by 10000 to 15000 (25%) and 1000 to 5000 (20%) respectively.

ESTABLISHMENT OF PROCESSING AND VALUE ADDITION UNIT

After acquainting with processing, preparation of value-added products and packing, farmers migrating for sugarcane cutting came forward to take up processing and value addition to millets as an entrepreneurial activity. They established units at home in available space and then purchased and installed the necessary equipments viz. solar dryer, pulveriser, weighing scale and sealing machine. KVK assisted them in procuring this equipment, installation and handling.

KVK Beed-I assisted the entrepreneurs to do registration of their unit with a brand name under Food Safety and Standards Authority of India 2006. The units were registered with brand names “Bajra Chunk” (FSSAI Reg. No. 21520314000052) and “Aditya Bajra Kharwadi” by the two entrepreneur self-help group. KVK also lend a hand in designing and printing of colourful labels and packing for product produced by these entrepreneurs facilitating better marketing opportunities.

Marketing of the value-added products: Initially the products produced by these women were sold under the brand name of KVK, Beed-I in different exhibitions in different locations. After creating a platform for these products, the products were placed in provision stores in Parali and Pune. An exclusive outlet for sale of these products was also opened in the vicinity of KVK for meeting the local demand. In addition to the regular market avenues, the Bajra Kharwadi included snacks to the participants of training programmes and other official meetings.

Table 3: Economics of Bajra Kharwadi: Year Wise Processing, Sale, Expenditure and Net Income

Year	*Food Item*	*Production*	*Expenditure*	*Gross Income*	*Net Income*
2020	Kharwadi	20 qtl	80000.00	216000.00	136000.00
2021	Kharwadi	116 qtl	464000.00	1184000.00	720000.00

Self-help Group produced 20 qt of Kharwadi as per demand and sold fried kharwadi @ 100 Rs/kg and raw kharwadi @140 Rs/kg. Non fried kharwadi fetches higher market price as it has good storage capacity compared to fried kharwadi. On the basis of two-year data, the economic analysis of the product Bajra kharwadi is presented in the Table 3. It is depicted from the table that in the year 2020 sale of kharwadi is around 20 qt with net profit of Rs. 136000.00 and in the year 2021 sale of bajra kharwadi is 116 qt with net profit of 720000.

Thus, a small intervention made by KVK, Beed in convergence with Infosys resulted in a sustainable income generation for poor rural women and enhanced their livelihood standards. Incidentally, the annual consumption of raw material by these value-added units annually is around 1000 kg of bajra, thereby creating market for rain-fed farmers of the region. This process can raise the income of farmers and will provide incentive for improving their management practices towards higher farm productivity. The income of the farmers can be enhanced by increasing production, value addition and better marketing options. An efficient value chain by linking small and marginal farmers to these value-added units will enhance the net returns for both mutually. This finding is in accordance with the other study, which shows that the pearl millet processing generates the additional employment and enhances income of the families. (P. Swarna, *et al.*, 2017 and Holvoet, N. 2005)

CONCLUSION

The interventions of Deendayal research institute Krishi Vigyan Kendra, Beed-I played a strategic role in increasing self-confidence among farm women in undertaking small scale food processing and value addition unit at their village level and reaching the market in urban area. The consolidated initiation of farm women on processing and preparation of value addition to pearl millets is a new way of self-reliance practice. The entrepreneurship activity focusing the millet products has not only generated the additional employment and enhanced income of the families but also saved the farm families from hunting of work to earn livelihood. Further availability of millet products helps in enhancing its consumption which in turn improve the nutritional intake of the consumers.

REFERENCES

[1] Arora, P.D., Fan, L., Sodek, J., Kapus, A. and McCulloch, C.A. (2003). Differential Binding to Dorsal and Ventral Cell Surfaces of Fibroblasts: Effect on Collagen Phagocytosis, *Exp. Cell Res.* 286:366–380. doi: 10.1016/S0014-4827(03)00096-X.

[2] Gopalan, C., Ramashastri, B.V. and Balasubramanian, S.C. (2004). Nutritive value of Indian Foods. NIN. Hyderabad, pp. 47–91.

[3] Holvoet, N. (2005). The impact of microfinance on decision making agency: Evidence from South India. *Dev. & Change*, 36(1): 75–10.

[4] Pattanashetti, S.K., Upadhyaya, H.D., Dwivedi, S.L., Vetriventhan, M. and Reddy, K.N. (2016). *Genetic and Genomic Resources for Grain Cereals Improvement.* Elsevier. *Amsterdam, The Netherlands: Pearl Millet,* pp. 253–289.

[5] Swarna, P.; Lakshmi, R. Prasanna; Reddy, P. Bala Hussain and Kumar, P. Ganesh (2017). Convergence led livelihood security: A case study in Chittoor district of Andhra Pradesh, AU Volume 12, Issue 4, 581–586.

[6] Ragaee, S., Abdelaal, E. and Noamam, M. (2006). Antioxidant Activity and Nutrient Composition of Selected Cereals for Food Use. *Food Chem,* 98:32–38.

[7] Singh, N., Singh, S.P., Kumar, M., Kanhiya, K. and Kumar, A. (2018). Nutri Cereal Pearlmillet: Way Forward. *Int. J. Curr. Microbiol. Appl. Sci.* 7:578–581. doi: 10.20546/ijcmas.2018.706.066.

[8] Taylor, J.R.N. and Emmambux, M.N. (2008). Technology of Functional Cereal Products. Elsevier; Amsterdam, The Netherlands: Products Containing Other Speciality Grains: Sorghum, the Millets and Pseudocereals, pp. 281–335.

Millets Policy of Maharashtra: A Content Analysis

Babasaheb Mundhe
Department of Political Science, Savitribai Phule Pune University, Pune, India

ABSTRACT: Maharashtra is a leading state in millets cultivation. But this is part of the story. Millets have lost out to other crops in the overall scheme of cultivation as a result of biased government policies, and their intrinsic problems. In the recent past, there is a realization among the policy makers and farmers about their dietary value and ability to withstand the uncertainty of climate change. Despite of these inherent qualities, millets have been conspicuously absent in the policy. Transitory efforts in the form of campaigns to promote the development of few select millets, jowar and bajri, have been the hallmark of the policy. The content analysis of the policy efforts in the state demonstrate that not only millets have been neglected all along in policy making, but those efforts have been found to be inadequate to promote millets' development. It is imperative to develop a robust policy framework that would create an environment conducive to cultivating millet and harnessing its social and economic potential.

Keywords: *Millets, Policy, Cultivation.*

INTRODUCTION

Millets have been back on the agenda of development thinkers and policy makers in India over last decade or so. Until then, they received scanty attention from policy makers and were often on the margins of policy agenda. Promoting the cultivation of millets, by some, is viewed as the only potent policy alternative to effectively deal with a series of insecurities—food, fodder, health, nutrition, livelihood and ecological—faced by humanity (Millet Network of India, 2008). The issues of hunger eradication, of climate change and predicted effects on farming and forest systems across the globe has brought millets in focus. However, the current debate and discussion over millets is disproportionaly skewed towards systematically documenting their superior nutritional value compared to fine cereals (NAAS, 2013; Singh A. *et al.*, 2020; Bhat S. *et al.*, 2018), overlooking their neglect in the agricultural development policy history. The agricultural development policy in India, including states' measures, is rarely subjected to systematic investigation from the millets perspective. This paper attempts to undertake the systematic analysis of contents of agricultural development policy of the State of Maharashtra with the purpose of developing the understanding of the extent of their adequacy to promote millets development.

A number of small-seeded annual grasses that are cultivated as grain crops, primarily on marginal lands in dry areas in temperate, subtropical and tropical regions, are collectively called millets (FAO, 2018). The most important species are pearl millet, finger millet, proso millet and foxtail millet, sorghum. They are grown worldwide in 131 countries. On the basis of grain size, these classified as major and minor millets. Little millet is a smart food- food that is good for consumers, hardy and thus good for farmers and environmentally sustainable. The various species differ in their physical characteristics, quality attributes, soil and climatic requirements and growth duration. However, they are most popular in developing countries from Asia and Africa. These countries account for about 74 percent of global output and 90 percent of world acreage (FAO STAT, 2021). The consumption of millet globally has declined at the rate of nearly one per cent and expected to witness positive movement during 2019–2024 (Anbukkani *et al.,* 2017). The significance of millet as food staples at global level has been declining due to demand and supply factors like rising incomes, urbanization, and government policies (King, 2017). In the recent past, global community has come realize the critical role millet can play in ensuring nutritional security, and sustainability of farming systems in the face of rising agricultural crisis and climate change. It is not surprising that conscious efforts are being made to re-launch millet agriculture and popularize global consumption.

While the production of millet in Asia restricted almost exclusively to two countries, India and China, small quantities are produced in Myanmar, Nepal and Pakistan. India is the world's leading producer of millets, harvesting over 17 million tons per year, about 20 percent of the world's output (FAO STAT, 2021). In India, millets are mostly cultivated in Karnataka, Andhra Pradesh, Tamil Nadu, Maharashtra, Odisha, Madhya Pradesh, Rajasthan and Uttarakhand states (Gowri M.U. and Shivakumar K.M., 2020). Pearl millet, which accounts for about two-thirds of India's millet production, is grown in the drier areas of the country, mainly in the states of Rajasthan, Maharashtra, Gujarat, Uttar Pradesh and Haryana. Finger millet is produced mainly in the state of Karnataka, but also in Orissa, Uttar Pradesh and Tamil Nadu. The productivity of millets is boosting through technologies and high yielding varieties. But shrinking acreage and production continues to be primary concern. Acreage under millet from 1965–70 to 2015–20 decreased by

56 percent (Department of Agriculture and Farmers Welfare, 2022). The area under small millets declined from 8 million hectare (1949–50) to 1.8 million hectare (2017–18) over last six decades. Likewise, the production of small millet showed the same trend, recording a decline from 4 million ton to 2.44 million ton in the respective year, predominantly loss occurred in all other small millets compared to finger millet (Shadang and Jaganathan, 2014). The importance of millet as food staples in India declined over the years (Michaelraj and Shanmugam, 2013). Millets constituted 20 percent share of the total food grains basket until 1965–70, but it is now reduced to merely 6 percent (Department of Agriculture and Farmers Welfare, 2022).

On the other hand, the history of the state of Maharashtra is not completely different from that of India, although the state is one of the major millet growers. The State has second largest area under millets (20.67 percent) in India (Assocham, 2022), whereas small millets are grown in over 8 percent of the area. They are also the main source of earning for small and marginal farmers. Millets are rich in nutritional value and bear characteristics like drought resistance, photo-sensitivity and resilience to climate change. Like India, the area in millet has decreased significantly over time in the state. Although area allocated to millets declined substantially overtime, the production increased mainly due to technological intervention. Increased use of high yielding/hybrid varieties of millets contributed to quantum jump in the productivity (Directorate of Millet Development, 2020). The empirical evidence demonstrates beyond doubt the importance and role of policy in development of millets.

EXISTING POLICY INITIATIVES

Millets[1] occupy an important place in socio-economic life of the State. This section provides an overview of the policy, focusing on State's initiatives for agricultural development in general and millets in particular. Their state is also analyzed. The State's millet development policy covered all main dimensions, namely, provision of core agricultural inputs, support services, output dimensions and policy awareness among coarse cereal farmers having a bearing upon agricultural development. Thus, the agricultural policy framework of the State appears to be elaborate and comprehensive governing all the aforesaid dimensions of agriculture (Dev, 1996:2). Core agricultural inputs include quality seeds, fertilizers, irrigation, crop protection and mechanization of agriculture.

Provision of quality seeds has been one of the main goals of the agricultural policy as they play a crucial role in increasing production and productivity of various crops. It is estimated that the use of quality seed accounts for 20–25 percent of agricultural production (India, Department of Agriculture and Cooperation, 2013: 51). Several policy initiatives were launched from time to time to ensure steady supply of quality seeds to farmers. Taluka seed multiplication farms scheme (1957), was launched by the Government of India, the 'Hybrid Seed Programme 1965–66', launched during the Fourth Five Year Plan to increase the acreage under improved and hybrid varieties of Jowar, Bajri, maize, etc., multi-agency approach by pressing into service the co-operative societies, zilla parishads and registered seed growers for distribution of seeds, involving the private sector this endeavor (Dev, 1996:10), Mini Seed Kit Scheme since 1970s, and training and orientation of farmers in production technology were some of policy instruments to popularize the modern seeds and technology.

The availability of organic manures and fertilizers in the required quantity, quality and at the right time is necessary for achieving higher yield (Jain 1965; Maharashtra, Finance Department, 1961). The policy initiatives introduced from time to time to improve the consumption of organic manures and chemical fertilizers include making available fertilizers at a uniform price throughout the country, developing a network of retail outlets and intensive use of fertilizers campaigns. These policy measures have helped to improve the consumption of fertilizers over time. A study of the State's agricultural policy has found that the farmers have benefited from the fertilizers pricing policy pursued since 1970s, but the benefits have been mostly cornered by the farmers with irrigation facility (Dev, 1996).

The protection of crops from insects, rodents, other animal pests, diseases, weeds and parasitic plants is of critical importance. The problem is serious because about 10 to 30 percent of crop is lost due pests and diseases in the country (India, Department of Agriculture and Cooperation, 2012). Necessary pesticides, manually operated dusters and sprayers have been supplied to farmers at concessional rates. The eco-friendly Integrated Pest Management (IPM) was adopted but the State seems to be not following IPM strategy widely even though it is ecologically sound, economically viable and socially acceptable.

The policy efforts with respect to agricultural credit or crop loans since independence have emphasized the need for increasing institutional lending to small and marginal farmers and other weaker sections (Rao and Jeromi, 2009). The policy recognized the necessity of an efficient and effective crop loan delivery system for providing timely, adequate and equitable access to credit for raising agricultural productivity and incomes. The State adopted a multi-agency approach to achieve these goals. It encouraged the development of agricultural primary cooperative societies to lend credit to farmers. Besides, the role of Regional Rural Banks and commercial banks was emphasized. The State is also pursuing the policy of making available crop loans at concessional or subsidized rates.

The extension services constitute another set of policy measures of the agricultural support policy. These services include information and support required to undertake agricultural activities. In the 1970s and 1980s, the basic principles of guiding extension policies changed from human and community development to transfer of technology. The main extension activities were continued to be input supplies. Training and Visit (T&V) mode was commonly used to disseminate new seeds of major cereal crops.

The central sector's 'mini kit seed programme' of pre-released varieties of coarse cereals introduced in 1974 and followed by launching its location specific programme in 1982 may be mentioned as millet production oriented programmes. Yet another was all-India coordinated Pearl Millet Improvement Project by ICAR launched in 1985. There were other special programmes executed in 1980s (India, Directorate of Millet Development, 2014:7). The Special Foodgrains Production Programme (SFPP) was introduced in 1988–89 with focus on improving productivity of low yield areas and maximizing production of high yielding areas (Planning Commission of India, 1992:2). Initially it covered rice and wheat, and later on expanded to Jowar, Bajri, maize and Ragi. It emphasized spreading improved technology; efficient delivery of inputs like seed, fertilizer, pesticides; and popularization of improved agricultural equipment. This contributed to an increase in the production of coarse cereals.

Agriculture output price policy is another policy instrument the State has relied upon to ensure millet growth. It is being implemented since 1965. Its main objectives are to ensure that the growers get remunerative prices for their produce and to encourage higher investment and production (Kalmakar, 2004:6). The Minimum Support Price (MSP) is announced by the Government for the principal agricultural products including Jowar and Bajri every year to achieve the above stated objectives. Prices are fixed by the Government on the recommendations of the Commission of Agricultural Costs and Prices (CACP). It is operationalized through the Agricultural Produce Marketing Committees (APMCs) across the State.

Some of the latest initiatives to popularize millets are nutria-cereals programme (2011), launching of national mission on millets, announcing and celebrating

2018 as the National Year for Millets, making Sub Mission on Millets as part of National Food Security Mission since 2018, encouraging state to launch state level mission on millets, inclusion of millets in Poshan Mission Abhiyan, extending support, financial and technological, to start-ups and technological development and so on. To generate awareness among farmers, the Department of Agriculture relies on media such as audio-video media, advertisements on radio and television sets, newspapers, exhibitions, organizing farmers' meets, etc. One of the popular modes of communication used in recent times is organizing farmers' fairs like *Krushi Pradarshan* or taking out *Krushi Dindi*.

METHODS AND MATERIALS

The paper relies on analytical framework to analyze millets policy in the state. The qualitative desk method is used as an analytical technique. Data for the study is drawn from various sources such as policy documents, reports, government publications, published research studies. The information thus collected is subjected to content analysis to ascertain extent of policy support enjoyed by millets.

RESULTS AND DISCUSSION

Millet development policy measures have been all along focused on improving the production. Only in not so distant past the focus has been shifted to their nutritional value. Other qualities of millets talked about are that they are climate change resilient, can grow in poor soil, good for nature and so on. However, the analysis of millets policy measures brings out gaps in the framework. These efforts are too broad based to make a long-lasting impact on the state of millets. Most of those efforts cater to the immediate production needs of major millets, neglecting altogether small/minor millets. Policy initiatives are also not provided adequate budgetary support (Bhide, 2011). Minimum support prices are announced on regular basis to boost the cultivation of millets. Only major millets are covered under the price policy. Even these efforts get thwarted in practice as a little or no procurement is undertaken. Thus, making these measures meaningless. The extension service is non-existent on the ground. The dysfunctional extension service deprives the small and marginal households of authentic advisory regarding use of inputs and marketing millets produce. The awareness among the farmers is very low about the policy. Overall, the analysis suggests that millets are most neglected crops in the realm of policy, contributing their marginalization.

CONCLUSION

The governments, development thinkers, and policy makers have renewed efforts to revive millets cultivation in the country and state. In reality, they suffer from numerous weaknesses. The initiatives by any standard are found to be inadequate in terms its focus and coverage. Many of those measures are transitory in nature raising questions over their longevity and sustainability. The existing policies thus make task of mainstreaming millets cultivation much more difficult and challenging. To achieve this goal, there is urgent need of developing a strong policy framework to address gamut of issues holding back the ability of government to harness the full potential of millets. A policy framework developed considering not only production and marketing needs but also processing aspects will go a long way in making them truly 'super food'. In absence of such a policy, millets will remain 'poor man's food'.

Note: *They were called coarse cereals until 2018, when the Government of India renamed them as nutri-cereals (Press Notification, 2018).*

REFERENCES

[1] Anbukkani, P., Balaji, S.J. and Nithyashree, M.L. (2017). Production and consumption of minor millets in India—A structural break analysis, *Ann. Agric. Res. New Series*, Vol. 38(4), pp. 1–8.

[2] Assocham (2022). Knowledge paper on Millets: The Future Super Food of India, The Associated Chambers of Commerce and Industry of India, New Delhi.

[3] Bhat, S. Nandini, C. Tippeswamy, V. and Prabhakar (2018). Significance of small millets in nutrition and health -A review, *Asian Journal of Dairy & Food Research*, Vol. 37(1), pp. 35–40.

[4] Bhide, S. (2011). The nutria-cereal challenge, The Hindu Business Line, March 9, 2011.

[5] Department of Agriculture and Farmers Welfare (2022). Annual Report 2021–22, Ministry of Agriculture and Farmers Welfare, Government of India, New Delhi.

[6] Dev, M.S. (1996). Agriculture Policy Framework for Maharashtra—Issues and Options, Proceedings/Project Reports No. 21, Indira Gandhi Institute of Development Research, Mumbai.

[7] DHAN Foundation (2012). Supporting millets in India: policy review and suggestions for action, *Part of Revalorizing Small Millets in Rainfed Regions of South Asia (RESMISA)*, pp. 5–11.

[8] Food and Agriculture Organization (FAO), (2021). The State of Food and Agriculture 2021, FAO, Rome.

[9] Gowri, M.U. and Shivakumar, K.M. (2020). Millet Scenario in India, Economic Affairs, 65(3), pp. 363–370.

[10] India, Department of Agriculture and Cooperation (2012). Annual Report 2011–12, Ministry of Agriculture, Government of India, New Delhi.

[11] India, Department of Agriculture and Cooperation (2013). Annual Report 2012–13, Ministry of Agriculture, Government of India, New Delhi.

[12] India, Directorate of Millets Development (2014). Status Paper on Coarse Cereals, Department of Agriculture and Cooperation, Ministry of Agriculture, Government of India, March, New Delhi.

[13] Jain, S.C. (1965). Agricultural Policy in India, Allied Publishers Private Limited, Bombay.

[14] Kalamkar S.S. (2006). Agricultural Development in Maharashtra, in Sthanumoorty (edited), Maharashtra Economy: Performance, Opportunities and Challenges, The ICFAI University Press, Hyderabad, pp. 64–86.

[15] King, E. (2017). Impact of Reduced drudgery of women in production and post-harvest processing of small millets, MSSRF Working paper, No. 9, Chennai.

[16] Maharashtra, Finance Department (1961). The Third Five-Year Plan Maharashtra, Government Central Press, Bombay.

[17] Michaelraj, P.S.J. and Shanmugam, A. (2013). A study on millets-based cultivation and consumption in India, *International Journal of Marketing, Financial Services and Management Research*, 2(4), pp. 49–58.

[18] Millets Network of India (2008). God's Own Crops—Report of a National Consultation on Millets, Booksline, Hyderabad.

[19] Mondal, A., King, I.O., Roy, S., Priyam, S., Meldrum, G., Padulosi, S. and Mishra, S. (2016). Making millets matter in Madhya Pradesh, *Farming Matters,* 32(2), pp. 10–13.

[20] National Academy of Agricultural Sciences (2013). Role of Millets in Nutritional Security of India, Policy Paper 66, National Academy of Agricultural Sciences, New Delhi.

[21] Planning Commission of India (1992). Eighth Five Year Plan (1992–1997), Government of India, New Delhi.

[22] Rao, V.M. and Jeromi, P.D. (2009). Modernizing Indian Agriculture in Krishna K.L. and Uma Kapila (edited), Readings in Indian Agriculture and Industry, Academic Foundation, New Delhi.

[23] Shadang, C. and Jaganathan, D. (2014). Millet—The Frugal Grain. *International Journal of Scientific Research and Reviews*, 3(4), pp 75–90.

[24] Singh, A. Kumar, M. and Shamim, M. (2020). Importance of minor millets (Nutri Cereals) for nutrition purpose in present scenario, *International Journal of Chemical Studies*, Vol. 8(1), pp. 3109–3113. (https://doi.org/10.22271/chemi.2020.v8.i1au.9226)

[25] Uma, Gowri, M. and Prabhu, R. (2017). Millet Production and Its Scope for Revival in India with special reference to Tamil Nadu, *International Journal of Farm Sciences,* 7(2), pp. 88–93.

Millets in India: An Annotated Bibliography of Essential Resources

Veena Thakare

Librarian, Ramnarain Ruia Autonomous College Matunga (East), Mumbai, India

ABSTRACT: Millets is one of the major food groups in India. There has been a multitude of research and scholarly literature published on various aspects of millets. This paper presents an annotated bibliography of all the open access resources that discuss millets in Indian context. The types of resources considered for this paper are doctoral theses, journal articles, and video lectures. To ensure the currency, only the resources published/theses submitted in last two years (2021 and 2022) are reviewed. In addition, a descriptive list of various organizations that specialize in millets research in India is constructed based on a brief description of websites. This detailed bibliography can serve as a one-stop solution for all the millets researchers to get an overview about the subject and current research trends in it.

Keywords: *Annotations, Bibliography, India, Millets, Open Access, Scholarly Resources, Websites.*

INTRODUCTION

According to Food and Agriculture Organization of the United Nations (FAO), Millet is a collective term referring to a number of small-seeded annual grasses that are cultivated as grain crops, primarily on marginal lands in dry areas in temperate, subtropical and tropical regions. The most important species are pearl millet, finger millet, proso millet and foxtail millet.‖[1] As per the statistics given by the United States Department of Agriculture's International Product Assessment Division, the highest i.e. 39% of the world production of millets is by India; and within India, maximum millet crops are taken by the state of Rajasthan.[2] Millets are considered as one of the major food groups and a base for preparing traditional alcoholic beverages, especially in the developing countries in Asia and Africa. However, in developed countries like the USA, millets are primarily considered as bird feed, and are not much consumed by humans. Being the highest producer of millets in the world, India celebrated 2018 as the National Year of Millets. At the behest of Government of India, the United Nations has commemorated 2023 to be the International Year of Millets, with the aim for—India to emerge as a Global Hub for Millets‖.

There has been a multitude of research done and scholarly literature published worldwide on the various aspects of millets, such as agricultural farming, mechanical processing, culinary recipes, dietetics, nutritional value, medicinal/ pharmaceutical uses, economics, etc. It was, therefore, observed that there should be a detailed bibliography that can inform the researchers about the essential resources on millets. This paper contains the annotated list of such scholarly resources on millets, curated with the goal to help the reader to get an overview about the current research trends in it.

SCOPE AND COVERAGE

The main scope of this bibliography is scholarly resources written/created on millets in Indian context. Further, the resources only with open access, with content in English language, and published in 2021 and 2022 are considered for this listing. The types of resources assessed are doctoral theses, journal articles, and video lectures. To ensure the currency, only the resources published/theses submitted/videos uploaded in last two years (2021 and 2022) are reviewed. In addition, a descriptive list of various organizations that specialize in millets research in India is constructed based on a brief description of the information provided on their websites.

METHODOLOGY

A literature search was carried out in open access databases such as, Directory of Open Access Journals (DOAJ), Multidisciplinary Digital Publishing Institute (MDPI), National Digital Library of India (NDLI), NISCAIR Repository, Open Access Publishing in European Networks (OAPEN), Research Gate, and Shodhganga. The keywords used for search were millet, millets, India, and Indian. The search was further refined using the delimiters specified in the scope and coverage, viz. language, publication year, and type of resource. The initial search retrieved 71 resources. Each of these was checked for the annotation points decided upon for this bibliography. They are as follows:

- **Title:** PhD thesis title/journal article title/video lecture title.
- **Authors' Expertise:** Author name/PhD researcher name; Guide name; Credentials (qualifications, affiliation).
- **Publisher's Credibility:** University of PhD thesis/Journal title; Country of publisher; ISSN/ISBN; Journal Impact Factor.
- **Content Coverage:** Core subject; keywords as provided in the resource.

- **Content Currency:** Publication year/Submission date/Video creation date.
- **Content Type:** Journal Article/PhD Thesis/Video Lecture.
- **Content Source:** Source of the open access content.

Link for Open Access

Based on the information available under above points, 46 resources were finalized for annotation. These 46 resources comprised of 25 PhD Theses, 15 journal articles, and 6 Video Lectures.

ANNOTATED BIBLIOGRAPHY OF 25 LATEST DOCTORAL THESES ON MILLETS

Arranged in descending order by thesis submission date:

1. **PhD Thesis Title:** Comparative Assessment of Phenolic Compounds Extracted from Millets for inhibiting the growth of colorectal and Breast carcinomas
   - **Researcher Name:** Mahadevaswamy, K.G.
   - **Guide Name(s):** M.V.S.S.T. SubbaRao
   - **University Name:** JSS Academy of Higher Education and Research, Mysore
   - **Core Subject:** Life Sciences
   - **Keywords as provided in the resource:** Anatomy and Morphology, Immunology, Life Sciences, Millets, Phytochemicals, Phenolic acids, Flavonoids, Breast cancer, Colorectal cancer
   - **Submission Date (Month & Year):** Dec. 2022
   - **Content Source:** Shodhganga
   - **Link for open access:** http://hdl.handle.net/10603/437231
2. **PhD Thesis Title:** Comparative Evaluation of Different Processing Methods on Nutritional Properties of Minor Millets
   - **Researcher Name:** Nithyashree, K.
   - **Guide Name(s):** Vijayalaxmi, K.G.
   - **University Name:** University of Agricultural Sciences, Bangalore
   - **Core Subject:** Food Science and Nutrition
   - **Keywords as provided in the resource:** Agricultural Sciences, Life Sciences.
   - **Submission Date (Month & Year):** Dec. 2022
   - **Content Source:** Shodhganga
   - **Link for open access:** http://hdl.handle.net/10603/440523

3. **PhD Thesis Title:** Studies on Kutki Millet panicum sumatrense starch hydrogel and its application as nutraceutical delivery in the food matrix
   - **Researcher Name:** Palak
   - **Guide Name(s):** Bera, Manab Bandhu and Panesar, Paramjit Singh
   - **University Name:** Sant Longowal Institute of Engineering and Technology, Punjab
   - **Core Subject:** Food Engineering and Technology
   - **Keywords as provided in the resource:** Agricultural Sciences, Food Science and Technology, Life Sciences
   - **Submission Date (Month & Year):** Nov. 2022
   - **Content Source:** Shodhganga
   - **Link for open access:** http://hdl.handle.net/10603/435253
4. **PhD Thesis Title:** Agronomic Investigations in Finger Millet Eleusine Coracana L Gaertn Adopting Sri Principles
   - **Researcher Name:** Shiva Kumar, H.D.
   - **Guide Name(s):** Kalyana Murthy, K.N.
   - **University Name:** University of Agricultural Sciences, Bangalore
   - **Core Subject:** Agronomy
   - **Keywords as provided in the resource:** Agricultural Sciences, Agronomy, Life Sciences
   - **Submission Date (Month & Year):** Nov. 2022
   - **Content Source:** Shodhganga
   - **Link for open access:** http://hdl.handle.net/10603/440303
5. **PhD Thesis Title:** Genetic Association Mapping to Identify qtls for Drought Tolerance and Expression Analysis of Selected Genes Under Drought Environment in Finger Millet Eleusine Coracana L. Gaertn
   - **Researcher Name:** Antre, Suresh Haribhau
   - **Guide Name(s):** Ravikumar, R.L.
   - **University Name:** University of Agricultural Sciences, Bangalore
   - **Core Subject:** Plant Biotechnology
   - **Keywords as provided in the resource:** Agricultural Sciences, Agriculture Multidisciplinary, Life Sciences
   - **Submission Date (Month & Year):** Aug. 2022
   - **Content Source:** Shodhganga
   - **Link for open access:** http://hdl.handle.net/10603/440469
6. **PhD Thesis Title:** The Molecular and Phenotypic Characterization of an Association Mapping Panel of Finger Millet Eleusine Coracana L Gaertn Germplasm for Drought Adaptive Traits

- **Researcher Name:** Manjugouda I. Patil
- **Guide Name(s):** M.S. Sheshshayee
- **University Name:** University of Agricultural Sciences, Bangalore
- **Core Subject:** Crop Physiology
- **Keywords as provided in the resource:** Agricultural Sciences, Agriculture Multidisciplinary, Life Sciences
- **Submission Date (Month & Year):** Jul. 2022
- **Content Source:** Shodhganga
- **Link for open access:** http://hdl.handle.net/10603/440898

7. **PhD Thesis Title:** Identification and Characterization of Donor Lines for Drought Tolerance in Finger Millet Eleusine coracana L. Gaertn
   - **Researcher Name:** Ramya, V.
   - **Guide Name(s):** Y.A. Nanja Reddy
   - **University Name:** University of Agricultural Sciences, Bangalore
   - **Core Subject:** Crop Physiology
   - **Keywords as provided in the resource:** Agricultural Sciences, Agriculture Multidisciplinary, Life Sciences
   - **Submission Date (Month & Year):** Jul. 2022
   - **Content Source:** Shodhganga
   - **Link for open access:** http://hdl.handle.net/10603/440888
8. **PhD Thesis Title:** Studies on species complex varietal Response host plant resistance and Management of earhead caterpillars in Finger millet eleusine coracana l Gaertn
   - **Researcher Name:** Chikkarugi, Nagesh
   - **Guide Name(s):** Vijayakumar, L.
   - **University Name:** University of Agricultural Sciences, Bangalore
   - **Core Subject:** Agricultural Entomology
   - **Keywords as provided in the resource:** Agricultural Sciences, Agriculture Multidisciplinary, Life Sciences
   - **Submission Date (Month & Year):** Apr. 2022
   - **Content Source:** Shodhganga
   - **Link for open access:** http://hdl.handle.net/10603/440491
9. **PhD Thesis Title:** Development of Value Added Probiotic Beverages from Millets
   - **Researcher Name:** Vinuta R. Yadagouda
   - **Guide Name(s):** Usha Ravindra
   - **University Name:** University of Agricultural Sciences, Bangalore
   - **Core Subject:** Food Science and Nutrition

- **Keywords as provided in the resource:** Agricultural Sciences, Food Science and Technology, Life Sciences
- **Submission Date (Month & Year):** Mar. 2022
- **Content Source:** Shodhganga
- **Link for open access:** http://hdl.handle.net/10603/440282

10. **PhD Thesis Title:** Glycemic index of recipes replaced with selected millets barnyard millet foxtail millet and kodo millet and their glycemic response in type II diabetics
    - **Researcher Name:** Rajeswari, N.
    - **Guide Name(s):** Priyadharsini, V. Premala
    - **University Name:** Avinashilingam Institute for Home Science and Higher Education for Women, Coimbatore
    - **Core Subject:** Food Service Management and Dietetics
    - **Keywords as provided in the resource:** Agricultural Sciences, Food Science and Technology, Life Sciences
    - **Submission Date (Month & Year):** Feb. 2022
    - **Content Source:** Shodhganga
    - **Link for open access:** http://hdl.handle.net/10603/421294
11. **PhD Thesis Title:** Seed Technological Studies in Little Millet Panicum Miliare L Foxtail Millet Setaria Italica Beauv and Proso Millet Panicum Miliaceum L
    - **Researcher Name:** Balaji D.S.
    - **Guide Name(s):** Sathiyanarayanan G.
    - **University Name:** Annamalai University, Chidambaram
    - **Core Subject:** Seed Science and Technology
    - **Keywords as provided in the resource:** Life Sciences, Agricultural Sciences, Food Science and Technology
    - **Submission Date (Month & Year):** Jan. 2022
    - **Content Source:** Shodhganga
    - **Link for open access:** http://hdl.handle.net/10603/410879
12. **PhD Thesis Title:** Integrated nutrient management in groundnut Arachis hypogaea L and its residual effect on finger millet Eleusine coracana L on brown forest soil of South Odisha
    - **Researcher Name:** Palai, Jnana Bharati
    - **Guide Name(s):** Malik, G.C.
    - **University Name:** Visva Bharti University, Kolkata
    - **Core Subject:** Agronomy
    - **Keywords as provided in the resource:** Agricultural Sciences, Life Sciences

- **Submission Date (Month & Year):** Oct. 2021
- **Content Source:** Shodhganga
- **Link for open access:** http://hdl.handle.net/10603/400781

13. **PhD Thesis Title:** Evaluation of stress alleviation in foxtail millet Setaria italic grown in Kashmir Himalaya and Ladakh
    - **Researcher Name:** Wasifa
    - **Guide Name(s):** Reiaz Ul Rehman, Inayatullah Tahir, and Manoj Prasad
    - **University Name:** University of Kashmir
    - **Core Subject:** Bioresources
    - **Keywords as provided in the resource:** Biology and Biochemistry, Developmental Biology, Life Sciences
    - **Submission Date (Month & Year):** Oct. 2021
    - **Content Source:** Shodhganga
    - **Link for open access:** http://hdl.handle.net/10603/402437
14. **PhD Thesis Title:** Social Practices in Conservation of Millets in Kolli Hills
    - **Researcher Name:** Jayashree, B.
    - **Guide Name(s):** Arul Aram I.
    - **University Name:** Anna University, Chennai
    - **Core Subject:** Faculty of Science and Humanities
    - **Keywords as provided in the resource:** Social Practices, Conservation, Millets
    - **Submission Date (Month & Year):** Oct. 2021
    - **Content Source:** Shodhganga
    - **Link for open access:** http://hdl.handle.net/10603/422474
15. **PhD Thesis Title:** Effects of Plant Growth Promoting Rhizobacteria Psedomonas and Bacillus strains on Finger Millet Eleusine coracana L Growth Yield and Seed Quality
    - **Researcher Name:** Rakesh Singh
    - **Guide Name(s):** Deepti Prabha
    - **University Name:** Hemwati Nandan Bahuguna Garhwal University, Uttarakhand
    - **Core Subject:** Seed Science
    - **Keywords as provided in the resource:** Agricultural Sciences, Agriculture Multidisciplinary, Finger Millet (Elesuine coracana L.), Growth, Yield and Seed Quality, Life Sciences, Plant Growth, Rhizobacteria (Pseudomonuas and Bacillus)

- **Submission Date (Month & Year):** Oct. 2021
- **Content Source:** Shodhganga
- **Link for open access:** http://hdl.handle.net/10603/442317

16. **PhD Thesis Title:** Evaluation of Microbial Consortia for Rapid Composting of Urban Waste and its Influence on Growth and Yield of Little Millet Panicum Miliaceum L
    - **Researcher Name:** Davalsab Javoor
    - **Guide Name(s):** B. Narayanaswamy
    - **University Name:** University of Agricultural Sciences, Bangalore
    - **Core Subject:** Agricultural Microbiology
    - **Keywords as provided in the resource:** Agricultural Sciences, Agriculture Multidisciplinary, Life Sciences
    - **Submission Date (Month & Year):** Sep. 2021
    - **Content Source:** Shodhganga
    - **Link for open access:** http://hdl.handle.net/10603/440835
17. **PhD Thesis Title:** Effect of Gamma Irradiation on Physico Chemical Characteristics of Kodo and Kutki Millets and In Vivo Studies for their Antidiabetic Potential
    - **Researcher Name:** Kaushik, Nidhi
    - **Guide Name(s):** Aggarwal, Manjeet and Chauhan, Komal
    - **University Name:** National Institute of Food Technology Entrepreneurship and Management—NIFTEM (K) Kundli, Haryana
    - **Core Subject:** Basic and Applied Science
    - **Keywords as provided in the resource:** Biology, Biochemistry, Life Sciences
    - **Submission Date (Month & Year):** Sep. 21
    - **Content Source:** Shodhganga
    - **Link for open access:** http://hdl.handle.net/10603/451996
18. **PhD Thesis Title:** Characterization of millets protein hydrolysates and peptides for functional food formulation
    - **Researcher Name:** Agrawal, Himani
    - **Guide Name(s):** Gupta, Mahesh
    - **University Name:** Academy of Scientific and Innovative Research (AcSIR), Chennai
    - **Core Subject:** Biological Sciences (CSIR-NBRI)
    - **Keywords as provided in the resource:** Biology, Biochemistry, Life Sciences
    - **Submission Date (Month & Year):** Jul–2021

- **Content Source:** Shodhganga
- **Link for open access:** http://hdl.handle.net/10603/395798

19. **PhD Thesis Title:** Effect of different levels of NPK on Browntop Millet and their residual Effect on field bean in paleustalfs of Eastern dry zone of Karnataka
    - **Researcher Name:** Anitha, K.V.
    - **Guide Name(s):** Chikkaramappa, T.
    - **University Name:** University of Agricultural Sciences, Bangalore
    - **Core Subject:** Soil Science and Agricultural Chemistry
    - **Keywords as provided in the resource:** Agricultural Sciences, Agriculture Multidisciplinary, Life Sciences
    - **Submission Date (Month & Year):** Jul. 2021
    - **Content Source:** Shodhganga
    - **Link for open access:** http://hdl.handle.net/10603/440526
20. **PhD Thesis Title:** Assessment of Seven Millets for Low Phosphate Stress Tolerance and Identification of Quantitative Trait Loci Qtl in Low Phosphate Stress Tolerance in Finger Millet Eleusine Coracana L Gaertn Using Linkage Mapping
    - **Researcher Name:** Maharajan, T.
    - **Guide Name(s):** Duraipandiyan, V.
    - **University Name:** University of Madras
    - **Core Subject:** Entomology
    - **Keywords as provided in the resource:** Clinical Pre-Clinical and Health, Clinical Medicine, Endocrinology and Metabolism tissue
    - **Submission Date (Month & Year):** Apr. 2021
    - **Content Source:** Shodhganga
    - **Link for open access:** http://hdl.handle.net/10603/392207
21. **PhD Thesis Title:** Effects of processing techniques on nutritional viscosity osmolality and sensory characteristics of millet based diarrheal replacement fluids and its effect on childhood diarrhea
    - **Researcher Name:** A. Shanmugapriya
    - **Guide Name(s):** P. Nazni
    - **University Name:** Periyar University, Salem
    - **Core Subject:** Food Science and Nutrition
    - **Keywords as provided in the resource:** Effects of processing techniques on nutritional viscosity osmolality and sensory characteristics of millet based diarrheal, Nutrition and Dietetics, replacement fluids and its effect on childhood diarrhea

- **Submission Date (Month & Year):** Mar. 2021
- **Content Source:** Shodhganga
- **Link for open access:** http://hdl.handle.net/10603/379189

22. **PhD Thesis Title:** Identification and Characterization of Alternaria Alternata Fr Keissler Inciting Leaf Blight in Little Millet
    - **Researcher Name:** Boda, Praveen
    - **Guide Name(s):** Nagaraja, A.
    - **University Name:** University of Agricultural Sciences, Bangalore
    - **Core Subject:** Plant Pathology
    - **Keywords as provided in the resource:** Agricultural Sciences, Agriculture Multidisciplinary, Life Sciences
    - **Submission Date (Month & Year):** Feb. 2021
    - **Content Source:** Shodhganga
    - **Link for open access:** http://hdl.handle.net/10603/437653

23. **PhD Thesis Title:** Effect of Kodo, Pearl and Foxtail millet incorporated cereal bars on the quality parameters and physical fitness among the selected sportspersons
    - **Researcher Name:** Kerenhappuch, Susan S.
    - **Guide Name(s):** P. Nazni
    - **University Name:** Periyar University, Salem
    - **Core Subject:** Food Science and Nutrition
    - **Keywords as provided in the resource:** Cereal bars on the quality parameters and physical fitness among the selected sportspersons, Effect of kodo pearl and foxtail millet incorporated, Food Science and Technology
    - **Submission Date (Month & Year):** Feb. 2021
    - **Content Source:** Shodhganga
    - **Link for open access:** http://hdl.handle.net/10603/379223

24. **PhD Thesis Title:** Transcript profiling of pearl mille Pennisetum glaucum L under drought stress and functional characterization of a stress responsive gene
    - **Researcher Name:** Radha Shivhare
    - **Guide Name(s):** Charu Lata
    - **University Name:** Academy of Scientific and Innovative Research (AcSIR), Chennai
    - **Core Subject:** Biological Sciences (CSIR-NBRI)
    - **Keywords as provided in the resource:** Biology, Biochemistry, Life Sciences

- **Submission Date (Month & Year):** Feb. 2021
- **Content Source:** Shodhganga
- **Link for open access:** http://hdl.handle.net/10603/394846

25. **PhD Thesis Title:** Processing Nutritional Quality and Value Added Products of Kodo Millet Paspalum Scrobiculatum
    - **Researcher Name:** Patil, Roopa B.
    - **Guide Name(s):** K.G. Vijayalaxmi
    - **University Name:** University of Agricultural Sciences, Bangalore
    - **Core Subject:** Food Science and Nutrition
    - **Keywords as provided in the resource:** Agricultural Sciences, Agriculture Multidisciplinary, Life Sciences
    - **Submission Date (Month & Year):** Jan. 2021
    - **Content Source:** Shodhganga
    - **Link for open access:** http://hdl.handle.net/10603/441014

ANNOTATED BIBLIOGRAPHY OF 15 LATEST JOURNAL ARTICLES ON MILLETS

Arranged in descending order by publication date:

1. **Article Title:** Leveraging Nutrient-Rich Traditional Foods to Improve Diets among Indigenous Populations in India: Value Chain Analysis of Finger Millet and Kionaar Leaves
    - **Author Name(s):** Shauna M. Downs, Ridhima Kapoor, Emily V. Merchant, Tamara Sullivan, Geetanjali Singh, Jessica Fanzo, Suparna Ghosh-Jerath
    - **Author Affiliation(s):** New Use Agriculture and Natural Plant Products Program, Department of Plant Biology, Center for Agricultural Food Ecosystems, The New Jersey Institute for Food, Nutrition and Health, Rutgers University, New Brunswick, NJ 08901, USA; The George Institute for Global Health India, New Delhi 110025, India; Department of Urban-Global Public Health, Rutgers School of Public Health, Newark, NJ 07102, USA; Department of Botany, Dr. Shyama Prasad Mukherjee University Ranchi, Ranchi 834008, India; Berman Institute of Bioethics, Nitze School of Advanced International Studies (SAIS), Bloomberg School of Public Health, Johns Hopkins University, Washington, DC 20036, USA.
    - **Journal Title & ISSN:** Foods ; 2304–8158 (Online)
    - **Country of Journal Publication:** Switzerland

- **Journal Impact Factor (2021) as given by Web of Science:** 5.561
- **Content Core subject**—Food and Nutrition
- **Keywords as provided in the resource:** finger millet, Indigenous foods, Koinaar leaves, nutrient-rich foods, traditional ecological knowledge, value chain analysis
- **Article Publication Month & Year:** Nov. 2022
- **Content Source:** Multidisciplinary Digital Publishing Institute (MDPI)
- **Link for open access:** https://www.mdpi.com/2304-8158/11/23/3774

2. **Article Title:** Impact of Precise Land Levelling on Pearl Millet-Based Cropping Systems
   - **Author Name(s):** Y.P. Singh, Sandeep Singh Tomar, Ravi Yadav, Ravi Gurjar, Deependra Sharma
   - **Author Affiliation(s):** Rajmata Vijayaraje Scindia Krishi Vishwa Vidyalaya, Gwalior, Madhya Pradesh, India–474 002
   - **Journal Title & ISSN:** Indian Farming (by Indian Council of Agricultural Research); 0019-4786 (Online)
   - **Country of Journal Publication:** India
   - **Journal Impact Factor (2021) as given by Web of Science:** Nil
   - **Content Core subject:** Agriculture
   - **Keywords as provided in the resource:** Cropping system, Economics, Precise leveling, Water productivity, Yield
   - **Article Publication Month & Year:** Aug. 2022
   - **Content Source:** ResearchGate
   - **Link for open access:** https://www.researchgate.net/publication/363587952_Indian_Farming_Impact_of_precise_ land_levelling_on_pearl_millet-based_cropping_systems

3. **Article Title:** Growth Dynamics and Forecasting of Minor Millets in India: A Time Series Analysis
   - **Author Name(s):** Prabhu, R.; Gowri, M.; Gayathri, R.; Govindaraj, M.; Manikandan, G.
   - **Author Affiliation(s):** Tamil Nadu Agricultural University, Coimbatore, India—School of Post Graduate Studies; Department of Agricultural Economics, Centre for Agricultural and Rural Development Studies, Department of Seed Science and Technology, Central Institute of Agricultural Engineering
   - **Journal Title & ISSN:** Journal of Applied and Natural Science; 2231-5209 (Online)

- **Country of Journal Publication:** India
- **Journal Impact Factor (2021) as given by Web of Science:** Nil
- **Content Core subject:** Agriculture
- **Keywords as provided in the resource:** Area, Forecasting, Minor millets, Production, Time series analysis
- **Article Publication Month & Year:** Jul. 2022
- **Content Source:** Research Gate
- **Link for open access:** https://journals.ansfoundation.org/index.php/jans/article/view/3600/2282

4. **Article Title:** A Natural Low Phytic Acid Finger Millet Accession Significantly Improves Iron Bioavailability in Indian Women
   - **Author Name(s):** Bellam H. Rajashekar Reddy, Prashanth Thankachan, Masoami Hatakayama, Netravati Hiremath, Diego Moretti, Yellodu A. Nanjareddy, Mathi B. Thumilan, Ramapura L. Ravikumar, Shamprasad Phadnis, Beena Bose, Lucy Poveda, Geetha Kalaiah, Michael B. Zimmermann, Kentaro K. Shimizu, Ralph Schlapbach, Anura V. Kurpad, and Sheshshayee M. Sreeman.
   - **Author Affiliation(s):** Division of Nutrition, St. John's Research Institute, St. John's National Academy of Health Sciences, Bengaluru, India; Department of Evolutionary Biology and Environmental Studies, University of Zurich, Zurich, Switzerland; Functional Genomics Center Zurich, Zurich, Switzerland; All India Coordinated Research Project (Foods and Nutrition), University of Agricultural Sciences, Bengaluru, India; Laboratory for Human Nutrition, Institute of Food, Nutrition and Health, ETH Zurich, Zurich, Switzerland; Department of Crop Physiology, University of Agricultural Sciences, Bengaluru, India; Department of Biotechnology, University of Agricultural Sciences, Bengaluru, India; Department of Physiology, St. John's Medical College, St. John's National Academy of Health Sciences, Bengaluru, India.
   - **Journal Title & ISSN:** Frontiers in Nutrition; 2296-861X (Online)
   - **Country of Journal Publication:** Switzerland
   - **Journal Impact Factor (2021) as given by Web of Science:** 6.590
   - **Content Core subject:** Nutrition and Sustainable Diets
   - Keywords as provided in the resource –
   - **Article Publication Month & Year:** Mar. 2022
   - **Content Source:** Research Gate
   - **Link for open access:** https://www.frontiersin.org/articles/10.3389/fnut.2021.791392/full

5. **Article Title:** Processing of Pearl Millet Grains to Develop Nutri-Smart Food
   - **Author Name(s):** Vinutha Thimmegowda; Suneha Goswami; Maharishi Tomar; Archana Sachdev; Veda Krishnan; Tara Satyavathi; Shelly Praveen; Aruna Tyagi
   - **Author Affiliation(s):** Indian Council of Agricultural Research (ICAR)
   - **Journal Title & ISSN:** Indian Council of Agricultural Research
   - **Country of Journal Publication:** India
   - **Journal Impact Factor (2021) as given by Web of Science:** Nil
   - **Content Core subject:** Agriculture
   - **Keywords as provided in the resource:** Pearl millets
   - **Article Publication Month & Year:** Mar. 2022
   - **Content Source:** Research Gate
   - **Link for open access:** http://www.aicpmip.res.in/tb01_2022.pdf
6. **Article Title:** Modern Processing of Indian Millets: A Perspective on Changes in Nutritional Properties
   - **Author Name(s):** N.A. Nanje Gowda, Kaliramesh Siliveru, P.V. Vara Prasad, Yogita Bhatt, B.P. Netravati, and Chennappa Gurikar
   - **Author Affiliation(s):** Department of Food Technology, Faculty of Life and Allied Health Sciences, Ramaiah University of Applied Sciences, Bangalore–560054, India; Department of Grain Science and Industry, Kansas State University, Manhattan, KS 66506, USA; Department of Agronomy, Kansas State University, Manhattan, KS 66506, USA
   - **Journal Title & ISSN:** Foods; 2304-8158 (Online)
   - **Country of Journal Publication:** Switzerland
   - **Journal Impact Factor (2021) as given by Web of Science:** 5.561
   - **Content Core subject:** Sustainable Diets and Agro-food Biodiversity
   - **Keywords as provided in the resource:** Millets, processing; nutrients, dietary fiber, pearl, foxtail
   - **Article Publication Month & Year:** Feb. 2022
   - **Content Source:** Multidisciplinary Digital Publishing Institute (MDPI)
   - **Link for open access:** https://www.mdpi.com/2304-8158/11/4/499
7. **Article Title:** Spatial Arrangement and Biofertilizers Enhance the Performance of Legume—Millet Intercropping System in Rainfed Areas of Southern India.
   - **Author Name(s):** Devesh Singh, Natarajan Mathimaran, Jegan Sekar, Prabavathy Vaiyapuri Ramalingam, Yuvaraj Perisamy, Kathiravan

Raju, Rengalakshmi Raj, Israel Oliver King, Thimmegowda Matadadoddi Nanjundegowda, Manjunatha Baiyapalli Narayanswamy, Bhavitha Nayakanahalli Chikkegowda, Savitha Matakere Siddegowda, Davis Joseph Bagyaraj, Paul Mäder, Thomas Boller, Ansgar Kahmen.

- **Author Affiliation(s):** Department of Environmental Sciences—Botany, University of Basel, Basel, Switzerland; Programme Area Ecotechnology, M.S. Swaminathan Research Foundation, Chennai, India; Dryland Agriculture Project, University of Agricultural Sciences, GKVK, Bengaluru, India; Centre for Natural Biological Resources and Community Development, Bengaluru, India; Department of Soil Sciences, Research Institute of Organic Agriculture (FiBL), Frick, Switzerland.
- **Journal Title & ISSN:** Frontiers in Sustainable Food Systems; 2571–581X (Online)
- **Country of Journal Publication:** Switzerland
- **Journal Impact Factor (2021) as given by Web of Science:** 5.005
- **Content Core subject**—Crop Biology and Sustainability
- **Keywords as provided in the resource:** Bengaluru and Kolli Hills, bioirrigation, drought, finger millet, intercropping, mycorrhiza, pigeon pea, spatial arrangement of component plants
- **Article Publication Month & Year:** Dec. 2021
- **Content Source:** Directory of Open Access Journals (DOAJ)
- **Link for open access:** https://www.frontiersin.org/articles/10.3389/fsufs.2021.711284/full

8. **Article Title:** Assessing Millets and Sorghum Consumption Behavior in Urban India: A Large-Scale Survey.
   - **Author Name(s):** Joanna Kane-Potaka, Seetha Anitha, Takuji W. Tsusaka, Rosemary Botha, Muralidhar Budumuru, Shweta Upadhyay, Parkavi Kumar, Karuna Mallesh, Raghavendra Hunasgi, Ashok Kumar Jalagam, and Swamikannu Nedumaran.
   - **Author Affiliation(s):** Smart Food Initiative, International Crops Research Institute for the Semi-Arid Tropics, Hyderabad, India; Organization for Advanced and Integrated Research, Kobe University, Kobe, Japan; Development Strategy and Governance Division, International Food Policy Research Institute, Lilongwe, Malawi; United Nations International Children's Emergency Fund, Lilongwe, Malawi; Marketing Leadership Institute, SR Global Media and Research Pvt. Ltd., Hyderabad, India; Thought Folks Digital,

Hyderabad, India; Innovation Systems for the Dry Land, International Crops Research Institute for the Semi-Arid Tropics, Hyderabad, India.
- **Journal Title & ISSN:** Frontiers in Sustainable Food Systems; 2571-581X (Online)
- **Country of Journal Publication:** Switzerland
- **Journal Impact Factor (2021) as given by Web of Science:** 5.005
- **Content Core subject:** Nutrition and Sustainable Diets
- **Article Publication Month & Year:** Aug. 2021
- **Content Source:** Research Gate
- **Link for open access:** https://www.frontiersin.org/articles/10.3389/fsufs.2021.680777/full

9. **Article Title:** Genome-Wide Assessment of Population Structure and Genetic Diversity of the Global Finger Millet Germplasm Panel Conserved at the ICRISAT Genebank.
   - **Author Name(s):** C. Backiyalakshmi, Mani Vetriventhan, Santosh Deshpande, C. Babu, V. Allan, D. Naresh, Rajeev Gupta, and Vania C.R. Azevedo.
   - **Author Affiliation(s):** Centre for Plant Breeding and Genetics, Tamil Nadu Agricultural University (TNAU), Coimbatore, India; International Crops Research Institute for the Semi-Arid Tropics (ICRISAT), Hyderabad, India.
   - **Journal Title & ISSN:** Frontiers in Plant Science; 1664-462X (Online)
   - **Country of Journal Publication:** Switzerland
   - **Journal Impact Factor (2021) as given by Web of Science:** 6.627
   - **Content Core subject:** Agriculture, Plant Breeding
   - Keywords as provided in the resource
   - **Article Publication Month & Year:** Aug. 2021
   - **Content Source:** Research Gate
   - **Link for open access:** https://www.frontiersin.org/articles/10.3389/fpls.2021.692463/full

10. **Article Title:** Comparative Study of Physical Properties of Whole and Hulled Minor Millets for Equipment Designing.
    - **Author Name(s):** Abhishek Gaurav, Rama Chandra Pradhan, and Sabyasachi Mishra
    - **Author Affiliation(s):** Food Process Engineering Department, National Institute of Technology, Rourkela, Odisha, India-769 008.

- **Journal Title & ISSN:** Journal of Scientific & Industrial Research; 0975–1084 (Online)
- **Country of Journal Publication:** India
- **Journal Impact Factor (2021) as given by Web of Science:** 0.555
- **Content Core subject:** Agricultural Technology
- **Keywords as provided in the resource:** Engineering properties, Gravimetric properties, Mechanical property, Minor millets, Whiteness index
- **Article Publication Month & Year:** Jul. 2021
- **Content Source:** NISCAIR
- **Link for open access:** http://op.niscair.res.in/index.php/JSIR/article/view/49358/465479731

11. **Article Title:** Iron-biofortified pearl millet consumption increases physical activity in Indian adolescent schoolchildren after a 6-month randomised feeding trial
    - **Author Name(s):** Laura M Pompano, Sarah V Luna, Shobha A Udipi, Padmini S Ghugre, Eric M Przybyszewski, and Jere Haas
    - **Author Affiliation(s):** Cornell University, Savage Hall, Ithaca, NY14853, USA; Harvest Plus, International Food Policy Research Institute, 1201 I St NW, Washington, DC 20005 USA; S.N.D.T. Women's University, 1, Nathibai Thackersey Road, Mumbai 400020, India
    - **Journal Title & ISSN:** British Journal of Nutrition; 1475–2662 (Online)
    - **Country of Journal Publication:** United Kingdom
    - **Journal Impact Factor (2021) as given by Web of Science:** 4.125
    - **Content Core subject:** Nutritional Science
    - **Keywords as provided in the resource:** Accelerometry; Adolescent health; Iron biofortification; Iron deficiency; Pearl millet; Physical activity; Sedentary behaviour
    - **Article Publication Month & Year:** Jun. 2021
    - **Content Source:** OAPEN
    - **Link for open access:** https://www.cambridge.org/core/journals/british-journal-of-nutrition/article/ironbiofortified-pearl-millet-consumption-increases-physical-activity-in-indian-adolescent-schoolchildren-after-a-6month-randomised-feeding-trial/3008220B89F25A6A1C6422F149DADA6C#

12. **Article Title:** Finger millet-based staple beverages consumed by the Gadaba ethnic community of Odisha, India: Preparation and nutritive characteristics.

- **Author Name(s):** Arup Kumar Mishra, Ranjana Bajpai, Amrita Swain
- **Author Affiliation(s):** Department of Applied Geography, and Department of Zoology, Ravenshaw University. Cuttack, Odisha 753003, India
- **Journal Title & ISSN:** Biodiversitas; 1412-033X (Print), 2085-4722 (Online)
- **Country of Journal Publication:** Indonesia
- **Journal Impact Factor (2021) as given by Web of Science:** 1.320
- **Content Core subject:** Food and Nutrition
- **Keywords as provided in the resource:** Finger millet, Gadaba, indigenous beverage, landa, nutritional parameter, pej
- **Article Publication Month & Year:** May. 2021
- **Content Source:** Directory of Open Access Journals (DOAJ)
- **Link for open access:** https://smujo.id/biodiv/article/view/8137

13. **Article Title:** Performance and Stability of Pearl Millet Varieties for Grain Yield and Micronutrients in Arid and Semi-Arid Regions of India
    - **Author Name(s):** P. Sanjana Reddy, C. Tara Satyavathi, Vikas Khandelwal, H.T. Patil, P.C. Gupta, L.D. Sharma, K.D. Mungra, Sumer P. Singh, R. Narasimhulu, H.H. Bhadarge, K. Iyanar, M.K. Tripathi, Devvart Yadav, Ruchika Bhardwaj, A.M. Talwar, V.K. Tiwari, U.G. Kachole, K. Sravanti, M. Shanthi Priya, B.K. Athoni, N. Anuradha, Mahalingam Govindaraj, T. Nepolean, Vilas A. Tonapi
    - **Author Affiliation(s):** Indian Institute of Millets Research, Hyderabad, India; All India Coordinated Research Project on Pearl Millet, Jodhpur, India; Bajra Research Scheme, College of Agriculture, Mahatma Phule Krishi Vidyapeeth (MPKV), Dhule, India; Agricultural Research Station, Swami Keshavanand Rajasthan Agriculture University (SKRAU), Bikaner, India; Rajasthan Agricultural Research Institute, Sri Karan Narendra Agriculture University (SKNAU), Jaipur, India; Pearl Millet Research Station, Junagadh Agricultural University (JAU), Jamnagar, India; ICAR-Indian Agricultural Research Institute (IARI), New Delhi, India; Agricultural Research Station, Acharya NG Ranga Agricultural University (ANGRAU), Ananthapuramu, India; National Agricultural Research Project, Vasantrao Naik Marathwada Krishi Vidyapeeth (VNMKV), Aurangabad, India; School of Genetics, Tamil Nadu Agricultural University (TNAU), Coimbatore, India; College of Agriculture, Rajmata Vijayaraje Scindia Krishi Vishwa

Vidyalaya (RVSKVV), Gwalior, India; Chaudhary Charan Singh Haryana Agricultural University (CCSHAU), Hisar, India; College of Agriculture, Punjab Agricultural University (PAU), Ludhiana, India; Agricultural Research Station, University of Agricultural Sciences (UAS), Raichur, India; Zonal Agricultural Research Station, Rajmata Vijayaraje Scindia Krishi Vishwa Vidyalaya (RVSKVV), Morena, India; Agricultural Research Station, Mahatma Phule Krishi Vidyapeeth (MPKV), Niphad, India; Regional Agricultural Research Station, Professor Jayashankar Telangana State Agricultural University (PJTSAU), Palem, India; Agricultural Research Station, Acharya NG Ranga Agricultural University (ANGRAU), Tirupati, India; Regional Agricultural Research Station, University of Agricultural Sciences (UAS), Vijayapura, India; Agricultural Research Station, Acharya NG Ranga Agricultural University (ANGRAU), Vizianagaram, India; International Crops Research Institute for the Semi-Arid Tropics (ICRISAT), Hyderabad, India.

- **Journal Title & ISSN:** Frontiers in Plant Science; 1664-462X (Online).
- **Country of Journal Publication:** Switzerland
- **Journal Impact Factor (2021) as given by Web of Science:** 6.627
- **Content Core subject:** Agriculture, Plant Breeding
- **Keywords as provided in the resource:** G × E, GGE-biplot, iron, zinc, grain yield, open pollinated varieties, representative study
- **Article Publication Month & Year:** May. 2021
- **Content Source:** Directory of Open Access Journals (DOAJ)
- **Link for open access:** https://www.frontiersin.org/articles/10.3389/fpls.2021.670201/full

14. **Article Title:** Our Grandmother Used to Sing Whilst Weeding': Oral histories, millet food culture, and farming rituals among women smallholders in Ramanagara district, Karnataka
    - **Author Name(s):** Sandip Hazareesingh
    - **Author Affiliation(s):** History Department, The Open University, UK
    - **Journal Title & ISSN:** Modern Asian Studies; 1469-8099 (Online)
    - **Country of Journal Publication:** United Kingdom
    - **Journal Impact Factor (2021) as given by Web of Science:** 1.075
    - **Content Core subject:** Agriculture
    - **Keywords as provided in the resource:** Changing Farming Lives in South India
    - **Article Publication Month & Year:** May. 2021

Annotated Bibliography of Six Video Lectures on Millets

Sr. No.	Title	Content Writer	Writer Affiliation	Uploaded by	Length of Video	Core Subject/ Keywords	Content Type	Content OA Source	Link for Open Access
1.	Millets—Types and Compositi on	Maria Joseph	Women's Christian College, Chennai	Inflibnet: e-PG Pathshala	26.38 mins	Food Science	Video Lecture	National Digital Library of India (NDLI)	*http://ndl.iitkgp.ac.in/document/WGo3Njhma012NldsbTlObUtjeVhCUFNWSHhkVENYV3l6b3FjZndxU0hGaWM1ZGJQdjh3akdwbWVidXd2czloVTJTN0hEak9yZTFsYjUwYng2NktzWnc9PQ*
2.	Cereals and Millets and their products	Maria Joseph	Women's Christian College, Chennai	Inflibnet: e-PG Pathshala	24.44 mins	Food Science	Video Lecture	National Digital Library of India (NDLI)	*http://ndl.iitkgp.ac.in/document/WGo3Njhma012NldsbTlObUtjeVhCUFNWSHhkVENYV3l6b3FjZndxU0hGaUN1Q3BZdW1Hay9qNTNuRHpUNGFVUnBjZkFWblJXQXN2WkU2MEJTbEFOaEE9PQ*
3.	Recipes with cereals and millets	P. A. Raajeswa ri	Avinashilingam Institute for Home Science and Higher Education for Women, Coimbatore	Inflibnet: e-PG Pathshala	26.32 mins	Food and Nutrition, Food Science	Video Lecture (with e- text and photographs)	National Digital Library of India (NDLI)	*http://ndl.iitkgp.ac.in/document/WGo3Njhma012NldsbTlObUtjeVhCUFNWSHhkVENYV3l6b3FjZndxU0hGaFlYQzdpQ1ppdW5JcTNubjlVTG1aOG02aTZrejFSc DFNcVJpQ3lQUm1NRGc9PQ*
4.	Processing products of millets	A Reni		Inflibnet: e-PG Pathshala	27.28 mins	Food, Nutrition, Food Science, Principles of food processing	Video Lecture	National Digital Library of India (NDLI)	*http://ndl.iitkgp.ac.in/document/WGo3Njhma012NldsbTlObUtjeVhCUFNWSHhkVENYV3l6b3FjZndxU0hGaFNvZzJmWXU4Qm5SRXd6czN3RnVUcjU0QkRna2c wa3cwdEgzVDVIbjZsQ3c9PQ*
5.	Sorghum crop production	Vinod Kumar	IIT Kanpur	NPTEL	5.09 mins	Agriculture	Video Lecture	National Programme on Technology Enhanced Learning (NPTEL)	*https://archive.nptel.ac.in/courses/126/104/126104001/*
6.	Pearl millet crop production	J.R. Yadav	Shekhar Azad University of Agriculture & Technology,	NPTEL	10.23 mins	Agriculture	Video Lecture		*https://archive.nptel.ac.in/courses/126/104/126104001/*

- **Content Source:** OAPEN
- **Link for open access:** https://www.cambridge.org/core/journals/modern-asian-studies/article/our-grandmother-used-to-sing-whilst-weeding-oral-histories-millet-food-culture-and-farming-rituals-among-women-smallholders-in-ramanagara-district-karnataka/53ECB0B7DCBC5D9EA7D4046FAAA34615

15. **Article Title:** Genetic Gains in Pearl Millet in India: Insights Into Historic Breeding Strategies and Future Perspective
    - **Author Name(s):** Om Parkash Yadav, S.K. Gupta, Mahalingam Govindaraj, Rajan Sharma, Rajeev K. Varshney, Rakesh K. Srivastava, A. Rathore, Rajendra Singh Mahala.
    - **Author Affiliation(s):** ICAR—Central Arid Zone Research Institute, Jodhpur, India; International Crops Research Institute for the Semi-Arid Tropics (ICRISAT), Patancheru, India; State Agricultural Biotechnology Centre, Centre for Crop and Food Innovation, Food Futures Institute, Murdoch University, Murdoch, WA, Australia; SeedWorks International Pvt. Ltd, Hyderabad, India.
    - **Journal Title & ISSN:** Frontiers in Plant Science; 1664-462X (Online)
    - **Country of Journal Publication:** Switzerland
    - **Journal Impact Factor (2021) as given by Web of Science:** 6.627
    - **Content Core subject:** Agriculture, Plant Breeding
    - **Keywords as provided in the resource:** Biofortification, disease resistance, drought tolerance, genetic gain, hybrid breeding, pearl millet
    - **Article Publication Month & Year:** Mar. 2021
    - **Content Source:** Directory of Open Access Journals (DOAJ)
    - **Link for open access:** https://www.frontiersin.org/articles/10.3389/fpls.2021.645038/full

INITIATIVES SPECIALIZING IN RESEARCH/INFORMATION ON MILLETS

Listed below are some of the Indian research institutes and initiatives that are dedicated to millets. Their development and the information provided on their websites are described here in brief. The first seven of these 10 websites are by the official departments of Government of India, while the remaining three are by either NGO or private companies.

1. **Indian Institute of Millets Research (IIMR)—https://www.millets.res.in:** IIMR is the premier organization researching millets. First

established under the Project on Intensified Research on Cotton, Oilseeds and Millets (**PIRCOM**) in 1958, it was named as IIMR in 2014. It has formed the—Society for Millets Research‖ (SMR)—to provide a common platform to all the individuals engaged in research and development efforts on Millets‖. IIMR website has many good and useful resources, such as—The Story of Millets‖ a 110-page booklet published in 2018 that gives information regarding various types of millets, their history, cultivation, nutritional value, utilization, etc. This site also contains freely downloadable millets recipe books in English, Hindi, Kannada, Telugu, and Japanese. It also hosts a special page for millet farmers, providing important information for them in various Indian languages. IIMR's annual reports and newsletters are also available for downloading.

2. **https://www.mygov.in/campaigns/millets:** This is a webpage on the Government of India's official website, devoted to International Year of Millets (IYoM) 2023. It is currently hosting various interactive activities that invite participation like, designing creative logos for IYoM, sharing millet recipes for Mega Food Event 2023, solving quizzes on millets, slogan contest on importance of millets for nutritious diet, comics story writing contest on millets for health, composing catchy songs on millets, creating informative documentaries on millets, and innovative ideas for start-ups to increase millet production and consumption.
3. **https://www.nutricereals.dac.gov.in:** This is a website created especially for IYoM 2023 by the Ministry of Agriculture & Farmers Welfare, Government of India. NFSM-Nutri Cereals is one of the sub-schemes under National Food Security Mission. This website gives detailed cultivation notes, varieties, and related information about each of eight millets acknowledged as Nutri-Crops viz., sorghum, pearl millet, finger millet, kodo millet, little millet, foxtail millet, barnyard millet, and proso millet. There are downloadable booklets as well on nutritional and health benefits of nutri cereals and healthy recipes in English/Hindi.
4. **All India Coordinated Research Project on Pearl Millet (AICRP-PM)—http://www.aicpmip.res.in/index.html:** Established in 1985 separately for Pearl Millet, this center is currently located at Jodhpur in Rajasthan—which occupies nearly half of pearl millet area‖ of India. Its website contains access to Pearl Millet news, an annual magazine; as well as other 100+ reports/publications.
5. **All India Coordinated Research Project on Small Millets (AICRP-SM)—https://smallmillets.naarm.org.in:** It was established in 1986 at Bengaluru—to plan, coordinate and execute the multi-locational research programmes to augment the production and productivity of the seven

small millets viz., finger millet, kodo millet, little millet, foxtail millet, barnyard millet, proso millet and browntop millet in the country‖. It has 13 centers located in various states, and maintains one of the largest collections of germplasm in India.

6. **Pearl Millet Research Station, Junagadh AgriculturalUniversity—** http://www.jau.in/index.php/resources/2-init-content/research/109-millet-research-station-jamnagar: It was formed in 1962 and has developed many new hybrids of pearl millets. Its main thrust area is to conduct location specific research regarding collection and utilization of genetic resources for development of high yielding hybrids with wide adaptability.
7. **Millet Network of India (MINI)—https://milletindia.org**—MINI was launched in 2007 by an organic farmer, Dr. Nammalwar, in Andhra Pradesh. Currently, it has more than 50 registered member organizations from nine Indian states. MINI's primary vision is to acknowledge and strengthen the millet-growing communities of India. To achieve this, they have conducted various activities like providing millet-based meals in *Aanganwadi* (children's day care) centers, cultivation of agricultural bio-heritage sites, and documentation to track and analyze trends in millets farming and consumption. MINI also provides job opportunities, especially for women, to work with farming communities in the arid and semi- arid communities of India.
8. **Indian Millets Association (IMA) https://www.indianmilletsassocia tion.com**: IMA was founded in 1940 in Hyderabad specifically for promotion activities and services of millets. They supply millets-based products, provide training, and support research of millets.
9. **The Millet Foundation (TMF)—https://themillet.org:** As noted on their website—The Millet Foundation is a public charitable trust registered in Bengaluru, India. It is an organization supporting players in the millet value chain seeking technical resources and promotional materials‖. It contains nearly 100 informative posts on various topics such as, agriculture and processing of millets, and millet foods.
10. **Let's Millet—https://letsmillet.org:** It gives AI accumulated information on various types of millet, its nutritional value, its market potential, its production and trends, various products of millets available, interesting millets recipes, and publications available for download.

Apart from the websites described above, there are a few other important websites that may not be dedicated fully to millets but are peripherally important for any millets researcher. A few of them are listed below.

- The International Crops Research Institute for the Semi-Arid Tropics (ICRISAT)—https://www.icrisat.org
- CSIR—Central Food Technological Research Institute—https://www.cftri.res.in
- ICMR—National Institute of Nutrition—https://www.nin.res.in
- Department of Agriculture & Farmers Welfare—https://agricoop.nic.in
- Indian Council of Agricultural Research—https://icar.org.in

CONCLUSION

The listed resources in this paper show that among the various types of millets, Finger Millet has been researched highest in India in last two years with eight doctoral theses focusing solely on it. In case of journal articles, five are on Pearl Millet, four on Finger Millet, and one article on sorghum. Next, it can be noted that the agriculture/plant breeding/crop processing aspects of millets have maximum research coverage with 10 doctoral theses; followed by seven theses on nutritional facets; and six theses on medicinal/pharmaceutical properties of millets. Among journal articles, maximum i.e. seven are dedicated to nutrition by millets; followed by six on agriculture-related; and two on social aspects of millets. The credibility of the authors of journal articles and creators of video contents is high as seen by their affiliations to the respective fields related to the topics discussed. Each article has at least one author from India or of Indian origin. Maximum i.e. 20 doctoral theses on millets out of the total 25 studied are from the universities in Southern India. This high level of research activity can possibly be due to many major millets research institutes including IIMR being located in that region. It can, therefore, be concluded that the research on millets in India is flourishing. The two major domains of research are the agricultural cultivation of millets and its related areas, and the nutritional and medicinal properties of millets. This research trend can further encourage a worldwide higher demand and more consumption of millets which can in turn benefit the economy of India and other millet-producing countries.

REFERENCES

[1] https://www.fao.org/3/w1808e/w1808e0c.htm

[2] https://ipad.fas.usda.gov/cropexplorer/cropview/commodityView.aspx?cropid=0459100

Reviewer's List

Editors

Prof. Anushree Lokur

Prof. Kamini J. Donde

Prof. Jessy Pius

Reviewer's List

Prof. (Dr.) Louiza Rodrigues

Dr. Bhavna R Narula

Dr. Urmila Moon

Dr. Madhavi Badole

Dr. Vaishali Dhabke

Dr. Mayuri Rege

Dr. Sachin Palekar

Dr. Nandini Girish

Dr. Madhura. Dhavale-Joshi

Plagiarism Check by

Dr. Veena Thakare

Author Index